THE CAMBRIDGE HANDBOOK OF ANTITRUST, INTELLECTUAL PROPERTY, AND HIGH TECH

This handbook, edited by Roger D. Blair and D. Daniel Sokol, brings together a group of world-renowned professors in the fields of law and economics to assess the theory and practice of antitrust, intellectual property, and high tech. With the increased globalization of antitrust, a better understanding of how law and economics shape this interface will assist academics, policymakers, and practitioners to understand the existing state of academic literature, its limits, and its relevance to real-world antitrust. The book will be an essential resource for anyone seeking to understand academic and policy considerations shaping the world of antitrust, intellectual property, and high tech.

ROGER D. BLAIR is Walter J. Matherly Professor of Economics at the University of Florida, where he has taught since 1970. He received his PhD from Michigan State University. Professor Blair is the author or co-author of numerous books, including *Proving Antitrust Damages*. Professor Blair has written more than 200 articles or chapters in professional economics journals, law reviews, and books.

D. DANIEL SOKOL is Professor of Law at the University of Florida. He focuses his teaching and scholarship on complex business issues from early stage start-ups to large multinational businesses, emphasizing antitrust issues. Sokol has published his work in law reviews, peer review journals in law and economics, books, and the popular press. *The Global Competition Review* named Sokol its Antitrust Academic of the Year in 2014.

The Cambridge Handbook of Antitrust, Intellectual Property, and High Tech

Edited by

ROGER D. BLAIR
University of Florida

D. DANIEL SOKOL
University of Florida

CAMBRIDGE
UNIVERSITY PRESS

University Printing House, Cambridge CB2 8BS, United Kingdom

One Liberty Plaza, 20th Floor, New York, NY 10006, USA

477 Williamstown Road, Port Melbourne, VIC 3207, Australia

314-321, 3rd Floor, Plot 3, Splendor Forum, Jasola District Centre, New Delhi - 110025, India

79 Anson Road, #06-04/06, Singapore 079906

Cambridge University Press is part of the University of Cambridge.

It furthers the University's mission by disseminating knowledge in the pursuit of education, learning and research at the highest international levels of excellence.

www.cambridge.org
Information on this title: www.cambridge.org/9781108722087
DOI: 10.1017/9781316671313

© Cambridge University Press 2017

This publication is in copyright. Subject to statutory exception and to the provisions of relevant collective licensing agreements, no reproduction of any part may take place without the written permission of Cambridge University Press.

First published 2017
First paperback edition 2019

A catalogue record for this publication is available from the British Library

Library of Congress Cataloging in Publication data
NAMES: Sokol, D. Daniel, editor. | Blair, Roger D., editor.
TITLE: The Cambridge handbook of antitrust, intellectual property, and high tech / Edited by D. Daniel Sokol, University of Florida, Roger D. Blair, University of Florida.
OTHER TITLES: Handbook of antitrust, intellectual property, and high tech
DESCRIPTION: New York, NY : Cambridge University Press, 2017. | Includes index.
IDENTIFIERS: LCCN 2016048781 | ISBN 9781107159136 (Hardback)
SUBJECTS: LCSH: Antitrust law. | Intellectual property. | High technology.
CLASSIFICATION: LCC K3850 .C36 2017 | DDC 343.07/21–dc23 LC record available at https://lccn.loc.gov/2016048781

ISBN 978-1-107-15913-6 Hardback
ISBN 978-1-108-72208-7 Paperback

Cambridge University Press has no responsibility for the persistence or accuracy of URLs for external or third-party internet websites referred to in this publication, and does not guarantee that any content on such websites is, or will remain, accurate or appropriate.

Contents

List of Figures		page ix
List of Tables		xi
List of Contributors		xiii
Preface		xvii
Roger D. Blair and D. Daniel Sokol		

PART I ECONOMICS OF ANTITRUST-IP　　　　　　　　　　　　　1

1 Economics of Innovation　　　　　　　　　　　　　　　　　3
   Joshua S. Gans

2 Antitrust and Intellectual Property: Developments in Pharmaceuticals　　21
   Sumanth Addanki

3 The Economics of the Internet　　　　　　　　　　　　　　43
   Babette E. Boliek

4 The Economics of FRAND　　　　　　　　　　　　　　　58
   Anne Layne-Farrar

PART II INSTITUTIONAL DESIGN: COUNTRY OVERVIEWS　　　　　79

5 Antitrust and Intellectual Property: A Brief Introduction　　　　　81
   Keith N. Hylton

6 The Antitrust and Intellectual Property Intersection in European
   Union Law　　　　　　　　　　　　　　　　　　　　　92
   Nicolas Petit

7	The IP–Antitrust Interface in China: Uncharted Territory Thomas K. Cheng	120
8	Intellectual Property and Antitrust in Japan Masako Wakui	138
9	Competition Law and Intellectual Property in Korea Hwang Lee	158

PART III MONOPOLIZATION 181

10	Is Pepsi Really a Substitute for Coke? Market Definition in Antitrust and IP Mark A. Lemley and Mark P. McKenna	183
11	Monopoly Power and Intellectual Property Roger D. Blair and Wenche Wang	204
12	Exploitative Abuses of Intellectual Property Rights Harry First	222
13	Patent Holdups Daryl Lim	245
14	*Walker Process* and Sham Litigation Leon Greenfield and Mark Ford	271
15	Does Antitrust Have a Role to Play in Regulating Big Data? D. Daniel Sokol and Roisin Comerford	293

PART IV COMPETITOR COLLABORATION 317

16	Drug Patent Settlements Michael A. Carrier	319
17	Copyright Licensing and the EU Digital Single Market Strategy Pablo Ibáñez Colomo	339
18	Patent Pools and Related Technology Sharing Erik Hovenkamp and Herbert Hovenkamp	358

PART V VERTICAL RELATIONS 377

19	Bundling and High Tech Industries Daniel J. Gifford and Robert T. Kudrle	379

20	Tying Arrangements and Intellectual Property Christopher R. Leslie	404
21	Online RPM John B. Kirkwood	425
	PART VI MERGERS IN HIGH TECHNOLOGY	443
22	US Merger Enforcement in the Information Technology Sector Jeffrey A. Eisenach	445
23	Competition Assessment of IPRs in China's Merger Control Liyang Hou	467
Index		489

Figures

1.1	Equilibrium rate of innovation	page 8
5.1	Welfare consequences of monopolization	83
8.1	Examination of IP-related practice under Article 21	144
11.1	An illustration of the Lerner Index to measure market power	214
23.1	Annual number of cases reviewed by Mofcom before 2016	470

Tables

22.1 Hard-Scott-Rodino filings, IT-sector transactions, 2010–2014 *page* 452
22.2 Major mergers and acquisitions reviewed by FCC, 2010–2015 453
23.1 Market structure in cases of structural divesture 477

Contributors

Sumanth Addanki specializes in antitrust, intellectual property, and the evaluation of commercial damages at NERA Economic Consulting, where he is a Senior Vice President.

Roger D. Blair is a Professor of Economics at the University of Florida.

Babette F Boliek, is an Associate Professor of Law at Pepperdine School of Law.

Michael A. Carrier is the Distinguished Professor of Law at Rutgers.

Thomas K. Cheng is an Associate Professor at the University of Hong Kong Faculty of Law.

Roisin Comerford is an associate in the antitrust group at Wilson Sonsini Goodrich and Rosati.

Jeffrey A. Eisenach is a Senior Vice President and Co-Chair of NERA's Communications, Media, and Internet Practice. He is also an Adjunct Professor at George Mason University Law School, where he teaches Regulated Industries, and a Visiting Scholar at the American Enterprise Institute, where he directs the Center for Internet, Communications, and Technology Policy.

Harry First is the Charles L. Denison Professor of Law at New York University School of Law and Co-Director of the law school's Competition, Innovation, and Information Law Program.

Mark Ford is a partner at WilmerHale in Boston.

Joshua S. Gans holds the Jeffrey C. Skoll Chair in Technical Innovation and Entrepreneurship and is a Professor, and Area Coordinator of Strategic Management at Rotman School of Management (with a cross-appointment in the Department of Economics) at the University of Toronto.

Daniel J. Gifford is the emeritus Robins Kaplan Professor of Law at the University of Minnesota Law School.

Leon Greenfield is a partner in the antitrust group at WilmerHale.

Liyang Hou is a Professor and Assistant Dean of International Relations at KoGuan Law School, Shanghai Jiao Tong University – KoGuan Law School.

Erik Hovenkamp is a PhD student in the Department of Economics at Northwestern University.

Herbert Hovenkamp holds the Ben and Dorothy Willie Chair at the University of Iowa College of Law.

Keith N. Hylton is the William Fairfield Warren Distinguished Professor of Law at Boston University Law School.

Pablo Ibáñez Colomo is Associate Professor of Law at the London School of Economics and Political Science.

John B. Kirkwood is a Professor of Law at Seattle University School of Law and a Senior Fellow of the American Antitrust Institute.

Robert T. Kudrle is the Orville and Jane Freeman Professor of International Trade and Investment Policy, Hubert H. Humphrey School of Public Affairs and the Law School, University of Minnesota.

Anne Layne-Farrar is a Vice President in the Antitrust & Competition Economics Practice of CRA and an Adjunct Professor at the Northwestern University Pritzker School of Law.

Hwang Lee is a Professor of Law at Korea University School of Law.

Mark A. Lemley is the William H. Neukom Professor of Law at Stanford Law School and the Director of the Stanford Program in Law, Science and Technology.

Christopher R. Leslie is the Chancellor's Professor of Law at the University of California, Irvine School of Law.

Daryl Lim is an Associate Professor and Director, Center for Intellectual Property, Information Technology & Privacy Law at John Marshall Law School.

Mark P. McKenna is Associate Dean for Faculty Research and Development, Professor of Law, and Notre Dame Presidential Fellow at Notre Dame Law School.

Nicolas Petit is a Professor at the Law School of the University of Liege (ULg) Belgium, and Visiting Professor at EDHEC Business School, France.

D. Daniel Sokol is the University of Florida Research Foundation Professor of Law at the University of Florida. He also serves as Senior of Counsel at Wilson Sonsini Goodrich & Rosati.

Masako Wakui is a Professor of Law at Osaka City University.

Wenche Wang is a research fellow in the Department of Economics at the University of Florida.

Preface

Antitrust has always faced issues of innovation and technology. Additionally, antitrust has had to confront its interface with issues that also receive intellectual property protection (of which patents and copyright loom large). Antitrust has proven adaptable to address issues involving monopoly power industries possessing characteristics of fast-moving high tech markets. Further, in the United States, the two federal antitrust enforcers have identified in a joint report that antitrust and intellectual property "share the same fundamental goals of enhancing consumer welfare and promoting innovation [,] ... work[ing] in tandem to bring new and better technologies, products, and services to consumers at lower prices."[1] As with so many issues in antitrust, understanding how these particular dynamics play out in both theory and practice requires a more in-depth analysis.

The purpose of this handbook is to bring together lawyers and economists to provide an assessment of the state of theory and practice, and to expose fault lines in the economics and legal doctrines surrounding the interface of antitrust, intellectual property, and high tech. With an increasing number of mergers and conduct cases that have global implications, a better understanding of law and economics will shape this interface. Although it is not possible for any handbook to be exhaustive, we have attempted to provide coverage of what we believe to be the major topics in the interface of antitrust, intellectual property, and high tech. We believe that further developments in law and economic analysis across the topics covered in the handbook will inform both academic and policy considerations.

Roger D. Blair and D. Daniel Sokol

---

[1] US Dep't of Justice & Fed. Trade Comm'n, *Antitrust Enforcement and Intellectual Property Rights: Promoting Innovation and Competition* (2007) 1.

PART I

Economics of Antitrust-IP

1

Economics of Innovation

Joshua S. Gans

1.1. INTRODUCTION

This chapter examines what lessons the economics of innovation hold for the analysis of antitrust violations. Of course, the relationship between innovation and competition is itself a complex question whose analysis goes back all the way to the work of Joseph Schumpeter. The purpose here is not to adjudicate this general issue but instead to consider how the usual static tools used by lawyers and economists to assess antitrust matters hold up when innovation plays an important role. That is, can static tools be applied when there are "dynamic considerations" that are evident and may play an important role.

The reason why such considerations may pose a challenge is that innovation can shift the locus of competition away from traditional *in the market* analysis (that holds the firm and product composition as fixed) to analysis where competition is *for the market* (as Evans and Schmalensee usefully distinguished).[1] So rather than markets being modeled where competition is largely, say, price based and in which static instruments can impact on market power, in some industries, and at some points in time, competition is more akin to a series of winner-take-all contests where the winner is determined by a race to have the "best" innovations.

The reason this distinction is important for the analysis of antitrust policy is that, in many cases, policy precedes via a two-step procedure in which a regulatory body determines whether a particular practice should be limited or deterred. First, regulatory bodies begin by examining whether the firm possesses monopoly, or at least a substantial degree, of market power. They then examine whether, the practice under examination (say, exclusionary contracts) would have been undertaken in the absence of market power, and whether this practice could potentially damage competition. Evans and Schmalensee are concerned that firms possessing a

---

Significant parts of this chapter are drawn directly from Gans (2010).
[1] Evans and Schmalensee (2002).

substantial degree of market power are, to use software lingo, a feature rather than a bug in some industries.[2] Therefore, prohibiting certain practices (for example, product tying or below cost pricing) by firms who have market power will necessarily inhibit and reduce the profitability of firms in those industries and, as a result, subvert the means by which dynamic competition operates; it will eradicate the high market prize associated with successful innovative activity, namely, the ability to displace incumbent monopolists.

The consequence of this line of reasoning is the emergence of a debate focused on the argument that when dynamic considerations (that is, the notion that the incumbency prize is a key driver of innovation) are taken into account, antitrust authorities should be more cautious about interventions, since such interventions might weaken the potential for long-run competition in the industry.[3] The argument that authorities should be more permissive of short-run exploitation of market power is based on the idea that it leads to continual and frequent changes in market leadership. This process requires a distribution of rent from the market leader to consumers and puts pressure on incumbents to invest in innovation so as to maintain their market leadership.[4]

Countering this is the concern, voiced most aggressively by antitrust authorities themselves, that great vigilance is needed when the source of innovative pressure in an industry is from new entrants rather than existing incumbents. The argument is that those entrants face significant hurdles and bear considerable risks in attempting to raise the required capital to introduce new products to markets, and that unfair behavior on the part of incumbents should be restricted so as to give entrants the greatest chance of success.

Below-cost pricing is a good example of a practice that creates this type of tension. On the one hand, such pricing is, under usual antitrust analyses, indicative of predatory behavior whereby an incumbent sets low prices upon entry in the hope of facilitating the exit of any new potential competitors and deterring any future entry (for example, by sending the signal that entry is unprofitable). Given the inherent risks associated with entrant innovation, antitrust authorities have long been concerned that aggressive post-entry behavior may exacerbate the already suboptimal levels of innovation.

On the other hand, it is argued that, in some industries, winner-take-all competition does not necessarily award the market to firms with the better product but may instead award it to those firms who build up market share the quickest. This can occur in markets for network goods, where consumer value for products depends not only on the intrinsic utility of the good, but also on how many other consumers are consuming the same or a similar product. Firms in these markets will be willing to

---

[2] Id.
[3] Gilbert (2006; 2007); Manne and Wright (2009).
[4] Gilbert and Newbery (1982).

"pay for market share," even if this involves below-cost pricing for a short period of time. Proponents of this view argue that even a monopolist may choose to set low prices for a short period of time so as to increase market share, thus, boosting the consumers' utility from the product and hence its profitability. Hence, it is argued that observed low pricing is not necessarily predicated on competition or the deterrence of it. Thus, to deny firms the ability to build markets will itself further reduce the incentives for new product innovation in such industries.

Notice that both sides of the argument essentially appeal to "dynamic considerations." Specifically, the incentives for innovation in industry are likely to be damaged should anticompetitive practices be either permitted or prohibited. Certainly, a similar tension appears in the purely static environment – that is, below-cost pricing may be entry deterring but it is also good for consumers who are able to purchase low-cost products. The issue for antitrust policy is to determine what tools are required to analyze these issues, and to determine whether a violation has taken place. In both cases, the proponents argue for less weight to be placed on static considerations (such as current and prospective monopoly power) and more weight to be placed on dynamic factors (such as the rate and sources of innovation).

Here I argue that recent economic theory implies that dynamic considerations can often be addressed and analyzed using the same tools we would use for static analysis. I base this argument on the application of a formal model of the dynamic impact of antitrust policies on innovation. While the formal model is not exposited here,[5] it should be noted that it is premised on the "dynamic considerations" that are put forward by those who think such factors should change the analysis.

The analysis here is based on Segal and Whinston,[6] hereafter, SW. SW argue that, in innovative industries, antitrust policies have two major consequences. First, if effective, antitrust policies are likely to prevent rents from flowing from entrants to incumbents and in the process, hopefully allow consumers to capture some of these rents. Second, the potential loss in rents will lower the value of incumbency. In an industry where competition is characterized by sequential monopolists rather than persistent rivalry, innovation is driven by the desire for incumbency profits.

These consequences of antitrust policy mirror both sides of the debate over "dynamic considerations." The first, that rents entrants receive immediately upon entry may be lost should a practice be permitted, is what most concerns antitrust authorities. The second, that prohibiting certain practices could devalue the role of incumbency, is what most concerns those in fear of placing excessive constraints on incumbents. Yet, SW note that *both* consequences drive innovation and, importantly, both interact with each another. After all, the value of being an incumbent is equal to the profits that a firm expects to make as the market leader, less the profit that it expects to make if it is a laggard. Thus, while antitrust policy might reduce the

---

[5] An interested reader can find that in Gans and Persson (2013).
[6] Segal and Whinston (2007).

profits of a market leader, it does so by increasing the profits of a laggard, making it hard, at first glance, to determine the net effect on the value of incumbency.

The outline of this chapter is as follows. In the next section, I will provide a non-technical exposition of SW's model. In Section 1.3, I show that, in many cases, this implies that the conclusions that can be drawn by taking into account dynamic considerations can be achieved using the same tools that we apply for static analysis. Section 1.4 then considers an extension to the SW framework to consider an aspect of firm behavior in innovative industries that is neglected by antitrust scholars – namely, that entrants often do not end up competing head to head with incumbents, but instead end up cooperating with them. I argue that this poses special issues for the application of static antitrust analysis and provide suggestions as to how we should evaluate the consequences of static market power. Indeed, this is an area that is likely to require new tools in order for proper antitrust analyses to be conducted. A final section concludes.

1.2. MODELING INNOVATION DYNAMICS

SW consider an environment in which developing product improvements in an industry leads to innovation. Examples include computer processors with increasingly superior performance, software with improved capabilities, or mobile phones with more features. While these product improvements could result from the R&D activities of incumbents my focus is on improvements that arise from entrant investment in R&D – specifically, R&D conducted in firms with little or no presence in the product market. Indeed, the easiest way to understand the SW framework is to begin by considering a situation in which all new products are the invention of entrants rather than incumbents.

R&D in new products is fundamentally a process of applying resources (in particular, capital and labor) in those activities that are most likely to increase the chances of generating a new product in a short amount of time. Of course, the sooner that a firm hopes to innovate, the costlier it is to achieve. But, the cost associated with bringing forward the innovation date will be worthwhile if the "prize" from innovating is large enough – delayed innovation will result in a delayed prize.

In what follows I describe the formal model that considers this tension. Let w denote the prize an entrant receives if it successfully innovates. The details of the prize are discussed below, but in the meantime, we can conceptualize a supply function for industry innovative activity, $S(w)$. This function, $S(w)$, is literally the likelihood that a new product generation is developed today, and is increasing in w. The logic behind this idea is that a higher prize will encourage more entrants to expend more resources trying to innovate more quickly, thus increasing the likelihood of an innovation appearing today. Importantly, the innovation supply function is driven purely by the costs associated with R&D. As we will see below, many practices that are of antitrust concern are not dependent on R&D costs, and as such do not impact upon this supply function. That said, the supply function does not depend only on the response of the winning innovator, but rather, depends on the response of all potential entrants. Thus, a higher prize could induce more research start-ups into the industry. One antitrust concern is that incumbent practices could deter these start-ups but, as we will see below, this concern has an effect on the level of the prize, w, but, *ceteris paribus*, does not change the nature of the innovation supply function itself.

1.2.1. *Determinants of the Innovation Prize*

One key question that needs to be addressed is: what determines the size of the innovation prize, w? In this model, the prize is simply equal to the profit that the entrant receives if it generates an innovation today. One component of this is the immediate post-entry profits of an entrant in competition with the incumbent. This includes any revenues that the entrant receives net of the costs associated with entry. The second component consists of the *additional* future profits that are associated with being the innovation leader, above and beyond those profits appropriated to the laggard in the industry, that is, the *incumbency advantage* or *IA* for short. I assume that the entrant receives this bonus because its innovation generates a product that is superior to the current incumbent's product, thus allowing the entrant to displace the existing incumbent. Antitrust policy will have an impact on both the immediate profits of the entrant and the *IA*. Generally, we focus our attention on how such policies favor the entrant's immediate profitability, increasing the short-term component of the prize; that is, we think of antitrust policy in a static sense. In contrast, the impact on *IA* – which captures the dynamic component – is subtle. Clearly, if antitrust policy were to increase (or weakly increase) the expected profits of the incumbent (rather than the expected profits of the entrant), then the impact of such policies on the prize associated with incumbency would be unambiguous. In order to determine precisely how antitrust policy affects expected profits, however, we need to understand what components make up the *IA*.

The first main driver of *IA* is the expected future rate of innovation. The rate at which future entrants choose to innovate depends on the prize that they expect from innovation. This implies that *IA* is decreasing in the rate of innovation; intuitively, the expected lifetime of incumbency is equal to the expected length of time between new product improvements. The higher the rate of innovation, the shorter the lifetime of incumbency. What this implies, is that w is decreasing in the rate of innovation. Specifically, the tradeoff associated with greater innovation is the reduction in the innovation prize. This market constraint on the rate of innovation and the prize associated with innovation is akin to the demand constraint firms face in the market; firms can only sell more units if they are willing to do so at a lower price. Here the market is unable to offer both a high rate of innovation and a high prize.

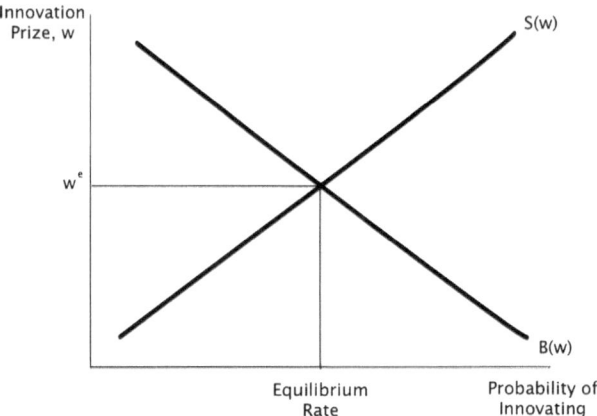

FIGURE 1.1: Equilibrium rate of innovation

This negative relationship between the benefits associated with innovation and the rate of innovation can be represented by a decreasing function, $B(w)$. We let $B(w)$ be the maximum likelihood of generating an innovation tomorrow in a market with innovation size w. Comparison of the supply and benefit functions associated with innovation highlights a fundamental tension. Since innovation supply is dictated by $S(w)$, a higher prize is needed if we want to encourage entrants to innovate more. In contrast, since $B(w)$ is decreasing in w, a lower prize is needed in order to sustain a higher level of innovation. SW note that the same tension between supply and demand exists in all markets. Moreover, as in any market, the innovation rate targeted must equal the innovation rate supplied, and so the intersection of $B(w)$ and $S(w)$ dictates the equilibrium level of w. Note from Figure 1.1, which illustrates this concept, if $w > w^e$, entrants want to supply a greater level of innovation than can be supported by the prize – hence, the prize will necessarily fall. In contrast, if $w < w^e$, the IA is too high and entrants do not want to supply too much innovation. Consequently, the innovation prize will rise to eliminate the shortage of innovation.

What is useful about this representation of antitrust policy is that it is relatively straightforward to examine the impact of policy changes on the equilibrium rate of innovation. For instance, if the only effect of antitrust policy was to increase immediate entrant profits, this would shift the B curve upwards and the new equilibrium point would result in a higher level of innovation. This is fairly intuitive as such profits are an important driver of the size of the innovation prize.

It is for this reason that it is so important that we understand *all* of the components that make up IA. SW use a dynamic equilibrium approach to analyze these components and the equations are presented in the appendix.[7] Here I will motivate the issues using a more intuitive approach by asking, *what is the maximum an entrant would be willing to pay to become an incumbent?*

---

[7] See Gans and Persson (2013) for details.

To begin with, note that the profit received by the incumbent in each period is less than it would receive were it a monopoly, since, in the presence of competition, it must compete with an entrant for profit. The maximum an entrant would be willing to pay to become the incumbent would be the additional amount it would earn if it were able to switch places with the incumbent. Were this possible, the entrant's payoff would rise by the difference between the incumbent's and its own profits in periods in which there is competition, and the monopoly profit in periods with no competition. There will always be competition in the period following the entrant's success in introducing a new product. Thus, the value of incumbency is a strict average (weighted by the probability of innovation in any given period) between the monopoly profit received when there is no competition (that is, when there is no innovation) and the difference between the incumbent and entrant profits when there is competition (that is, when an innovation occurs). In other words, the IA is a function of incumbency profits, weighted by the probability of entry.

1.2.2. *Impact of Antitrust Policy*

Two features of antitrust policy become particularly interesting once we view IA in this light. First, antitrust policy that increases immediate entry profits may lead to a change in w. As has already been noted, an increase in immediate entry profits raises the immediate payoff to the entrant from innovation. Note however, that a rise in immediate entry profits also reduces the incumbency advantage because the payoff associated with being an entrant also rises. Since the benefit attributed to the entrant is necessarily incurred in future periods, and thus discounted, whereas the rise in incumbency profits are immediate, the second effect is outweighed by the first. Importantly, however, this analysis demonstrates that once dynamic considerations are taken into account, the quantitative impact of antitrust policy may differ from the estimated cost or benefit derived from a static analysis.

Secondly, antitrust policy can affect the expected immediate payoff to the incumbent of innovation. Since the incumbent only receives this increase in profits in the absence of entry, the IA depends on the probability that the incumbent is not overtaken by a competitor. In this regard, if there is a practice that can reduce the probability that an entrant innovates, the incumbent will be willing to accept a reduction in its expected immediate payoff in order to reduce the probability of entry. That is, in an attempt to retain its incumbency advantage, the incumbent is willing to invest today in R&D deterrence tomorrow.

Putting these two features together generates an important result:

Outlawing any incumbent practice whose profitability is dependent on a reduction in entrant innovation will increase the equilibrium rate of entrant innovation.

It is important to note that this argument assumes that prohibiting the practice will raise immediate entrant profits. Similarly, it is critical to acknowledge the fact that, while incumbents might engage in practices that raise their profits, this does not necessarily mean that such practices raise the incumbency advantage – the driver of entrant innovation. So, even though the prize for entrant innovation is dependent on the expected profits of an incumbent, practices that are themselves only profitable should future entrant innovation be reduced will, it turns out, lead to outcomes that erode the incumbency advantage.

This result is consistent with Ordover and Willig's[8] definition of predatory behavior as any behavior that eliminates existing rivals. SW essentially extends this definition to include behavior that reduces the likelihood of innovative entry.[9]

1.3. USING STATIC ANALYSIS

According to Evans and Schmalensee,[10] in industries where dynamic considerations are important, competition for the market is more important for welfare than competition within the market. An interpretation of this is that when investigating industries in which dynamic considerations are important, antitrust authorities can be somewhat relaxed about practices that allow dominant firms in the market to increase their profits since increased incumbent profits will serve to stimulate innovative entry.

The SW framework both captures and refines this argument. For example, if prohibiting a practice causes a disproportionately large fall in expected incumbent profits per period relative to the increase in immediate entrant profits, then the rate of entrant innovation may fall as a result of antitrust policy. In order to identify which policies will lead to a fall in entrant innovation, we must first examine in more detail a range of different policies.

Evans and Schmalensee[11] point to the case of Microsoft and the District Court's decision that Microsoft's promotion of Internet Explorer as a competitor to Netscape was, in fact, anticompetitive. Their argument relied on the conjecture that Microsoft would not have expended such a large quantity of resources into the promotion of Internet Explorer had it not come to the conclusion that it was in a "winner-take-all" race to be the dominant browser. Thus, the District Court concluded that the profitability of Microsoft's investment was contingent on Netscape's exit.

At first glance, this argument contradicts the result of SW described in detail above. By considering the broader case as put forward by the Department of Justice and its economic expert Franklin Fisher,[12] however, it becomes apparent that these

---

[8] Ordover and Willig (1981).
[9] For a discussion of how this framework can be applied in a real litigation matter see Gans (2013) in relation to the Intel antitrust suits brought by the EC and FTC.
[10] Evans and Schmalensee (2002).
[11] Id.
[12] Fisher (2000).

two views are not, in fact, contradictory. The Department of Justice argued that Microsoft engaged in a variety of practices to promote Internet Explorer (including pre-installation and integration with their operating system), and that the profitability of each of these practices was contingent on deterring entry and innovation in its operating system and related products, rather than simply in the market for web browsers.

In this context, the difference between the two arguments put forward rests on the definition of the relevant market in which Microsoft was deemed to have been dominant. In Evans and Schmalensee's description,[13] the prohibition on activities that lead to an accelerated development of Internet Explorer would make little sense if the market in which Microsoft was deemed to be creating market power was simply the market for web browsers. Indeed, SW, who include an extension of their model to involve incumbent innovation, would argue that such innovation would not be a problem for competition and, moreover, that its promotion should be a goal of antitrust policy.

According to Fisher, however, Microsoft's investments in the development of Internet Explorer (and here I am leaving aside other related issues such as tying) would only be profitable in markets where it unambiguously held a dominant position – that is, in markets which extended beyond the market for web browsers.

The point here is that the evaluation of these competing arguments rests not so much on the notion that monopoly power is the prize associated with "winner-take-all" markets. Instead it relies on both the definition of which markets are relevant when evaluating anticompetitive practices (in this case, the relevant market might be the market for browsers, or a somewhat broader market), and also the possibility that, contingent on the presence of monopoly power in at least one of the relevant markets, the practice under consideration could raise barriers to entry The fact that the browser market may have been a "winner-take-all" market was important for understanding whether achieving market dominance would raise barriers to entry into the market for operating systems. In other words, the risk that prohibiting Microsoft's development of Internet Explorer would damage innovation in the web browser market did not depend on whether the market was a "winner-take-all" market or not. The dynamic considerations advanced by Evans and Schmalensee[14] did not appear to come into play.

A similar argument can be made in relation to other practices that were under investigation in the Microsoft case. Consider, as an example, the investigation into Microsoft's decision to tie or bundle Internet Explorer with the Windows operating system (either deeply or through pre-installation[15]). Evans and Schmalensee[16]

---

[13] Evans and Schmalensee (2002).
[14] Id.
[15] Carlton, Gans, and Waldman (2010) argue that pre-installation can lead to welfare reductions even when deep integration would be of benefit to consumers.
[16] Evans and Schmalensee (2002).

argued that "the analysis of tying claims in new-economy industries must consider the ubiquity of integration as a competitive strategy and the extreme risk of having judges and juries second-guess product design decisions" (p. 34). This argument, however, is not a dynamic one in which competition is primarily "for the market," but instead a static argument based on the idea that efficiencies are associated with certain product bundles that serve to reduce price or increase consumer value in the context of "within the market" competition. For SW, this same tradeoff is borne out if it was the case that prohibiting product integration substantially reduced ongoing monopoly profits, once again focusing attention on the static drivers of dynamic competition and innovation.

In summary, SW's analysis teaches us that when evaluating antitrust policies and their impact on innovation in a dynamic environment, we must focus on the impact of antitrust policies on period-by-period profits under both monopoly and competition. Traditional tools of static market analysis can, therefore, be applied to analyze whether a practice will harm entrant profits – often by raising standard entry barriers – in a way that is not justified by other efficiencies, or in a way that reflects behavior that would normally take place in an industry that does not face the prospect of entrant innovation. Whether the market is "winner-take-all" is not something that will necessarily prove decisive in such an analysis.

1.4. ANTITRUST AND COOPERATIVE COMMERCIALIZATION

Much antitrust analysis, including the discussion and papers referred to above, assume that entrant innovators commercialize new technologies primarily by entering product markets and competing head to head with incumbent firms. In actuality, this is not the dominant path for commercialization in many industries. Rather, entrant innovators engage in cooperative commercialization, contracting in some manner with incumbent firms. For example, they might license their patents to established firms, joint venture with them, or be acquired.

1.4.1. *Rationale for Cooperative Commercialization*

The rationale for the pursuit of such cooperation is straightforward. First, entry into a product market following successful innovation can be costly. In particular, incumbents may already have complementary assets (in marketing, distribution, and regulation) that would involve large sunk expenditures to replicate. Moreover, these complimentary assets would themselves be devalued by duplicative entry. As Teece argues,[17] such duplication can be avoided through contracting between entrants and incumbents.

---

[17] Teece (1987).

Second, and of particular relevance to this discussion, contracting can allow entrants and incumbents to avoid head-to-head competition – even if only temporary – and consequent rent dissipation.[18] Again, this is a mutual benefit that can be realized through some cooperative arrangement.

Avoiding duplication and rent dissipation are two static reasons why cooperative commercialization might arise. From an antitrust perspective, evaluating the deals themselves would be similar to analyzing the consequences of a merger – specifically, we are interested in whether the savings associated with not having to duplicate fixed assets outweigh the potential harm to consumers that arises from a lack of competition. The question here, however, is whether there are dynamic considerations that make it necessary to analyze such transactions differently to the static analysis of mergers. Similarly, we are concerned with whether incumbent behavior influences the terms of transactions differently under static and dynamic considerations.

To understand how this question can be answered, consider the total payoff to the incumbent and an innovating entrant if they cooperate immediately after a new product is generated. This is given by:

Per Period Monopoly Profit + Market Leader's Future Profits + Entrant's Future Profits as a Non-Producer

That is, the incumbent and the innovating entrant will, despite the new innovation, jointly earn a monopoly profit, the incumbent will remain the market leader and the entrant will earn future profits as a continuing innovator (but non-producer) in the industry. In contrast, if they fail to cooperate, they will return to the competitive scenario as examined earlier. In this case, their total payoff will be:

Immediate Profits in Competition + Market Leader's Future Profits + Incumbent's Future Profits as a Follower

That is, in the absence of cooperation, the incumbent and entrant will remain in immediate competition, the entrant will become the market leader, and each party will earn their respective future profits. In this scenario, the incumbent will take on the role of follower.

In comparing these outcomes, notice that, under cooperation, the incumbent continues to earn market leader profits in the future, whereas under cooperation, after an initial bout of competition, the innovating entrant takes the market leader position. Thus, a cooperative deal will be struck if the total payoff from cooperation exceeds the total payoff from competition; that is:

Immediate Gain in Monopoly Rents > Incumbent's Future Profits as a Follower – Entrant's Future Profits as Non-Producer

---

[18] Gans and Stern (2000).

Notice that the market leader's future profits do not play a role in determining whether a cooperative deal is struck since these profits will be realized regardless. Notice, however, that in order for cooperation to occur, the static drivers of cooperation – namely, saving duplicative assets and avoiding rent dissipation – must be such that future profits (if any) of the incumbent outweigh what the entrant would receive (by never being the incumbent; that is, not producing). Consequently, it is possible that a deal might not be struck if, in the absence of market leadership, the incumbent was able to earn a high level of future profits (say, by bouncing back and regaining that leadership quickly), while the entrant was not. Thus, while in a static environment, the case for licensing is compelling, in an environment in which dynamic considerations play an important role, the case for licensing is less conclusive.

1.4.4.1. Determinants of the Innovation Prize

What we are interested in, of course, is what drives innovation when licensing is expected. In order to examine the determinants of the innovation prize, suppose that the condition above that supports a licensing outcome holds. If this is the case, the entrant's innovation prize is simply the license fee (or other payment) it expects to receive as a result of innovating, plus the difference between its future profits with and without an immediate innovation. Thus, by innovating today, the entrant will benefit from a license payment, will "lose" the profits they would have received had they not generated an innovation, and will gain the profits they would receive in the post-innovation environment. It is likely that these post-innovation future profits are less than or equal to the profits that the entrant would expect to receive from its innovative efforts in the current generation.

To understand the nature of this prize, the first task is to pin down the license fee. Here I use the Nash bargaining solution that posits that the license fee will be that which equates each party's payoff less their outside option. In this case[19] the prize becomes:

$$w = \frac{1}{2} \text{(Static Gains from Licensing)} + IA$$

$$+ \frac{1}{2} \text{ Entrant's Future Profits with Innovation under Competition}$$

$$- \frac{1}{2} \text{ Incumbent's Future Profits without Innovation under Competition}$$

Notice that the prize has three components. The first is a share of the static gains from cooperation. The second term is the pre-innovation incumbency advantage of being the incumbent rather than the entrant innovator; this concept is the same as the one that played a role in determining the prize under competitive commercialization.

---

[19] As is demonstrated in Gans and Persson (2013).

The final term is a share of the advantage (if any) in terms of profits that the entrant has over the incumbent if it displaces the incumbent as the market leader, that is, the post-innovation incumbency advantage.

1.4.2. Impact of Antitrust Policy

To understand the impact of antitrust policy, we need to understand the primitive drivers of the innovation prize under cooperation. It is demonstrated[20] that w is increasing in the difference between (a) the per period profits of a competitive entrant and (b) the per period profits of an incumbent. w is also increasing in discounted per period monopoly profits. Thus, in contrast to the case of competitive commercialization, when cooperation is expected any antitrust policy that shifts rents under competition from the incumbent to the entrant (that is, increases entrant profits while decreasing incumbent profits) will increase the innovation prize, so long as the reduction in ongoing monopoly profits is not too large.

What is significant is that, compared with the competitive case, under cooperative commercialization, any negative impact of antitrust policies on incumbent profits under competition stimulate rather than reduce incentives for entrant innovation. This is because the entrant shares in the incremental benefits that the incumbent receives from avoiding competition and preserving monopoly. Thus, antitrust authorities should be primarily concerned with actions that both shift rents from the incumbent to the entrant and take place prior to or immediately following an entrant innovation, but before any licensing deal is signed.

One such example is exclusive customer contracts. SW analyzed these in the context of competitive commercialization and argued that, should an entrant innovate, an incumbent has an incentive to provide customers with discounts if they agree not to purchase from the entrant. Exclusive contracts have the effect of limiting the share of customers that the entrant can compete for (thus reducing immediate entrant profits under competition), but as SW show, after the initial period, the incumbent ceases to provide consumers with a discount. Consequently, restricting such contracts improves entrant profits while leaving expected incumbent profits per period (after the initial entry period) unchanged. Thus, such a prohibition is likely to increase the rate of entrant innovation.

The level of the discount required to entice consumers to sign an exclusive contract is related to the rate of entrant innovation. The sooner consumers expect entrant innovation, the larger the discount necessary in order to compensate consumers for the prospective inability to purchase the improved product in the future. When there is cooperative commercialization in the industry, it is the incumbent rather than the entrant who provides new products to the market following entrant innovation. Knowing this, consumers incur no costs associated with signing an

[20] Id.

exclusive deal with the incumbent. This shrinks the size of the entrant's potential customer base, should a licensing deal not be signed, thus lowering the entrant's outside option, and hence lowering the license fee agreed upon between the entrant and the incumbent. As a result, exclusive contracts have the effect of increasing incumbent profits under competition and reducing ongoing monopoly profits. A prohibition on exclusive customer contracts would have the reverse effect. Thus, exclusionary practices are less costly for incumbents when there is cooperative commercialization. Such a prohibition would likely increase the innovation prize.

Another example of a practice that should alert antitrust authorities is accelerated product development by incumbents. These practices are designed to reduce license or acquisition fees by encouraging customer purchases of products that are known to be inferior to the entrant's new product. Of course, this suggests that if antitrust authorities observe intense competition between the incumbent and the entrant, a possible enforcement mechanism would be to prohibit cooperative activities. Knowing this, however, the incumbent would refrain from intense competition in an attempt to avoid prohibition.

Nonetheless, playing these games is difficult and risky. For instance, the incumbent might refrain from product development for fear of being prohibited from cooperative deals, even when this product development is efficient. In addition, if the entrant is unlikely to be successful in penetrating the market a cooperative deal may be the only means by which the entrant's innovation can be commercialized.[21] If this is the case, the prohibition may be more harmful to the entrant than the incumbent.[22]

The broader point here is that when cooperative commercialization is a means by which entrant innovation is stimulated in an industry, the evaluation of those cooperative deals from an antitrust perspective does not currently move beyond a static analysis. It would be better to permit such deals on the basis of dynamic considerations – since the overall impact may be a rise in the rate of innovation. Thus, antitrust policy should be permissive of cooperative commercialization deals, but should remain aggressive with respect to competitive practices and should directly limit these practices rather than resort to indirect punishments.

1.4.3. Merger Analysis

The above analysis suggests that vigilant antitrust enforcement of practices that shift rents from entrants to incumbents in periods of innovative entry will have the additional impact of stimulating entrant innovation in industries where cooperative commercialization is the norm. This lends support to the notion that dynamic considerations strengthen the case for strong antitrust action.

---

[21] Of course, in that situation, the incumbent may not want to discourage or to even be seen to discourage entrant innovation (Gans and Stern 2003).

[22] Of course, such factors could protect incumbents in useful ways (Rasmussen 1988).

But what effect do other antitrust policies, say, in relation to the cooperative commercialization deals, have on entrant innovation? I suggested earlier that if such deals were primarily associated with preventing rent dissipation, static merger analysis could be applied, but this would overestimate consumer surplus today and fail to acknowledge any potential reduction in innovation in the future. That is, static analysis of these situations focuses on "in the market" competition rather than "for the market" competition.

But what if competition is really "winner-take-all" in the sense that Evans and Schmalensee envisage it,[23] and that consumers do not actually benefit in the short run? Were this the case, the entrant would displace the incumbent as the monopolist immediately, and there would be no instantaneous competitive profits. Yet, the question still begs to be asked: does this necessarily imply a permissive licensing or merger policy?

These issues are addressed in further detail in Gans,[24] but I nonetheless briefly highlight the main results here. Consider a cooperative licensing arrangement. As noted above, when a licensing deal is struck, the incumbent retains its role as market leader and the entrant is free to develop future innovations. Thus, in determining the rules of the licensing agreement, the parties are effectively negotiating over which role each will take in expanding R&D resources for the next generation. Given their current position, both the incumbent and the entrant have a joint interest in ensuring that resources are provided sparingly rather than excessively.

As is well known,[25] one problem that arises is that the incumbent has less incentive than the entrant to expend resources in R&D since innovation hastens the cannibalization of the incumbent's existing product. Thus, if the incumbent and entrant are asymmetric with respect to their capabilities in research, the next incumbent will be the firm with the greatest propensity to innovate. The conditions for dynamic economic efficiency, however, are the converse.

This suggests two somewhat contradictory statements: firstly, that when parties have an incentive to license we should prohibit them from doing so, and secondly, when parties do not wish to license, we should encourage them to do so. While the economic argument for such a policy is straightforward (indeed, it was simply explained above), it is easy to see that such a policy would be hard to enact. Nonetheless, a prohibition on licensing, which is possibly an easier policy to implement than one that encourages licensing, would still provide some welfare improvement according to this logic, since it would prevent licensing when firms wish to do so, and would have little effect on welfare when firms do not seek out licensing possibilities.

Mergers are different from licensing deals in that both firms become the incumbent in the former but not the latter. When the incumbent has fewer research

---

[23] Evans and Schmalensee (2002).
[24] Gans (2009).
[25] See Arrow (1962); Reinganum (1989).

opportunities than the entrant, a merger will be mutually advantageous since it allows the consolidation of research capabilities. In contrast, should this consolidation of research capabilities make a third party, and stronger, research organization likely to innovate more intensely than if the two were separated, the incumbent and potential entrant will not choose to merge. The end result is that, when pursued, mergers are likely to result in dynamic inefficiencies.

These issues have been recognized by the Department of Justice in their proposed new Merger Guidelines. They write that the merger may reduce the incentive to introduce new product development and this

> ... effect is most likely to occur if at least one of the merging firms has capabilities that are likely to lead it to develop new products in the future that would capture substantial revenues from the other merging firm. The Agencies therefore also consider whether a merger will diminish innovation competition by combining two of a very small number of firms with the strongest capabilities to successfully innovate in a specific direction.[26]

Their focus is on the innovative capabilities that exist before and after the merger. As discussed above, in Gans (2009), I emphasize the importance of the incentives that are matched with those capabilities. Specifically, high capabilities in the hands of those who are not the market leader is likely to increase the rate of innovation in an industry. Mergers impact on this matching as does merger policy.

The purpose of this argument is not to suggest that licensing and merger deals are undesirable; the environment considered here is unique, and one in which many other considerations may come into play. Rather, the point that I wish to raise is that it is far from obvious that dynamic considerations lead us to a more permissive antitrust stance on cooperative arrangements. It is certainly true that dynamic considerations are a key issue that needs to be addressed, but it is not clear that we currently possess the tools to adequately address these concerns in general. In the meantime, while we wait for these tools to be developed, a case-by-case approach has some advantages.

1.5. CONCLUSIONS

This chapter uses the recent analysis of SW to analyze how antitrust authorities should alter their evaluation procedures and which tools should be used to conduct this analysis when dynamic considerations (specifically, the importance of competition for the market) play an important role. I argue that, in many cases, the tools associated with static analysis are the appropriate tools to apply even when dynamic considerations are an important factor. This is because the primitives upon which this analysis is based – market definition and practices that raise entry barriers – drive dynamic innovation "for the market" just as they drive competition "in the market."

---

[26] Department of Justice (2010) 23.

Nonetheless, in some industries entrant innovators commercialize via contracting (for example, licensing) or by being acquired by existing incumbents (through mergers). In these cases, applying static tools to evaluate antitrust practices poses special difficulties; the cooperation in these industries is itself a difficult issue for antitrust authorities to analyze since it drives innovation in the industry. Complications also arise because these practices are, to some extent, salient in situations where cooperative commercialization breaks down, and so it is the threat of such practices rather than the ability to enforce these practices that is the key issue. Static tools are simply not equipped to address these concerns. Moreover, it is not clear whether antitrust policy should be permissive or vigilant.

While dynamic considerations have played an important role in the evaluation of antitrust matters in the past (for instance, for IBM in the 1980s and Microsoft in the 1990s), they are being raised today as a routine matter in antitrust. This includes the evaluation of mergers such as the industry maneuvers between Google and Microsoft over Yahoo!, the interaction of intellectual property law and antitrust in the Google Books settlement deal and the evaluation of exclusionary practices by firms (most recently, the EU and FTC investigations of Intel). In each case, there is a tension between those who utilize static tools and those who argue for more complicated dynamic analysis. While the appropriate antitrust policy to be applied will always be a matter of identifying tradeoffs, the analysis here suggests that the static tools will likely do a reasonable task in fulfilling that function.

Hopefully, future research will be able to further untangle these issues. In particular, the role of intellectual property protection in providing a separate instrument to stimulate innovation is an important issue that needs to be addressed. Brennan argues that if intellectual property law can be used to preserve innovative rewards, there is no reason why antitrust authorities should be less vigilant in pursuing activities than they would be were they simply to rely on static analysis to draw conclusions regarding optimal policy choice.[27] A similar argument is put forward by SW, who provide an example in which the optimal patent policy completely eliminates any ambiguity as to whether prohibiting certain practices will be harmful to innovation. The interaction of IP protection and antitrust policy is therefore likely to be an important focus of future research in this area.

REFERENCES

Arrow, KJ. 1962. Economic Welfare and the Allocation of Resources for Invention. In *The Rate and Direction of Inventive Activity*. Princeton, NJ: Princeton University Press, pp. 609–25.

Brennan, TJ. 2007. Should Innovation Rationalize Supra-Competitive Prices? A Skeptical Speculation. In A. Fredenberg (ed.), *The Pros and Cons of High Prices*. Stockholm: Konkurrensverket/Swedish Competition Authority, pp. 88–127.

---

[27] Brennan (2007).

Carlton, D., JS. Gans, and M. Waldman. 2010. Why Tie a Product Consumers Do Not Use? *American Economic Journal: Microeconomics*, 2(3), 85–105.

Department of Justice. 2010. Horizontal Merger Guidelines: Proposed. Washington, DC: Mimeo.

Evans, DS. and R. Schmalensee. 2002. Some Economic Aspects of Antitrust Analysis in Dynamically Competitive Industries. In A.B. Jaffe, J. Lerner and S. Stern (eds.), *Innovation Policy and the Economy*, Vol. 2, Cambridge, MA: MIT Press, chapter 1.

Fisher, FM. 2000. The IBM and Microsoft Cases: What's the Difference? *American Economic Review*, 90(2) 180–3.

Gans, JS. 2009. *Negotiating for the Market*. Melbourne: Mimeo.

2010. When is Static Analysis a Sufficient Proxy for Dynamic Considerations? Reconsidering Innovation and Antitrust. In J. Lerner and S. Stern (eds.), *Innovation Policy and the Economy*, Vol. 11. Cambridge, MA: NBER.

2013. Intel and Blocking Practices (2010). In J. Kwoka and L. White (eds.), *The Antitrust Revolution*, 6th edition. Oxford University Press.

Gans, JS. and L. Persson. 2013. Entrepreneurial Commercialization Choices and the Interaction between IPR and Competition Policy. *Industrial and Corporate Change*, 22(1), 131–51.

Gans, JS. and S. Stern. 2000. Incumbency and R&D Incentives: Licensing the Gale of Creative Destruction. *Journal of Economics and Management Strategy*, 9(4), 485–511.

2003. The Product Market and the "Market for Ideas": Commercialization Strategies for Technology Entrepreneurs. *Research Policy*, 32(2), 333–50.

Gilbert, R. 2006. Looking for Mr. Schumpeter: Where Are We in the Competition Innovation Debate? In Adam B. Jaffe, Josh Lerner, and Scott Stern (eds.), *Innovation Policy and the Economy*, 6, National Bureau of Economic Research.

2007. Holding Innovation to an Antitrust Standard. *Competition Policy International*, 3, 47–77.

Gilbert, R. and D. Newbery. 1982. Preemptive Patenting and the Persistence of Monopoly. *American Economic Review*, 72(3), 514–26.

Manne, G. and J. Wright. 2009. "Innovation and the Limits of Antitrust," George Mason Law & Economics Research Paper No. 09–54; Lewis & Clark Law School Legal Studies Research Paper No. 2009–26. Available at SSRN: http://ssrn.com/abstract=1490849.

Ordover, J. and R. Willig. 1981. An Economic Definition of Predation: Pricing and Product Innovation. *Yale Law Journal*, 9, 8–53.

Rasmussen, E. 1988. Entry for Buyout. *Journal of Industrial Economics*, 36(3), 281–99.

Reinganum, J.F. 1989. On the Timing of Innovation. In R. Schmalansee and R. Willig (eds.), *Handbook of Industrial Organization*, Vol. 1. Amsterdam: Elsevier, pp. 849–908.

Segal, I. and M. Whinston. 2007. Antitrust in Innovative Industries. *American Economic Review*, 97(5), 1703–30.

Teece, DJ. 1987. Profiting from Technological Innovation: Implications for Integration, Collaboration, Licensing, and Public Policy. In DJ. Teece (ed.), *The Competitive Challenge: Strategies for Industrial Innovation and Renewal*. Cambridge, MA: Ballinger, pp. 185–220.

2

Antitrust and Intellectual Property: Developments in Pharmaceuticals

Sumanth Addanki

2.1. INDUSTRY BACKGROUND: WHAT MAKES IT SPECIAL?

2.1.1. *Overview*

In an industry in which firms compete in large part by developing new products, where the average cost to develop and bring a new product to market is well over $2.5 billion, and where firms spend upwards of 11 percent of their sales revenue on research and development (R&D), it is no surprise that firms devote a good deal of time and resources to creating, protecting, and generally worrying about intellectual property (IP).[1] The vast bulk of R&D expenditures in the pharmaceutical industry is undertaken by innovator or branded drug manufacturers seeking to invent and commercialize new drugs or find new uses for existing drugs. These branded drug products embody a broad array of intellectual property: patents covering various aspects of the drug, from the underlying molecule through manufacturing process innovations to delivery forms and even methods of use; trademarks and trade dress associated with the drug and its packaging; and the clinical research and data associated with the development, approval, and sales of the drug.

As with any IP-intensive industry, issues relating to IP often loom large in antitrust analysis pertaining to pharmaceuticals, but analysis of the pharmaceutical industry is also further affected – perhaps uniquely so – by the industry's institutional and regulatory idiosyncrasies. There are at least three such factors that are important to keep in mind. First, there is the gatekeeping role played by the primary government regulator – the Food and Drug Administration (FDA) in the United States and analogous regulatory bodies abroad – among whose functions is to limit market participation to those pharmaceutical products that have been demonstrated (to the

---

I am indebted to Alan Daskin, Stephanie Demperio, Bryan Ray and Claire Xie for valuable comments and discussion.

[1] See Lamberti and Getz (2015). Also, National Science Foundation (2015).

agency's satisfaction) to be safe and efficacious. The second factor is the disconnect among the parties involved in consuming pharmaceuticals: the decision maker choosing which product will be prescribed (typically the physician); the entity paying for that product (generally some combination of the patient and a third-party payer); and the actual consumer (the patient). This disconnect makes analysis of the market very different from, say, the market for cars or bread, where one entity – a household or individual consumer – chooses the product, pays for it, and consumes it. The third factor is the regulatory overlay governing the patent protection and other types of exclusivity granted to innovator (i.e., branded drug) manufacturers as well as the "rules of the road" governing competition between branded drugs and their generic substitutes. I will discuss each of these factors in more detail and then devote the remainder of this chapter to a discussion of interplay between IP and these factors in analyses of some of the antitrust questions that are of current interest in the industry.

2.1.2. The Role of the FDA in Approving Brand and Generic Drugs: A Brief Overview

Many of the interesting antitrust issues that arise in pharmaceuticals arise in the context of the entry of (and subsequent competition engendered by) new drugs, so understanding the role played by regulatory bodies such as the FDA in this regard can be of critical importance.[2] A manufacturer wishing to introduce a new branded drug into the marketplace must first obtain FDA approval for doing so. The process leading to approval, which starts with filing an Investigative New Drug application (IND) and culminates in a New Drug Application (or NDA), involves several stages of clinical testing of the proposed product, starting with small-scale evaluations of its safety and proceeding to controlled clinical trials to assess its safety and efficacy, eventually involving hundreds or thousands of subjects. The manufacturer typically works closely with the FDA in designing and monitoring these trials, and the whole process leading up to approval (if granted) can take ten to fifteen years and today costs, on average, well over $2.5 billion.[3] Given the time, resources, and uncertainty involved in this process, manufacturers ensure that their branded drugs in development are generally protected by one or more patents, so that if the product is eventually approved for sale, they will have an opportunity to recoup these investments without the threat of immediate competition.[4]

[2] For brevity and convenience, I will frame my discussion here in terms of US regulatory structures, but there are generally analogous structures in many other jurisdictions, and their impact on antitrust analyses are broadly similar.
[3] Of course, these are average numbers and there is considerable variation across drugs. See Lambert and Getz (2015).
[4] The cursory treatment here is not intended to be an exhaustive description of the NDA process but is merely meant to highlight its important features. For instance, when a new drug also contains a molecule that has not been available to the market before, a so-called New

In contrast to the prolonged and expensive route to approval under an NDA, the FDA offers a greatly abbreviated pathway to market entry for a manufacturer seeking to introduce a generic copy of an existing approved branded drug. By submitting data demonstrating that its proposed generic copy is *bioequivalent* to the branded drug at issue, i.e., contains the same active ingredient in the same proportions and is metabolized in the same fashion within the human body as is the branded drug, the generic manufacturer can submit an Abbreviated New Drug Application or ANDA to the FDA. If there are unexpired patents covering the branded drug at issue (which the brand manufacturer is legally obliged to list in an FDA publication known as the Orange Book), the ANDA filer has to certify, for each such patent, either that it will only enter the market upon expiration of the patent or that the patent is invalid and/or will not be infringed by the proposed generic product. The latter certification creates a technical act of infringement, which entitles the patent holder to sue the ANDA filer. It also automatically places the ANDA on "hold" at the FDA for a thirty-month period, upon the expiration of which the FDA may approve the ANDA if the filer satisfies all the other requirements involved, *even if there has been no final adjudication of the patent litigation, if any.* Thus, the ANDA process provides a swift path to market for generic competition and leaves open the possibility that a generic competitor may be approved and enter the market even while patent litigation remains pending between the brand and generic manufacturers. As we will see below, this regulatory structure has significant implications for the antitrust analysis of firm conduct.

2.1.3. *Third-party Payers and the Principal/Agent Problem*

For most products and markets with which we are generally familiar, the decision maker choosing whether to buy a product (or which product to buy) is also the economic entity – an individual, household, or firm – paying for the product and consuming or otherwise using it. The situation with prescription pharmaceuticals is very different. In the typical pharmaceutical purchasing situation, the primary decision maker selecting the product that will be consumed is the physician, who has no pecuniary interest in the outcome of the selection: he/she pays nothing toward the cost of the drug and makes no profit from its sale or consumption. The actual consumer – the patient – typically pays a small co-payment or co-insurance amount. The bulk of the cost is borne by a third-party payer (TPP), frequently an insurer. Depending upon the degree to which a given drug has effective therapeutic substitutes, the TPP may well be able to demand, and obtain, sizeable rebate payments from the manufacturer of the drug.

> Chemical Entity or NCE, upon approval it automatically obtains five years of exclusivity, whether or not it is also protected by patents. For a more complete discussion of the NDA process, see the FDA's website US Food and Drug Administration, *Applications for FDA Approval to Market a New Drug,* available at www.accessdata.fda.gov/scripts/cdrh/cfdocs/cfcfr/CFRSearch.cfm?CFRPart=314.

It should be obvious, then, that the situation is very unlike, say, the purchasing decision for white bread, where a 5 or 10 percent coupon on Brand A might well prompt a consumer who generally buys Brand B to opt for the discounted Brand A offering instead. There are few such short-run price incentives in pharmaceuticals, largely because of the disjunction among decision maker, consumer, and payer. Thus, there is no point along the distribution chain where one might observe short-run price changes that could be expected to engender changes in consumption patterns.

Note that this is *not* to say that the demand for a particular drug is entirely unresponsive to price, or even necessarily inelastic, as some have suggested. In fact, various institutional features of the industry often do constrain the pricing of pharmaceuticals. The prevalence of TPPs in health care has resulted – as economics predicts – in a tendency to over-consume health care products and services.[5] The TPPs have responded to this, in part, by trying to control reimbursement (i.e., their costs) in a number of ways. One important vehicle for such cost control in the pharmaceutical area is the use of *formularies* and tiered reimbursement for drugs. Tiered reimbursement simply means that, in each therapeutic category, the drugs available are organized for reimbursement purposes into groups, with the most preferred (Tier 1) group containing those drugs for which the TPP incurs the lowest net reimbursement costs, either because the drugs are available cheaply to pharmacies from a number of competing suppliers or because the manufacturers of the drugs pay substantial rebates to the TPP. These Tier 1 drugs are also associated with the lowest out-of-pocket expenditure by the insured patient. Successively higher-tier products require greater co-payments by the insured patient and, in some instances, the highest tier may allow for no reimbursement at all by the TPP. The formulary is just the grouping of drugs into these tiers. The grouping into tiers, as discussed above, is based on the relative costs incurred by the TPP in reimbursing pharmacies for dispensing the drugs, with drugs in higher tiers costing the TPP relatively more to reimburse. The tiers encourage the use of drugs that result in the lowest net reimbursements by the TPP within the relevant therapeutic class.[6]

Not surprisingly, manufacturers of branded drugs respond to this by jockeying for the most preferred formulary status that they can achieve, generally by offering rebates to the TPPs based on the volume of sales/reimbursements of their drugs. If a manufacturer of a drug attempted to increase its effective price by scaling back the

---

[5] "The way we pay for health care encourages both patients and physicians to overuse resources. On the average, every time a patient spends a dollar on health care, 79 cents of it is paid by employers, insurance companies, government and charitable giving. That encourages patients to purchase services that they would not purchase if they paid the full bill" (Robbins, Robbins and Goodman 1994). Also see "Heading for the Emergency Room," *The Economist*, June 25, 2009.

[6] A number of mechanisms affect this: patients may ask physicians to prescribe preferred drugs where they will do as well as non-preferred alternatives; physicians, aware of formulary lists, might spontaneously prescribe preferred drugs; and pharmacies do their part by dispensing generic products (usually Tier 1) wherever possible.

TPP rebates it paid, the TPPs would respond by placing the drug on a less preferred formulary tier, typically resulting in fewer prescriptions for that drug (assuming that there are therapeutic substitutes available). In other words, the quantity demanded *will* respond to the net price charged by the manufacturer.

Thus, the problem is not that demand is unresponsive to price. Rather, the problem facing the economic analyst is a logistical one – there is no convenient observation point (akin to the supermarket scanner) at which one could observe frequent price movements and their impact on consumption decisions. Given the unavailability of suitable data, methods that have become commonplace elsewhere in antitrust, such as the use of econometric demand analyses to define markets or to characterize the closeness of competition between products in the market, are generally not feasible, and the analyst seeking answers on market definition and substitutability needs to look elsewhere, as I discuss in more detail below.[7]

2.1.4. *Regulations Governing Brand/Generic Substitution*

The regulations and institutional arrangements governing the substitution between branded pharmaceuticals and their generic equivalents result in competitive environments rather different from those that obtain in most markets. As noted above in the section on FDA drug approval regulations, generic competition can be introduced via the ANDA process, which, when successfully concluded, results in the generic entrant's receiving so-called "AB rating" to the branded product, reflecting the FDA's determination that the brand and generic are bioequivalent. Also, in order to qualify as a generic substitute, the entrant must set its list price at a discount to the brand version's list price. Once the entry of this AB-rated generic version occurs, a combination of state regulations and financial incentives ensure that when a prescription is presented to a pharmacy for the branded drug, the pharmacist either is required by law to fill it with the generic product or has strong financial incentives to do so (or both).[8]

Moreover, generic drugs are generally on the most preferred tier (Tier 1) of a formulary, which generally means that they require the lowest co-pay amount from

---

[7] I should stress that it is the econometric demand analysis of the sort frequently done in consumer goods mergers – which generally relies on substantial price variation over time and across geographies – that is infeasible. It is entirely possible, even in pharmaceutical markets, that history provides natural experiments that can be exploited via econometric analysis to shed some light on questions of substitutability among products.

[8] In some states, such substitution is "automatic" (i.e., mandatory) unless the physician writes "DAW" (dispense as written) on the prescription; in other states, such substitution by the pharmacist is permissible. For further discussion of state laws regarding substitution, see Vivian (2008). ("All states in the U.S. have laws addressing generic substitution to one degree or another. There are 'positive formulary' states, which identify generics that can be substituted, and there are 'negative formulary' states, which list drugs that cannot be substituted. There are also states that do not refer to *Orange Book* standards and have ne[i]ther a positive nor a negative formulary, and where pharmacists are permitted to perform generic substitution so long as the drugs are pharmaceutically equivalent.")

the patient, which in turn means that even if state law or the pharmacist do not mandate or recommend generic substitution, the patient has the incentive to request it of his or her own accord. Consequently, the entry of an AB-rated generic will necessarily take sales away from the branded product and, because the generic must be priced lower in order to qualify as a generic substitute, the weighted average cost to the pharmacy will be lower after such entry, *entirely regardless of whether the branded product faced vigorous competition from a dozen other therapeutic substitutes or was the sole therapy available to treat a particular condition prior to generic entry.* As we will see below, this fact has important implications for the analysis of monopoly power.

2.2. ANTITRUST ANALYSIS OF COORDINATED AND SINGLE FIRM CONDUCT: ASSESSING MONOPOLY POWER AND COMPETITIVE EFFECTS

2.2.1. *Introduction: Two Interesting Types of "Conduct" Cases*

The impact that the institutional peculiarities of the pharmaceutical industry discussed above can have on antitrust analysis can be particularly acute in analyses of potentially anticompetitive joint or single firm conduct.[9] I will discuss two types of such inquiries, both of which have attracted a good deal of recent interest: so-called "reverse payment" or "pay-for-delay" cases (which allege joint anticompetitive conduct) and "product hopping" cases (which allege unilateral anticompetitive conduct). "Pay-for-delay" cases, which are now attracting interest in jurisdictions all over the world, have a long and somewhat circuitous jurisprudential history in the United States, culminating in the Supreme Court's decision in *Actavis*. I will not dwell here on either this history or the *Actavis* decision itself, both of which have been exhaustively covered in the literature,[10] except to point out that *Actavis*'s primary conclusion – that such cases should be tried as rule of reason cases – is the right outcome from the standpoint of economics.[11]

Reverse payment cases arise in the context of patent litigation between the seller of a branded drug covered by a patent (the NDA holder) and a would-be generic competitor who seeks to enter the market with its generic substitute either by invalidating the patent at issue or by establishing that its generic product does not infringe (the ANDA filer who has made a so-called Paragraph IV certification). Before final adjudication of patent validity or infringement issues, the parties settle the lawsuit, with the ANDA filer agreeing not to enter until some

---

[9] To be sure, merger analysis is not immune to the problems created by these factors, as I discuss briefly at the end of this chapter.
[10] Addanki and Butler (2014).
[11] It is also the framework I recommended and used in 2001, when I served as an economic expert in the FTC's case against Schering-Plough and Upsher Smith.

future date before patent expiration and the patent holder (the branded drug supplier), according to the plaintiffs in these cases, making a payment of some kind to the ANDA filer. The plaintiffs' antitrust theory is that the "reverse payment" ("reverse" because most patent infringement suits are settled by a payment from the alleged infringer to the patentee, rather than the other way around) is intended to – and does – delay entry by the generic beyond the date on which it would have insisted but for the payment.[12] In effect, according to the plaintiffs, in exchange for preserving its "monopoly profits" longer, the brand manufacturer agrees to share some of those added profits in a collusive scheme with the generic.

Product hopping cases also arise in the ANDA context, but they do not involve allegations of collusion between brand and generic; in fact, in at least one recent case, the generic firm involved was actually a plaintiff.[13] In these cases, the branded drug manufacturer facing the prospect of the expiration of the patent(s) protecting its drug (i.e., those listed in the Orange Book for this drug), and the subsequent competition that might be expected from generic entry, introduces a new formulation of the drug, often protected by newer patents, markets the new version to physicians and de-emphasizes the marketing of (or even entirely discontinues supplying) the original formulation.[14] Extended or delayed release versions, tablets instead of capsules, and different strengths or dosages are some of the variations that have been involved in these cases. The theory of competitive harm here is that the innovation represented by the new formulation (and the patent(s) covering it, if any) is trivial or pretextual, and that the only reason to "move the market" to the new formulation is to avoid competition from generic substitutes. But for the move, according to this theory, prescriptions for the original branded drug would have been filled with the AB-rated generic product once it became available. However, if physicians were persuaded to switch their prescribing to the new formulation, pharmacists would not be able to dispense the generic in its place, because the generic product whose entry is imminent would not be AB-rated to the new formulation. Thus, the plaintiffs allege, generic competition has been illegally thwarted.

---

[12] Recognizing, perhaps, that this assumes that a settlement without a payment would have been possible, the plaintiffs sometimes argue in the alternative that the payment results in entry later than the entry date which represents the mathematical expected outcome of litigation had it been pursued to a final resolution, i.e., the weighted average of the likely generic entry date should the ANDA filer prevail in the lawsuit and the likely entry date (presumably post-patent-expiration) should it lose, weighted by the probabilities of each of those outcomes.
[13] See Mylan Pharmaceuticals, Inc. v. Warner Chilcott plc, et al. Civ. No. 12–3824 (E.D. Pa. Apr. 16, 2015).
[14] While this is the "standard" fact pattern, product hopping allegations have also been made where there are no unexpired patents at issue and the event precipitating the "hop" was the filing of an ANDA on the original formulation.

2.2.2. *The Rule of Reason Is the Appropriate Mode of Analysis*

In both types of cases, it would be a mistake from the economic standpoint to view the alleged conduct as illegal per se. The reverse payment case, after all, represents a settlement of patent litigation that allows for entry (typically free of any royalty obligation) at some date earlier than the expiration of the patent(s) at issue. Should the underlying patent(s) be valid and infringed by the proposed generic, the settlement is unequivocally *procompetitive*, because it provides for entry sooner than the patent would have allowed. Similarly, because the alleged reverse payment is frequently embedded in a separate business deal (often for co-promotion, licensing other products, etc.) between the parties, the question of whether a reverse payment was even made has to be addressed through fact-specific inquiry.

The product hopping cases, for their part, are similar in many respects to cases alleging predatory innovation of one kind or another. As with those cases, it would make no sense to bar product improvements and innovations outright. Rather, in both types of cases, the conduct should be assessed under the rule of reason, with the ultimate question being one of competitive effects: on balance, did the conduct actually harm consumers?

As with any rule of reason case, the analysis properly begins with a monopoly power screen: if the branded drug manufacturer in these cases has monopoly power, the inquiry then turns to the impact of the conduct at issue: did the settlement agreement or the product innovations, as the case may be, improperly maintain that monopoly power without any offsetting procompetitive benefits? On the other hand, if the screen establishes that the branded drug manufacturer does *not* possess monopoly power, the inquiry need proceed no further: If there is no monopoly power to protect or maintain, as in any rule of reason analysis, the conduct can be presumed to be neutral or procompetitive.[15]

2.2.3. *Applying the Monopoly Power Screen*

Application of the monopoly power screen in pharmaceuticals can be complicated, and there are traps for the unwary. As discussed earlier, it is generally not feasible to carry out econometric studies of demand to inform a market definition/monopoly power analysis. Rather, these questions are best addressed via close inquiry into the behavior of market participants – manufacturers, physicians, patients, and TPPs – as reflected in documents, data, and testimony, as I describe toward the end of this section. First, however, I will deal with a seductively simple – but ultimately incorrect – approach to assessing monopoly power that some have proffered.

---

[15] Again, as noted above, the reverse payment case involves a settlement of litigation, one that permits entry and competition. Likewise, the product hopping case involves innovation and product development, again generally a pro-consumer activity.

This so-called "direct test" has been espoused by the US Federal Trade Commission (FTC) and several plaintiffs in reverse payment and product hopping cases, and it runs roughly as follows. In most cases of entry of the first AB-rated generic to a branded product, price falls. Even if the branded product does not drop its price in response to the entry (and it generally does not), because the generic enters at a lower price and takes away sales from the brand, the average price paid for the product, across brand and generic, is lower after generic entry. Therefore, the argument runs, the branded product must have enjoyed monopoly power.[16]

But, as I have explained above, the fact that the generic *must* enter at a discount to its corresponding branded product, and the fact that prescriptions for the brand *will* be filled with the generic in many cases, are simply institutional facts of the industry. They will *always* be true, regardless of whether the branded molecule at issue was the only treatment available for a rare form of melanoma or was one of twenty-five equally effective oral contraceptive products. These facts – and the reduced average price they necessarily engender – are institutional features of the industry that, by themselves, tell us *nothing* useful about monopoly power or the lack thereof.

To see this, consider generic entry in two very different situations. First, consider a branded drug that confers genuine therapeutic advantages over any existing formulation. In that case, once the manufacturer has made medical practitioners aware of those advantages, the drug's therapeutic benefits will sustain its marketplace success, even at a price premium, as long as no equivalent substitute product is available. When an AB-rated generic enters in this case, providing the same product at a lower price, the effect may well be to drive increased use of the advantageous therapy. In effect, the market will "move down the demand curve" with a larger quantity purchased at reduced prices. If there is a significant increase in quantity, the branded manufacturer may, in fact, have enjoyed monopoly power prior to generic entry: the price premium it enjoyed may have represented a restriction of output and reduction in consumer welfare.

Now consider an entirely different case: suppose that the branded drug, in fact, confers little or no material therapeutic benefit over existing alternatives. Rather, the brand-name drug uses an alternative (technically unique) delivery mechanism. That mechanism can be exploited, let us say, by creative marketing and brand-building activities. In that event, such premium as the branded manufacturer can extract depends entirely upon the brand awareness created by its advertising and marketing efforts. It is well known, of course, that such premiums can and do exist. Our daily lives abound with examples of the premiums that brands may confer; the branded loaf of white bread is more expensive than most private labels, even if the private label loaf is equal or even greater in objective quality. But in the overwhelming

---

[16] Brief for the Federal Trade Commission as *Amicus Curiae*, pp. 15–16, *Mylan Pharmaceuticals, Inc. v. Warner Chilcott plc, et al.*, available at www.ftc.gov/system/files/documents/amicus_briefs/mylan-pharmaceuticals-inc.v.warner-chilcott-plc-et-al./151001mylanamicusbrief.pdf. Also see Cramer and Berger (2004), among others.

majority of cases, this premium has nothing whatsoever to do with monopoly power. It is merely the economic return to the advertising, promotional, and other brand-building efforts undertaken by the manufacturer.

Note that generic entry in this latter case will have exactly the same price effect as in the former case. The effects on *output*, however, are very different. In the second case, generic entry destroys the branded manufacturer's incentive to engage in any advertising, promotion, or brand-building activity. That is because the branded supplier can no longer reap the benefit of such demand-building efforts: pharmacists will simply substitute the lower-priced generic in place of the branded product. Predictably, the branded supplier will simply cease to promote its product under these circumstances. When these advertising, promotion, and other demand-building activities cease, the entire demand curve will, in effect, shift inwards. In other words, at any given price there will simply be less demand for the branded product than there had been before. Where, as in this latter case, the manufacturer of the branded drug relies on marketing and similar demand-building efforts, and where its product confers little or no therapeutic advantage, the disappearance of advertising and demand-building efforts is likely to cause substantial demand erosion. Even at the reduced price charged by the generic firm, there may be little or no increase in the total quantity demanded of the branded drug (and its generic counterpart). Indeed, quantity may even decline. In such a case, there was no output restriction, no exercise of monopoly power before generic entry.

Rather, market success (including, possibly, price premiums) that is purely the result of brand and demand-building efforts is, as pointed out earlier, simply the economic return earned by those efforts and does not connote monopoly power. Thus, the so-called "direct test" of monopoly power is, in the context of brand/generic pharmaceutical competition, nothing of the kind. The fact that the generic product enters at a lower price than the brand-name drug and captures sales does not provide evidence of prior monopoly power but simply reflects unique institutional aspects of the pharmaceuticals market. The analyst seeking answers on market definition and substitutability needs to look elsewhere.[17]

Fortunately, the multiple layers of competition in pharmaceuticals provide rich sources of information for such an exercise. To begin with, of course, the products have FDA labels, which describe the indications for which they are approved for use, together with other prescribing information, data on side-effects, and so on. However, the labels are only a starting point because many (if not most) pharmaceutical products are used for off-label purposes, sometimes quite widely.[18] Thus, data on

---

[17] As noted above, while econometric demand analysis of the sort frequently done in consumer goods mergers is infeasible, the analyst should certainly explore whether there were natural experiments that could, through the judicious use of econometrics, shed light on these questions.

[18] There is nothing improper or illegal about such use if it represents the physician's best professional judgment on the appropriate treatment for the patient. For example, "[g]ood medical practice and the best interests of the patient require that physicians use legally available drugs,

the actual conditions for which drugs are prescribed can be very useful. In some instances, these data may be available from surveys conducted by the manufacturers themselves. In others, the data may be available from public sources such as IMS in the United States, in its National Disease and Therapies Index (NDTI) dataset.

There are several other potentially rich sources of information on the degree to which the various players in the market – manufacturers, physicians, and TPPs – believe that (or behave as if) different therapies competitively constrain one another. Because the decision maker in the first instance is the physician (operating, no doubt, under whatever constraints might be imposed by the TPP system), pharmaceutical manufacturers devote considerable resources to advertising and promotional activities to physicians. Frequently, the documents and data associated with these efforts provide useful insights into the set of therapies viewed by the manufacturer as being the "competitive set." Similarly, manufacturers' brand plans and competitive analyses often pay particular attention to a small group of products as a competitive set. Again, such a competitive set may well be a good starting point for a candidate-relevant market.

As noted above, formulary status can also be very important in driving prescribing decisions, so manufacturers also devote considerable efforts to negotiating and striking deals with TPPs (often with their pharmacy benefit managers, or PBMs). The competitive sets referred to in these negotiations can often inform our understanding of the set of products that are viewed as genuine therapeutic substitutes. Likewise, the actions PBMs or TPPs take in replacing one product with another as the preferred therapy in a particular category can be valuable in the same way.

In some therapeutic categories, a patient may need to be placed on long-term therapy following initial intervention in a hospital setting. For instance, patients admitted to a hospital for treatment of a stroke are likely to be asked to continue taking a blood thinner after their discharge; likewise, patients treated in a hospital for gastro-intestinal reflux disorder (GERD) are often prescribed proton pump inhibitor (PPI) therapy after discharge. In such instances, the specific therapy administered in the hospital is often the one prescribed for long-term use (possibly in different form, i.e., oral rather than intravascular). Pharmaceutical manufacturers in these

---

biologics and devices according to their best knowledge and judgment. If physicians use a product for an indication not in the approved labeling, they have the responsibility to be well informed about the product, to base its use on firm scientific rationale and on sound medical evidence, and to maintain records of the product's use and effects. Use of a marketed product in this manner *when the intent is the 'practice of medicine'* does not require the submission of an Investigational New Drug Application (IND), Investigational Device Exemption (IDE) or review by an Institutional Review Board (IRB). However, the institution at which the product will be used may, under its own authority, require IRB review or other institutional oversight" (US Food and Drug Administration (2014) (emphasis in the original). "Many experts will use an unlicensed medication if they think the medication is likely to be effective and the benefits of treatment outweigh any associated risk" (UK National Health Service (2014)).

therapeutic categories often offer heavily discounted pricing to hospitals in the hope that their own therapy will be the one administered – and the one prescribed for the long term. Documentation and data surrounding these activities – particularly any references to the competitive set of products – can be useful sources of information about the scope of the market and about competition within it.[19]

In sum, the unique institutional features of the pharmaceutical industry and markets can effectively preclude the use of some otherwise well-accepted analytic tools in assessing monopoly power (econometric demand estimation in particular) and also rule out so-called "direct tests" of monopoly power or competitive effects. However, those same institutional features result in a range of competitive efforts by manufacturers that can effectively inform our understanding of the scope of the relevant market(s) involved as well as the extent of monopoly power, if any.

2.2.4. Competitive Effects in Reverse Payment Cases

Let us assume that monopoly power cannot be ruled out in a reverse payment case and that the fact of a payment from brand to ANDA filer can be established. Does that necessarily mean that the agreement was anticompetitive? One school of thought, prominently represented by the FTC among others, says, unequivocally, "yes."[20]

The proponents of this view argue as follows. To begin with, they argue that the appropriate measure of any "anticompetitive effect" of a given settlement agreement is the amount of time by which it delays entry relative to alternative settlements or litigation, because consumers are better off the sooner the entrant enters the market.[21] They then argue that settlements that involve payments from the patentee to the alleged infringer are necessarily anticompetitive, because, if the parties could reach a settlement without a side payment, the settlements reached with side

---

[19] Manufacturers often sponsor clinical research to explore the efficacy of their products in the treatment of off-label indications. That is because physicians using the drugs off-label rely ultimately on the results of such clinical research in informing their prescribing decisions. Therefore, the conditions being targeted in such research, the alternative therapies used in the clinical trials (if any) and the results of the research can sometimes provide valuable insights as well.

[20] See, e.g., *Schering-Plough Corp. v. FTC*, 402 F.3d 1056, 1060 (11th Cir. 2005). Among others supporting this view, see Leffler and Leffler (2002); Hirsh and Dorfman (2005) and Cotter (2004).

[21] This formulation is not strictly correct; risk aversion and discounting (the economic reality that a dollar today is worth more than a dollar payable in the future, even setting aside inflation), among other things, mean that a "date certain" entry four years in the future, for instance, is not equivalent, from the consumer's standpoint, to a lawsuit under which the expected outcome is an entry date four years into the future. It is certainly entirely possible to incorporate these features into an economic model, which I have done elsewhere, but I will not here, because their inclusion greatly complicates the exposition without materially changing the qualitative results. In any event, much of the public debate has been framed (simplistically) in terms of entry dates.

payments are "more anticompetitive," i.e., result in later entry, than the settlement that those same parties would have reached otherwise.[22]

On the other hand, the argument goes, when payments are necessary for settlement even to be feasible, such payments in the "wrong" direction, from incumbent to entrant, lead to outcomes "more anticompetitive" – i.e., later entry dates – than either party expects under litigation.[23] This conclusion rests on the following argument: Suppose for simplicity that the litigation has reached a stage where discovery is complete, so that the parties have learned all that they could expect to learn prior to trial about their odds of winning at trial; suppose further that both parties agree that each one's probability of prevailing in the litigation is roughly 50 percent. Then each party expects that, if they continued to litigate, the probability that the defendant will prevail and entry will occur virtually immediately is 50 percent, while the probability that the patentee will prevail and entry will be delayed until expiration of the patent is also 50 percent.[24] Therefore, the argument goes, the "expected" time to entry under litigation (i.e., the probability-weighted average of the two entry dates under the two alternative outcomes) is approximately one-half of the term remaining on the patent.[25] Any settlement that results in an entry date later than this benchmark would then be deemed anticompetitive.

The argument further holds that if the parties agreed that their respective odds of prevailing were 50 percent each, neither side would agree, absent side payments, to any settlement that specified an entry date different from this benchmark date; the patentee, according to this view, would accept no date earlier than the benchmark, whereas the entrant would accept no date later than the benchmark, each party reasoning that it could expect to do at least as well should it pursue the litigation to its conclusion. Therefore, the argument concludes, any payment from patentee to entrant must necessarily be a "bribe" to persuade the entrant to delay its entry.[26]

While the foregoing argument has considerable appeal if only for its simplicity, it is not usable, because it relies on assumptions that are not generally valid. For example, one crucial flaw in the chain of reasoning lies in the assertion that the

---

[22] "The issue of exclusion payments has been the subject of significant debate, but the Commission's position is clear. Where a patent holder makes a payment to a challenger to induce it to agree to a later entry than it would otherwise agree to, consumers are harmed *either* because a settlement with an earlier date might have been reached, *or* because continuation of the litigation without settlement would yield a greater prospect of competition" *Barriers to Generic Entry: Hearing Before the Sen. Special Comm. on Aging*, 109th Cong. 18 (2006) (statement of FTC) (footnote omitted).
[23] See Elhauge and Krueger (2012) and Edlin et al. (2013).
[24] For simplicity only, assume that the outcome of the trial will be made known relatively quickly, so that, should the alleged infringer prevail, its entry would not be subject to any additional delay. Assume, too, for simplicity, that there is no additional litigation cost.
[25] For instance, if the patent has eight years to run, the probability of instantaneous entry is 50 percent, but the probability that entry would be deferred for eight years is also 50 percent, so the expected time to entry under litigation is four years (50 percent probability of zero and 50 percent probability of eight years).
[26] Shapiro (2003).

patentee would not settle for an entry date earlier than the benchmark (i.e., the expected, or probability-weighted average, date of entry under litigation). The logical flaw stems from the implicit assumption that the patentee would view a date certain entry of, say, four years in the future as exactly equivalent to engaging in litigation whose expected entry date is also four years in the future (because, for example, it offers equal odds of entry today or entry eight years hence, upon patent expiration).

The problem with this assumption is that it is frequently violated in practice. There are many sound economic reasons why a patentee may be willing to settle for an entry date earlier than that expected under litigation.[27] Among these are people's attitudes toward risk. Economists have long understood that most individuals are "risk averse," in that they value outcomes that are inherently uncertain less than outcomes that can be known with certainty. Everyday experience is replete with examples. Companies whose fortunes are more risky have to offer higher expected returns to their investors than do companies that are less risky. The interest rates on corporate bonds reflect the same reality: Companies whose prospects are regarded as more risky (and whose ratings by bond rating services like Moody's reflect that assessment) have to offer higher interest rates in order to attract investor interest than do companies that are regarded as less risky. The immediate implication, of course, is that an individual who is risk averse might well be willing to sacrifice some portion of his expected return from a venture if, in exchange, he could reduce the uncertainty associated with that venture. A patentee who has built a substantial business around a patent may well be risk averse in exactly that fashion: when choosing between a settlement and pursuing litigation to its final outcome, the patentee would recognize that the nonzero probability associated with "losing it all" creates very real risk, regardless of the expected value associated with litigation. If, as in our example above, the expected date of entry associated with litigation were four years (equal likelihood of immediate entry or entry after eight years, upon patent expiration), the risk-averse patentee would be willing to sacrifice some of that expected value in exchange for reducing the uncertainty attendant upon litigation.

In other words, the risk-averse patentee would be willing to settle for entry by the would-be entrant at a date certain earlier than the expected date under litigation. In effect, the patentee's risk aversion could make the settlement more favorable to consumers than the expected outcome under litigation. Of course, such a settlement could also be attractive to the entrant, because it would permit entry sooner than might have been expected under litigation. The problem is that the would-be infringer may well also find that its liquidity position does not permit it to "wait out" the period until that entry date. In other words, while attractive, the settlement may not be feasible for the entrant without some sort of cash infusion that would help it

---

[27] Among other sources, see Addanki and Daskin (2008); Orszag (2013); Langenfeld (2013); Harris et al. (2014); Padilla and Meunier (2015).

to survive until the entry date at issue (even though that date is earlier than the expected date of entry under litigation). In this situation, the only path to a settlement could well be one in which the patentee provides such a cash infusion. Without the infusion, even though the patentee would be willing to entertain a definite entry date earlier than the expected outcome of litigation, that earlier date would remain infeasible for the entrant. Any date that the entrant would regard as feasible (absent the cash infusion) would be too early for the patentee to accept, given its odds of prevailing in the lawsuit (even allowing for risk aversion). Thus, the only alternative to the settlement with a cash payment might, in fact, have been litigation.

Note that this does not mean that the resulting date of entry would be later than the expected outcome of the litigation. In fact, the date agreed upon by parties – even with the cash payment – may well be earlier than the date that might be expected under litigation. That, of course, is the crucial question if one assumes that earlier generic entry is always better for consumers: Is the entry date specified in the settlement earlier or later than the benchmark entry date that might be expected under litigation?[28] In this example, whether it is earlier than the benchmark date depends upon the degree of risk aversion of the patentee, the amount of the payment required and the returns that each party expects to earn under the alternatives.[29]

The simple argument above, therefore, cannot be used by itself to establish that a reverse payment settlement is necessarily anticompetitive in its effect. Its critical assumption that the patentee would never agree to a settlement that embodied an entry date earlier than the date that might be expected under litigation is fundamentally invalid. Moreover, the risk aversion discussed above represents only one of several possible reasons why the argument's key underlying assumptions could easily be invalid. For instance, the patentee might simply be unduly pessimistic about its case; the judge or magistrate may have placed particular pressure on the patentee to settle; or litigation costs, including out-of-pocket costs as well as the significant opportunity costs that litigation imposes on senior management time and attention, could be a factor.[30] There are certainly other reasons as well why the assumptions

---

[28] The assumption, sometimes implicit and sometimes made explicit, that earlier generic entry is unequivocally better for consumers, does not hold if, for instance, the level of, and growth in, demand for the branded drug is largely sustained by advertising and promotional efforts on the part of its manufacturer. Also see the Memorandum Opinion of the Court in *Meijer, Inc., et al vs. Barr Pharmaceuticals, et al*, U.S. District Court for the District of Columbia, August 11, 2008.

[29] Shapiro (2003).

[30] To be sure, some authors have suggested that reverse payments up to the anticipated litigation costs may not be anticompetitive, but it is notoriously difficult to measure the sometimes substantial economic costs of litigation beyond the obvious out-of-pocket expenses. In any event, as other authors have shown, settlements that involve reverse payments that exceed litigation costs can nevertheless be procompetitive (see, e.g., Harris *et al.* 2014).

may be violated.[31] Therefore, contrary to the argument above, agreements that provide for payments from the patentee to the ANDA filer could, in fact, be procompetitive.[32]

One strand of economic literature has attempted to assess whether such agreements are anticompetitive on average by using "event study" methodology to assess whether the stock market rewards brand pharmaceutical companies for entering into reverse payment settlements. The underlying idea is that if it does, one could then infer that the excess return earned by companies entering into such settlements flows from the preservation/enhancement of monopoly profits, implying that, at least on average, such settlements are anticompetitive. Thus, for example, Drake et al. analyze a sample of sixty-eight brand/generic settlements from 1993 to 2013, categorizing twenty-seven as containing terms they regard as consistent with a reverse payment and forty-one as not containing such terms. They find that the excess returns following the settlement announcement were significantly greater for the "reverse payment" subset than for the others. From this they infer that such settlements, at least on average, may be anticompetitive.[33]

Space does not permit a complete critique of their work here, but four important limitations or shortcomings are worth mention. First, as they are quick to acknowledge, whatever the results may prove on average, the study cannot shed light on whether a *given* agreement is anticompetitive in its effect. Second, and again acknowledged by the authors, they had to rely on public information about whether each settlement might or might not have included a reverse payment. The mere fact of a side deal's having accompanied a settlement (which was essentially their mechanism to identify their reverse payment settlements) does not tell us anything about whether the side deal actually constituted net payment flowing from brand to generic or was, rather, an arm's-length transaction mutually beneficial to both parties. Thus, settlements that the authors characterize as reverse payment deals may in fact have been nothing of the kind, which means that their central result may well be spurious. Third, even if the result is not in fact spurious, and the market rewards settlements with side deals more than settlements without such deals, it is not at all clear what "signal" evokes that response. The authors hypothesize that it is the market recognizing an anticompetitive deal that preserves monopoly profits, but it could just as well be other things, such as the market recognizing that lawsuits that *needed* side deals in order to be settled were more contentious and less likely to settle; all else equal, the resolution of a fractious, contentious lawsuit could well be viewed by the market as a "more positive" event than the resolution of a lawsuit that was relatively simple to settle.

---

[31] Among other things, there might be antitrust counterclaims that would be disposed of concurrently with the patent litigation, which could bear on the parties' incentives to settle.
[32] See the Appendix in Addanki and Daskin (2008) for the algebraic demonstration of this.
[33] Drake, Starr, and McGuire (2014).

Finally, even assuming that the authors are correct in their hypothesis that the market views the existence of a side deal (perhaps including a reverse payment) as a signal that the entry would be later than it might have been absent such a payment, it remains unclear whether *any* of the "reverse payment" settlements in their sample were actually anticompetitive. As I have discussed above, a hypothetical settlement without a side deal may not even have been possible, which means that the appropriate test is whether the actual settlements were anticompetitive relative to the litigation alternative, i.e., whether *any* of the settlements specified an entry date that was less favorable to customers than the expected outcome of litigation. Given the complexity of patent litigation – and the unavailability of important litigation-related information to the outside investor – it is implausible that the market could have a sufficiently accurate estimate of the likely outcome of litigation as to reward a settlement for "doing better" than litigation.[34] Thus, in light of these shortcomings, this line of research is unlikely to shed useful light on the topic even in general and certainly would not help the analyst in a specific case.

What, then, is the analyst to do, assuming that monopoly power and the fact of a reverse payment have been established? The appropriate test is whether customers are better off under the settlement than they would have been (in expectation terms) under litigation. In evaluating a settlement agreement with a so-called reverse payment and an agreed date of entry, in some cases, the appropriate test may simply be whether the settlement resulted in an agreed-upon entry date later than what might have been expected under litigation.

To answer this question, one must evaluate the likely outcomes of the patent case as well as each party's odds of prevailing in litigation. These facts would help establish what the expected outcome would have been under litigation. A few points are worth noting in this connection. First, the analyst or fact finder needs to determine the likely outcomes of the patent case and the objective odds that each party will prevail in the litigation, not the parties' subjective estimates of those odds. Consequently, there would generally be no need to examine privileged documents to estimate those odds. Second, it is not generally necessary to estimate those odds with tremendous precision. If the proposed settlement splits the remaining patent term in half, for example, the inquiry need only determine if the expected time to entry under litigation would have been longer, not whether the patentee's probability of prevailing in the litigation is 60, 70, or 75 percent.[35]

---

[34] As a threshold matter, it is important to note, too, that Drake, Starr, and McGuire (2014) implicitly assume that every brand firm in their sample enjoyed monopoly power in the sale of its branded drug.

[35] Many authors who argue in favor of a presumption of illegality for reverse payment settlements offer, in support of their position, the presumed advantage that such treatment obviates the need to debate the merits of the patent case, because the settlement including the reverse payment would necessarily have been worse for consumers than an alternative settlement that did not include such a payment. The problem, as discussed above, and as has been demonstrated before,

In this connection, it is important to recall that the assumption underlying these discussions is that entry would be virtually instantaneous should the entrant prevail in the litigation. In actual fact, even a victory by the entrant could result in deferred entry, either because of appeals or because the entrant's entry would be delayed by the need to undertake various investments or seek regulatory approvals, among other sources of delay. In that case, the expected time to entry would exceed one-half of the time remaining on the patent even if the probability of the entrant's prevailing were 50 percent. Therefore, any empirical evaluation of whether or not a given agreement involving reverse payments is anticompetitive requires that we inquire not only about the odds of each party prevailing, but also about likely entry dates under alternative litigation outcomes.

2.2.5. *Competitive Effects in the Product Hopping Context*

Let us assume that we are dealing with a case in which the brand manufacturer has developed and patented an alternative formulation (an extended release version, say) to its existing drug and is in the process of promoting the extended release version to physicians as the better drug to prescribe. The manufacturer has also discontinued all marketing and promotion of the original formulation, although it remains available. Plaintiffs, including the generic manufacturer who filed an ANDA referring to the original brand formulation, have sued, alleging that the claimed improvements embodied in the new formulation are a sham and that the only reason for the switch was to thwart generic competition to the detriment of consumers.

Clearly, again, the monopoly power screen is an important first step in the analysis. If there is no monopoly power, the analysis need proceed no further. If monopoly power has not been (or cannot be) ruled out definitively, the potential for competitive harm from the "product hop" has to be weighed against the potential for consumer benefit from the improvements embodied in the new formulation, whatever they may be. This is a fact-specific inquiry that will necessarily involve medical as well as economic inputs and insights.

That said, in a case where the generic manufacturer was free to enter with the AB-rated generic version of the original product, and where the brand manufacturer discontinued all marketing and promotion of the original formulation but did not discontinue supplying it before the ANDA was approved, the path to demonstrating

---

is that no such alternative settlement was agreed to, and one might not have even been feasible (see Addanki and Daskin 2008, for example). Any attempt to construct a hypothetical alternative settlement and to justify it as being one to which the parties would rationally have agreed can only be done *in the context of the merits of the underlying lawsuit*. Thus, for example, to suggest that patentee and ANDA filer would each have been irrational not to agree to entry on a given date necessarily is to make a statement about the odds of each party prevailing in the patent suit – quite apart from any other considerations such as liquidity, discount rates, etc., which may have rendered the hypothesized settlement infeasible in any event.

competitive harm is perhaps a difficult one because of the internal tension inherent in a product hopping case.

The claim by the plaintiffs in the case would presumably be that the defendant's de-emphasis of the "old" formulation and its promotion of "new" strengths or formulations – the so-called "product hopping" – resulted in the disappearance of demand for the old formulation. In short, the defendant stopped promoting the old product, so the demand for it disappeared, to the alleged detriment of the various plaintiffs. But if the demand for the old formulation was entirely driven by promotion, it could scarcely have been in a position to exercise monopoly power by restricting output and raising price. The plaintiffs' contention that promotional activities were needed to prop up demand is curiously inconsistent with the assertion that the old formulation enjoyed monopoly power. It is certainly not evident how one can have meaningful monopoly power in the sale of a product for which there is no autonomous demand.

Of course, it is possible that demand for the old formulation withered because the new formulation has therapeutic benefits that are sufficiently great that physicians no longer saw any reason to prescribe the old one. But in that case, the alleged product hopping caused no harm to competition or to consumers, representing, as it did, real innovation and improvement.

In any event, in cases in which the plaintiffs acknowledge, at least implicitly, that demand for the old strengths/formulations was overwhelmingly, if not entirely, the result of promotional activities and disappeared on cessation of those activities, the tension is real. If there is no meaningful demand for the old products absent promotional efforts, how could meaningful monopoly power inhere in them? And, if there is no monopoly power to be protected, how could the hop engender competitive harm?

That being said, as noted above, there could certainly be the possibility of competitive harm in situations in which there was monopoly power, in which the original formulation that conferred the monopoly power was discontinued before generic entry, and where prescriptions were "hard switched" to the new formulation. If the new formulation conferred little or no therapeutic benefit over the original version, it would certainly be possible that monopoly power was being preserved with no material offsetting procompetitive benefit.

2.3. CONCLUSION

2.3.1. A Word About Merger Analysis

Although there are certainly exceptions, mergers in the industry generally involve firms that sell several products, often spanning a range of chemical compositions and therapeutic applications. Thus, as with most such "portfolio" mergers, the competitive analysis of the transaction necessarily proceeds on a product-by-product (and application-by-application) basis. It seems eminently reasonable that treatments for, say, asthma, are unlikely, except by the purest coincidence, to be suitable treatments for strep infections.[36] Moreover, in carrying out this "market-by-market" assessment in an industry in which an important aspect of competition is the development and introduction of innovative new products, potential future competitive pressure from products in the merging firms' R&D pipelines may represent as significant an antitrust issue as competition provided by their existing products. Thus, competitive issues could arise if both firms sell treatments for asthma today, *or* Firm A sells it today while Firm B has a product in development, *or* neither sells one today but both have products in development.

There is generally ample information about pipeline products because of the highly regulated process for reviewing and approving a new drug. This means that if both the merging firms are developing treatments for asthma, details about the nature of the products being developed, their mechanisms of action (MOA) and their stage(s) of development would be available to the antitrust reviewer, although, to be sure, the timeline for final approval and, indeed, the question of whether one or both drugs would actually even be approved, would be subject to perhaps considerable uncertainty.

This uncertainty notwithstanding, the antitrust review might need to assess whether the merger could foreclose important future competition (i.e., between the yet-to-be-introduced therapies). The analysis would need to explore whether there are already numerous effective therapies available, whether one or both of the prospective therapies offer any unique attributes that would sharply differentiate them from the existing ones, whether the prospective therapies' MOAs are very similar, or whether the prospective therapies would compete particularly closely with one another (and less so with the existing therapies).

We have already discussed the difficulties that arise in defining relevant markets and assessing competitive conditions in pharmaceuticals even where the products exist today (or the nature of the potential competition, as with AB-rated generics, is well known). These difficulties are greatly compounded when attempting to assess potential competition with future products. To begin with, it is not clear if a given development effort will even result in an approved product; if a product is approved, it is not clear what its labeled indication will be; whatever the label may say, it is not clear how the product might actually be used. Thus, whatever the avowed goal of a development effort might be, the nature and extent of the competitive pressure that would actually be provided by the end product of the program, *if any*, are notoriously difficult to predict. The risk-averse path to enforcement, then, may be simply to demand divesture of a development program as a condition of approving a merger

---

[36] To be sure, an enterprising plaintiff attempted to sue to overturn the Pfizer/Wyeth merger (after it had been approved by the FTC), alleging that it would harm pharmacies in an alleged relevant market for all prescription pharmaceuticals; however, the case was dismissed and the dismissal affirmed by the Ninth Circuit. *Golden Gate Pharmacy Services, Inc. v. Pfizer, Inc.*, No. 10–15978 (9th Cir. May 19, 2011).

if there is even the *possibility* of future competition being diminished. One concern with such an approach is that R&D programs pursuing similar goals using different means can benefit each other in synergistic ways, so an overly conservative enforcement strategy could result in such synergies being forgone.

2.3.2. Closing Thoughts

The pharmaceutical industry is both important and complex in many ways, and its institutional idiosyncrasies can significantly complicate antitrust analysis. As is true of many R&D-intensive industries, antitrust analysis in pharmaceuticals frequently involves IP in a fundamental way. I have tried to provide some flavor of the role that industry features such as its regulatory structure and the principal–agent problems created by third-party-payer arrangements, among other factors, play in complicating the antitrust analysis of IP issues. As I have shown, these complications manifest themselves to a considerable degree in reverse payment and product hopping cases, but they are generally ubiquitous in antitrust inquiries involving this industry.

REFERENCES

Addanki, Sumanth and Henry N. Butler. 2014. Activating Actavis: Economic Issues in Applying the Rule of Reason to Reverse Payment Settlements. *Minnesota Journal of Law, Science & Technology*, 15(1), 77–94.

Addanki, Sumanth and Alan J. Daskin. 2008. Patent Settlement Agreements. *ABA Section of Antitrust Law*, 3, 2127–53.

Cotter, Thomas F. 2004. Antitrust Implications of Patent Settlements Involving Reverse Payments: Defending a Rebuttable Presumption of Illegality in Light of Some Recent Scholarship. *Antitrust Law Journal*, 71(3), 1069–97.

Cramer, Eric L. and Daniel Berger. 2004. The Superiority of Direct Proof of Monopoly Power and Anticompetitive Effects in Antitrust Cases Involving Delayed Entry of Generic Drugs. *University of San Francisco Law Review*, 39, 81–140.

Drake, Keith M., Martha A. Starr, and Thomas McGuire. 2014. Do "Reverse Payment" Settlements of Brand-Generic Patent Disputes in the Pharmaceutical Industry Constitute an Anticompetitive Pay for Delay? NBER Working Paper No. 20292. July.

Edlin, Aaron, Scott Hemphill, Herbert Hovenkamp, and Carl Shapiro. 2013. Activating Actavis. *Antitrust Magazine*, 28(1), 16–23.

Elhauge, Einer and Alex Krueger. 2012. Solving the Patent Settlement Puzzle. *Texas Law Review*, 91, 283–330.

Federal Trade Commission. 2006. Prepared Statement of the Federal Trade Commission Before the Special Committee on Aging of the United States Senate on Barriers to Generic Entry. Available at www.ftc.gov/sites/default/files/documents/public_statements/prepared-statement-federal-trade-commission-generic-drug-entry/p052103barrierstogenericentrytestimonysenate07202006.pdf.

    2015. Brief for Amicus Curiae Federal Trade Commission Supporting Plaintiff-Appellant in Mylan Pharmaceuticals, Inc. v. Warner Chilcott plc, et al.. Available at www.ftc.gov/system/files/documents/amicus_briefs/mylan-pharmaceuticals-inc.v.warner-chilcott-plc-et-al./151001mylanamicusbrief.pdf.

Golden Gate Pharmacy Services, Inc. v. Pfizer, Inc., No. 10–15978 (9th Cir. May 19, 2011).

Harris, Barry C., Kevin M. Murphy, Robert D. Willig, and Matthew B. Wright. 2014. Activating Actavis: A More Complete Story. *Antitrust Magazine*, 28(2), 83–9.
Heading for the Emergency Room. 2009. *The Economist*, June 25. Available at www.economist.com/node/13899647.
Hirsh, Merrill and Dan Zoloth Dorfman. 2005. I Didn't Say Orphan Often: The Benefits of a Bright-Line Rule Barring Brand to Generic Payments in Hatch-Waxman Patent Settlements. *Antitrust Health Care Chronicle*, 19(2).
Lamberti, Mary J. and Kenneth Getz. 2015. *Profiles of New Approaches to Improving the Efficiency and Performance of Pharmaceutical Drug Development*. Tufts Center for the Study of Drug Development White Paper. Available at http://csdd.tufts.edu/files/uploads/CSSD_PhRMAWhitePaperNEWEST.pdf.
Langenfeld, James. 2013. Evaluating the Size of "Reverse Payments" in Light of the Supreme Court's Decision in FTC v. Actavis. *CPI Antitrust Chronicle*. Available at www.competitionpolicyinternational.com/evaluating-the-size-of-reverse-payments-in-light-of-the-supreme-court-s-decision-in-ftc-v-actavis-2/.
Leffler, Keith B. and Cristofer I. Leffler. 2002. Want to Pay a Competitor to Exit the Market? Settle a Patent Infringement Case; An Argument for Per Se Condemnation of Payments by the Patent Holder. *ABA Section of Antitrust Law Economics Committee Newsletter*, Spring, 26–35.
Meijer, Inc. v. Barr Pharmaceuticals, Inc., 572 F. Supp. 2d 38 (D.C. 2008).
Mylan Pharmaceuticals, Inc. v. Warner Chilcott plc, et al. Civ. No. 12–3824 (E.D. Pa. Apr. 16, 2015).
National Science Foundation. 2015. Business Research and Development and Innovation: 2012. Available at www.nsf.gov/statistics/2016/nsf16301/.
Orszag, Jonathan. 2013. Statement of Jon Orszag Before the Committee on Senate Judiciary Subcommittee on Antitrust, Competition Policy and Consumer Rights. Available at www.judiciary.senate.gov/imo/media/doc/CHRG-113shrg87818.pdf.
Padilla, Jorge and Valerie Meunier. 2015. Should Reverse Payment Patent Settlements Be Prohibited Per Se? Available at http://ssrn.com/abstract=2604071.
Robbins, Gary, Aldonna Robbins, and John Goodman. 1994. Inefficiency in the U.S. Health Care System: What Can We Do? NCPA Policy Report No. 182. Available at www.ncpa.org/pdfs/st182.pdf.
Schering-Plough Corp. v. FTC, 402 F.3d 1056, 1060 (11th Cir. 2005).
Shapiro, Carl F. 2003. Antitrust Limits to Patent Settlements. *RAND Journal of Economics*, 34 (1), 391–411.
UK National Health Service. 2014. Medicines information – Licensing. Available at www.nhs.uk/conditions/Medicinesinfo/pages/safetyissues.aspx.
US Food and Drug Administration. 2014. "Off-Label" and Investigational Use of Marketed Drugs, Biologics, and Medical Devices – Information Sheet. Available at www.fda.gov/RegulatoryInformation/Guidances/ucm126486.htm.
  2015. Applications for FDA Approval to Market a New Drug. Available at www.accessdata.fda.gov/scripts/cdrh/cfdocs/cfcfr/CFRSearch.cfm?CFRPart=314.
Vivian, Jesse C. 2008. Generic-Substitution Laws. *U.S. Pharmacist*. Available at www.uspharmacist.com/content/s/44/c/9787/.

3

The Economics of the Internet

Babette E. Boliek

As a matter of first principles, the economics of the Internet, and of much of the current high tech industries, are largely the same as for all other industries – consumers choose a bundle of goods and services that maximize their objectives; if the price of a preferred good or service increases, the quantity demanded by consumers decreases. If demand is particularly strong (inelastic) for a good or service, and the prices charged for that item are high or even supracompetitive (reflective of the offering firm exercising market power) competitors and innovators will attempt to take part of the rival's market to the extent to do so is profitable.[1] But although still subservient to basic economic principles high tech industries – in particular those related to the Internet ecosystem – are susceptible to certain market tendencies and forces dictated by such industries' defining characteristics.

In a review of the economic and legal literature of the high tech and Internet ecosystem these defining characteristics can be separated into three broad categories: (1) two-sided markets and network effects; (2) rapid innovation; and (3) regulatory impacts. The following sets forward a review of the literature that corresponds with each category and a description of how that category tends to define industries and may ultimately affect robust, antitrust analyses of the market.

3.1. TWO-SIDED MARKETS AND NETWORK EFFECTS

Simply stated, the study of network effects focuses on the "adoption by users and optimal network size."[2] In network economies such as the Internet, "consumers place greater value on larger networks than small ones."[3] In turn, a two-sided market is "one in which 1) two sets of agents interact through an intermediary platform, and

---

[1] Fisher (1978).
[2] Rysman (2009).
[3] Shapiro (1999).

2) the decisions of each set of agents affects the outcomes of the other set of agents, typically through an externality."[4] In the study of two-sided markets, the focus is on the conduct of the intermediary, and specifically, pricing choices.[5] The study of network effects and two-sided markets may be separate but the research is intertwined because the adoption by users will affect how much an intermediary will charge for a good or service. In a two-sided market such as the payment card market, for example, potential merchants and card-users likely take into consideration which network has the largest user base when selecting a payment card.[6] Below is a review of general research on two-sided markets and network effects followed by a discussion of network effects and antitrust as well as network effects in the analysis of net neutrality regulation.

3.1.1. *General*

Focusing on two-sided markets rather than network effects, Rysman speaks to the pricing decision of the profit maximizer in a two-sided market. He posits that "pricing to one side of the market depends not only on the demand and costs that those consumers bring but also on how their participation affects the participation on the other side and the profit that is extracted from that participation."[7] It follows from this argument that an increased demand for a product in a network may raise the price for other products within the same network. Alternatively, the intermediary "might charge a price below cost on one side if those agents have a large price elasticity and their participation attracts a large number of participants on the other side who are relatively price inelastic (and hence have a high mark-up)."[8] These pricing realities may explain some of the zero-pricing models so often found in the Internet ecosystem. As Parker and Alstyne argue, a firm can rationally invest in a product that it gives away for free, so long as the giveaway increases demand in the complementary premium market that covers the cost of the goods given away for free.[9]

Parker and Alstyne note also the importance of determining which side of a two-sided market receives a discount or zero-price.[10] This determination "depends on cross-price elasticities as well as the relative sizes of the two-sided network effects."[11] Parker and Alstyne go on to distinguish indirect network effects from true network effects. "Indirect network effects are consumption externalities from purchasing compatible products such as hardware and software."[12] In contrast, the authors argue

---

[4] Rysman (2009).
[5] *Id.*
[6] *Id.*
[7] *Id.*
[8] *Id.*
[9] Parker and Alystyne (2005).
[10] *Id.*
[11] *Id.*
[12] *Id.*

that a true network effect occurs when one population's choice of product or service affects another population's choice of a different good or service.[13]

Armstrong joins the study of competition and pricing in two-sided markets, and sets forth three factors that determine prices offered to the different groups in the market.[14] The first factor is the relative sizes of cross-group externalities.[15] To compete within a two-sided market, platforms on both sides of the market must be competitive.[16] As a result, there is "downward pressure on the prices to both sides compared to the case where no cross-group network effects exist."[17] Armstrong's second factor is the presence of fixed fees or per-transaction charges because, he finds, such fees mitigate network effects. For example, a platform may want to mitigate network effects by "charg[ing] for their services on a lump sum basis, so that an agent's payment does not explicitly depend on how well the platform performs on the other side of the market."[18] Armstrong goes as far to say that "[b]ecause network effects are lessened with per-transaction charges, it is plausible that platform profits are higher when this form of charging is used."[19] The exception, Armstrong believes, is when a monopoly exists because "the incumbent's profits typically *increase* with the size of the network effects since entrants find it hard to gain a toehold even when the incumbent sets high prices."[20] However, Shapiro contends that two-sided markets in high tech markets are unlikely to be monopolized because "alliances form and the original developer of a new technology makes it 'open' in order to gain acceptance and build an installed base."[21]

Armstrong's, third and final factor affecting price is whether an agent is "single-homing" or "multi-homing."[22] Whereas single-homing is when an agent uses only one platform, multi-homing is when an agent uses several platforms.[23] An example of multi-homing would be a consumer that used one Internet Service Provider (ISP) at home, another at work, and a third on their mobile device.

3.1.2. Network Effects and Antitrust

As previously mentioned, Shapiro suggests that markets subject to network effects are unlikely to be monopolized.[24] To support this conclusion, Shapiro relies on the

---

[13] Id.
[14] Armstrong (2006).
[15] Id.
[16] Id.
[17] Id.
[18] Id.
[19] Id.
[20] Id.
[21] Shapiro (1999).
[22] Armstrong (2006).
[23] Id.
[24] Shapiro (1999).

premise that original developers will make new technology open in order to capture the network effects and acquire a wide user base.[25] Likewise, Shapiro argues that any alliances formed between developers will be procompetitive.[26] However, Shapiro states that when a dominant firm engages in exclusionary conduct, there may be harm to competition as well as future innovation.[27]

Exclusionary conduct in network economies, as with any market, may stifle innovation by excluding potential entrants. Shapiro contends that consumers will select products based on the strength of that product's network effects.[28] Therefore, firms in a dominant network might more easily exclude potential entrants than in non-networked markets by refusing potential entrant access to the dominant network. Nevertheless, Shapiro puts forth procompetitive effects of strengthening network industries, such as "differentiating products and networks, encourage investment in these networks, and to overcome freeriding."[29] Despite the possible harm to competition, Shapiro is reluctant to force network incumbents to open their networks to potential entrants.[30] This hesitation is consistent with empirical studies that suggest forced sharing decreases competition and investment.

3.1.3. Net Neutrality, Two-Sided Markets, and Network Effects

A significant movement based on the potential downsides of network effects in the presence of dominant incumbents has been in the move for net neutrality regulation. Discussed further in the regulation literature discussed in Section 3.3, proponents of network neutrality are "concerned that cable modem and DSL systems will use their control of the 'last mile' of the network to block or slow access to content and applications that threaten their proprietary operations."[31] However, cable modem and DSL providers contend that they "have not blocked access to any content or applications and that competitive forces would preclude any future attempt to do so."[32] Yoo suggests that, in accordance with antitrust policy, government should not prohibit a practice unless it has evaluated the practice's effect on competition.[33] Net neutrality, similarly to the over-regulation of antitrust laws, stifles future innovation.[34] Owen contends that such pre-emptive regulation is unnecessary because the courts may intervene through antitrust law once ex post evidence of

---

[25] Id.
[26] Id.
[27] Id.
[28] Id.
[29] Id.
[30] Id.
[31] Yoo (2005).
[32] Id.
[33] Id.
[34] Crandall and Jackson (2011).

harm to competition arises.³⁵ Moreover, because the Internet is a two-sided market, pre-emptive regulation may have a negative effect on both markets.³⁶

Owen, like other authors, characterizes net neutrality as a debate on the competitive benefits of vertical integration. Vertical integration describes a firm's control of sequential steps of production.³⁷ Because the chain of production is only as efficient as the weakest link, Owen urges regulators to focus on "the link that is the most concentrated and the most protected by entry barriers and design regulations to increase its competitiveness."³⁸ Proponents of net neutrality believe that vertical integration "may give firms both the opportunity (through denial of access or price discrimination) and incentive (increased profit) to restrict competition."³⁹ However, Owen contends that "[b]oth vertical integration and changes in the extent of vertical integration are benign characteristics of efficient, dynamic, competitive markets."⁴⁰ Furthermore, Yoo claims that the establishment of network neutrality may "impede the emergence of competition in the last mile by reinforcing the economic characteristics that drive markets for telecommunications networks toward natural monopoly (i.e., high up-front costs and network economic effects)."⁴¹

3.2. INNOVATION ECONOMICS

Almost by definition, high tech and Internet technology markets exemplify almost incessant innovation. In such industries, antitrust enforcers have been cautious to intervene lest they interfere with the innovative process which in and of itself may quickly resolve any perceived anticompetitive practices. There is an extensive literature on the factors promoting innovation and its implication for policy makers. Described below is a sampling of that literature, including research within the Innovation Economics school of thought, a Schumpeterian view of innovation, limits placed on innovation by antitrust and intellectual property economics, and research on innovation in the Internet broadband markets.

3.2.1. Antitrust and Intellectual Property Law as Limits to Innovation

There is a delicate balance between incentivizing innovation and overprotecting innovators. As Fisher stated the case long ago, even in the presence of monopoly profits (a sign of limited competition) antitrust intervention may not be necessary.⁴² He warns that aggressively enforcing antitrust laws to protect innovators may result in

---

[35] Owen (2011).
[36] Id.
[37] Id.
[38] Owen (2011).
[39] Id.
[40] Id.
[41] Yoo (2005).
[42] Fisher (1978).

false-positives, which would stifle innovation in dynamic markets. Indeed monopoly profits themselves may incite procompetitive innovation.[43] As Fisher lays out the case, "if there are no barriers to imitation, other firms, through reverse engineering, for example, will learn how to make the new product and begin bringing it out."[44] In turn, the new firm will likely offer the new product at a lower price than the incumbent in order to lure customers away.[45] Antitrust intervention or regulatory attempts to control or lower the monopoly price risks stifling this cycle of innovation and product differentiation. Fisher's view is rejected entirely by those who question the societal benefits of consumer-driven innovation.[46]

A cautious approach is consistent with the view expressed by Atkinson and Audretsch, who state that "[i]nnovation economics holds that the policy priority is long-term economic growth and that major drivers of growth are productive efficiency and adaptive efficiency."[47] Adherents of the Innovation Economics School believe that the primary purpose of antitrust is to encourage productivity and innovation, not analyzing allocation efficiency or distributional effects.[48] Moreover, the Innovation School sees inter-firm interaction as a positive sign, rather than a troublesome indicator of potential collusion.[49] The Innovation School suggests that antitrust enforcers should treat industries differently[50] and recommends that enforcers focus "on the pragmatic issues surrounding each issue, and judge it based on the extent to which [a given practice] spurs innovation and productivity."[51]

Similarly, intellectual property laws may stifle innovation by "granting intellectual property holders far beyond what is necessary to create appropriate incentives to innovate."[52] Further, Hovenkamp notes that innovation is not embodied in a patent or copyright merely because such ownership rights are the result of relentless lobbying to the legislature.[53] Nevertheless, Hovenkamp believes that protecting innovation in dynamic markets should be primarily the function of patent law, not antitrust law.[54] This is because Hovenkamp sees antitrust as a mechanism to correct private markets.[55] But Hovenkamp values the use of antitrust laws to punish firm conduct that is directed solely at spoiling rival innovation.[56]

---

[43] Id.
[44] Id.
[45] Id.
[46] Pasquale (2010).
[47] Atkinson and Audretsch (2011).
[48] Id.
[49] Id.
[50] Id.
[51] Id.
[52] Id.
[53] Hovenkamp (2012b).
[54] Id.
[55] Id.
[56] Id.

Hovenkamp suggests that the antitrust and intellectual property laws are not necessarily in conflict. For example, Hovenkamp argues that "[m]any intellectual property practices, such as tying arrangements, may increase the returns to innovation or licensing without doing any harm whatsoever to competition or consumer welfare."[57] Rather, both laws condemn conduct that unreasonably limits innovation or competition.[58]

Manne and Wright agree that antitrust intervention may stifle innovation in dynamic markets and illustrate their concern by employing an error-cost framework.[59] The authors use historical evidence, such as the damage to Microsoft after it lost its antitrust case, to demonstrate how intervention may stifle innovation.[60] Hazlett also demonstrated that, even excluding Microsoft, strong antitrust measures lead to declining equity values in the computer sector.[61] Nevertheless, the authors make clear that antitrust has a role in preventing anticompetitive behavior, and therefore antitrust analysis should "account for [the] special features of IT – just like it ought to account for special features of any industry under consideration."[62]

Easterbrook, like Manne and Wright, also employs an error-cost framework to illustrate two related, and opposing concerns with antitrust enforcement. As characterized by Manne and Wright, (1) false-positives are more harmful than false-negatives because the market may self-correct in a false-negative, but not in a false-positive; and (2) both types of errors will inevitably occur because it is very difficult to distinguish procompetitive conduct from anticompetitive conduct.[63] In addition to guiding antitrust intervention, Manne and Wright believe the error-cost framework "is also useful for providing a taxonomy of antitrust errors in creating mistaken likelihoods about the competitive impact of a particular practice . . . ."[64]

3.2.2. Schumpeterian Competition

Katz and Shelanski put forward the Schumpeterian argument that "competition primarily occurs through cycles of innovation, rather than through static price output or competition."[65] As a result, "firms do not compete simultaneously for a share of the market, but rather sequentially for the market as a whole."[66] In this "Game of Thrones," a firm with a significant market share may be overthrown at any

---

[57] Id.
[58] Id.
[59] Manne and Wright (2010).
[60] Id.
[61] Hazlett and Weisman (2009)
[62] Manne and Wright (2010).
[63] Id.
[64] Id.
[65] Katz and Shelanski (2005).
[66] Id.

moment by a firm that develops a superior product. (But note how these effects may be countered or minimized by exclusionary conduct in a network economy.)

Schumpeterians push for the systematic retreat of antitrust intervention by the courts in dynamic markets.[67] Intervention in dynamic markets may stifle innovation by "distorting the reward structure for risky R&D (i.e. by impeding temporary monopoly returns) or by preventing dynamically beneficial mergers (i.e. transactions that might speed innovation by aggregating complementary assets)."[68] The examination of pricing and output by a firm is a short-term view of things, whereas scrutinizing the potential R&D benefits is more of a long-term view.[69] The Schumpeterian view focuses on the latter. For example, a merger, "might increase prices in the short run but, by bringing together complementary assets needed to develop or exploit new technology, might increase firms' abilities to innovate."[70]

There are some members of the Schumpeterian School that believe that current market share is a poor proxy for market power and future market performance in dynamic markets in innovative cases.[71] Because dynamic markets focus on R&D, dominant incumbents may be dethroned at any time. However, the Schumpeterian School concedes that market share is a practical proxy for market power in antitrust enforcement.[72] Katz and Shelanski agree that Schumpeterians "are right that significant innovation complicates the use of current market shares to determine market power and future market performance and that antitrust enforcers must be sensitive to these complications."[73] However, Katz and Shelanski believe that "Schumpeterians are wrong when they assert that the process of innovation-based competition and the relevant economic evidence counsel the presumption that concentration and market power promote innovation and consumer welfare and that a systematic retreat from antitrust enforcement is warranted."[74]

Katz and Shelanksi differ in this conclusion from Evans, who puts forth a Schumpeterian argument for allowing firms to use network effects to create a monopoly.[75] He contends that a firm that uses network effects to create higher barriers to entry may nonetheless use the revenues from the monopoly to innovate new products resulting in large benefit to social welfare. Moreover, the network effects in IT markets means that error-costs are much higher; thus a false-positive would have crippling effects on a firm and detriment to the social welfare.[76] Lastly,

---

[67] Id.
[68] Id.
[69] Id.
[70] Id.
[71] Id.
[72] Id.
[73] Id.
[74] Id.
[75] Evans (2009).
[76] Id.

Evans puts forth the Schumpeterian view that "dominant IT firms get displaced pretty regularly and often from surprising directions."[77]

Evans also discusses, but disagrees with, the notion that antitrust should be removed from information technology.[78] The notion is that antitrust fails to consider how dynamic IT markets are.[79] Evans strikes down this notion, explaining that antitrust law in the United States allows for monopolies to exist.[80] Next, Evans contends that "there's no reason to believe that dominant firms in the information technology industries are less able or have fewer incentives to engage in anti-competitive practices than dominant firms in other industries."[81] However, it seems logical that monopolies are more likely to engage in anticompetitive conduct because of how dynamic IT markets are. Despite this, Evans suggests, in contrast to the Innovation School, that IT markets be regulated no more, and no less, than any other market.[82]

3.2.3. *Innovation in the Broadband Markets*

Given the important role of broadband markets in current policy debates, a study examining innovation in these markets is worth mention. The broadband communication services market is "characterized by rapid innovation, declining costs, product differentiation, competitive price discrimination, network effects and 'multisidedness.'"[83] Eisenach states that "broadband is treated differently from other IT industries when it comes to competition policy: competition in the rest of the IT sector is subject to scrutiny under antitrust laws, while broadband is regulated by the FCC."[84] However, Eisenach argues that the competitive dynamics of the broadband market are now substantially similar to other markets within the Internet ecosystem and that the "broadband markets should therefore be brought into conformity with the ex post, case-specific approach applied to the other IT markets."[85]

In regulating IT markets such as the broadband market, the consensus is that policing potentially harmful conduct should be conducted on a case-by-case basis.[86] One side argues that "competition in IT markets is so naturally intense, or that the risk of policy error are sufficiently high, that enforcers should apply a reduced level of antitrust scrutiny."[87] Others contend that "IT markets are in some respects more

---

[77] *Id.*
[78] *Id.*
[79] *Id.*
[80] *Id.*
[81] *Id.*
[82] *Id.*
[83] Eisenach (2012).
[84] *Id.*
[85] *Id.*
[86] *Id.*
[87] *Id.*

prone to market failure than more traditional markets and hence deserve enhanced scrutiny."[88]

Governmental regulators continue to use, as a basis of their analysis, the structural presumption, which involves the likelihood of a firm to exercise market power in concentrated markets.[89] Proponents of heightened scrutiny of the broadband markets prefer the structural approach because the broadband market is a concentrated market, and may be seen as a "cozy duopoly."[90] Eisenach contends that despite the broadband market's concentration, the rapid rates of innovation in the market "erode[] market power, and perhaps reduces the need for antitrust enforcement in general."[91]

3.2.4. A Note on Patent Law

Whereas antitrust policy is generally aimed at increasing consumer welfare by increasing competition within a market, intellectual property policy goals are specifically targeted to incentivize innovation by rewarding innovators with above-cost returns.[92] However, antitrust enforcers will intervene if an intellectual property holder uses their rights in an anticompetitive way, such as using a patent to restrict entry into the market.

Hovenkamp notes that unlike most goods, intellectual property rights are non-rivalrous, meaning one's use of the right will not deplete the supply.[93] Hovenkamp sees intellectual property markets as collaborative markets.[94] Hovenkamp suggests collaborating in high tech markets is often procompetitive because tech products have high fixed costs and usually require networking and interconnectivity with other products.[95] Thus, the procompetitive effects of the collaborating should likely keep antitrust enforcers at bay.

3.3. REGULATION AND THE ECONOMICS OF THE INTERNET

No description of the economics of high technology and the Internet would be complete without reference to applicable regulations. In the United States, both regulation and antitrust laws share a common purpose: "to protect consumers and promote allocative efficiencies in production."[96] However, "regulated industries are carved out from the rest of the economy and are subject to proactive, regulatory

---

[88] Id.
[89] Id.
[90] Id.
[91] Id.
[92] Hovenkamp (2012a).
[93] Id.
[94] Id.
[95] Id.
[96] Boliek (2011).

intervention that goes above and beyond antitrust enforcement measures."[97] In the communications context, for example, Title II of the Communications Act grants the Federal Communications Commission (FCC) extensive powers to regulate common carriers, thereby limiting antitrust enforcement measures.[98] Large on the regulatory agenda of US and international regulators has been the passage of so-called net neutrality rules. The tradeoff between net neutrality regulation and antitrust enforcement is so important that examples of research in that area are listed in a separate section below. Also set forth is literature examining more generally the benefits of ex post antitrust enforcement versus ex ante regulation. And finally, the unique legal impact of regulation on US antitrust enforcement is considered.

3.3.1. *The FCC and Net Neutrality*

As described briefly above, network neutrality advocates voice concerns that (1) ISPs will use their market position to favor affiliate application providers and block or raise costs for rival application providers; and (2) allowing ISPs to charge for services currently priced at zero would stifle innovation.[99] To address these concerns, the FCC enacted rules that prohibit ISPs from blocking or unreasonably discriminating in the transmission of lawful network content.

Because the FCC's mandate is so broad, Boliek contends that it allows the "FCC to expand its jurisdictional reach far beyond the industries and problems within the contemplation of the original, legislative draftsmen."[100] Moreover, Boliek, Yoo and other opponents of network neutrality regulation suggest that policymakers should take a more passive role and allow network diversity "[u]nless such experimentation poses potential harms that are catastrophic or irreversible."[101] Yoo argues that the effects of a diverse network are not pernicious and unjustifiable as to warrant such burdensome regulation.[102] Rather, network diversity has some procompetitive effects.

Yoo contends that structure of the telecommunications industry indicates that price discrimination will occur, but that it is beneficial.[103] For example, the telecommunications industry contains an enormous barrier to entry, the up-front investment needed to create a telecommunications network, which as a result transformed telecommunications carriers into natural monopolies.[104] Yoo explains that for a monopolist to recoup its startup costs, it must charge equal to or above the average

---

[97] *Id.*
[98] *Id.*
[99] Hemphill (2008).
[100] Boliek (2011).
[101] Yoo (2006).
[102] *Id.*
[103] *Id.*
[104] *Id.*

cost.[105] By charging at the average cost, the monopolist is charging at inefficient levels, thus creating inefficient shortfalls in production.[106] Therefore, to maximize economic welfare, a monopolist must offer some consumers the "service at a price below average cost and mak[e] up the difference by charging other customers a price that exceeds average cost."[107] In deciding how to discriminate, the monopolist will "allocate fixed cost[s] in inverse proportion to the elasticities of demand and [] fund fixed costs by transferring the surplus that would otherwise be captured by inframarginal consumers to the producers."[108] Thus, allowing network diversity would eliminate inefficiencies in production and forgone welfare gains.

Further complicating the discussion of net neutrality has been the proper definition of relevant markets, in particular the economic discussion of whether, and to what extent, wireless broadband is a substitute for landline broadband.[109] Although similar in other aspects – both mobile and landline broadband are in two-sided markets and are subject to similar regulations – the technological limitations and competitive environments are arguably distinct and merit separate antitrust evaluation.[110]

3.3.2. Ex Post Regulation

In striking a regulatory–antitrust balance, Shelaski contends that generally,

> ex ante regulation that depends for its rationale on monopoly market structure should give way to ex post intervention against specific, anticompetitive acts on the model of conventional antitrust and competition policy, with resort to ex ante rules only where experience provides a compelling case that such rules are necessary to protect consumer welfare.[111]

Shelanski notes that some regulations, such as network interconnection rules, are important to advancing telecommunications policy goals.[112]

Shelanski also posits that "the benefits of regulation diminish as markets become competitive, while the costs of regulation remain and even increase as that transition occurs."[113] This view is also held by Alfred Kahn, who argued that diverse platforms in the telecommunications market, for example, were sufficient to "envision deregulated competition as the general rule and continued regulation the exception."[114] In

---

[105] *Id.*
[106] *Id.*
[107] *Id.*
[108] *Id.*
[109] Hahn, Litan, and Singer (2007).
[110] *Id.*
[111] Shelanski (2007).
[112] *Id.*
[113] *Id.*
[114] Kahn (2006).

competitive markets, regulated prices that are too high allow firms to set prices that are higher than the prices would be if the market was unregulated.[115] For example, Shelanski noted that when AT&T was the price leader in the telephony market, its competitors "knew in advance what AT&T's prices would be and had incentive to follow just under the 'umbrella' of AT&T's prices rather than aggressively cutting prices themselves."[116] In contrast, when the regulators set prices too low, at a "level below that which provides the return competitors need to attract investment and profitably enter the market [this] will deter competition and impede the benefits it would provide to consumers."[117]

As James Prieger found, the telecommunications market today is very competitive.[118] Prieger found that competition in the ISP markets drove up Internet speeds.[119] Specifically, "[w]hen a cable ISP is present in a market, the probability that the Incumbent Local Exchange Carrier (ILEC) offers high-speed DSL (the standard ILEC broadband technology) goes up dramatically. This probability jumps even higher when the cable ISP offers speeds exceeding 50 Mbps."[120] However, Internet speeds did not significantly improve when Competitive Local Exchange Carrier (CLEC) competitors were given low-priced access to ILEC networks.[121] The shift in the telecommunications industry from a monopoly-dominated market to a competitive market demands a shift in regulatory policy.

Shelanski recommends that regulators adopt an ex post regime analogous to the rule of reason analysis used in antitrust cases.[122] Although antitrust law serves as base in preventing anticompetitive practices, Shelanski contends that antitrust laws should not necessarily establish the limit to competition policy.[123] "For example, the absence of any duty to deal or of a meaningful essential facilities doctrine in US antitrust law (see the *Trinko* case) might hinder effective competitive enforcement in an industry in transition from regulation to unregulated competition."[124] Nevertheless, the shortcoming of current antitrust law does not justify the enactment of ex ante net neutrality regulations.[125] Rather, Shelanski argues that Congress should "articulate a standard of network competition and conduct that the FCC and the US antitrust agencies can enforce without being blocked by contrary precedent from general antitrust law."[126]

---

[115] Shelanski (2007).
[116] Id.
[117] Id.
[118] Prieger, Molner, and Savage (2014).
[119] Id.
[120] Id.
[121] Id.
[122] Shelanski (2007).
[123] Id.
[124] Id.
[125] Id.
[126] Id.

3.3.3. Implied Immunity from Antitrust Law

One of the most interesting aspects of regulatory economics is that in certain arenas, regulation may not only preclude, but prohibit antitrust oversight. As Daniel F. Spulber and Christopher S. Yoo point out, Congress can (and has) passed laws which repeal antitrust laws with regard to certain conduct.[127] For instance, Congress granted the FCC exclusive authority to regulate common carriers within Title II.[128] Indeed, a key issue of the recent net neutrality debate is that moving broadband from Title I to Title II regulation removes jurisdiction from the Federal Trade Commission (FTC). To the extent that net neutrality is primarily an argument over "vertical leveraging," the removal of such antitrust expertise is problematic[129]

Even without express immunity, however, Spulber and Yoo note that defendants may still argue that an implied immunity from antitrust laws applies.[130] Nonetheless, trial courts "have rejected claims that the federal regulatory scheme overseen by the FCC immunizes telephone companies' refusals to allow other companies to interconnect with their networks from antitrust scrutiny."[131] Spulber and Yoo contend that state regulation may also grant antitrust immunity.[132] There are two requirements to establish state action immunity: "First, the challenged restraint must be 'one clearly articulated and affirmatively expressed as state policy'; second, the policy must be 'actively supervised' by the State itself."[133]

REFERENCES

Armstrong, M. 2006. Competition in Two-sided Markets. *The RAND Journal of Economics*, 37(3), 668–91.
Atkinson, R.D. and D.B. Audretsch. 2011. Economic Doctrines and Approaches to Antitrust. *Information Technology & Innovation Foundation*, Summer, 6.
Boliek, B. 2011. FCC Regulation Versus Antitrust: How Net Neutrality is Defining the Boundaries. *Boston College Law Review*, 52, 1628–9.
Crandall, R.W. and C.L. Jackson. 2011. Antitrust in High-Tech Industries. *Review of Industrial Organization* 38(4), 319–62.
Eisenach, J.A. 2012. Broadband Competition in the Internet Ecosystem. *AEI Economic Studies*, October, 1.
Evans, D. 2009. The Middle Way on Applying Antitrust to Information Technology. *Competition Policy International* , November (2), 2.
Fisher, Franklin M. 1978. *Diagnosing Monopoly*. Working Paper Department of Economics, MIT No. 226.

---

[127] Spulber and Yoo (2007).
[128] Id.
[129] Nuechterlein (2009).
[130] Spulber and Yoo (2007).
[131] Id.
[132] Id.
[133] Id.

Hahn, Robert W., Robert E. Litan and Hal J. Singer. 2007. The Economics of "Wireless Net Neutrality." AEI-Brookings Joint Center Working Paper No. RP07–10.

Hazlett, T.W. and D. Weisman. 2009. Market Power in U.S. Broadband Services. George Mason Law and Economic Research Paper Series 1–32.

Hemphill, C. Scott. 2008. Network Neutrality and the False Promise of Zero-Price Regulation. *Yale Journal on Regulation*, 25, 135.

Hovenkamp, H. 2012a. Antitrust and the Movement of Technology. *George Mason Law Review*, 19, 1119.

2012b. Antitrust and Innovation: Where We Are and Where We Should Be Going. University of Iowa Legal Studies Research Paper No. 12–03, 749.

Kahn, Alfred E. 2006. Telecommunications: The Transition from Regulation to Antitrust. *Journal on Telecommunications & High Technology Law*, 5, 159.

Katz, M.L. and H.A. Shelanski. 2005. Schumpeterian Competition and Antitrust Policy in High Tech Markets. *Competition*, 7, 4.

Manne, G.A. and J.D. Wright. 2010. Innovation and the Limits of Antitrust. *Journal of Competition Law & Economics*, 6(1), 156–7.

Nuechterlein, Jonathan E. 2009. Antitrust Oversight of an Antitrust Debate: An Institutional Perspective on the Net Neutrality Debate. *Journal on Telecommunications & High Technology Law*, 7, 19.

Owen, B.M. 2011. Antitrust and Vertical Integration in "New Economy" Industries with Application to Broadband Access. *Review of Industrial Organization* 38(4), 363–86.

Parker, G.G. and M.W. Alstyne. 2005. Two-Sided Network Effects: A Theory of Information Product Design. *Management Science*, 51(10), 1494–504.

Pasquale, Frank. 2010. Beyond Innovation and Competition: The Need for Qualified Transparency in Internet Intermediaries. *Northwestern University Law Review*, 104, 105.

Prieger, James E., Gabor Molnar, and Scott J. Savage. 2014. *Quality Competition in the Broadband Service Provision Industry*. Paper presented at 2014 TPRC/42nd Research Conference on Communication, Information and Internet Policy.

Rysman, M. 2009. The Economics of Two-Sided Markets. *Journal of Economic Perspectives*, 23(3), 125–30.

Shapiro, C. 1999. Exclusivity in Network Industries. *George Mason Law Review*, 7, 673.

Shelanski, H.A. 2007. Adjusting Regulation to Competition: Toward a New Model for U.S. Telecommunications Policy. *Yale Law Journal on Regulation*, 24, 57.

Spulber, D.F. and C.S. Yoo. 2007. Mandating Access to Telecom and the Internet: The Hidden Side of Trinko. *Columbia Law Review*, 107.

Yoo, C. 2005. Beyond Network Neutrality. *Harvard Journal of Law & Technology*, 19, 1–7.

2006. Network Neutrality and the Economics of Congestion. *Georgetown Law Journal*, 94, 1851–2.

4

The Economics of FRAND

Anne Layne-Farrar

4.1. INTRODUCTION

Since the issue first emerged in the policy arena in the early 2000s, economists have been debating the meaning and implications of FRAND licensing commitments within cooperative technology standard setting organizations (SSOs). Today the issue is global, with scholars and policymakers in Europe, Korea, Japan, China, Taiwan, India, and Brazil all weighing in. Most SSOs around the world ask their members to commit to offer any patents that might be needed to implement a standard in commercial products and services (that is, any patents that might be "essential" for the practice of the standard) on Fair, Reasonable and Non-Discriminatory (FRAND) terms.[1] But what exactly does it mean to license a patent on fair, reasonable, and non-discriminatory terms and conditions? Does promising to do so come with other, implicit, obligations as well, such as forgoing seeking an injunction which is otherwise an option for patent holders? And more fundamentally, why do SSOs ask their patent-contributing members to commit to FRAND?

In this chapter, I review the academic literature on FRAND licensing. My review is intended to discuss the economic underpinnings of FRAND, but it would be incomplete without discussing court rulings to date, as FRAND court decisions provide real-world boundaries to interpretations of FRAND licensing. That being said, my summary of the court cases focuses on the economic interpretation of FRAND and the practical implications for expert analysis, and does not cover any legal assessment (for which I am unqualified). While the underlying economics of FRAND are universal, legal and institutional factors affect court outcomes and policy interests; I therefore restrict my discussion to the United States.

---

[1] The commitment is sometimes referred to as RAND, without "fair"; there is no meaningful difference between RAND and FRAND however. Royalty-free licensing can be viewed as a form of FRAND and is typically referred to as FRAND-RF.

To put the discussion in context, I start with consensus views on SSOs' motivations for including FRAND commitments within their regulations. I then move to the evolution of substantive interpretations of what FRAND commitments mean in practice. I cover early theoretical views of FRAND as well as litigation and agency investigation matters that have involved allegations of a breach of FRAND, from the earliest cases brought by the US Federal Trade Commission (FTC) through to the most recent court decisions (as of publication).[2] The chapter closes with a discussion of the open issues prevalent in FRAND disputes that have yet to be fully resolved by either the economic literature or court rulings.

4.2. THE IMPETUS FOR FRAND

At the simplest level, a FRAND commitment is a pledge made by a patent holder to an SSO[3] that it will license patented technology it has submitted for inclusion in a standard on fair, reasonable, and non-discriminatory terms. The typical FRAND pledge contains two key elements: (1) ready access to patented technology for implementers of the standard, and (2) a reasonable price for that access. No SSO provides a specific definition of what fair or reasonable means, or how to assess whether any particular terms and conditions are "discriminatory." This lack of specificity is, in fact, the reason that scholars and practitioners have been debating the interpretation of FRAND commitments for so long. Before reviewing how FRAND has been construed over the years, however, it is important first to understand why SSOs include FRAND commitments in their intellectual property rights (IPR) policies in the first place, and why those commitments have continued to remain (with a few exceptions, as explained below) imprecise.

The key impetus for FRAND commitments stems from the process of standard setting. New interoperability standards are initiated when industry participants perceive a technical problem that cannot be solved by a single firm or when they desire to reach a new technical goal that requires industry cooperation. Participants in standard setting efforts – who are largely engineers and other technical people, not lawyers[4] – can contribute existing proprietary technologies or develop new

---

[2] One strand of the debate focuses on whether antitrust intervention is ever warranted, as FRAND disputes may be better handled by the courts under contract or patent law. See, e.g. Brooks and Geradin (2011). I do not go into the details of this topic as it has more of a legal bent.

[3] FRAND commitments are also made unilaterally outside of SSOs. See, e.g., The Program for Information Justice and Intellectual Property, at Washington College of Law, which maintains a database of over 150 public non-SSO patent pledges: www.pijip.org/non-sdo-patent-commitments/. Contreras (2015) and Elhauge (2015) discuss the legal basis for enforcing FRAND commitments made outside of SSOs. For other discussions of non-SSO patent pledges, see Harkrider (2013) and Layne-Farrar (2014).

[4] The typical composition of members is one reason why most SSOs do not define in detail what they mean by FRAND. Working groups within an SSO focus on technical solutions, not commercial dealings, and tend to avoid things that will delay standard development or commercialization.

technical components. SSO members then vote on which of the proposed technologies to include in the standard.

Developing technical standards can take considerable time. For example, working groups for the UMTS (3G) mobile standard were first formed within the European Telecommunications Standards Institute (ETSI) in December of 1998, the first release of the UMTS standard specification came in December of 1999, and the first commercial release of UMTS products and services was not until late 2001.[5] The vast majority of licenses are negotiated just before or around the time of commercialization, primarily due to uncertainty as to the direction the standard will ultimately take and which patents may end up covering technologies essential for practice of the standard. Put differently, uncertainty over the standard as it evolves makes licensees reluctant to enter into licensing talks until there is greater certainty about which specific contributors they need a license from. Similarly, standard essential patent (SEP) holders may not know the entities that will implement a standard until commercialization announcements are made. As a result of these practicalities, there is typically a time gap between when a patented technology is proposed for a standard, when it is selected (or rejected) for inclusion in a standard, and when it is ultimately licensed to standard implementers (assuming inclusion).

The time gap between a patent's inclusion in a standard and the likely point of license negotiations poses a potential economic problem. Namely, once a patent is deemed "essential" for compliance with a standard, that patent holder may be able to "holdup" licensees, using either the implementer's inability to move to alternative technologies and still comply with the standard or the cost of redefining the standard as a whole as leverage to obtain royalty payments that exceed some objectively determined fair and reasonable amount.[6]

Empirical research also clearly establishes that being included in a standard can affect a patent's value. Rysman and Simcoe find both nature and nurture play a role.[7] First, they find that SSOs tend to identify promising technical solutions by including patents that already have significant indications of value and importance at the time of adoption (nature). Second, they find that SSOs can enhance the value and importance of included patents by promoting their adoption and diffusion (nurture). The nurture effect would raise the ability of a SEP holder to practice holdup, while the nature effect indicates the SEP holder can charge a relatively higher royalty simply due to the contribution the SEP makes to the standard and to implementers' products.

While mentions of "patent holdup" are rife in any discussion of FRAND, it is important to understand that the potential for holdup is not specific to patents and is

---

[5] Layne-Farrar (2011).
[6] Shapiro (2001); Lemley and Shapiro (2007).
[7] Rysman and Simcoe (2008).

in fact a well-studied phenomenon throughout the economy.[8] For the potential holdup risk to translate into actuality, two conditions must be met: (1) asset specificity or "lock-in" (e.g., the patent is essential for the practice of a standard) and (2) self-interest seeking with guile or deception (e.g., the SEP holder attempts to exploit switching costs ex post as a means of charging supra-FRAND rates).[9] In plainer language, if firms planning to implement a standard know of the patents at the time the standard is developed and have a reasonable expectation of the royalty rates they will be charged, they can plan accordingly and cannot be held up. These factors may explain the lack of empirical evidence establishing that patent holdup is prevalent in the economy, as opposed to being an isolated occurrence.[10] However, if the SEP holder surprises or deceives licensees, say with a hidden patent not properly disclosed during the standard's development[11] or by seeking royalties in excess of any reasonable expectations after the standard is codified, then the licensee may be held up (though licensees have some recourse, like suing the SEP holder for breach of a FRAND commitment, e.g., *Microsoft v. Motorola* 2013).

By urging patent holders to commit to licensing their essential patents on FRAND terms and conditions during standard setting, SSOs deter essential IPR holders from exercising any market power they might gain through the standardization process – as opposed to market power gained through the value and importance of their patented technologies as a contribution to products complying with the standard. In most technology markets, licensee-implementers must commit their own resources to the commercialization of products and services compliant with the standard. SSO requests for FRAND commitments provide assurances to these implementers that their commercialization investments will not be exploited and that essential standard technologies will be available for licensing on fair, reasonable, and non-discriminatory terms and conditions.[12]

The metes and bounds of FRAND commitments do differ across standards, however with a couple of exceptions most have remained imprecise or flexible. One exception is VITA, the standards body that promulgates standards for VME bus technology. In 2007 VITA imposed additional restrictions on its FRAND pledge to prevent ex post opportunistic behavior: all VITA members must disclose the maximum rates and terms they intend to seek for any essential patents before any relevant technology vote takes place. Patent holders not complying with the during-standard-development rate disclosure rule are then restricted to royalty-free licensing. Put differently, VITA imposes a "disclose it or lose it" rule.

Early rate disclosure might appear an obvious solution to maintaining FRAND licensing terms and conditions, but such early licensing commitments pose practical

---

[8] Williamson (1985).
[9] Kieff and Layne-Farrar (2013).
[10] Galetovic, Haber, and Levine (2015).
[11] E.g., *Rambus v. FTC* (2008).
[12] Geradin (2014).

problems given the technical uncertainty prevalent during the development of a standard, as noted above. Mandatory rate disclosure is also ill-suited for patent holders that do not intend to enforce their patent rights, but who do not want to exclude the licensing option irrevocably. Any rate such firms disclose will be misleading, as no enforcement is planned, but announcing a zero rate can be detrimental strategically, such as for defensive cross-licensing or countersuits.[13]

Less severe than VITA's mandatory disclosure rule, IEEE instituted an optional rate disclosure on its patent disclosure form (the Letter of Assurance), also in 2007. After eight years in place, however, the option has rarely been taken. More recently, in 2015 IEEE instituted a series of specific restrictions within its FRAND pledge, including the preclusion of seeking an injunction on any disclosed patents and a commitment to use the "smallest saleable patent practicing unit" ("SSPPU") as the royalty base in any FRAND license (IEEE IPR policy). The policy change is quite controversial, with several IEEE members stating that they will follow IEEE's prior policy but not the new one.[14] Nevertheless, the Department of Justice provided IEEE with a favorable Business Review Letter.[15]

To put the FRAND obligations into perspective, recall that outside of standard setting contexts, patent holders can refuse to license any of their patents and can seek whatever terms and conditions they desire (though, of course, whether the sought rates and terms actually will be paid by anyone depends on the market's view of the value of the patented technologies). Seen in the broader context of patent licensing, then, the FRAND commitment can be thought of as a quid pro quo: in exchange for inclusion in a standard, the patent holder commits to granting licenses on FRAND terms and conditions.

4.2.1. FRAND Implies Balance

Thus far, we have spoken only of the SEP holder's obligations under FRAND, but is FRAND a one-way street? No, as the section title above indicates. Economic reasons, explained below, make clear the need for balance. And some SSOs make the need for balance in SEP licensing explicit. One example is ETSI's IPR policy. Section 2.1 of that policy includes the following language on one-half of the bargain:

> ETSI IPR POLICY seeks to reduce the risk to ETSI, MEMBERS, and others applying ETSI STANDARDS, that investment in the preparation, adoption and application of STANDARDS could be wasted as a result of an ESSENTIAL IPR for a STANDARD being unavailable.

In other words, implementers need access to essential patents, as discussed above. Section 2.2 of ETSI's policy then addresses the other side of the bargaining table:

---

[13] Geradin and Layne-Farrar (2007).
[14] Lloyd (2015).
[15] Department of Justice (2015).

"IPR holders whether members of ETSI and their AFFILIATES or third parties, should be adequately and fairly rewarded for the use of their IPRs in the implementation of STANDARDS." In other words, SEP holders will provide access, but should be paid fairly for it.

While patent holdup theory clarifies the risk of SEP holders upsetting the balance with excessive royalty rates, reverse holdup or holdout theories clarify the symmetric risk of licensees upsetting the balance with demands of below-FRAND royalties. Reverse holdup is the other side of the patent holdup coin: while holdup leverages implementers' costs of switching away from technologies codified in a standard, reverse holdup leverages SEP holders' limited (sometimes non-existent) market opportunities to license standard essential technologies outside of the standard.[16] Concerns over a buyers' cartel, where implementers bargain as a group to keep licensing rates below FRAND levels, are a key reason SSOs have not incorporated FRAND rate setting mechanisms as part of the standardization process.[17] As yet another reason for keeping licensing negotiations separate from standard setting activities, Lerner and Tirole find that "even balanced SSOs will put pressure on IP owners to accept low prices in exchange for their functionalities' being selected into the standard."[18]

An additional risk for SEP holders is licensee holdout. In this scenario, the licensee simply refuses to take any license, FRAND or otherwise.[19] The SEP holder therefore has to choose between letting the implementer infringe versus expending the resources to litigate over patent infringement. The standard setting context imposes two important differences for FRAND litigation as compared to traditional patent infringement. First, all essential patents must be licensed for compliance with a standard, which implies SEP portfolio licensing, as inventive SSO participants contributing technologies to a standard tend to contribute more than one patent. But at least without dramatic changes from current practice, infringement litigation does not allow portfolio assertions: courts restrict patent holders to a handful of asserted patents at most.[20] As a result, unless the litigation leads the parties to a broad portfolio settlement, multiple litigations will be required to cover the SEP portfolio. Second, the FRAND commitment imposes a cap on royalties, in contrast to traditional infringement cases in which damages are "no less than a reasonable royalty" (i.e., the reasonable royalty sets a pricing floor). Relatedly, while damages set in non-FRAND-related cases may be increased to reflect the court's finding of validity and infringement (as compared to the uncertainty over validity and infringement affecting arm's-length licensing negotiations), the courts have capped FRAND

---

[16] Geradin (2010).
[17] Swanson and Baumol (2005).
[18] Lerner and Tirole (2015).
[19] Chien (2014).
[20] See, for example, *Realtek v. LSI*, which covered only two SEPs, though LSI held far more.

rates at those granted in comparable arm's-length agreements, with no enhancement for validity and infringement.

While litigation imposes costs on both parties, this expense may reflect the price of an "option" for certain licensees: there is some probability that the patent holder will not sue, there is also a nonzero probability that if sued the licensee will win, and even if the licensee loses damages will not exceed the FRAND rates that would have otherwise been paid. Taking the expected value across these possible outcomes could well suggest that licensee holdout is a very attractive strategy for some implementers. Litigation therefore entails asymmetric risks for SEP holders: the patent holder can only assert its SEP portfolio piecemeal and risks a finding that the patents are not essential, not infringed, or not valid, which eliminates revenues for those patents against the party sued plus all others; but if the SEP holder wins the litigation, it will get in damages no more than it could have earned through bilateral negotiations, though less the cost of litigation.

Considering the above, it is easy to understand why "fair and reasonable" needs to apply to both SEP holders and licensees. If implementers cannot count on access to essential technologies for compliance with the standard, and at reasonable rates, then any standard developed will die on the vine.[21] But equally important, if SEP holders cannot expect adequate compensation for their risky and often large up-front investments in inventions that move the evolution of standards forward, there would be little incentive for such innovative firms to make such investments or to participate in cooperative standard setting efforts.[22] In short, technical interoperability SSOs form platforms for the two-sided process of standard setting. Both sides are critical to a commercially successful standard, and FRAND licensing is a critical element in achieving balance. Thus, SSO IPR policies often recognize two different goals: (1) to fairly and adequately compensate patent holders who participate in SSOs by contributing valuable technologies, and (2) to make essential patents available on reasonable terms to those wishing to implement the standard to ensure its commercial success.

4.3. THE EVOLUTION OF FRAND INTERPRETATIONS

4.3.1. *An Early Emphasis on Fair and Reasonable*

With the above understanding of why SSOs ask for FRAND commitments, we can now turn to what FRAND actually means in practice. One of the earliest concerns was patent ambush, a violation of a FRAND commitment involving deception-enabled holdup.[23] Under ambush, a patent holder participates in a standard

---

[21] Lemley and Shapiro (2007).
[22] Layne-Farrar, Llobet, and Padilla (2014a).
[23] Farrell et al. (2007).

development effort, perhaps even encouraging the adoption of its technology at SSO working group meetings, but without disclosing that it holds patents on that technology. At the commercialization stage, the patent holder then reveals its patent rights, demanding rates that exploit implementers' investments in producing standard-compliant products.

The definition above describes the FTC's allegations in its long-running case against Rambus (FTC Notice filed in 2002). The FTC argued that Rambus had illegally acquired monopoly power by failing to disclose its relevant patent rights, which deceived JEDEC members. The FTC argued that the JEDEC standards body members likely would have chosen different technological components, and not Rambus' technologies, had they known that Rambus would attempt supra-FRAND licensing rates after the standard's codification. While the FTC won early rounds of the case, on appeal the court found that the FTC had not sufficiently established that the but-for world without Rambus' deception would have resulted in another technology being chosen for the standard over Rambus' technology, eviscerating the FTC's market power arguments. The Supreme Court then declined to take up the FTC's appeal.

Despite its ultimate loss, however, the publicity that attended the seven-year-long Rambus investigation (along with two earlier similar FTC investigations – *UNOCAL* and *Dell*) threw a bright spotlight on patent disclosure and deceptive practices at SSOs. In the wake of these investigations, a number of SSOs modified their IPR policies to clarify their patent disclosure rules[24] and patent disclosures at certain prominent SSOs have since skyrocketed (see, e.g. data available on SSOPatents.org).

More generally, even when potential SEPs are clearly disclosed, the question of fair and reasonable rates remains because the SEP holder may act with guile, reneging on its FRAND commitment, or more simply and less nefarious, because (unsurprisingly) licensors and licensees disagree over what constitutes a fair and reasonable price and thus need some objective metric to settle disputes. Swanson and Baumol provide an early – and influential – interpretation of benchmarking for FRAND licensing. Swanson and Baumol posited that "the concept of a 'reasonable' royalty for purposes of RAND licensing must be defined and implemented by reference to ex ante competition, i.e., competition in advance of standard selection."[25] The insight is an important one: during standard development, patented technologies may be competing with one another (or with other non-patented technologies), so using this "ex ante" period as a reference point captures market forces that keep prices fair and reasonable.[26]

---

[24] Layne-Farrar (2014).
[25] Swanson and Baumol (2005).
[26] Note that the period during a standard's development is ex ante to implementers' investments for commercialization, but is ex post to the R&D investments made by SEP contributors.

Farrell et al.[27] developed the notion of "incremental value" laid out in Swanson and Baumol,[28] which in turn was adapted to patent licensing from the discussions of incremental value in traditional manufacturing contexts (e.g., Farrell and Katz[29]). The idea is that FRAND royalties should be capped at the added value a SEP contributes to products implementing the standard. Under traditional model assumptions, competition among as few as two technology rivals during the development of a standard can lead to theoretical prediction of a near-zero royalty for the winning technology. However, Layne-Farrar and Llobet call into question a number of those traditional assumptions. In particular, they find that

> ...when patented technologies must be weighed on numerous factors, and not simply one-dimensional cost-savings [a situation that is prevalent in real world standard setting], there is unlikely to be a single incremental value that can be agreed upon by all relevant parties. Furthermore, ex ante competition fails to select the efficient technologies by penalizing the more versatile ones.[30]

Mariniello expanded on the ex ante competition concept in a different regard, defining a safe harbor for FRAND licensing. He observed that "while it is difficult to assess whether a FRAND commitment has been breached, assessing whether a FRAND commitment has not been infringed can be easier."[31] He then defined a number of conditions that can act as a screening mechanism, all of which have to hold before a more detailed investigation of any breach of FRAND would commence. His conditions are: (1) a credible alternative to the technology must have existed during the development of the standard; (2) licensees could not reasonably anticipate the rates that the SEP holder seeks ex post; (3) the SEP holder seeks more onerous terms ex post than ex ante (to test this condition, at least one license for the SEPs in question must have been signed before the standard was codified); and (4) the licensee must be locked in to using the SEP. If any one of these conditions does not hold, the SEP holder cannot have violated its FRAND commitment by charging "excessive" royalties; if, however, all four conditions hold, then a detailed investigation should be opened to assess whether a breach has occurred.

As noted above, the screening test proposed by Mariniello[32] only works in the presence of at least one license concluded ex ante, which is not often the case. Thus, turning the competition over technologies to define a standard into a practicable ex post licensing rule has remained elusive. Swanson and Baumol proposed an auction during a standard's development in which patented technologies would be auctioned off at licensing royalty rates, with the auction winner then defining the

---

[27] Farrell et al. (2007).
[28] Swanson and Baumol (2005).
[29] Farrell and Katz (2005).
[30] Layne-Farrar and Llobet (2014).
[31] Mariniello (2011).
[32] Id.

standard and licensing at the "winning bid."[33] Later papers,[34] however, explained why such auctions would not be feasible in practice and would not yield the desired outcome at any rate, so the quest then turned to other means for establishing ex ante pricing benchmarks.

As a simpler approach than licensing auctions, some industry participants proposed joint negotiations among SSO members over patent rates.[35] Under this approach, the SSO would sponsor, or at least provide a forum for members to collectively bargain over licensing terms during the standard development phase, much along the lines of labor union bargaining. In fact, the Department of Justice issued a statement that it would evaluate joint bargaining within SSOs under a rule of reason,[36] signaling to SSOs that such rules would *not* be treated as per se buyer cartels and that instead collective bargaining with appropriate safeguards in place could be approved. To evaluate whether such an approach would be a good idea, Gilbert compared bilateral SEP licensing against joint negotiations at a theoretical level, under a Nash bargaining framework.[37] He found that joint negotiations would indeed lead to lower royalties than bilaterally negotiated agreements, but that the reduction in rates could be the result of overly aggressive ex ante group negotiations that push rates to a level that deters innovation. Thus, he concluded that joint negotiations should only be explored and adopted when a significant risk of ex post patent holdup exists and when there is clear evidence that the SSO lacks meaningful market power. Thus far, no SSO has adopted a joint negotiation approach.

In an effort at identifying a more practical approach, Layne-Farrar, Padilla, and Schmalensee[38] tied FRAND determinations to the framework established for assessing reasonable royalties in traditional patent infringement cases, as defined in the seminal case *Georgia Pacific* (*Georgia-Pacific Corp. v. U.S. Plywood Corp.* 1970).[39] While it is unwieldy, and can be abused as an unbounded checklist, the fifteen factors enumerated in *Georgia Pacific* do encompass a number of economic aspects commonly part of arm's-length licensing negotiations, such as the royalties earned by the same patents in other licenses and the advantages of the patented technology over old modes of operation. For application in FRAND determination, some factors are dropped, such as whether the patent holder and the licensee compete with one another, but the remainder of the factors are easily modified to account for the limitations imposed by a FRAND commitment. In fact, this approach was adopted in the first district court case to consider a claimed breach of a FRAND promise (*Microsoft v. Motorola* 2013).

---

[33] Swanson and Baumol (2005).
[34] E.g., Geradin, Layne-Farrar, and Padilla (2008),
[35] Ohana, Hansen, and Shah (2003).
[36] Department of Justice (2006).
[37] Gilbert (2011).
[38] Layne-Farrar, Padilla, and Schmalensee (2007).
[39] Layne-Farrar, Padilla, and Schmalensee (2007) also developed a Shapley Value method for valuing FRAND patents, though this was more of a thought exercise than a practical proposal.

4.3.2. *Dispute Resolution: Contracts and Arbitration*

Taking an entirely different tack, some scholars have argued that FRAND commitments form a contract – mainly between the SEP holder and the SSO, but with implementing members of the SSO garnering third-party benefits. Under this framework, "the role of a court is not to determine what 'fair and reasonable' terms would be, but whether the terms offered, taking into account all of the specific circumstances between the parties and prevailing market conditions, fall outside the range of reasonableness contemplated by the FRAND commitment."[40] Geradin softened this position somewhat, arguing that FRAND commitments do create a contract, but an incomplete one given that prices and other terms are not specified.[41] The incomplete nature of the contract then affects its legal interpretation and enforcement. As a result, Geradin concluded that "while the determination of contractual terms are much better left to the parties, there is no reason to believe that courts should be unable to determine the terms 'fair and reasonable' when asked to do so."[42] Several courts have accepted a contractual reading of FRAND commitments and have proceeded to determine what fair and reasonable rates would be given the facts presented in the case (see, e.g., *Microsoft v. Motorola* 2013).

More recently the debate over FRAND rate determination has centered on alternative dispute resolution mechanisms that avoid the need for court determinations, namely arbitration. Lemley and Shapiro propose that SSOs include provisions that bind members to "baseball-style" arbitration, should a FRAND dispute arise after a standard is set. In brief, their argument is as follows:

> ... if a standard-essential patent owner and an implementer of the standard cannot agree on licensing terms, the standard-essential patent owner is obligated to enter into binding baseball-style (or "final offer") arbitration with any willing licensee to determine the royalty rate. This obligation may be conditioned on the implementer making a reciprocal FRAND Commitment for any standard-essential patents it owns that read on the same standard. If the implementer is unwilling to enter into binding arbitration, the standard-essential patent owner's FRAND commitment not to go to court to enforce its standard-essential patents against that party is discharged.[43]

Larouche et al. counter that binding "baseball-style" arbitration could well lead to outcomes opposite of those desired. In particular, they argue that baseball arbitration would reduce parties' incentives to negotiate bilaterally in good faith and would instead lead to increased adjudication, potentially with biased results in relation to

---

[40] Brooks and Geradin (2011).
[41] Geradin (2014).
[42] Id.
[43] Lemley and Shapiro (2013).

fair compensation for SEP holders.[44] They further point out that arbitration is not necessarily more efficient or cost-effective than litigation. It also has the downside of being confidential, limiting the licensing information that makes its way into the public sphere.

Some SSOs do mention arbitration as an option, if both parties agree (see, e.g., IEEE's IPR Policy), however I have not seen any SSO mandate binding arbitration with no appeal, as suggested by Lemley and Shapiro.[45] Even IEEE, an aggressive first mover among SSOs in terms of adding more precise bounds to the FRAND commitment, states in its 2015 IPR Policy that arbitration is by "the parties' mutual agreement."

4.3.3. The Non-discrimination Prong of FRAND

Far less attention has been paid to the "ND" element of FRAND, with only a handful of papers addressing what it means to offer non-discriminatory licensing rates and terms for SEPs. Swanson and Baumol provide a benchmarking framework for non-discriminatory licensing akin to transfer pricing: a SEP holder can charge no more for its essential patents than it implicitly charges itself for downstream production.[46] Of course, this rule only applies to SEP holders that also implement the standard in downstream markets (i.e., for vertically integrated firms); the benchmark offers no help for nonpracticing entities with SEPs.

In a broader assessment of the ND commitment, Layne-Farrar starts by reviewing the large economics literature on price discrimination in markets for traditional goods to glean teachings relevant for patent licensing under FRAND commitments.[47] The bottom line in that paper is that "discrimination" in and of itself is not harmful, and in some circumstances can even be procompetitive. What is important for FRAND assessments, therefore, is consideration of core antitrust principles, first and foremost establishing that the SEP holder has market power (something that cannot be presumed simply as resulting from SEP ownership). Once that threshold has been met, the licensees' relative positions are the key element of analysis. Similarly situated licensees should be treated similarly, with any differences in rates or terms explained and assessed for their impact on the competitive position of the party receiving the less favorable terms.

Gilbert[48] and Carlton and Shampine[49] do not provide explicit analysis of non-discriminatory licensing, but rather argue that the ND commitment within FRAND can be used as a tool to prevent patent holdup. That is, license agreements that are

---

[44] Larouche, Padilla, and Taffet (2014).
[45] Lemley and Shapiro (2013).
[46] Swanson and Baumol (2005).
[47] Layne-Farrar (2010).
[48] Gilbert (2011).
[49] Carlton and Shampine (2014).

clearly FRAND (say, because they were negotiated prior to the codification of the standard, and thus before any market power could have been gained through the patent's inclusion in the standard) can be used as benchmarks for other agreements, to restrict the terms and conditions to fair and reasonable for all licensees.

4.3.4. *Royalty Stacking*

Closely related to the literature on FRAND licensing is the literature on royalty stacking. More precisely, if a standard incorporates a number of different technologies with patents held by a large number of entities, the royalty payments to access each set of SEPs can create a "stack" that in the aggregate renders commercialization of products compliant with the standard uneconomical or unprofitable. Thus, a royalty stack could exaggerate any holdup rates set by individual SEP holders. On the basis of this possibility, some have argued that in order for a license to comport with FRAND, it should not contribute to any "royalty stack."[50] In other words, while royalty stacking is an aggregate licensing concept, individual SEP holders should actively work to prevent a stack in order to comply with their FRAND commitments. Some court decisions (e.g., *Microsoft v. Motorola* 2013) have thus argued that individual SEP holders need to account for their place within the set of all SEPs to be licensed, such that the potential for a royalty stack to emerge is part of the FRAND rate determination for individual patents.

While the double marginalization theory that royalty stacking is based upon has a long and well-respected history,[51] the application of that theory to patent licensing – and more importantly, the conclusion that royalty stacking is so prevalent as to be a mandatory part of individual SEP holders' FRAND license determinations – has been controversial. For example, Elhauge assesses the Lemley-Shapiro model and finds that even under their specification, the resulting royalty rate is likely to fall *below* the optimal rate, not above it. While Elhauge notes that "[n]one of the analysis that follows denies that there are some cases in which extreme holdup problems could result in royalties that exceed the socially optimal rate," he concludes that economic theory does not support the inference that royalty stacking is a systematic problem, and instead suggests that under-compensation of SEPs is more likely.[52]

As another consideration, the royalty stacking theory assumes that each SEP holder will set its royalty rate without any consideration of the rates charged by other SEP holders. Layne-Farrar and Wong-Ervin question that assumption, observing that SEP holders interact closely with one another during the development of the standard.[53] Given the joint production nature of standard setting, there is little justification for

---

[50] Lemley and Shapiro (2007).
[51] Cournot (1838).
[52] Elhauge (2008).
[53] Layne-Farrar and Wong-Ervin (2015).

the assumption that each SEP holder will act independently when it comes time to a license, ignoring all other SEP holders in setting its own royalty rates.

Since theory does not provide a definitive answer to the actual prevalence of royalty stacking, the Federal Circuit has called for case-specific evidence, demanding that "[c]ertainly something more than a general argument that these phenomena are possibilities is necessary" (*Ericsson v. D-Link* 2014). To meet the burden of providing "something more," the Federal Circuit ruled that the actual cumulative royalty paid by a particular implementer must be proven and assessed during the case in order to establish that it is excessive.

4.4. LITIGATING SEPS: OPEN ISSUES

As the discussion above illustrates, a number of FRAND's parameters have become clearer over time, through both the literature and court rulings. That being said, a number of open issues remain, particularly in relation to FRAND litigation. I close this chapter with a brief review of those topics.

4.4.1. Injunctions

The question here is whether or not a FRAND commitment precludes a SEP holder from ever seeking an injunction – either preliminary or permanent. The debate over injunctive relief ties directly back to the debate over patent holdup. That is, some have expressed concern that the ability to threaten a licensee with an injunction or exclusion order provides a SEP holder with undue market power that can be used to extract supra-FRAND royalty payments.[54] Knowing that the SEP holder can foreclose it from the market entirely, the implementer may be more likely to accept unreasonable licensing terms and conditions. Thus, this camp argues that a FRAND commitment prevents a SEP holder from ever seeking an injunction.[55]

On the other side of the debate are those who argue that without the ability to at least seek an injunction (as a separate issue from when an injunction should actually issue and be enforced), SEP holders have no means of pushing infringers to the negotiation table or to avoid licensee holdout.[56] Put differently, standards are published publicly, or are available for at least all SSO members, so that the patented technologies comprising a standard are easy to infringe. A firm can implement a standard, following the published guidelines, without taking a license to any of the relevant SEPs, forcing the SEP holders to file patent infringement litigation. But, as discussed above (see the discussion of licensee holdout), this route provides inadequate remediation for SEP holders since patent litigation is limited to a handful of patents at a time, not the patent

---

[54] Lemley and Shapiro (2007).
[55] Miller (2006).
[56] Geradin (2010).

holder's full SEP portfolio. Moreover, the best the SEP holder can get if it wins is a FRAND rate – FRAND caps royalties – but this amount ignores the costly efforts that the SEP holder had to make to earn any royalty at all.

Until 2015, no SSO explicitly addressed injunctive relief. That changed with the IEEE IPR policy change, which, as explained above, imposed significant modifications. In particular, the IEEE IPR policy now states that SEP holders making a FRAND commitment may not seek, or threaten to seek, injunctions against potential licensees who are willing to negotiate for licenses. IEEE is in the minority thus far, however, as no other SSO has enacted any rule preventing SEP holders from ever seeking injunctive relief.

The Federal Circuit has not issued any definitive rules yet, but has thus far upheld rulings that prevent SEP holders from obtaining an injunction. For example, in *Microsoft v. Motorola*, the Court ruled that a FRAND commitment can preclude an injunction: "the sweeping promise of Motorola's RAND agreements, at least arguably a guarantee that the patent-holder will not take steps to keep would-be users from using the patented material" implies that it will refrain from "seeking an injunction, but will instead proffer licenses consistent with the commitment made." The Court justified its conclusion on the basis of evidence in the record that it felt made clear Motorola's use of an injunction as a tool of holdup: "The district court identified the testimony of five different experts from which the jury could conclude that Motorola's actions were intended to induce hold-up, i.e., to pressure Microsoft into accepting a higher RAND rate than was objectively merited, and thereby to frustrate the purpose of the contract."

What remains to be seen is whether there are any circumstances under which a SEP holder may seek an injunction. Certainly the antitrust agencies allow for the possibility. In a 2013 statement, after noting that granting an injunction on a FRAND-encumbered patent may harm competition and consumers, the US Department of Justice (DOJ) made the following clarification:

> This is not to say that consideration of the public interest factors set out in the statute would always counsel against the issuance of an exclusion order to address infringement of a F/RAND-encumbered, standards-essential patent. An exclusion order may still be an appropriate remedy in some circumstances, such as where the putative licensee is unable or refuses to take a F/RAND license and is acting outside the scope of the patent holder's commitment to license on F/RAND terms. For example, if a putative licensee refuses to pay what has been determined to be a F/RAND royalty, or refuses to engage in a negotiation to determine F/RAND terms, an exclusion order could be appropriate. Such a refusal could take the form of a constructive refusal to negotiate, such as by insisting on terms clearly outside the bounds of what could reasonably be considered to be F/RAND ... This list is not an exhaustive one.[57]

---

[57] Department of Justice and U.S. Patent Office (2013).

4.4.2. The Influence of the SEP Holder's Business Model

Another evolving debate regarding FRAND centers on the business practices of the SEP holder. In 2006, the Supreme Court issued its opinion in *eBay*, known primarily for its rule that all patent holders seeking an injunction in federal court must pass a four-part test, but which also included a concurring opinion from Justice Kennedy that lamented that "[a]n industry has developed in which firms use patents not as a basis for producing and selling goods but, instead, primarily for obtaining licensing fees." In the wake of the *eBay* decision, a significant strand of literature emerged investigating patent licensing incentives and practices among entities not practicing their patents directly, so-called NPEs or, more pejoratively, "patent trolls."[58]

The broader debate over NPEs and how they operate under different motives than practicing entities then led to a debate over NPEs and FRAND commitments. Starting in 2008, the FTC held a series of workshops related to this topic, culminating in one in 2012 exploring the relationship between patent assertion entities (PAEs, a subcategory of NPE) and FRAND promises.[59] Lemley and Melamed[60] expressed concerns over the sale of patents to NPEs or the use of "privateering" contracts – where a practicing entity transfers patents to an NPE who then acts as an agent for the original patent owner, sharing the returns of enforcement[61] – as a means of evading FRAND commitments. Though they use the term "hybrid PAE" rather than privateer, Scott Morton and Shapiro further explore these issues.[62] Thus far, the literature has raised a number of questions and potential areas of concern for PAEs and FRAND commitments, but no consensus has emerged on the need for any action (aside from perennial calls for general patent reform).

4.4.3. Damages for Infringement of SEPs

To close out this chapter, consider damages calculations for SEPs with FRAND obligations. The issues here are also hot topics in traditional (non-FRAND-related) patent infringement cases, but take on a new twist with a FRAND-committed patent. Indeed, different views regarding the strength of patents overall, and SEPs in particular, can have a profound impact on the balance of power between SEP holders and standard implementers, not to mention strategic choices in terms of negotiation and licensing practices, participation in standard setting efforts, and the business models adopted or abandoned by industry participants.

---

[58] See, e.g., Golden (2007); Lemley and Shapiro (2007); Geradin, Layne-Farrar and Padilla (2011).
[59] Mintzer and Munck (2014).
[60] Lemley and Melamed (2013).
[61] Ewing (2011).
[62] Scott Morton and Shapiro (2014, 2015).

The underlying question in litigation damages is where in the production chain a SEP holder should license. As noted earlier, the consensus view is that FRAND prevents a SEP holder from refusing to license. That does not necessarily mean licensing at all levels in the production chain, however. For example, non-assertion practices or covenants not to sue may meet the FRAND obligation without the SEP holder needing to negotiate specific licenses with all parties. At least at a theoretical level, Layne-Farrar, Llobet, and Padilla find that the aggregate royalty level is unaffected by licensing to downstream manufacturers alone versus splitting royalties up across production levels.[63]

This issue fits within the larger debate over the SSPPU that arose out of traditional patent infringement cases (e.g., *VirnetX v. Cisco* 2014). The basic concept is that the royalty base should be limited to the smallest product component that has a separate price (to ensure objectivity) and that also fully captures the benefits of the patented technology. This principle makes sense when dealing, say, with a patent on a narrow software program running on a laptop PC. In that case, calculating royalties for the software program on the basis of revenues for the entire PC is unlikely to make economic sense and is instead likely to inflate damages beyond the value of the patent.

A literature applying this topic to FRAND-encumbered patents has yet to emerge, but the debate has been active within policy circles and led IEEE to change its IPR policy. Among the changes IEEE made (as discussed above), the policy now specifies that "[t]he value that the functionality of the claimed invention or inventive feature within the Essential Patent Claim contributes to the value of the relevant functionality of the smallest saleable Compliant Implementation that practices the Essential Patent Claim." In other words, IEEE is mandating the royalty base used to calculate royalty payments within good faith bilateral licensing negotiations, outside of any litigation.

On its face, an SSPPU restriction might seem like a good idea, as it can work to limit royalties to the specific benefits the SEP provides to products implementing the standard. However, industry practice in high tech sectors has been to license on end product prices, for a host of practical and transactional cost reasons. Thus, hard-line insistence on an SSPPU approach would drive a wedge between litigation damages and arm's-length licensing. Moreover, inserting rules into how private parties negotiate contracts is highly controversial.

In its *Ericsson* (2014) ruling, the Federal Circuit made it clear that an SSPPU approach is not the only way to arrive at a rationally apportioned reasonable royalty; the use of comparable licenses is still a viable approach. Then in a December 2015 ruling, the Federal Circuit expounded on this view in a case involving SEPs without a FRAND commitment attached to them (*CSIRO v. Cisco*). The Court

---

[63] Layne-Farrar, Llobet, and Padilla (2014b).

found "[t]he rule Cisco advances – which would require all damages models to begin with the smallest salable patent-practicing unit – is untenable. It conflicts with our prior approvals of a methodology that values the asserted patent based on comparable licenses." In another interesting and related aspect of this same ruling, the Court found that even though CSIRO had not made a FRAND commitment on its SEPs, nevertheless, "reasonable royalties for SEPs generally – and not only those subject to a RAND commitment – must not include any value flowing to the patent from the standard's adoption." It remains to be seen exactly how this Federal Circuit ruling will be interpreted by the lower courts.

As this chapter makes abundantly clear, reaching consensus on any aspect of FRAND is a daunting task. It is a complicated subject requiring a great deal of nuanced thinking. Nevertheless, the economics literature has shown at least one thing: the Gaia principle is alive and well in all things related to standards. FRAND licensing is but one aspect of larger commercial dealings among a diverse ecosystem of participants, running the gamut from pure-licensing implementers to upstream innovation specialists, with all manner of hybrid forms in between. In whatever is to come for FRAND, be it from the courts or the competition agencies, we can only hope that the many interdependencies are kept front of mind so as to limit undesirable unintended consequences.

REFERENCES

Brooks, Roger and Damien Geradin. 2011. Taking Contracts Seriously: The Meaning of the Voluntary Commitment to License Essential Patents on "Fair and Reasonable" Terms. In S. Anderman and A. Ezrachi (eds.), *Intellectual Property and Competition Law: New Frontiers*. Oxford University Press.

Carlton, Dennis W. and Allan L. Shampine. 2014. Identifying Benchmarks for Applying Non-Discrimination in FRAND. *Competition Policy International Chronicle*, 1, 2.

Chien, Colleen. 2014. "Holding up" and "Holding Out". *Michigan Telecommunications & Technology Law Review*, 21, 1.

Contreras, Jorge. 2015. Market Reliance and Patent Pledges. *Utah Law Review*, 2.

Cournot, Augustin. 1838. *Researches into the Mathematical Principles of the Theory of Wealth*. Nathaniel T. Bacon trans. New York: Macmillan (1987).

CSIRO v. Cisco. Federal Circuit. December 1, 2015.

Department of Justice. 2006. Letter from Assistant Attorney General Thomas O. Barnett to Robert A. Skitol, October 30. Available at www.justice.gov/atr/response-vmebus-international-trade-association-vitas-request-business-review-letter.

2015. Letter from Acting Assistant Attorney General Renata B. Hesse to Michael A. Lindsay, February 2. Available at www.justice.gov/atr/response-institute-electrical-and-electronics-engineers-incorporated.

Department of Justice and U.S. Patent Office. 2013. Policy Statement on Remedies For Standards-Essential Patents Subject To Voluntary F/Rand Commitments. Available at www.uspto.gov/about/offices/ogc/Final_DOJ-PTO_Policy_Statement_on_FRAND_SEPs_1-8-13.pdf.

Elhauge, Einer. 2008. Do Patent Holdup and Royalty Stacking Lead to Systematically Excessive Royalties? *Journal of Competition Law & Economics*, 4, 535.
   2015. Treating RAND Commitments Neutrally. *Journal of Competition Law and Economics*, 11, 1.
Ericsson v. D-Link, 773 F.3d 1201 (Fed. Cir. 2014).
ETSI IPR Policy. 2014. Available at www.etsi.org/images/files/ipr/etsi-ipr-policy.pdf.
Ewing, Tom. 2011. Indirect Exploitation of Intellectual Property Rights By Corporations and Investors. *Hastings Science & Technology Law Journal*, 4.
FTC. 2007. Opinion of the Commission on Remedy in the Matter of Rambus Inc. by Chairman Deborah Platt Majoras.
Farrell, Joseph and Michael Katz. 2005. Competition or Predation? Consumer Coordination, Strategic Pricing and Price Floors in Network Markets. *Journal of Industrial Economics*, 53(2), 203.
Farrell, Joseph, John Hayes, Carl Shapiro, and Catherine Sullivan. 2007. Standard Setting, Patents, and Hold-Up. *Antitrust Law Journal*, 74, 603.
Galetovic, Alexander, Stephen Haber, and Ross Levine. 2015. An Empirical Examination of Patent Holdup. *Journal of Competition Law & Economics*, 11(3), 1.
Georgia-Pacific Corp. v. U.S. Plywood Corp., 318 F. Supp. 1116, 1120 (S.D.N.Y.) 1970.
Geradin, Damien. 2010. *Reverse Hold-ups: The (Often Ignored) Risks Faced by Innovators in Standardized Areas. The Pros and Cons of Standard Setting* 2010. Swedish Competition Authority.
   2014. The Meaning of "Fair and Reasonable" in the Context of Third-Party Determination of FRAND Terms. *George Mason Law Review*, 21, 919.
Geradin, Damien and Anne Layne-Farrar. 2007. The Logic and Limits of Ex Ante Competition in a Standard-Setting Environment. *Competition Policy International*, 3(1).
Geradin, Damien, Anne Layne-Farrar, and Jorge Padilla. 2008. The Ex Ante Auction Model for the Control of Market Power in Standard Setting Organizations. *European Competition Journal*, 4(8).
   2011. Elves or Trolls? The Role of Nonpracticing Patent Owners in the Innovation Economy. *Industrial and Corporate Change*, June, 1.
Gilbert, Richard. 2011. Deal or No Deal? Licensing Negotiations by Standard Setting Organizations. *Antitrust Law Journal*, 77, 855.
Golden, John. 2007. Commentary: "Patent Trolls" and Patent Remedies. *Texas Law Review*, 85, 2111.
Harkrider, John. 2013. REPs Not SEPs: A Reasonable and Non-Discriminatory Approach to Licensing Commitments. *Antitrust Chronicle*, 10.
IEEE IPR Policy. Available at https://standards.ieee.org/develop/policies/bylaws/approved-changes.pdf.
Kieff, F. Scott and Anne Layne-Farrar. 2013. Incentive Effects from Different Approaches to Holdup Mitigation Surrounding Patent Remedies and Standard-Setting Organizations. *Journal of Competition Law & Economics*, 9(4), 1–33.
Larouche, Pierre, Jorge Padilla, and Richard Taffet. 2014. Settling FRAND Disputes: Is Mandatory Arbitration a Reasonable and Non-Discriminatory Alternative? *Journal of Competition Law and Economics*, 10, 581.
Layne-Farrar, Anne. 2010. Nondiscriminatory Pricing: Is Standard Setting Different? *Journal of Competition Law & Economics*, 6(4), 1.
   2011. Innovative or Indefensible? An Empirical Assessment of Patenting within Standard Setting. *International Journal of IT Standards and Standardization Research*, 9, 1–18.

2014. Moving Past the SEP RAND Obsession: Some Thoughts on the Economic Implications of Unilateral Commitments and the Complexities of Patent Licensing. *George Mason Law Review*, 21.

Layne-Farrar, Anne and Gerard Llobet. 2014. Moving Beyond Simple Examples: Assessing the Incremental Value Rule within Standards. *International Journal of Industrial Organization*, 36, 57.

Layne-Farrar, Anne and Koren Wong-Ervin. 2015. An Analysis of the Federal Circuit's Decision in Ericsson v. D-Link. *CPI Antitrust Chronicle*, March.

Layne-Farrar, Anne, Gerard Llobet, and Jorge Padilla. 2014a. Payments and Participation: The Incentives to Join Cooperative Standard Setting Efforts. *Journal of Economics & Management Strategy*, 23, 24–49.

2014b. Patent Licensing in Vertically Disaggregated Industries: The Royalty Allocation Neutrality Principle. *Communications and Strategies* 5.

Layne-Farrar, Anne, Jorge Padilla, and Richard Schmalensee. 2007. Pricing Patents for Licensing in Standard Setting Organizations: Making Sense of FRAND Commitments. *Antitrust Law Journal*, 74, 671.

Lemley, Mark and Douglas Melamed. 2013. Missing the Forest for the Trolls. *Columbia Law Review*, 113, 2117.

Lemley, Mark and Carl Shapiro. 2007. Patent Holdup and Royalty Stacking. *Texas Law Review*, 85, 1989.

2013. A Simple Approach to Setting Reasonable Royalties for Standard-Essential Patents. *Berkeley Technology Law Journal*, 28, 1135.

Lerner, Joshua and Jean Tirole. 2015. Restoring Competition in Standard-essential Patents. *Journal of Political Economy*, 123(3), 547.

Lloyd, Andrew. 2015. Ericsson and Nokia the latest to confirm that they will not license under the new IEEE patent policy. IAM, blog. Available at www.iam-media.com/Blog/Detail.aspx?g=d07d0bde-ebd6-495a-aa72-4eecb9dac67d.

Mariniello, Mario. 2011. Fair, Reasonable and Non-Discriminatory (FRAND) Terms: A Challenge for Competition Authorities. *Journal of Competition Law & Economics*, 7, 523.

Microsoft v. Motorola. 2012. Federal Circuit. 696 F.3d 872 (9th Cir.).

2013. Federal Circuit. Case No. C10-1823JLR.

Miller, Joseph Scott. 2006. Standard Setting, Patents, and Access Lock-In: RAND Licensing and the Theory of the Firm. *Indiana Law Review*, 40.

Mintzer, Erica and Suzanne Munck. 2014. The Joint U.S. Department of Justice and Federal Trade Commission Workshop on Patent Assertion Entity Activities – "Follow The Money." *Antitrust Law Journal*, 79, 423.

Ohana, Gil, Marc Hansen and Omar Shah. 2003. Disclosure and Negotiation of Licensing Terms Prior to Adoption of Industry Standards: Preventing Another Patent Ambush. *European Competition Law Review*, 24, 644.

Rambus v. FTC. 2008. 522 F.3d 456 Nos. 07-1086, 07-1124. United States Court of Appeals, District of Columbia Circuit.

Realtek v. LSI. 2014. District Court, Northern District of California. Case No. C-12-3451-RMW.

Rysman, Marc and Tim Simcoe. 2008. Patents and the Performance of Voluntary Standard-Setting Organizations. *Management Science*, 54(11), 1920.

Scott Morton, Fiona and Carl Shapiro. 2014. Strategic Patent Acquisitions. *Antitrust Law Journal*, 79(2) 463.

2015. *Patent Assertions: Are We Any Closer to Aligning Reward to Contribution?* NBER Working Paper.

Shapiro, Carl. 2001. Navigating the Patent Thicket: Cross Licenses, Patent Pools, and Standard-Setting. In Adam Jaffe, Joshua Lerner, and Scott Stern (eds.), *Innovation Policy and the Economy*, Vol. I. Cambridge, MA: MIT Press.

Swanson, D. and W. Baumol. 2005. Reasonable and Nondiscriminatory (RAND) Royalties, Standards Selection, and the Control of Market Power. *Antitrust Law Journal*, 73, 727.

VirnetX, Inc. and Science Applications International Corporation v. Cisco Systems, Inc. and Apple Inc. 2014. Court of Appeals of the Federal Circuit.

Williamson, Oliver E. 1985. *The Economic Institutions of Capitalism*. New York: Free Press.

PART II

Institutional Design: Country Overviews

5

Antitrust and Intellectual Property: A Brief Introduction

Keith N. Hylton

5.1. INTRODUCTION

Intellectual property law and antitrust have been described as conflicting bodies of law, and the reason for such a characterization is easy to see. Antitrust law, described simply, aims to protect consumers from the consequences of monopolization. Intellectual property law, in contrast, seeks to enhance incentives to innovate by granting monopolies in ideas or expressions of ideas. This is an overly simplistic description of both types of law, and the apparent tension I have described is due in part to simplistic framing.

More subtle descriptions of the law have suggested that the conflict between antitrust and intellectual property is mostly superficial. Perhaps the most prominent of these more sophisticated descriptions argues that the conflict between antitrust and intellectual property is just a short-run phenomenon. Antitrust and intellectual property are in conflict in the short run, under this view, but not in the long run. In the long run, both areas of the law attempt to maximize the set of choices available to consumers.

Another effort to reconcile antitrust and intellectual property rejects the short-run versus long-run distinction and holds that the tensions result mostly from misapplications of the law, even in the short run. Under this view, there should be no conflict in an ideal legal system. Both types of law strive toward optimal tradeoffs between the maintenance of innovation incentives and the protection of consumers. If those tradeoffs are managed correctly in each case, in a manner that maximizes society's welfare, the same results should be achieved under both types of law. Since law is far from perfect in application, however, conflicts arise.

The purpose of this chapter is to explore the purported conflict between antitrust and intellectual property. The chapter is largely descriptive, and focuses on current or developing litigation rather than historical controversies. Many of the modern examples of conflict can be attributed to problems of classification. The general trend has been to reclassify issues that were once considered solely within the province of patent law as antitrust issues.

In Section 5.2 below, I present a straightforward discussion of the economics of intellectual property and antitrust law. I show that the tension between antitrust and intellectual property law can be viewed within a common economic framework. Section 5.3 presents modern examples of the patent–antitrust conflict.

5.2. ECONOMICS

I will start with an exploration of the basic economics of antitrust and intellectual property, separately. My hope is to use economics to offer a simple and reasonably rigorous account of the potential scope of conflict between antitrust and intellectual property.

5.2.1. Antitrust

To simplify matters, I will treat monopolization as the same whether it occurs through unilateral action or through cartel activity. When a single firm monopolizes, it excludes competitors and takes the market to itself, raising price to the monopolistic price level and cutting output back to the monopolistic quantity. When a group of firms forms a cartel, the effect is the same. The cartel raises price and cuts output to the monopolistic price-output combination.

Figure 5.1 shows the familiar welfare consequences of monopolization. The downward sloping line represents the demand for the good, with quantity measured along the horizontal access and price measured along the vertical axis. The marginal cost of production is shown by the flat line – for simplicity, I assume constant marginal cost. In a competitive market, price would equal marginal cost. The diagram shows that the monopolizing firm exploits its monopoly power by cutting output and increasing price from the competitive level, p_1, to the monopoly level, p_2. As a result, society loses the surplus from goods that are no longer sold to consumers, represented by the area labeled Social Loss. Also, some of the surplus that had been enjoyed by consumers under competition is transferred to the monopolist, shown by the area labeled Transfer.

In view of the welfare consequences of monopolization, antitrust laws have been justified by two arguments. One, the most important, holds that antitrust is necessary to prevent the waste of resources that monopoly generates. In terms of the diagram (Figure 5.1), then, the purpose of antitrust is to prevent the area labeled Social Loss from arising. Under this view, antitrust is justifiable because it enhances society's welfare (or wealth), if employed judiciously. An alternative justification for antitrust emphasizes the distribution of wealth. Since the monopolizing entity transfers part of the surplus that would have gone to consumers to itself, monopolization transfers wealth between groups in society. Assuming consumers to be more numerous than

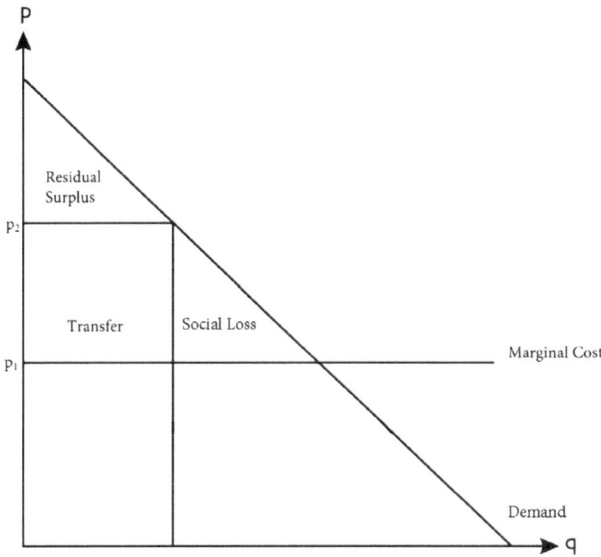

FIGURE 5.1: Welfare consequences of monopolization

the owners of the firm, the transfer of surplus enriches a relatively small class (owners) at the expense of a larger class (consumers). There is no clear theoretical basis for preferring one distribution of resources to another, as long as the total amount of resources is the same. However, a simple Benthamite approach to utility, assuming everyone has the same marginal utility of income, suggests that welfare would be enhanced by shifting the money back from owners to consumers.

I will adhere to the generally accepted approach and treat total resources, rather than the distribution of resources, as the major concern of antitrust law. Thus, monopolization is undesirable because it reduces society's total wealth, not because of its distributional impact.

5.2.2. Intellectual Property

Figure 5.1 can be used to set out the economic basis for intellectual property laws. Consider a product that has been patented. The patent allows the patentee to exclude competition. Since competition is excluded by the patent, the patentee can raise price above the competitive level. If the competitive level would be p_1, the patentee may raise the price to p_2. The result is the same as in the monopolization scenario described earlier. Some of the welfare of consumers is transferred to the patentee (Transfer). Some of the welfare is forfeited (Social Loss).

Of course, the immediate question this generates is: Why ever award a patent given the harmful effects just described? The answer is that the patent, under the appropriate conditions, incentivizes the producer to introduce the product into the market. In other words, without the patent grant, the good would not exist on the market. If this is valid, then how does society gain from the patent grant? Society's *potential* gain from the patent grant is the whole surplus triangle shown in Figure 5.1, which is the sum of the Social Loss, Transfer, and Residual Surplus. Once the patent is granted and the firm responds by raising its price, the amount society actually gains is the sum of the Transfer and the Residual Surplus. The patentee takes the entire transfer as his own, and consumers receive the residual surplus.

Note that under this theory, the social loss identified as the key justification for enforcing the antitrust laws does not exist as a practical matter in the patent setting. The reason is that in the absence of the patent, the surplus would not have been available to consumers at all. Hence the real gain to society from the patent is the sum of the transfer and the residual surplus. Society has not suffered a loss as a result of patent monopolization because the additional surplus represented by the Social Loss triangle would not have been available to consumers in the absence of the patent grant.

One might object to this argument by noting that it might be possible to have the patent grant without the patent-induced price rise. Supposing that it is possible to have the patent without the price increase – that is, the patentee might be awarded a patent and forgo the increase in price from p_1 to p_2 – then it would seem to follow that the entire loss in welfare resulting from the price rise should be counted as a loss to society. Under this assumption, the analysis of welfare under intellectual property is the same as that under antitrust. But this view ignores the underlying premise that the patent is necessary to generate the innovation. If the patent is necessary to generate the innovation, then in the absence of the prospect of a patent-induced price rise, the incentive to patent would not have existed. The patent and the resultant increase from p_1 to p_2 cannot be separated under the starting premise of this analysis.

A compromise between these conflicting views – one viewing intellectual property as entirely beneficial and the other viewing intellectual property as a harmful form of monopolization – might be reached by modifying the premise of this discussion slightly and assuming that a particular price increase was necessary to bring the innovation to market, but that the increase from p_1 to p_2 is greater than that necessary price increase. Under this view, part of the area labeled Social Loss could then be described as a waste of society's resources. The patent was too generous. The patentee would have brought the innovation to market with a smaller reward. The loss in welfare resulting from the differential between p_2 and the necessary price (or minimum necessary innovation-inducing price) would then represent a genuine loss to society's welfare. Under this view, one can argue that there is a tradeoff between consumer welfare protection and incentivizing innovation – and indeed a conflict between antitrust and patent law.

Many scholars have addressed the tradeoff between consumer welfare protection and incentivizing innovation, mostly in connection with the patent–antitrust conflict. Carrier provides a useful survey,[1] and groups the arguments into three general categories. One is the position taken by Ward Bowman,[2] which is that legal policy should aim to maximize the reward generated by the competitive advantage provided by the patent. This view is consistent, very generally, with permitting the patentee to obtain the entire transfer, and perhaps the entire potential surplus in a system of perfect price discrimination. A second view, attributed to Baxter,[3] seeks to minimize the social loss due to patent protection. Thus, as long as the reward is sufficient to bring the innovation to market, the goal should be to minimize the social loss resulting from the patent-induced price increase. The third position, attributed to Kaplow,[4] holds that policy should aim to vary the extent of antitrust regulation of patentee conduct as the ratio of the reward from the patent (that is, the transfer) to the social loss changes. Thus, as the reward increases relative to the social loss, patent protection should be given more weight as a policy goal relative to the consumer protection goal. Carrier noted, interestingly, that patent protection should also vary with the responsiveness of innovation to such protection – which implies an industry-specific approach to trading off patent and antitrust law.[5]

5.3. APPLICATIONS

In the abstract, it seems that the conflict between patent and antitrust law should be pervasive. Under the general theory outlined above, the controversy potentially exists as long as the patentee receives a reward that is greater than the minimum necessary for the patentee to bring the innovation into existence. However, this condition is likely to hold for many if not all patents. The ideas that do not come to market as patents fail because the patent reward was less than the cost of creation. The ideas that succeed, by becoming valuable patents, are those for which the patent reward exceeds the cost of creation. It should be the rare instance where the reward was just sufficient to cover the cost of creation. Given this, it would appear that almost every patent represents a potential instance of a patent–antitrust conflict, because the reward probably could have been reduced at least slightly while maintaining the incentive to bring the innovation to market.

Although this view appears to be well received, it is somewhat static and backward looking. Suppose, for example, the reward is fixed at $100, and the cost of innovation is random. Innovation occurs whenever the realized innovation cost is less than the fixed reward. Under this simple model of innovation, any reduction in the reward

---

[1] Carrier (2002).
[2] Bowman (1973).
[3] Baxter (1966).
[4] Kaplow (1984).
[5] Carrier (2002) at 790 (noting that there should be "differentiated analysis between industries in which patents are essentials for innovation (e.g. chemicals) and industries in which competition (and not patents) is essential for innovation (e.g. software)").

will generate less innovation. Although some innovations would have occurred with a smaller reward – for example, the cost of innovation was only $50 – there would always exist innovation that occurs on the innovation–incentivizing frontier – that is, where the cost of innovation is $99.99. If one could identify the innovations that would have occurred with a smaller reward and force a price reduction for those patents only, consumer welfare could be enhanced. But the difficulty is identifying the subset of innovations that fall into this category. An error in identifying this subset of cases would generate a general reduction in the incentive to innovate.

Still, accepting this theory of tradeoff as the fundamental basis for using antitrust law to control the conduct of patentees, there are major concepts in antitrust law that limit the scope of antitrust law's application to the conduct of a patentee. For unilateral conduct, there are two general antitrust "constraints" of this sort. The first is the general idea of market power, and the second is the distinction between exclusion of competition and exploitation of monopoly power. For collusive conduct, antitrust requires proof of agreement.

The concepts of market power and exploitation are most important in this context. In general, for unilateral conduct to be unlawful under Section 2 of the Sherman Act, the defendant firm must have monopoly power (*U.S. v. Grinnell Corp.*, 570 ("The offense of monopoly under s 2 of the Sherman Act has two elements: (1) the possession of monopoly power in the relevant market and (2) the willful acquisition or maintenance of that power . . . .").[6] Moreover, it is lawful for a monopolist to exploit its monopoly by charging the monopoly price and producing the monopolistic quantity. The monopolist violates antitrust law only by excluding competitors.

Market power is a rather elusive concept in antitrust. Generally, it is considered the power to raise price significantly above the competitive level. The definition implies that competition is too weak to prevent the firm from imposing its desired price increase on consumers. Market power, in short, is the power to raise price above the competitive level without having to worry too much about the constraints imposed by competition. A monopolist can raise price at will, but will restrain himself to some degree by the loss in sales that results as fewer consumers prefer to purchase the product. A firm in a perfectly competitive market can raise price, but will lose all of its sales to rivals. Monopoly power is the intermediate status, reached before full monopolization, where the firm that raises its price anticipates losing consumers mostly because of their willingness to pay and not because of their ease of switching to a rival with a reasonable substitute.

The market power requirement clearly should constrain the extent to which antitrust law applies to intellectual property. Most patents exclude competition within the scope of the patent without creating monopoly power. The reason is that, for such patents, there are substitute technologies that a potential implementer

---

[6] See generally, Hylton (2003) at 186–202.

or consumer could turn to if the patentee tries to raise the price. These substitutes constrain the patentee's ability to profitably impose a price increase on implementers or consumers.

The Supreme Court in *Illinois Tool Works Inc. v. Independent Ink, Inc.* held that in the tying context no violation of the antitrust laws could be shown in the absence of proof of market power in the patented product. The opinion rejects the pre-existing presumption that a patent confers market power on the patentee. The reasoning of *Independent Ink* implies more generally that market power is a necessary condition for a monopolization claim to be successful based on a theory that the patentee leveraged the exclusionary power of the patent.

The Supreme Court's abandonment of the patent-market-power presumption in *Independent Ink* implies that unilateral conduct claims founded on the theory that the patentee leveraged the exclusivity of the patent to exclude a rival or to constrain competition in some other market would be unsuccessful under the Sherman Act. In spite of this, at least one special rule originally based on the leverage theory, though falling under the label "patent misuse," was recently upheld by the Court. In *Kimble v. Marvel Entertainment*, the Court reaffirmed the rule of *Brulotte v. Thys* (31, patentee's use of a royalty agreement that projects beyond the expiration date of the patent is unlawful per se), under which a patentee cannot charge royalties that extend beyond the term of the patent. *Kimble* argued that the *Brulotte* rule reflected a settled judicial interpretation of the Patent Act, and that such an interpretation should remain in force unless strong reasons could be marshaled for overturning it. The majority in *Kimble* believed that strong reasons had not been provided for overturning *Brulotte*. The error in *Kimble* was the Court's failure to see that the leveraging theory upon which *Brulotte* had been based had been firmly rejected in *Independent Ink*.

The more significant issue generated by *Kimble* is one of categorization. Although *Independent Ink* limits the scope of antitrust theories based on leveraging the exclusionary power of the patent, courts may fail to notice or explicitly acknowledge when such a theory is brought before them. The majority in *Kimble* had been persuaded that the core issue of the case was one of statutory interpretation, and viewed Congress's failure to explicitly overturn *Brulotte* as an implicit legislative enactment of the *Brulotte* rule. Such categorization gamesmanship can always be played by litigants. This leaves open the question of how far courts will go in applying *Independent Ink* to theories of leverage directed toward patentees. The Court may continue to claim fidelity to *Indendent Ink* while at the same time opting to avoid the leveraging question in particular patent cases and to focus on some tangential legal issue (such as statutory interpretation).

A more direct antitrust attack on patenting activity is represented by the developing law on "product hopping."[7] The product hopping charge is asserted against a

---

[7] Carrier (2010).

pharmaceutical firm with a patented drug on the theory that the firm sought or obtained a new patent based on a minor reformulation of the drug for the purpose of eliminating generic competition. The monopoly power of the pharmaceutical drug is likely to be easily established, so the rule of *Independent Ink* would not present an obstacle to such a claim.[8] Moreover, unlike the leveraging cases, which are based on the theory that the monopolist seeks to extend monopoly power into a related market, the product hopping claims assert that the pharmaceutical firm has violated the antitrust law by abusing its monopoly power, within the monopolized market, by enforcing its new patent.

At present, the most important court decision on product hopping is *New York v. Actavis*. *New York v. Actavis* applies the balancing framework for monopolization claims established in *U.S. v. Microsoft (Microsoft III)*. Under the *Microsoft III* balancing test, the Court must compare the anticompetitive effects of the defendant's conduct to the procompetitive or efficiency benefits resulting from it. Using this reasoning, the Second Circuit held that a "hard switch" – that is, a change in the formulation of the drug which accompanies a termination of sales of the original formulation – violates the antitrust laws if the change in the formulation appears to offer little that was not provided by the original formulation. The Second Circuit rejected the alternative approach of the Ninth Circuit in *Allied Orthopedic Appliances Inc. v. Tyco Health Care Group LP.*, which refuses to balance competitive effects in the context of a predatory innovation claim. Under *Tyco*, which rejects the *Microsoft III* balancing approach, any significant enhancement of utility to consumers would be sufficient to defeat an antitrust claim based on a theory of predatory innovation.

Although the product hopping case-law is still in development, and the issue has not yet reached the Supreme Court, it represents the most direct conflict one could imagine between the patent and the antitrust laws. Under the balancing test adopted in *New York v. Actavis*, courts are, in effect, directed to balance the interests of the patent laws against the interests of the antitrust laws. One could argue that the proper forum for striking such a balance would be within a claim for patent infringement. The defendant in such a case would typically argue that the patent is invalid. One reason a court might find a patent invalid is that it fails the novelty or nonobviousness requirements.[9] If a reformulation of a drug is a trivial alteration of the original formulation, then the new patent should be found invalid because of its obviousness. In this sense, patent law has always incorporated a direct mechanism for examining antitrust-based theories. However, *New York v. Actavis* essentially

---

[8] Of course, the monopoly power of the drug may not be easily proven in some cases. For example, if the particular drug faces competition from substitutes that may be nearly as effective for the particular treatment required, then a defendant pharmaceutical firm may be able to present sufficient evidence of competition to avoid a finding of monopoly power. On the analysis of monopoly power, see Hylton (2003) at 232–9.

[9] See Cass and Hylton (2013) at 49–75.

gives the infringing party an alternative route to challenge the patent and to seek a treble damages award from the patentee.

One can view *New York v. Actavis* as raising a similar categorization question to that generated by *Kimble*. A decision by a patentee to abandon exploitation of a previous patent and to put his entire resources into the exploitation of a new patent would seem to be well within the rights protected by the patent laws. Indeed, the patent-as-property theory adopted by the Supreme Court in *Bement v. National Harrow* would seem to provide sufficient legal basis for such a decision. However, the product hopping cases view the decision to abandon an old patent and shift resources to a new one as simply a competitive stratagem falling well within the scope of the antitrust laws.

Apart from unilateral conduct antitrust theories, antitrust law has also been used to attack settlements of patent infringement litigation on the ground that the settlements amounted to collusive market-sharing agreements. The most important area of case-law on this matter today consists of reverse payment settlement cases. In the typical case, a pharmaceutical firm (pioneer firm) sues a generic for infringement. The two firms enter into a settlement agreement in which the pioneer pays the generic to stay off of the market until some date before the expiration of the patent.

The Supreme Court held in *FTC v. Actavis* that such agreements may violate the antitrust laws and should therefore be analyzed under the rule of reason. While this may seem to be unexceptional at first glance, it was a substantial change of the preexisting law. The rule in effect before *Actavis*, the so-called "scope of the patent test," immunized settlement agreements from antitrust attack as long as the terms of the agreement did not effectively extend the duration or scope of the patent.[10] The typical reverse payment settlement agreement, which generally permits generic entry before expiration of the patent, would be lawful under the scope-of-the-patent test. Under *Actavis*, such an agreement would be upheld only if a court concluded that the rule of reason balancing test indicated that the anticompetitive effects of the settlement were outweighed by procompetitive benefits.

At present, the precise contours of *Actavis*' rule of reason test are unclear. The most important question in litigation currently is whether *Actavis* requires that the reverse payment be in cash for its restrictions to apply.[11] If courts ultimately conclude that *Actavis* does not require a cash transfer, then the rule's scope will extend to a

---

[10] See, e.g., *id.* at 199.
[11] Compare *In re Lamictal Direct Purchaser Antitrust Litigation*, 18 F. Supp. 3d 560, 569 (D. N.J. 2014) (refusing to extend *Actavis* to include non-monetary payments) with *In re Niaspan Antitrust Litigation*, 42 F. Supp. 3d 735, 751 (E.D. Penn. 2014) (holding that "reverse payment" is not limited to a cash payment (citing *Black's Law Dictionary* 1309 (10th edn. 2014))) and *In re Lipitor Antitrust Litigation*, 46 F. Supp. 3d 523, 543 (D. N.J. 2014) (holding that non-monetary payments may trigger antitrust review but must be converted to a reliable estimate of its monetary value in order to survive a motion to dismiss).

large set of patent infringement dispute settlements, and perhaps nearly all. Few cash payments have been observed since the Supreme Court's decision in *Actavis*, reflecting the general awareness that antitrust litigation could easily follow any cash settlement. However, patent infringement lawsuits continue to be settled, many involving complicated deals. Still, since each of these deals arguably involves some transfer of resources from the patentee to the generic (see *Asahi Glass Co. v. Pentech Pharm., Inc.*, 994 ("But *any* settlement agreement can be characterized as involving 'compensation' to the defendant, who would not settle unless he had something to show for the settlement. If any settlement agreement is thus to be classified as involving a forbidden 'reverse payment,' we shall have no more patent settlements.") (emphasis in original)), each arguably falls under the restrictions of *Actavis*. And since each settlement is arguably subject to scrutiny under *Actavis*, each settlement could give rise to a challenge from a third party (or a federal enforcement agency) based on *Actavis*.

In addition to the general possibility of an *Actavis*-based challenge to *any* settlement of a patent infringement lawsuit in the pharmaceutical sector, the potential costs of such challenges are advanced further by the unusual incentives of challengers. The patentee and generic, having settled, have no interest in continuing litigation. The third-party challenger, by contrast, has no interest in terminating litigation as long as (1) the potential payoff, in the form of a damages or settlement payment from the patentee and generic firms, is high (which is often true in pharmaceutical patent litigation), and (2) a possibility exists that a court might find an antitrust violation under the *Actavis* balancing test.

The characterization question that plagues so many of the recent decisions on the patent–antitrust conflict is very much in evidence in *Actavis*. Viewing a patent as a monopoly, traditional antitrust doctrine presumably would hold that the monopolist may exploit the patent in full without violating the antitrust laws.[12] The decision to settle a patent infringement lawsuit along terms that are consistent with the scope of the original patent is arguably just one of many ways of exploiting the monopoly power of the patent. The decision to pay to settle a patent infringement lawsuit is a decision to pay to eliminate a challenge to the title. It is, in effect, a sale of a limited right in the patent itself to the challenger, which is a view that provides support to the traditional scope-of-the patent test (overturned by *Actavis*). However, under the theory of *Actavis*, a settlement is not properly viewed as a decision to sell a limited right in the patent, but an agreement between two potential competitors to divide the market within the patent.

The characterization problem becomes obvious in the extreme scenarios. The view adopted by *Actavis* would imply that any decision to sell a right in a patent to a potential competitor might violate the antitrust laws. Indeed, the decision to sell the

---

[12] See, e.g., Cass and Hylton (2013) at 188–92.

patent in full to a potential competitor might violate the antitrust laws. Under this reasoning, any sale of a patent could be subjected to antitrust scrutiny.

5.4. CONCLUSION

The aim of this short chapter is to explain the basic economics of the patent–antitrust conflict and to illustrate its implications for antitrust law. This is a fertile source of litigation. The antitrust laws are relatively clear, and patent rights have been clear for a long time. The conflicts arise where courts have difficulty deciding which rights should dominate. The general trend over time has been one of courts finding more ways in which antitrust laws control the rights of patentees.

REFERENCES

Baxter, William F. 1966. Legal Restrictions on Exploitation of the Patent Monopoly: An Economic Analysis. *Yale Law Journal*, 76, 267.

Bowman, Ward S. 1973. *Patent and Antitrust Law: A Legal and Economic Appraisal*. University of Chicago Press.

Carrier, Michael A. 2002. Unraveling the Patent-Antitrust Paradox. *University of Pennsylvania Law Review*, 150, 761.

2010. A Real-World Analysis of Pharmaceutical Settlements: The Missing Dimension of Product Hopping. *Florida Law Review*, 62, 1016–20.

Cass, Ronald A. and Keith N. Hylton. 2013. *Laws of Creation: Property Rights in the World of Ideas*. Cambridge, MA: Harvard University Press.

Hylton, Keith N. 2003. *Antitrust Law: Economic Theory and Common Law Evolution*. Cambridge University Press.

Kaplow, Louis. 1984. The Patent-Antitrust Intersection: A Reappraisal. *Harvard Law Journal*, 97, 1813.

Cases Cited

Allied Orthopedic Appliances Inc. v. Tyco Health Care Group LP., 592 F.3d 991 (9th Cir. 2010).
Asahi Glass Co. v. Pentech Pharm., Inc., 289 F. Supp. 2d 986 (N.D. Ill. 2003) (Posner, J.).
Bement v. National Harrow, 186 U.S. 70 (1902).
Brulotte v. Thys, 79 U.S. 29 (1964).
FTC v. Actavis, 133 S.Ct. 2223 (2013).
Illinois Tool Works Inc. v. Independent Ink, Inc., 547 U.S. 28 (2006).
In re Lamictal Direct Purchaser Antitrust Litigation, 18 F. Supp. 3d 560 (D. N.J. 2014).
In re Lipitor Antitrust Litigation, 46 F. Supp. 3d 523 (D. N.J. 2014).
In re Niaspan Antitrust Litigation, 42 F. Supp. 3d 735 (E.D. Penn. 2014).
Kimble v. Marvel Entertainment, 135 S.Ct. 2401 (2015).
New York v. Actavis, 787 F.3d 638 (2d Cir. 2015).
U.S. v. Grinnell Corp., 384 U.S. 563 (1966).
U.S. v. Microsoft, 253 F.3d 34 (D.C. Cir. 2001).

6

The Antitrust and Intellectual Property Intersection in European Union Law

Nicolas Petit

6.1. INTRODUCTION

In European legal scholarship, many articles discuss the equilibrium reached in the case-law of the Court of Justice of the European Union (CJEU) when the EU antitrust prohibitions apply to, and restrain, the free and ordinary use of intellectual property rights (IPRs).[1] I call this the antitrust–IP intersection.

The most interesting feature of this literature is perhaps the common assumption that a unifying substantive perspective, vision, or theory on IPR underpins the intersection point reached by the antitrust case-law. The theory of "absolutism" can be quickly disposed of – it treats IPRs as statutorily crafted islands of monopoly in a sea of competition and grants IPR owners absolute freedom to exploit their rights without any risk of antitrust liability.[2] Several other theories are often advanced to rationalize the antitrust–IP intersection. The "theory of inherency" is one of them. Drexl explains that the theory of inherency "allow[s] competition-law application only if the right is used to restrain competition outside the scope of the exclusive right."[3] Liannos and Dreyfuss write that it "protects the practices inherent to the exercise of the IP right from the application of competition law."[4]

An alternative theory that some scholars and practitioners extract from the case-law is the "theory of exceptionalism," which says that antitrust liability for

---

I am grateful to D. Auer and J. Marcos Ramos for helpful comments on an earlier draft of this chapter.

[1] This research is the by-product of the CJEU judgments, which often use abstract concepts and do not clearly articulate the elements of this supposed theory. Its judgments distil an "it-is-so-because-we-says-so" approach, in the words of Antonin Scalia. See *Webster v. Reproductive Health Servs.*, 492 U.S. 490, 552 (1989). A related doctrine is the limited license theory, whereby the fact that an IPR owner has granted a license on restrictive terms cannot give rise to antitrust liability, for the simple reason that the IPR owner could have instead decided *not to* provide a license.

[2] Käseberg (2012).

[3] Drexl (2008) at 36.

[4] Liannos and Dreyfuss (2013).

IPR-related conduct can only be found in "exceptional circumstances."[5] The substantive foundations of exceptionalism are unclear, and can be traced to natural rights, utilitarianism or consequentialism.[6] At a more pedestrian level, exceptionalism entails setting the rules of engagement of antitrust liability at a level that is, at least on paper, higher in IPR cases than in non-IPR cases.

Lastly, the "theory of complementarity"[7] views antitrust rules and IP protection as a complementary set of market institutions which converge toward a common goal. This view has become "mainstream" in recent years[8] and seems to be the one favored by policymakers. Its main implication is that ex post antitrust enforcement is never perceived as illegitimate when it seeks to correct certain ex ante defects of the IP system, and in particular when it attacks "improvidently defined IPRs."

Against this backdrop, this chapter is one of a legal realist. It submits that the antitrust–IP intersection does not rest on any unitary theory, which, in turn, bespeaks the Court's vision of the social function of IPRs. Instead, the main feature of the CJEU case-law is that a specific methodology is applied to antitrust cases with IPR ramifications. The CJEU deals with most of such cases under a rule-based approach, as opposed to a standard-based approach. By rule-based approach, I refer to the ex ante setting of structured tests of liability or justifiability, by opposition to ex post case-by-case resolution on grounds of a predetermined, general standard (e.g., reasonableness, competition on the merits, efficiency, fairness, equity, etc.).

As will be seen below, this approach has many virtues, not least in terms of legal certainty. But it also has a major qualification. While the Court has consistently defined rules of liability and justifiability at the antitrust and IP intersection, it has at the same time often embedded abstract standards within those rules. The implications of this mixed approach remain unclear.

To address those issues, this chapter proceeds by elimination. I first show that the theory of inherency has not played a major role in the CJEU antitrust case-law (Section 6.2). I then move on to track the prominence gained by the theory of exceptionalism in modern case-law. I demonstrate that it is a vacuous concept, devoid of substantive density (Section 6.3). I turn then to the mainstream theory of complementarity, and find somewhat surprisingly that the EU Courts make little reference to it in their case-law (Section 6.4). I then pause for a short while to examine less influential theories which were rejected by the CJEU (Section 6.5). This brings us to the last section, which claims that the main systemic feature of the CJEU antitrust case-law in IPR cases may not be substantive, but methodological. In a significant proportion of the IP-related antitrust cases issued to date, the Court has almost invariably resorted to a rule-based approach, instead of a standard-based approach (Section 6.6).

[5] Ahlborn, Evans, and Padilla (2004).
[6] Epstein (2016).
[7] Kolstad (2008).
[8] Peeperkorn and Paulis (2005).

6.2. INHERENCY

The theory of inherency is an unconvincing rationalization of the antitrust–IP intersection. The point here is not that it does not exist in EU law. It does. Instead, it is that this theory has not exerted much influence in the review of antitrust allegations (as compared to other allegations of EU law infringements).

This can be seen in the treatment of the inherency concept in early antitrust case-law. While the EU Courts have introduced inherency-colored concepts like the "existence v. exercise" dichotomy or "specific subject matter" and "essential function" of IPRs (Section 6.2.1), those concepts have mostly served rhetorical purposes in relation to antitrust allegations (Section 6.2.2). Moreover, the irrelevance of those concepts has been precipitated in the modern era, with the EU Courts progressively phasing out any reference to them in their case-law (Section 6.2.3).[9]

6.2.1. *Emergence of Inherency Concepts in Formative Case-law*

A noteworthy feature of the early antitrust case-law of the EU Courts lies in the introduction of a number of inherency-related concepts which all seem to accept the existence of IPRs, and to limit antitrust intervention to certain residual acts of IPR owners which go beyond what is deemed to be reasonable.

The existence v. exercise dichotomy is one of them. In *Consten and Grundig v. Commission* – the first antitrust infringement case ever decided by the Commission – the Court was invited to pass judgment on the antitrust–IP intersection.[10] Grundig, a manufacturer of radio receivers, recorders, dictaphones, and television sets, had granted exclusive territorial protection to Consten for the distribution of its products in France.[11] In addition, Grundig as the owner of an international trademark, had authorized Consten to register in France the trademark GINT (for Grundig International), so that Consten could block parallel imports of GINT labeled products coming from other countries. The Commission affirmed liability under Article 101 Treaty on the Functioning of the European Union (TFEU), finding that the agreement between Grundig and Consten that authorized the latter to register the GINT trademark under its own name was unlawful.[12]

On appeal, the Court confirmed in full the Commission's finding of infringement. On this occasion, it made a number of seminal substantive pronouncements,

---

[9] It is difficult to provide a date for the beginning of what we call the modern era in EU competition law, but a possible starting point is early 2004, following the major institutional reform triggered by the adoption of Regulation 1/2003.
[10] Ebb (1967).
[11] CJEU, Joined Cases 56 and 58/64, *Etablissements Consten SARL and Grundig-Verkaufs-GmbH v. Commission*, ECLI:EU:C:1966:41.
[12] The Court found this agreement unlawful because it reinforced the exclusive territorial protection afforded to the retailer. A similar finding is made in CJEU, 28/77, *Tepea v. Commission*, ECLI:EU:C:1978:133.

which remain good law today. What may be less appreciated, however, is the CJEU dictum that the decision of the Commission could be deemed lawful because it did "not affect the grant of those rights but only limits their exercise to the extent necessary to give effect to the prohibition under Article 85 (1) [now Article 101(1) TFEU]."[13] With this, the Court inaugurated the idea that competition law must operate on the basis of the inherently legitimate existence of IPRs.

This initial case paved the way to the development of an important stream of judgments where the Court invariably held that the "existence" of IPRs could not be affected by the prohibitions contained in Articles 101 and 102 TFEU. Only the "exercise" of IPRs could, in contrast, come within the ambit of the antitrust rules. In what will later become a leitmotiv in the case-law, the Court repeated the existence v. exercise dichotomy in relation to almost all forms of IPRs, including patents (*Parke Davis*),[14] copyrights (*Deutsche Grammophon*),[15] designs (*Keurkoop v. Nancy Kean Gifts*)[16] etc.

In other cases, the Courts carried the inherency language even further. In *Parke Davis*, the Court said that not any use of IPRs is potentially anticompetitive, but that it had to "degenerate" into an infringement. In *Hoffmann-La Roche v. Centrafarm*, it held that the exercise of trademarks was not unlawful in itself, unless the IPR was used as "an instrument for the abuse" of a dominant position.[17]

In parallel to the existence v. exercise dichotomy, the CJEU further introduced other inherency-colored concepts in the antitrust case-law, like the concepts of "specific-subject matter" and of "essential functions" of IPRs. Enchelmaier defines them as follows: "while the specific subject matter tells us what owners of IPRs are allowed to do based on these rights, the essential function tells us the economic or other policy reasons why the legal system allows them to do so."[18]

Windsurfing is one of those cases. Here, the Commission had declared contrary to Article 101 TFEU several clauses of licensing agreements between Windsurfing, which owned patents over some parts of sailboards, and sailboards manufacturers. The impugned agreements included a clause requiring prior approval by Windsurfing of the boards used by licensees as well as a non-challenge clause. In relation to both clauses, the Court noted that they did not fall within the "specific subject matter of the patent," and therefore confirmed that they constituted unlawful restrictions of competition.[19]

---

[13] Id.
[14] CJEU, 24/67, *Parke, Davis & Co. v. Probel, Reese, Beintema-Interpharm and Centrafarm*, ECLI: EU:C:1968:11, p. 72: "the existence of the rights granted by a Member State to the holder of a patent is not affected by the prohibitions contained in Articles 85(1) and 86 of the Treaty."
[15] CJEU, 78/70, *Deutsche Grammophon Gesellschaft mbH v. Metro-SB-Großmärkte GmbH & Co. KG*, ECLI:EU:C:1971:59, §§6–7.
[16] CJEU, 144/81, *Keurkoop v. Nancy Keane Gifts*, ECLI:EU:C:1982:289, §27.
[17] CJEU, 102/771, *Hoffmann-La Roche v. Centrafarm*, ECLI:EU:C:1978:108, §16.
[18] Enchelmaier (2010) at p. 411.
[19] CJEU, 193/83, *Windsurfing International v. Commission*, ECLI:EU:C:1986:75.

6.2.2. Operational Irrelevance of Inherency Concepts in Formative Case-law

One claim that is often encountered is that the abovementioned inherency concepts have played (and still play) a role in the antitrust case-law of the CJEU. For instance, Korah has written that "the [CJEU] looks to the specific subject matter of the particular kind of IPR when applying not only the rules for free movement, but also the competition rules."[20]

This position cannot be sustained. If they really had operational relevance, the protective inherency concepts should have resulted in exoneration of antitrust liability in a significant number of cases. On the contrary, in most of the cases where the Court had to decide on the merits of the competition allegations, it affirmed liability.

The literature also occasionally obscures the point that the case-law on "specific-subject matter" and "essential function" is not, in reality, competition case-law. While it is true that the cases in discussion certainly involved allegations of competition law infringements, those allegations were systematically preceded by allegations of violations of the free trade provisions of the EU Treaties. And it is during the review of those allegations that the Court employed the inherency concepts, arguably to soothe Member States' concerns over the division of competence between the European Union and its Member States.[21]

Lastly, in the rare cases where the inherency concepts were put into practice, they proved very poor at guiding the antitrust inquiry, and in particular at discriminating between antitrust-reprehensible and antitrust-immune conduct. This can be seen by comparing two cases, *Sirena* and *EMI v. CBS*. In *Sirena*, a US producer of cosmetic and medicinal cream had assigned trademark rights to distinct manufacturers in Germany and Italy.[22] The Italian manufacturer had sought to rely on the trademark to restrict imports of the German-made cream on domestic territory. The CJEU held that Article 101 TFEU was applicable to the extent to which "trademarks were invoked so as to prevent imports of products" across distinct Member States.[23] Several years later, the Court reversed this case-law in *EMI v. CBS*,[24] insisting on the proof of an additional agreement or concerted practice for a finding of Article 101 TFEU liability, and stressing that the "mere exercise of the national trade mark rights" could not be akin to an infringement.[25]

With this background, the rationalization of the early case-law on the antitrust–IP intersection as inherency spirited resembles a legal fable. The inherency concepts appear mostly to have been used as prose introduced in judgments to dispel

---

[20] Korah (2002) at p. 125.
[21] Liannos and Dreyfuss (2013).
[22] CJEU, 40/70, *Sirena v. EDA*, ECLI:EU:C:1971:18, §9.
[23] *Id.*, §11.
[24] CJEU, 86/75, *EMI Records Limited v. CBS Grammofon A/S*, ECLI:EU:C:1976:86.
[25] *Id.*, §§24–9.

concerns of antitrust activism in IPR-related matters. Instead, the formative era pictures the Court stepping over the very existence of IPRs in several cases. In *Windsurfing International* for instance, the Court held that "although the Commission is not competent to determine the scope of a patent, it is still the case that it may not refrain from all action when the scope of the patent is relevant for the purposes of determining whether there has been an infringement of Articles [101] or [102 TFEU]."[26] And in *Sirena*, the Court daringly affirmed that the public interest protected by trademarks is lower than that protected by other IPRs.

6.2.3. Phasing Out of Inherency Concepts in Modern Case-law

Liannos and Dreyfuss argue that since the Court's 1982 decision in *Coditel II*, the existence v. exercise dichotomy has never featured as an important element of the Court's reasoning in competition cases involving IP rights.[27] This finding is not entirely right, at least time wise. In 1988 the Court held in *AB Volvo and Erik Veng* that a car manufacturer could lawfully refuse to grant licenses of its design rights on car parts, even in exchange for a reasonable royalty. In turn, the Court reiterated that the "exercise" of this exclusive right could be prohibited by Article 102 TFEU, if it involves "certain abusive conduct."

That said, Liannos and Dreyfuss' point is certainly valid in relation to the other inherency concepts. The seminal *Magill* case of 1995 best evidences this. At issue was the conduct of British and Irish TV channels, which had relied in parallel on their respective copyrights over weekly schedules to prevent a third party from editing a novel, comprehensive TV magazine covering all channels' programs. The Commission found an infringement of Article 102 TFEU. On appeal, the General Court (GC) upheld the Commission's reasoning, relying heavily on the inherency notion of "essential function." It held in particular that the channels' conduct went "beyond what is necessary to fulfil the essential function of the copyright as permitted in Community law."[28]

The judgment was further appealed before the CJEU. During the proceedings, Advocate General Gulmann used the inherency concepts to advise the Court to vacate the appeal. He noted that "the right to refuse licences forms part of the specific subject matter of copyright."[29] And he further lambasted the GC for its failure to extend antitrust immunity to conduct falling within the scope of the "essential function of copyright."

---

[26] CJEU 193/83, *Windsurfing International v. Commission*, ECLI:EU:C:1986:75, §26.
[27] Liannos and Dreyfuss (2013).
[28] GC, T-69/89, *Radio Telefís Eireann v. Commission*, ECLI:EU:T:1991:39, §73.
[29] Opinion of Advocate General Gulmann in Joined cases C-241/91 P and C-242/91 P, *Radio Telefís Eireann (RTE) and Independent Television Publications Ltd (ITP) v. Commission*, ECLI:EU:C:1994:210, §§38 and 70.

In its final determination, the CJEU took a different tack, and refused to review the case through the "specific subject matter" and/or "essential function" of copyrights. Instead, the judgment soberly stated that "the exercise of an exclusive right by the proprietor may, in exceptional circumstances, involve abusive conduct."[30] As Czapracka rightly observes, proceeding on the basis of the specific subject matter would have restricted the Court's ability to hold the TV channels liable of an abuse, as the TV channels had strictly behaved as ordinary IP owners.[31]

In retrospect, *Magill* signed the death knell of inherency concepts in antitrust case-law. As will be seen in the next section, no judgment of the CJEU has since then relied on "specific subject matter" or "essential function" as a decisive parameter of antitrust liability. Instead, the Court has embraced a novel concept of "exceptional circumstances," whose relevance now deserves to be discussed.

6.3. EXCEPTIONALISM

Several authors consider that the antitrust–IP intersection is governed by a theory of "exceptionalism." Under this theory, the exercise of IPRs is deemed presumptively lawful under the antitrust provisions, save in "exceptional circumstances." With this, the rules of engagement of antitrust liability are allegedly set at a level that is higher than in non-IPR cases. The theory of exceptionalism would have been the one followed by the CJEU and the lower courts in most antitrust cases with IP ramifications since *Magill*, though with variations. In the following sections, I track the evolution of this theory since *Magill* (Section 6.3.1) and discuss whether it is indeed the theory that today governs the antitrust–IP intersection (Section 6.3.2). I conclude with a discussion on the scholarly interpretation of the meaning of exceptionalism (Section 6.3.3).

6.3.1. *Cases*

The novel concept of "exceptional circumstances" introduced in *Magill* was not immediately picked upon by the lower courts, and a period of fluctuation appeared at the GC level. In the 1997 *Tiercé Ladbroke SA v. Commission* judgment, the GC refused to discuss the existence of "exceptional circumstances" despite the fact that the dispute concerned very similar allegations, namely that a French horse trading company had refused to license its copyrights over televised pictures and sound commentaries of French races to Ladbroke, whose business consisted in operating horse betting outlets in Belgium. The Court confirmed the Commission's decision to reject Labroke's complaint, and noted that *Magill* was highly specific and limited to scenarios of anticompetitive leveraging where the dominant firm also exploits the

---

[30] CJEU, C-241/91 and C-242/91, *RTE and ITP Ltd v. Commission*, ECLI:EU:C:1995:98.
[31] Czapracka (2007).

IPRs, and uses them to harm competitors.[32] Two years later, the General Court would, however, endorse *Magill*'s "exceptional circumstances" in *Micro Leader Business v. Commission*. The case focused on Microsoft's attempts to restrict Canadian wholesalers from exporting copyrighted computer software to the European Union. The GC stressed that Microsoft could lawfully enforce its copyrights in Europe in order to prevent imports of products first sold in Canada. However, it added, obiter dictum, and quoting *Magill*, that:

> [I]t is clear from the case-law that whilst, as a rule, the enforcement of copyright by its holder, as in the case of the prohibition on importing certain products from outside the Community in to a Member State of the Community, is not in itself a breach of Article 86 [now 102 TFEU] of the Treaty, such enforcement may, in *exceptional circumstances*, involve abusive conduct (emphasis added).[33]

In the next case to reach the EU Courts' docket, the CJEU made clear that the concept of "exceptional circumstances" introduced in *Magill* was not a *passade*. In *IMS Health*, a marketing data firm had designed a very granular map which could be used by stakeholders of the pharmaceutical industry to track sales over the German territory. The map soon became an industry standard, not least because it had been developed in cooperation with the pharmaceutical industry. When a former employee of IMS Health set up a new marketing data company (NDC) and designed its own competing map, he did not manage to attract clients as most prospective users were accustomed to IMS Health maps. NDC thus proceeded to use IMS Health's type of maps. A dispute ensued. IMS Health, which held copyrights over the maps, sought and obtained a prohibitory injunction against NDC before the German courts. NDC reciprocated by launching antitrust proceedings before the Commission alleging that IMS Health was guilty of abusive refusal to supply. The Commission affirmed liability, but the operation of the Commission's decision was suspended on appeal. In parallel, as the initial injunctive relief case moved through the German appeals system, a preliminary reference was eventually sent to the CJEU.

The wording of the judgment leaves no shred of a doubt on the Court's resolve to cement the *Magill* case-law. Quoting *Magill*, the Court held generally that "exercise of an exclusive right by the owner may, in exceptional circumstances, involve abusive conduct."[34] This statement is not specifically confined to refusals to license. In fact, the Court quotes as further authority a paragraph of *AB Volvo and Erik Veng* where it is said that pricing levels of IPR-protected goods can fall foul of Article 102 TFEU. All this notwithstanding, what is perhaps more important in *IMS Health* is, however, not explicit in the judgment. In the scholarship, some had adhered to the view that *Magill* was confined to its own facts, and that the main driver behind the

---

[32] See GC, T-504/93, *Tiercé Ladbroke SA v. Commission*, ECLI:EU:T:1997:84, §130.
[33] See GC, T-198/98, *Micro Leader Business v. Commission*, ECLI:EU:T:1999:341, §§34 and 56.
[34] See CJEU, C-418/01 *IMS Health*, ECLI:EU:C:2004:257, §35.

finding of antitrust liability was to correct an anomaly in Irish IP law, namely that ludicrous IPRs could be improvidently granted for TV schedules.[35] *IMS Health* proved this reading wrong. *Magill* was not an anecdotal judgment. Unlike in *Magill*, where the IPRs in question might have seemed improvidently awarded, the IPRs in dispute were of the ordinary garden variety.

Since *IMS Health*, two sets of cases have seemed to further the theory of "exceptionalism." First, exceptionalism has been the approach ordinarily followed in refusal to license cases, not least in the *Microsoft I* and *II* judgments of the GC where it was held that Microsoft had unlawfully withheld essential interoperability information from rivals, in a bid to leverage its dominant position on the market for operating systems (OS) for PCs toward the adjacent market for work group servers' OS.[36] In the two *Microsoft* cases, the GC recalled that such exceptional circumstances were met, even though *Microsoft I* promoted a somewhat controversial interpretation of *Magill* and *IMS Health*.

Second, the theory of exceptionalism has also been applied to IP remedy cases. In *Protégé International v. Commission*,[37] Pernod Ricard SA, who owned the "Wild Turkey" trademark for whiskey, had commenced opposition proceedings with several trademark offices upon learning that Protégé International had applied for the registration of the "Wild Geese" trademark for whiskey. Protégé International reciprocated by lodging an antitrust complaint with the EU Commission, alleging that the initiation of opposition proceedings by Pernod Ricard SA was an anticompetitive abuse. The Commission dismissed the complaint, and the GC affirmed. It observed that because access to justice is a fundamental right, it is only in "wholly exceptional circumstances" that the pursuit of legal remedies – including on the basis of IPRs – can be deemed abusive. In the GC's view, such circumstances are present when the proceedings cannot (i) be "considered as an attempt to establish its rights and can therefore only serve to harass the opposite party"; and (ii) are conceived as a "framework of a plan whose goal is to eliminate competition."[38]

6.3.2. Is the Antitrust–IP Intersection Governed by the Theory of Exceptionalism?

With this background, the following sections seek to understand if the theory of exceptionalism can be deemed to be the lynchpin of the antitrust–IP intersection in modern competition law. The evidence is mixed. I expose hereafter validating (Section 6.3.2.1) and invalidating arguments (Section 6.3.2.2).

---

[35] Czapracka (2009).
[36] See GC, T-201/04, *Microsoft v. Commission* (Microsoft I), ECLI:EU:T:2007:289, §331; GC, T-167/08, *Microsoft Corp. v. Commission* (Microsoft II), ECLI:EU:T:2012:323, §§139–40.
[37] See GC, T-119/09, *Protégé International Ltd v. European Commission*, ECLI:EU:T:2012:421.
[38] *Id.*, §49.

6.3.2.1. Yes

The 2012 judgment of the CJEU in *AstraZeneca v. Commission* corroborates particularly well the hypothesis that the theory of exceptionalism is the lynchpin of the antitrust–IP intersection.[39] This judgment is often – and understandably – overlooked in the antitrust and IP literature because the impugned conduct did not involve an IP-instrumented strategy. This notwithstanding, *AstraZeneca v. Commission* contains important dicta on the antitrust–IP intersection. The *AstraZeneca* Court had to review previous decisions that had held a dominant drug manufacturer liable for unlawful abusive tactics aimed at delaying generics competition. Among other things, the dominant firm had acted before three national agencies to obtain deregistration of its marketing authorizations over capsules of its blockbuster drug Losec. In turn, the dominant firm had withdrawn all existing Losec capsules from those markets, and launched Losec in tablet format. The Commission found, and the GC confirmed, that those measures sought to prevent rival generic manufacturers from relying on the marketing authorization to speedily release Losec capsules.

In the course of the proceedings, AstraZeneca sought to draw an analogy with the *IMS Health* case, and argued that it is only in "exceptional circumstances" that its statutory right to free deregistration could be undermined by Article 102 TFEU. With this, AstraZeneca hoped to have the Court declare that the Commission and the GC had applied an insufficiently stringent liability test.

The Court dismissed AstraZeneca's claim, and confirmed the Commission's and GC's findings of unlawful abuse. However, an often-missed point is that the Court's judgment unequivocally endorsed the exceptionalism hypothesis. The reasoning is convoluted, yet the point is clear. The Court first held that antitrust law can bring limitations to the exercise of a statutorily granted right to deregistration, and that a finding of abuse under such circumstances is in no way an "exceptional case."[40] The Court then moved on to add an important statement: The situation in the case at hand "does not justify a derogation from Article [102 TFEU], unlike a situation in which the unfettered exercise of an exclusive right awarded for the realisation of an investment or creation is limited."[41] What this means does not deserve long explanation. The Court mutters that IPRs – it talks of "exclusive rights" – are subject to a derogatory regime under Article 102 TFEU. Their "unfettered exercise," says the Court implicitly, can only be restricted in "exceptional" cases. And this is due to the fact that they are the reward of an investment of creation, unlike the right to deregistration.

Upon review of the Court's entire antitrust case-law, I believe that no judgment better carries the theory of exceptionalism than that in *AstraZeneca v. Commission*.

---

[39] See CJEU, C-457/10, *AstraZeneca v. Commission*, ECLI:EU:C:2012:770.
[40] *Id.*, §150.
[41] *Id.*

The wording of the judgment – which I assume was carefully chosen by the judges – is not restricted to refusal to license or IP remedies, and talks generally of the "exercise of an exclusive right."

In the literature, some scholars consider that exceptionalism is the transversal principle behind the antitrust–IP intersection, at least in so far as Article 102 TFEU is concerned. Anderman and Schmidt note that "[b]roadly, one can talk of an 'exceptional circumstances' test under Article [102 TFEU], which operates to limit its application to the exercise of IPRs quite explicitly in the case of the abuse of refusal to supply and implicitly in the case of the abuse of excessive pricing." Further, they observe that "both the language of Article [102(b) TFEU] and the [EU] judgment in IMS offer good grounds for concluding that other types of abuses can also fall within the category of 'exceptional circumstances'."[42]

6.3.2.2. No

At the same time, other contemporary pronouncements of the EU Courts seem to discredit the claim that the theory of exceptionalism is the lynchpin of antitrust jurisprudence in IPR cases. I review the various arguments in turn.

6.3.2.2.1. EXCEPTIONALISM IS NOT IPR-SPECIFIC. The theory of exceptionalism is not an idiosyncrasy of the IPR-related antitrust case-law. On the contrary, the theory of exceptionalism is the one conventionally applied in all refusal to deal cases, regardless of whether the asset to which the dominant firm refuses access is IPR protected or not. In *Bronner v. Mediaprint*, the Court's seminal case on refusal to deal cases, the Court applied the concept of "exceptional circumstances" previously affirmed in *Magill* to a situation where a dominant firm had refused access to a pedestrian facility, namely a distribution system for daily newspapers.[43]

Similarly, the concept of "wholly exceptional circumstances" found in *Protégé International v. Commission* is not specific to IPR remedy cases. The GC had previously applied this strict version of the theory of exceptionalism in *ITT Promedia*, a case where the Belgian telecoms incumbent had sought ordinary judicial remedies against a publisher of directories that had used its non-IPR-protected subscriber data.[44]

6.3.2.2.2. EXCEPTIONALISM IS NOT SYSTEMIC IN IPR ANTITRUST LAW. If the theory of exceptionalism were the regulating concept of the antitrust–IP intersection, then one would expect to observe it in all IPR-related cases. However, in the high-profile *DSD* case of 2009, the Grand Chamber of the Court dispensed with

---

[42] Anderman and Schmidt (2007).
[43] CJEU, C-7/97, *Bronner v. Mediaprint*, ECLI:EU:C:1998:569, §26.
[44] GC, T-111/96, *ITT Promedia NV v. Commission*, ECLI:EU:T:1998:183, §60.

laboring "exceptional circumstances." In dispute was whether DSD, a dominant German waste collection system, had committed an abuse by requiring from its customers the payment of a fee for all packaging bearing its trademark, even if those customers had not used the DSD collecting service. The Court affirmed antitrust liability, and made no single reference to "exceptional circumstances."[45] Instead, the Court framed the case as a conventional exploitative abuse case, and referenced non-IP precedents in support of its reasoning.

In addition to this, the theory of exceptionalism has mostly been used in abuse of dominance cases, and comparatively little, if at all, in anticompetitive coordination cases.

Surely, this discrepancy may be explained by the dearth of IPR-related Article 101 TFEU cases in recent years, which in turn originates in a variety of reasons. Interestingly, the Commission implicitly rebuffed the applicability of the theory of exceptionalism under Article 101 TFEU in its infringement decision in *Lundbeck*.[46] With patent settlement agreements in sight, it held that "[s]uch agreements are fully subject to the discipline of competition law."[47] And it added that "this also applies to agreements whose purpose is to put an end to or otherwise deal with patent litigation or, more broadly, patent disputes."[48] This broad conception remains to be scrutinized by the EU Courts, but it suggests that the limiting principles of the Article 102 TFEU case-law may not apply in the anticompetitive coordination space.

6.3.2.2.3. EXCEPTIONALISM AND PARTICULARISM? In *Huawei v. ZTE*, the CJEU introduced a novel possibility to find antitrust liability under Article 102 TFEU in the presence of "particular circumstances."[49] In the case in point, the legal issue was whether the holder of a patent that has been declared essential to a collaborative technology standard (SEPs), and which it has pledged to license on Fair, Reasonable and Non-Discriminatory (FRAND) terms, could be deemed guilty of unlawful abuse when it applies for a court injunction and/or product recall against unlicensed implementers of its technology.

The Court held that such conduct could be deemed abusive not in "exceptional" but in "particular circumstances," presumably recognizing that the facts before it could not properly be considered as "exceptional" given the widespread nature of patent disputes in standardized industries.[50] Such particular circumstances could be deemed to occur, according to the Court, when (i) the patent at issue is "essential ... rendering its use indispensable to all competitors which envisage manufacturing products that comply with the standard [...]"; and (ii) the SEP

---

[45] See CJEU, C-385/07 P, *Duales System Deutschland v. Commission*, ECLI:EU:C:2009:456, §143.
[46] See Commission Decision, Case AT.39226 – Lundbeck C(2013) 3803 final (19.06.2013), §§598 and following.
[47] Id., at §600.
[48] Id.
[49] CJEU, C-170/13, *Huawei v. ZTE*, ECLI:EU:C:2014:2391, §48.
[50] Rato and English (2016).

holder has given an "irrevocable undertaking [...] to the standardisation body in question, that it is prepared to grant licences on FRAND terms."[51]

In the scholarship, some practitioners have argued that *Huawei v. ZTE* sets a "different standard" which strays from the theory of exceptionalism.[52] However, the same authors concede that the new concept "resemble[s] the 'exceptional circumstances' of the conventional case law" so that *Huawei v. ZTE* may be seen as the progeny of *Magill* and *IMS Health*. This impression is further supported by the case-law cited by the Court, which refers to *Volvo, Magill*, and *IMS Health*.

That said, the right reading is in my view that *Huawei v. ZTE* sets a new rule of engagement of Article 102 TFEU liability that is distinct from the theory of exceptionalism. As the Court explicitly affirms, "the particular circumstances of the case in the main proceedings distinguish that case from the cases which gave rise to the case-law" in *Volvo, Magill* and *IMS Health*.[53] The Court's reference to the conventional case-law is a formal trick. As is well known, the CJEU hates to explicitly admit that it adapts, improves, or reverses conventional case-law. What *Huawei v. ZTE* thus confusingly suggests is that the theory of exceptionalism is not the whole and sole threshold for the engagement of antitrust liability against IPR strategies.

6.3.2.2.4. OPEN-ENDED CONTENT OF EXCEPTIONALISM? I believe that there is one last argument which flouts the idea that the theory of exceptionalism is the lynchpin of the antitrust–IP intersection. The concept of "exceptional circumstances" has been interpreted in the case-law to be open ended, in violation of the general principle of law that exceptions ought to be interpreted restrictively. As a result, if the theory of exceptionalism exists as a formal legal construct, its scope seems so elusive that its substantive value can be called into question.

Antitrust-savvy readers will recall that in *IMS Health*, the CJEU had considered that the presence of three market circumstances was "sufficient" to create "exceptional circumstances," i.e. that the refusal to license (i) prevents the emergence of a new product; (ii) has no objective justification; and (iii) eliminates all competition on a secondary market.[54] During the pleadings of the *Microsoft* case, the Commission interpreted the wording of the CJEU literally and argued that the "exceptional circumstances" found in *IMS Health* were not exhaustive. Accordingly, proof of the three criteria designated in *IMS Health* was not a necessary requirement. The Commission conceivably understood that the main weakness of its case was that it had not established that Microsoft's withholding of interoperability information had prevented the launch of new products. The GC backed the Commission,

---

[51] CJEU, *Huawei v. ZTE*, §59.
[52] See Rato and English (2016) at p. 107.
[53] *Huawei v. ZTE*, §48.
[54] See CJEU, *IMS Health*, §38: "in order for the refusal by an undertaking which owns a copyright to give access to a product or service indispensable for carrying on a particular business to be treated as abusive, it is sufficient that three cumulative conditions be satisfied ..."

contemplating the possibility to assess other "particular circumstances" if "one or more of those circumstances [identified in Magill and IMS Health] are absent."[55] This notwithstanding, the GC found that "the circumstance relating to the appearance of a new product [was] present in this case."[56]

6.3.2.3. Summation

The above analysis hints that one should not read too much into the concept of "exceptional circumstances." The theory of exceptionalism lacks substance. Several versions of it coexist in the case-law. The classic one, elaborated in *Magill* and *IMS Health*, applies to refusal to license cases. A stricter version applies in IP remedies cases, like *Protégé International*. And as an exception to this, an even looser version governs IP remedies cases like *Huawei v. ZTE*.

Besides, the theory of exceptionalism has not cross-fertilized the Article 101 TFEU case-law.

With all this, the theory of exceptionalism seems again to be less of a theory than a semantic device, used by the Court to convince readers of its judgments (and maybe itself?) that it is not an antitrust activist.

6.3.3. Is Exceptionalism a Pro- or Anti-IP Doctrine?

In the scholarship, there seems to be a consensus that the theory of exceptionalism is protective of IPR owners. Within the IPR community, Vaver writes that the use of concepts like "exceptions" is skewed toward IPR owners, because it treats what "IP owners can do as rights" and "what everyone else can do as indulgences, aberrations from some preordained norm, activities to be narrowly construed and not extended."[57] Similarly, in the competition community, Anderman and Schmidt stress that the "exceptional circumstances" test "represents an important acceptance by competition law that IPRs are not the same as all other forms of property rights" and the "corollary proposition that the 'normal' exercise of IPRs will not abuse a dominant position."[58] Lastly, seasoned antitrust economists seem to consider that "exceptional circumstances" constitute an optimal legal standard that adequately protects IPR owners and socially beneficial investments.[59]

This, however, is an illusion. A quick look at the case-law shows that exceptionalism has entitled the Commission and the Courts to affirm antitrust liability in cases where inherency theory would have commanded a finding of antitrust immunity. In *Magill*, the EU Courts applied the exceptionalism framework and concluded that

---

[55] See GC, *Microsoft I*, §336.
[56] *Id.*, §665.
[57] Vaver (2009) at p. 159.
[58] Anderman and Schmidt (2007) at p. 107.
[59] Ahlborn, Evans, and Padilla (2004).

an unlawful abuse existed. In contrast, the application of inherency concepts like "specific subject matter" and "essential function" endorsed by Advocate General Gulmann would have led to the TV channels being exonerated from antitrust liability as their conduct would have remained within the scope of copyright protection.

6.4. COMPLEMENTARITY

In 2004, the EU Commission introduced a novel idea into soft law instruments: that antitrust and IPR policies share complementary goals. Without more qualifications, this idea means that competition law can apply with full force to IPR conduct. Pushed to the extreme, virtually any ex post antitrust intervention into IPR strategies is legitimate.

In this section, I expose (Section 6.4.1) the source of the theory of complementarity, and proceed to evaluate whether it is borne out by the case-law of the CJEU (Section 6.4.2).

6.4.1. *Exposition*

In 2004, a new theory of complementarity was introduced into a soft law instrument of the EU Commission entitled "Guidelines on Technology Transfer Agreements." Pursuant to this theory, which has been reproduced verbatim in the latest update of the Guidelines in 2014:

> [B]oth bodies of law share the same basic objective of promoting consumer welfare and an efficient allocation of resources. Innovation constitutes an essential and dynamic component of an open and competitive market economy. Intellectual property rights promote dynamic competition by encouraging undertakings to invest in developing new or improved products and processes. So does competition by putting pressure on undertakings to innovate. Therefore, both intellectual property rights and competition are necessary to promote innovation and ensure a competitive exploitation thereof.[60]

The introduction of this new theory seems to pair with an implicit rejection of the theory of exceptionalism. The Guidelines nowhere refer to "exceptional circumstances." Neither do they mention the protective inherency concepts adumbrated in prior case-law.

Drexl offers the following reading of the 2004 Guidelines: in his view they "refe[r] to a concept of complementarity of the two fields of intellectual property law and competition law."[61]

---

[60] 2004 Commission Notice – Guidelines on the application of Article 81 of the EC Treaty to technology transfer agreements, 2004/C 101/02, §7.
[61] See Drexl (2008) at p. 53.

At the operational level, the Guidelines' theory of complementarity entails that IPR agreements must be scrutinized pursuant to Article 101 TFEU under a "balancing" approach. The idea is to weigh the private interest of safeguarding IPR owners' freedom against the public interest of protecting undistorted competition.

Moreover, the Commission's interpretation of the public interest is not limited to undistorted competition, but seems instead to be an open-ended one. In relation to non-challenge clauses, the Commission suggests that eradicating weak IPRs belongs to the public interest protected by the competition rules:

> The public interest of strengthening the incentive of the licensor to license out by not being forced to continue dealing with a licensee that challenges the very subject matter of the licence agreement has to be balanced against the public interest to eliminate any obstacle to economic activity which may arise where an intellectual property right was granted in error.[62]

With all this, it is obvious that the theory of complementarity is less deferential to the interests of IPR owners than the theory of exceptionalism that prevails under Article 102 TFEU. Drexl, who seems wary of excessive IPR protection advocates that "[t]he Commission would be well advised to extend its concept of complementary goals of IP law and competition law from the field of technology transfer agreements to the area of abuse of dominance (Article 82 EC) and merger control."[63]

6.4.2. Evaluation

The theory of complementarity has made forays into the Court's contemporary case-law. *Scarlett Extended* – a non-antitrust case occasionally mentioned in EU Commission decisions as supportive authority for its enforcement initiatives[64] – suggests that a degree of balancing is appropriate between the private interest of IP owners and the freedom of third parties to conduct their own business:

> The protection of the right to intellectual property is indeed enshrined in Article 17(2) of the Charter of Fundamental Rights of the European Union ("the Charter"). There is, however, nothing whatsoever in the wording of that provision or in the Court's case-law to suggest that that right is inviolable and must for that reason be absolutely protected.[65]

Even more clearly, the Court of Justice has seemed to endorse the theory of complementarity in its 2015 judgment in *Huawei v. ZTE*, holding that the "Court must strike a balance between maintaining free competition – in respect of which primary law and, in particular, Article 102 TFEU – and the requirement to safeguard

---

[62] Guidelines, 2014/C 89/03 §138.
[63] Drexl (2008) at p. 53.
[64] See Commission Decision, Case AT.39939 – *Samsung* C(2014) 2891, (29.04.2014), at footnote 44.
[65] See Case C-70/10, *Scarlet Extended*, ECLI:EU:C:2011:771.

that proprietor's intellectual-property rights and its right to effective judicial protection guaranteed by Article 17(2) and Article 47 of the Charter."[66]

Besides those judicial pronouncements, the theory of complementarity seems particularly popular with antitrust enforcers. On the conference circuit, agency officials in charge of selling antitrust policy routinely use it to appease critiques of antitrust dogmatism.

If anything, this emerging case-law suggests that the Commission's soft law instruments can progressively be given legal force through judicial endorsement. However, the nature of the complementarity relationship that governs the antitrust and IP intersection is left entirely open by the Court of Justice. The CJEU case-law does not specify the respective weight to be ascribed to the antitrust and IPR variables in the balancing equation. Therefore, the theory of complementarity says nothing more than that antitrust law can apply to IP rights.

More generally, I would like to stress that in a relationship of complementarity, components can and do generally occupy distinct hierarchical positions. To take a graphic analogy, ethologists often explain that complementary species in a same ecosystem occupy high and low places in the food chain, the lion and the antelope being a perfect example. With this background, in *Huawei v. ZTE*, the Court has ordered the competition and IP variables in a way that suggests that property and judicial rights may occupy a higher position than antitrust protection. The operative part of the judgment, which sets out a safe harbor for patent owners willing to avoid antitrust liability, clearly confirms this reading.

Last, I note that the theory of complementarity has not been introduced into the legislative instrument that the Guidelines purport to accompany, namely the Block Exemption Regulation on technology transfer agreements. This silence may be interpreted as a sign of the lawmakers' discomfort with the theory of complementarity.

6.5. REJECTED THEORIES

In a 2010 paper on compulsory licensing, Temple Lang noted that "the law has developed piecemeal without any explicit framework or general principle."[67] His paper, which had the ambitioned of curing this defect, advanced the proposition that IPR conduct should never be deemed unlawful in itself, unless it is accompanied by an "additional abusive conduct."

On careful consideration, this reading of the case-law seems, however, doubtful. In *AB Volvo and Erik Veng*, the Court simply held that the exercise of an exclusive right may be caught by Article 102 TFEU if "certain abusive conduct" is present.[68]

---

[66] See CJEU, *Huawei v. ZTE*, §42.
[67] Temple Lang (2010).
[68] CJEU, C-238/87, *AB Volvo v Erik Veng*, ECLI:EU:C:1988:477, §9.

In my view, the element of "additional" abuse thus constitutes a misleading extrapolation. Nowhere does it appear in the judgment.

Another possible critique is that the Court's ruling seems to say something slightly different than what Temple Lang argues in his paper. In essence, the Court notes that the exclusive prerogatives attached to IPRs can blossom into several types of anticompetitive abuses (e.g., outright refusal to license, excessive licensing conditions, etc.). In the Court's view, IPR conduct and antitrust abuse are one and the same thing, and there is no need for additional conduct to trigger the prohibition rule of Article 102 TFEU.

6.6. THE RULE-BASED APPROACH

We now approach the main finding of the chapter. In my view, the most noteworthy feature of the CJEU case-law on the antitrust–IP intersection is methodological, not substantive. In almost all cases, the CJEU promotes a rule-based approach, instead of a standards-based approach (Section 6.6.1). Given, however, that this case-law has essentially developed under Article 102 TFEU, an open question is whether a rule-based approach also prevails under Article 101 TFEU. Notwithstanding the fact that the CJEU's pronouncements in that field are rarer, there is credence to the argument that Article 101 TFEU cases are also scrutinized under a rule-based approach (Section 6.6.2). In my opinion, the CJEU's systemic commitment to a rule-based approach in such cases denotes a plausible concern that discretionary antitrust intervention against IPRs on the basis of *post hoc* standards would send adverse signals to IPR owners, and disincentivize the production of socially beneficial innovation (Section 6.6.3).

6.6.1. *Rule-based Approach in Abuse of Dominance Cases Involving IPRs*

6.6.1.1. Hypothesis

Crane defines rules as "specifications of liability criteria in formal, seemingly precise, and usually short directives."[69] And Posner writes that a rule "singles out one or a few facts and makes it or them legally determinative."[70] In essence, a rule-based approach entails the ex ante setting of structured tests of liability.

By contrast, a standard-based approach consists in a flexible resolution of each case on grounds of "open ended, multi-factor and post hoc" considerations such as reasonableness, consumer welfare, competition on the merits, efficiency, fairness, equity, etc.[71]

---

[69] Crane (2007) at p. 55.
[70] Posner (2001) at p. 39.
[71] Crane (2007) at p. 57.

Rules can be found piecemeal across various areas of EU competition law. The justifiability segment found in Article 101 paragraph 3 TFEU is a good illustration of a rule-based system. A four-pronged test must be met to trigger the benefit of an exemption. In Article 102 TFEU, predatory pricing is subject to a rule-based approach. The *AKZO* case-law holds that prices below average variable costs are presumptively abusive, while prices above average total costs are presumptively lawful.[72] Finally, rules can also be encountered in merger law. The *Airtours* judgment requires the fulfillment of three cumulative conditions for the assessment of coordinated effects.[73]

6.6.1.2. Verification

In this section, I show that the Court has conventionally followed a rule-based approach in the various categories of cases submitted to its scrutiny.

6.6.1.2.1. REFUSAL TO LICENSE IPRS. Refusal to license cases clearly carry the point, though with a somewhat convoluted history. In *AB Volvo and Erik Veng*, the Court initially embraced a standard-based approach, which left entirely open the conditions under which it could be deemed abusive for an IPR holder to refuse to license its technology. This judicial policy was abandoned in *Magill*, when the Court promulgated the "exceptional circumstances" rule whereby a refusal to license is abusive if it prevents "the appearance of a new product," without justification, and this "reserves for [the dominant firm] the secondary market and exclu[des] all competition on that market."[74]

Magill was not, however, the end of the story. In *IMS Health*, the next compulsory licensing case to reach the CJEU, a "serious dispute" occurred as to whether the satisfaction of either of the *Magill* circumstances was sufficient to affirm liability pursuant to Article 102 TFEU.[75] The *IMS Health* Court resolved the issue, holding all three conditions to be cumulative, and not alternative.

Interestingly, the rule-based approach in compulsory licensing cases was not instantly sustained by the GC. In *Microsoft*, the GC used a standard-based approach to review the Commission decision that had declared the dominant OS maker guilty of unlawful refusal to supply interoperability information. The judgment found that Microsoft's conduct ran counter to the abstract wording of Article 102(b) TFEU, which talks of a "limitation [of] technical development."[76] As was previously argued, the Commission's decision did not meet the requirements of the *Magill/IMS Health* rule, and especially the "new product" and elimination of "all competition"

---

[72] CJEU, C-62/86, *AKZO v. Commission*, ECLI:EU:C:1991:286, §§71–2.
[73] GC, T-342/99, *Airtours*, ECLI:EU:T:2002:146, §§159 and 210.
[74] CJEU, *RTE and ITP*, §§52–6.
[75] Case T-184/01 R, *Order of the President of the Court of First Instance*, ECLI:EU:T:2001:259, §100.
[76] GC, *Microsoft I*, §647.

conditions. The GC held that this did not matter, for the Commission had met the "limitation of technical development" standard.

In 2012, the GC nevertheless performed a spectacular U-turn. In *Microsoft II* – a case about the remedial consequences of the Commission's infringement decision – the GC no longer talked to the "limitation of technical development" standard. In an obiter dictum, the GC explicitly refers to the "three cumulative conditions" of *IMS Health* as the test applicable to refusals to license IPRs.[77]

6.6.2. *IPR Remedy Cases (1)*

IPR remedy cases have also been dealt with under a rule-based approach. While there is no CJEU case-law on this, the judgment of the GC in *Protégé International* held that a finding of abuse can only occur in "wholly exceptional circumstances" and this requires the showing of two conditions: "the action (i) cannot reasonably be considered as an attempt to establish the rights of the undertaking concerned and can therefore only serve to harass the opposite party; and (ii) it is conceived in the framework of a plan whose goal is to eliminate competition."[78] Those conditions are cumulative and sequential: the second condition is only examined if the first condition is satisfied.[79] And given the fundamental right of access to court, the two conditions "must be construed and applied strictly."[80]

6.6.3. *IPR Remedy Cases (2)*

The CJEU again endorsed a rule-based approach in cases in which owners of FRAND-pledged SEPs had sought injunctions against unlicensed implementers of their technology.

In *Huawei v. ZTE*, the Court held that antitrust liability could be affirmed against IPR owners' enforcement actions in the presence of two particular circumstances. First, the "patent at issue is essential to a standard established by a standardisation body."[81] Second, "the patent at issue obtained SEP status only in return for the proprietor's irrevocable undertaking, given to the standardisation body in question, that it is prepared to grant licences on FRAND terms."[82] According to the Court, such conditions create "legitimate expectations on the part of third parties" that holders of FRAND-pledged SEPs will openly license their technology.[83]

---

[77] GC, *Microsoft II*, §139.
[78] *Id.*, §49.
[79] *Id.*
[80] *Id.*
[81] *Id.*
[82] *Id.*, §51.
[83] *Id.*, §53.

Closely related to this is the fact that *Huawei v. ZTE* also sets a rule of exoneration, besides the liability rule. The *Huawei v. ZTE* court explains the conditions under which a dominant FRAND-pledged SEP holder that applies for an injunction can avoid a finding of abuse. First, the patent owner must "alert the alleged infringer of the infringement complained about by designating that SEP and specifying the way in which it has been infringed."[84] Second, the patent owner must "present to that alleged infringer a specific, written offer for a licence on FRAND terms, in accordance with the undertaking given to the standardisation body, specifying, in particular, the amount of the royalty and the way in which that royalty is to be calculated."[85] Third, at each stage, the patent owner must leave to the implementing infringer a sufficient time to react – declaration of willingness to transact, acquiescence, or counter-offer – prior to contemplating the seeking of a court injunction. If the patent owner complies with those procedural "conditions," he "does not abuse its dominant position ... by bringing an action for infringement."[86]

6.6.3.1. Pricing (and Terms) of IPR Licenses

The *DSD* judgment is also rule based, though less explicitly. The Grand Chamber of the CJEU upheld the GC and the Commission decisions on the ground that they had correctly applied the rule set in the case-law on unfair pricing. As the Court recalls, a firm "abuses its dominant position where it charges for its services fees which are disproportionate to the economic value of the service provided," a finding that is not left to full agency discretion, but that must be established by reference to price-cost metrics.[87] On the facts, the CJEU seemed satisfied that the Commission had carried out – though in a crude way – the price-cost rule required for exploitative abuses cases.[88]

6.6.4. *Rule-based Approach in Coordinated Conduct Cases?*

If it is true that the CJEU follows a rule-based approach in antitrust cases with IP ramifications, then one should expect to observe it also in the Article 101 TFEU case-law. Unfortunately, however, no such observation can be made. For the reasons previously explained, the CJEU has produced little case-law on the intersection

---

[84] *Id.*, §61.
[85] *Id.*, §63.
[86] *Id.*, §71.
[87] CJEU, DSD, §142; GC, T-151/01, *Duales System Deutschland v. Commission*, ECLI:EU:T:2007:154, §121.
[88] Commission Decision 2001/463/EC of 20 April 2001, Case COMP D3/34.493 – DSD, [2001] OJL 166, p. 1, §111.

between Article 101 TFEU and IPRs. Until new cases are brought to the Court, this question will remain unresolved.

This notwithstanding, the submission of this chapter is that the CJEU need not introduce (or could introduce fewer) judge-made "rules" in its Article 101 TFEU case-law for the simple reason that such rules are already hard wired into the wording of Article 101 TFEU. Unlike Article 102 TFEU, the letter of Article 101 TFEU imposes a very structured process and several substantive conditions before a finding of liability can be attained. The first paragraph of Article 101 TFEU imposes the characterization of an "agreement" and a manifestation of anticompetitive impact by "object" or "effect." The demonstration of anticompetitive effect, in turn, requires proof of an adverse, plausible, and appreciable impact on competition. If all those conditions are met, the third paragraph of Article 101 TFEU then gives defendants an opportunity to rebut antitrust liability if they bring proof of four cumulative conditions, namely that their agreement: (i) "promo[tes] the production or distribution of goods or to promoting technical or economic progress"; (ii) "allow[s] consumers a fair share of the resulting benefit"; (iii) does not impose "restrictions which are not indispensable"; and (iv) does not lead to the "possibility of eliminating competition." By design, Article 101 TFEU cases thus enshrine a rule-based approach.

In addition to this, the scant case-law on the antitrust–IP intersection under Article 101 TFEU hints at adherence to the rule-based framework. To start, let us look at one contemporary case where the CJEU was invited to move away from the rule-based framework of the Treaty, and embrace a standard-based approach. In *GSK v. Commission*, the pharmaceutical company GSK had been found guilty of a restriction of competition by object on the ground that it had applied a system of dual pricing which penalized Spanish buyers who engaged in parallel exports. On appeal, the GC considered that the existence of a restriction by object had to be characterized by recourse to a standard, namely consumer harm.

On further appeal, the CJEU vacated the GC judgment. It noted that "neither the wording of Article [101(1)] TFEU nor the case-law lend support to such a position" and that the GC had committed an error of law by "requiring proof that the agreement entails disadvantages for final consumers as a prerequisite for a finding of anti-competitive object."[89] What seemed to drive the Court's analysis is the imperative of sticking to the clear text of the Treaty: "there is nothing *in that provision* to indicate that only those agreements which deprive consumers of certain advantages may have an anti-competitive object"[90] (emphasis added).

In all events, many older judgments of the Court at the antitrust–IP intersection suggest a commitment to a rule-based approach in Article 101 TFEU cases. This is

---

[89] CJEU, C-501/06 P, C-513/06 P, C-515/06 P and C-519/06 P, *GSK v. Commission*, ECLI:EU:C:2009:610, §§62–4.
[90] *Id.*, §63.

true of the case-law on territorial restrictions in licensing agreements, where the Court has affirmed a per se prohibition rule. The scope of this per se rule is often confusingly presented as a blanket prohibition of all licensing restrictions which limit "parallel trade" in the European Union. This reading is wrong. The Court's judgments set out a structured liability test which only apprehends a subset of licensing restrictions that exhibit certain features. Those conditions are the following: (1) the licensing restrictions must compartmentalize national markets, by opposition to licensing restrictions that insulate regional or local markets; (2) the licensing restrictions must provide "absolute" protection to the licensee on a national territory, and not any kind of protection from competitive imports.

Let us sift through the main cases. The per se prohibition rule was affirmed in *Consten and Grundig v. Commission*. As previously explained, Grundig, a manufacturer of radio receivers, recorders, dictaphones, and television sets, had granted exclusive territorial protection to Consten for the distribution of its products in France. Under the contract, Consten would be prohibited from actively and passively exporting machines outside the contract area, and Grundig had imposed a similar prohibition on its sole distributors in other countries and on its German wholesalers. Moreover, Grundig would not sell machines to other distributors in France. On appeal, the CJEU confirmed the existence of an infringement, noting that the arrangement gave rise to "absolute territorial protection" and resulted in the "isolation of the French market."

Since then, *Consten and Grundig* has been repeated on many occasions, and in relation to a variety of IPRs. To take a few examples, in *Nungesser and Eisele*, the French National Institute for Agricultural Research ("INRA") had licensed breeders' rights over maize seeds to a supplier of seeds in Germany. The licensing agreement provided that INRA would refrain from issuing other licenses in Germany and from supplying seeds itself in Germany. Moreover, there was a contractual commitment by INRA and the licensee to use all possible means to ensure that third parties would not import from other Member States into Germany, and not export from Germany to other Member States. When parallel importers introduced INRA seeds for sale to German buyers, INRA sought to exert pressures and commenced legal proceedings. The dispute was finally resolved by the Court, which held that: "*absolute territorial protection* granted to a licensee in order to enable parallel imports to be controlled and prevented results in the artificial maintenance of *separate national markets*, contrary to the Treaty" (emphasis added).[91]

Last, in *Football Association Premier League Ltd. (FAPL)*, the Court was questioned on the legality of the exclusive licensing agreements between FAPL and a number of national broadcasters over Europe. Those agreements included a general obligation on each national broadcaster to ensure that broadcasts could not be "received outside that territory" and a specific "prohibition from supplying

---

[91] CJEU, C-258/78, *Nungesser v. Commission*, ECLI:EU:C:1982:211, §61.

decoding devices that allow their broadcasts to be decrypted ... outside the territory."[92] Though not strictly about copyright, the case shocked the European broadcasting industry because the Court affirmed that "an agreement which might tend to restore the divisions *between national markets* is liable to frustrate the Treaty's objective of achieving the integration of those markets" and "must be regarded, in principle, as agreements whose object is to restrict competition" (emphasis added).

Clearly, the wording of those judgments bespeaks a per se prohibition *rule* against licensing restrictions which create "export bans," "partition national markets," and ultimately give rise to "absolute territorial protection." The Court's case-law, by contrast, does not endorse a loose, abstract, and general *standard* like the limitation, distortion, or restriction of parallel trade.

To close, we can observe that a rule-based approach also seems to prevail in relation to arguments made by defendants willing to redeem anticompetitive licensing restrictions. *Ottung* and *Bayer AG v. Süllhöfer* provide two possible illustrations. While the Court admitted that some licensing clauses could be dealt with under a standards approach (the Court pointed to an analysis of the "economic and legal context"), both judgments are essentially known for the rule of immunity, and the accompanying structured test, set out in relation to certain anticompetitive licensing restrictions. In *Ottung*, the Court considered that an obligation to pay royalties for an indeterminate period that possibly extends beyond patent expiry was per se lawful, as long as the licensee kept the ability to terminate the licensing agreement by giving reasonable notice.[93] This rule of legality was recently recalled by Advocate General Wathelet in his Opinion in *Genentech Inc. v. Hoechst GmbH*.[94] In *Bayer AG v. Süllhöfer*, the Court ruled that non-challenge clauses fall short of Article 101(1) TFEU when the agreement provides for a free license or relates to an outdated process that the licensee did not use.[95]

Even more clearly, the judgment in *Erauw-Jacquery v. La Hesbignonne* conveys a rule-based approach.[96] Here, the CJEU held that liability for absolute territorial protection can be eschewed if this is needed to protect (i) a supplier who has incurred "considerable financial commitment"; (ii) against risks of "improper handling" of his input.

---

[92] CJEU, C-403/08 and C-429/08, *Football Association Premier League Ltd. (FAPL)*, ECLI:EU:C:2011:631, §35.
[93] CJEU, 382/87, *Kai Ottung v. Klee & Weilbach A/S and Thomas Schmidt A/S*, ECLI:EU:C:1989:195, §13.
[94] Opinion of AG Wathelet in C-567/14, *Genentech Inc. v. Hoechst GmbH*, not yet published.
[95] CJEU, 65/86, *Bayer AG and Maschinenfabrik Hennecke GmbH v. Heinz Süllhöfer*, ECLI:EU:C:1988:448, §§17 and 18.
[96] CJEU, 27/87, *SPRL Louis Erauw-Jacquery v. La Hesbignonne SC.*, ECLI:EU:C:1988:183, §10.

6.6.5. Rationale for a Rule-based Approach in Antitrust Cases Involving IPRs?

All readers of EU case-law are familiar with the CJEU long-standing practice of stating what the law is, without articulating why this law is justified. In the IP field, Korah once talked about the Court's "frequent unwillingness to analyze theoretically, but rather to rule on results."[97] In this section, I therefore attempt to uncover the intellectual foundations of the rule-based approach.

At an abstract level, there are several explanatory factors for a rule-based approach in antitrust cases with IPR ramifications. Given the perils of ex post rationalization, I decline to choose among them.

First, a well-known advantage of structured tests of liability, as opposed to loose standards, is their predictability. In contrast, standards leave more space for discretionary intervention.

With this background, it is conceivable that the Court adheres to the idea that legal uncertainty disincentivizes investments, and that this risk is even more compounded in areas with high IPR density where the strength of intellectual property exerts an indeterminate – yet certain – influence on investments in innovation. I concur in this respect with Regibeau and Rockett, who suggest that the adoption of clear rules acts as a remedy against regulatory opportunism into IPRs.[98]

Second, the rule-based approach may be a signaling device, used to reassure the Member States that the European Union will not discretionarily encroach upon their exclusive competence in relation to property rights.

Third, judicial economy considerations may also have exerted influence on the Court's case-law. Given the frequency of IPR transactions, the additional cost of designing a rule (which is borne only once) is likely to be lower than the costs of applying a standard on a case-by-case basis. The same is true for users of the law (be they competition agencies, national courts, or practitioners) who incur only once the cost of figuring out what the law is, and can then spread this fixed cost over an indefinite number of cases.

Last, the rule-based approach may again be rationalized on the basis of the well-known aversion of the CJEU to the use of economics in the adjudication of cases. And while the Court's distaste for economics in genuine antitrust cases is well known, its discomfort with the economics of IP seems at least equal. We certainly have no direct proof of this, but some indirect pronouncements provide some evidence. In its case-law, the Court has made many conflicting pronouncements on the relationship between ownership of IPRs and market power. While in *Magill*, the Court asserted that "[s]o far as dominant position is concerned, it is to be remembered at the outset that mere ownership of an intellectual property right

---

[97] Korah (2002).
[98] Regibeau and Rockett (2007).

cannot confer such a position," it contradicted itself in *Sot Lelos kai Sia EE and Others v. GlaxoSmithKline* with the formidable – and flawed – statement that "a medicine is protected by a patent which confers a temporary monopoly on its holder."[99] If anything, the spectacular degree of confusion of the Court on this basic issue strongly suggests a discomfort with the economics of IPRs.

6.7. CONCLUSION

This chapter has provided a survey of the case-law and of the scholarship on the antitrust–IP intersection. Its main conclusion is that no substantive theory drives the relationship between ex ante IPR policy and ex post competition enforcement. Instead, the singularity of the case-law in this field is that the CJEU predominantly approaches antitrust cases with IPR ramifications under a rule-based approach.

This methodological choice certainly has many merits. But it begets a fundamental question: that of the optimal design (content) of the legal rules applicable at the antitrust–IP intersection.

Moreover, when looked at more closely, the rules defined by the CJEU often seem to embed standards. Put differently, the case-law at the antitrust–IP intersection formulates structured tests of liability or justifiability that often resort to abstract concepts. For instance, the *DSD* judgment talks of fees which are "disproportionate" to the economic value of the service. Similarly, the four-prong test of Article 101(3) talks of "improving," "fair," "indispensable," or "substantial."

With this qualification, our conclusion ought to be that the Court approaches the antitrust–IP intersection with a rule-based spirit, but when it comes to giving content to a structured test of liability and justifiability, it resorts to abstract concepts. The implications of this judicial policy are not entirely clear. Does this correct the traditional risks of over- and under-inclusion (or both) encountered with the rules-based approach? If this is the case, does this come at the expense of legal certainty?

Future research in this field is needed. As often in antitrust law, the input of economics will surely be of great assistance.

REFERENCES

Ahlborn, C., D.S. Evans, and A.J. Padilla, 2004. The Logic and Limits of the "Exceptional Circumstances Test" in Magill and IMS Health. *Fordham International Law Journal*, 28 (4), 1108–56.
Anderman S.D. and H. Schmidt. 2007. EC Competition Policy and IPRs. In S.D. Anderman (ed.), *Intellectual Property Rights and Competition Policy*. Cambridge University Press.
Crane, D.A. 2007. Rules Versus Standards in Antitrust Adjudication. *Washington & Lee Law Review*, 64(1), 49–110.

---

[99] CJEU, C-468/06 to C-478/06, *Sot. Lélos kai Sia EE and Others v. GlaxoSmithKline AEVE*, ECLI:EU:C:2008:504, §64.

Czapracka, K. 2007. Where Antitrust Ends and IP Begins – on the Roots of the Transatlantic Clashes. *Yale Journal of Law and Technology*, 9(1), 44–108.
  2009. *Intellectual Property and the Limits of Antitrust – A Comparative Study of US and EU Approaches*. Cheltenham: Edward Elgar Publishing.
Drexl, J. 2008. *Research Handbook on Intellectual Property and Competition Law.*, Munich: Max Planck Institute for Innovation and Competition.
Ebb, L.F. 1967. The Grundig-Consten Case Revisited: Judicial Harmonization of National Law and Treaty Law in the Common Market. *University of Pennsylvania Law Review*, 115 (6), 855–9.
Enchelmaier, S. 2010. Intellectual Property, the Internal Market and Competition Law. In J. Drexl (ed.), *Research Handbook on Intellectual Property and Competition law*. Cheltenham: Edward Elgar Publishing.
Epstein, R.A. 2016. The Constitutional Foundations of Intellectual Property: A Natural Rights Perspective (Book Review). *Federalist Society Review*, 17(1), 54–6.
Käseberg, T. 2012. *Intellectual Property, Antitrust and Cumulative Innovation in the EU and the US*. Oxford/Portland, OG: Hart Publishing.
Kolstad, O. 2008. Competition Law and Intellectual Property Rights – Outline of an Economics-based Approach in J. Drexl (ed.), *Research Handbook on Intellectual Property and Competition Law*. Munich: Max Planck Institute for Innovation and Competition.
Korah, Valentine. 2002. The Interface Between Intellectual Property and Antitrust: The European Experience. *Antitrust Law Journal*, 69(3), 801–39.
Liannos, I. and R.C. Dreyfuss. 2013. *New Challenges in the Intersection of Intellectual Property Rights with Competition Law – A View from Europe and the United States*. CLES Working Paper Series 4/2013.
Peeperkorn, L. and E. Paulis. 2005. Competition and Innovation: Two Horses Pulling the Same Cart. In P. Lugard, and L. Hancher (eds.), *On the Merits: Current Issues in Competition Law and Policy: Liber Amicorum Peter Plompen*. Antwerp: Intersentia.
Posner, R. 2001. *Antitrust Law* (2nd edn.) University of Chicago Press.
Rato, M., and M. English. 2016. An Assessment of Injunctions, Patents, and Standards Following the Court of Justice's Huawei/ZTE Ruling. *Journal of European Competition Law & Practice*, 7(2), 103–12.
Regibeau, P. and K. Rockett. 2007. The Relationship Between Intellectual Property Law and Competition Law" In S.D. Anderman (ed.), *Intellectual Property Rights and Competition Policy*, Cambridge University Press.
Temple Lang, J. 2010. European Competition Law and Intellectual Property Rights – A New Analysis. *ERA Forum*, 11(3), 411–37.
Vaver, D. 2009. Reforming Intellectual Property Law: An Obvious and Not-so-obvious Agenda. *Intellectual Property Quarterly*, 2, 143–6.

Cases and Official Documents

Almunia, J. speech, IP Summit 2013 (Paris), 9 December 2013.
Case T-184/01 R, Order of the President of the Court of First Instance, ECLI:EU:T:2001:259.
Case C-70/10, *Scarlet Extended*, ECLI:EU:C:2011:771.
Commission Decision 2001/463/EC of 20 April 2001, Case COMP D3/34.493 – *DSD*, [2001] OJL 166.
Commission Decision, Case AT.39226 – *Lundbeck*, C(2013) 3803 final (19.06.2013).
2004 Commission Notice – Guidelines on the application of Article 81 of the EC Treaty to technology transfer agreement, 2004/C 101/02.

Commission Regulation (EU) No. 316/2014 of 21 March 2014 on the application of Article 101(3) of the Treaty on the Functioning of the European Union to categories of technology transfer agreements, OJ L 93, 28.3.2014.
CJEU, Joined Cases 56 and 58/64, *Etablissements Consten SARL and Grundig-Verkaufs-GmbH v. Commission*, ECLI:EU:C:1966:41.
CJEU, 24/67, *Parke, Davis & Co. v. Probel, Reese, Beintema-Interpharm and Centrafarm*, ECLI:EU:C:1968:11
CJEU, 78/70, *Deutsche Grammophon Gesellschaft mbH v. Metro-SB-Großmärkte GmbH & Co. KG*, ECLI:EU:C:1971:59.
CJEU, 40/70, *Sirena v. EDA*, ECLI:EU:C:1971:18, §9.
CJEU, 86/75, *EMI Records Limited v. CBS Grammofon A/S*, ECLI:EU:C:1976:86.
CJEU, 102/771, *Hoffmann-La Roche v. Centrafarm*, ECLI:EU:C:1978:108.
CJEU, 28/77, *Tepea v. Commission*, ECLI:EU:C:1978:133.
CJEU, 144/81, *Keurkoop v. Nancy Keane Gifts*, ECLI:EU:C:1982:289.
CJEU, C-258/78, *Nungesser v. Commission*, ECLI:EU:C:1982:211.
CJEU, 193/83, *Windsurfing International v. Commission*, ECLI:EU:C:1986:75.
CJEU, 65/86, *Bayer AG and Maschinenfabrik Hennecke GmbH v. Heinz Süllhöfer*, ECLI:EU:C:1988:448.
CJEU, 238/87, *AB Volvo v Erik Veng*, ECLI:EU:C:1988:477, §9.
CJEU, 27/87, *SPRL Louis Erauw-Jacquery v. La Hesbignonne SC.*, ECLI:EU:C:1988:183
CJEU, 382/87, *Kai Ottung v. Klee & Weilbach A/S and Thomas Schmidt A/S*, ECLI:EU:C:1989:195.
CJEU, C-62/86, *AKZO v. Commission*, ECLI:EU:C:1991:286.
CJEU, C-241/91 and C-242/91, *RTE and ITP Ltd v. Commission*, ECLI:EU:C:1995:98.
CJEU, C-7/97, *Bronner v. Mediaprint*, ECLI:EU:C:1998:569.
CJEU, C-418/01 *IMS Health*, ECLI:EU:C:2004:257.
CJEU, C-468/06 to C-478/06, *Sot. Lélos kai Sia EE and Others v. GlaxoSmithKline AEVE*, ECLI:EU:C:2008:504
CJEU, C-385/07 P, *Duales System Deutschland v. Commission*, ECLI:EU:C:2009:456.
CJEU, C-501/06 P, C-513/06 P, C-515/06 P and C-519/06 P, *GSK v. Commission*, ECLI:EU:C:2009:610.
CJEU, C 403/08 and C 429/08, *Football Association Premier League Ltd. (FAPL)*, ECLI:EU:C:2011:631.
CJEU, C-457/10, *AstraZeneca v. Commission*, ECLI:EU:C:2012:770.
CJEU, C-170/13, *Huawei v. ZTE*, ECLI:EU:C:2014:2391.
GC, T-69/89, *Radio Telefis Eireann v. Commission*, ECLI:EU:T:1991:39.
GC, T-504/93, *Tiercé Ladbroke SA v. Commission*, ECLI:EU:T:1997:84.
GC, T-111/96, *ITT Promedia NV v. Commission*, ECLI:EU:T:1998:183.
GC, T-198/98, *Micro Leader Business v. Commission*, ECLI:EU:T:1999:341.
GC, T-342/99, *Airtours*, ECLI:EU:T:2002:146.
GC, T-151/01, *Duales System Deutschland v. Commission*, ECLI:EU:T:2007:154.
GC, T-201/04, *Microsoft v. Commission (Microsoft I)*, ECLI:EU:T:2007:289.
GC, T-167/08, *Microsoft Corp. v. Commission (Microsoft II)*, ECLI:EU:T:2012:323.
GC, T-119/09, *Protégé International Ltd v. European Commission*, ECLI:EU:T:2012:421.
Opinion of Advocate General Gulmann in Joined cases C-241/91 P and C-242/91 P, *Radio Telefis Eireann (RTE) and Independent Television Publications Ltd (ITP) v. Commission*, ECLI:EU:C:1994:210.
Webster v. Reproductive Health Servs., 492 U.S. 490 (1989).

7

The IP–Antitrust Interface in China: Uncharted Territory

Thomas K. Cheng

7.1. INTRODUCTION

One aspect of the Chinese Anti-Monopoly Law (AML) that attracted much attention early on was the treatment of intellectual property rights under the law. Intellectual property rights were specifically mentioned in the AML. Article 55 somewhat cryptically notes that while the lawful exercise of intellectual property rights is protected by the AML, the abuse of intellectual property rights is prohibited. The fact that the treatment of intellectual property rights occupies a full article out of a fifty-seven-article law underscores the weight attached to the issue. The IP–antitrust interface is particularly important in the Chinese context given the fact that Chinese companies are often the licensees of foreign technology and China itself is emerging as a center of innovation. The AML's treatment of patent licensing will have a significant impact on the licensing relationship between Chinese licensees and foreign patentees. The fulcrum of Article 55 is the concept of abuse. It is what determines whether a particular exercise of intellectual property rights is legal or not. The concept, however, is not self-explanatory. The law itself provides no further explanation or elucidation of the concept. The expectations were that the two enforcement authorities responsible for non-merger enforcement of the AML, the National Development and Reform Commission (NDRC) and the State Administration of Industry and Commerce (SAIC) would issue guidelines to provide further guidance. However, no guidelines were forthcoming for years. In the meantime, only a handful of cases that implicate the IP–antitrust interface have been decided by the courts. The treatment of the IP–antitrust interface under the AML has remained shrouded in mystery.

Some clarity has finally emerged. The SAIC issued regulations on the IP–antitrust interface, the Provisions on Prohibiting the Abuse of Intellectual Property Rights to Exclude and Restrain Competition (hereinafter the "Provisions") in April 2015. Although the Provisions are brief, they nonetheless provide some guidance on the

SAIC's attitude toward some patent exploitation practices. In February 2015, the NDRC concluded the highest-profile enforcement action by a Chinese authority against a foreign company, the NDRC's investigation of Qualcomm. The NDRC fined Qualcomm close to US$1 billion and imposed a range of behavioral remedies. The decision sheds light on the NDRC's stance on a number of fairly common patent licensing practices.

This chapter provides an overview of the recent development regarding the IP–antitrust interface under the AML, examining the SAIC Provisions and the NDRC *Qualcomm* decision, together with another important case and other relevant laws and administrative regulations.

7.2. THE ANTI-MONOPOLY LAW AND THE IP–ANTITRUST GUIDELINES

Article 55 provides the basis for the treatment of the IP–antitrust interface under the AML. It states that "[t]his Law is not applicable to the lawful exercise of intellectual property rights in accordance with the laws and administrative regulations pertaining to intellectual property rights. However, this Law applies to the abuse of intellectual property rights by business operators to eliminate or restrict competition."[1] This provision itself reveals little. We do know that the AML can apply to the exercise of intellectual property rights; China has not exempted the exercise of intellectual property rights from the purview of its competition law, as in the case of Jamaica.[2] But the provision does not provide much guidance on what constitutes an abuse of intellectual property rights. Early on, the expectations were that the enforcement authorities would issue guidelines to clarify the permissible scope for the exercise of intellectual property rights. Over the years, there were rumors that guidelines were being prepared by the NDRC and the SAIC. Multiple drafts were circulated among experts and those with close connections with the enforcement authorities. However, none was officially issued and the public remained in the dark about these guidelines. In April 2015, the wait was finally over when the SAIC issued the Provisions.

Despite the long wait, the Provisions still leave many questions unanswered. The Provisions do not lay out in detail the analytical approach of the SAIC with respect to various patent exploitation practices. They seem to adopt a fairly categorical approach, asserting whether a particular practice is prohibited outright without explaining what factors will be taken into account in the analysis. Moreover, for a wide variety of practices, the Provisions state that the practices are prohibited absent justifications. There is no guidance, however, on what kind of justifications will be accepted and whether any kind of balancing will be undertaken between the

---

[1] National People's Congress Standing Committee (2007).
[2] Jamaica Fair Trade Commission (2006).

purported anticompetitive effects and the procompetitive justifications. The relevance of these Provisions is further undermined by the announcement of the NDRC in August 2015 that it is in charge of drafting another set of IP–antitrust guidelines on behalf of the Anti-Monopoly Commission (AMC), the State Council body that oversees AML enforcement by the three enforcement authorities. It is unclear at this stage whether these AMC Guidelines will supersede the SAIC Provisions or whether the Provisions will continue to guide SAIC enforcement. It is technically possible for the NDRC and the SAIC to adopt different guidelines, as they have done in the past, due to the demarcation of jurisdiction between them. The NDRC is in charge of enforcement against price-related violations while the SAIC is responsible for non-price-related violations.

It is worth examining the Provisions to see what sort of guidance has been provided by the SAIC. If the two enforcement authorities were to take their division of labor literally, there may not be much that falls within the NDRC's jurisdiction. Apart from royalty-related issues and RPM, most of the issues arising from patent licensing are not directly price related. The kind of tying that is common in patent licensing situations is not directly price related. The same applies for exclusive dealing, package licenses, blanket licenses, block booking, unilateral refusal to license, and reverse payments. Therefore, the SAIC will likely have a significant role in shaping the law on the IP–antitrust interface in China.

Articles 1 and 2 provide some comfort to intellectual property owners in that they affirm the importance of encouragement of innovation as far as AML enforcement against the exercise of intellectual property rights is concerned. Article 1 states that the Provisions are formulated with a view to protecting fair competition and encouraging innovation.[3] Article 2 reaffirms the importance of encouraging innovation by emphasizing that the AML and intellectual property rights share the common objectives of promoting competition and innovation, improving the efficiency of economic operations, and safeguarding consumer and public interests.[4] Article 2 downplays the occasional conflict between antitrust and intellectual property law that has been noted by some commentators[5] and chooses to emphasize the shared objectives of the two bodies of law.[6] However, it does muddy the water by introducing public interest into the equation, which could be interpreted to mean general societal interest or national technological development that may detract from the focus on consumer welfare.

Article 15 sets out the analytical framework for determining whether the exercise of intellectual property rights is abusive.[7] First, the SAIC will determine the nature and manifestation of the exercise of the intellectual property rights by the business

---

[3] State Administration of Trade and Commerce (2015).
[4] Id.
[5] Kaplow (1984).
[6] Hovenkamp *et al.* (2009).
[7] State Administration of Trade and Commerce (2015).

operator. It is not clear what is meant by the manifestation of the exercise of intellectual property rights. Second, the SAIC will ascertain the nature of the relationship between the business operators that have exercised intellectual property rights. This is only relevant if multiple business operators are involved in the exercise of an intellectual property right. Third, the SAIC will define the relevant market. Fourth, the SAIC will determine the market position of the business operators involved. Fifth, the SAIC will assess the impact of the exercise of intellectual property rights on competition in the relevant market. In Article 16, the SAIC identifies eight factors that are relevant to the determination of the competitive impact of the exercise of an intellectual property right.[8] These are: (1) the respective market position of the business operator and its transaction counterparty, (2) the degree of concentration in the relevant market, (3) barriers to entry to the relevant market, (4) industry practices and the stage of industrial development, (5) the effectiveness in terms of time and scope of the restriction brought about by the exercise of intellectual property right on output, geographic area, and consumers, (6) the impact on innovation and promotion of technology, (7) the innovative capacity of the business operator and the speed of technological change, and (8) other relevant factors. Again, there is recognition of the importance of innovation incentives. This is a rather comprehensive list of well-balanced factors that should provide sufficient guidance to the SAIC.

Regarding market definition, the Provisions state that it will generally be done in accordance with the Anti-Monopoly Commission Guidelines on Market Definition. The Provisions do distinguish situations involving intellectual property rights licensing, and suggest that in such circumstance the relevant market may be a technology market or a product market containing a specific intellectual property right.[9]

Regarding monopolistic agreements, Article 5 provides safe harbors for horizontal and vertical agreements respectively.[10] Unless there is evidence to the contrary, an agreement among competitors whose collective market share in the relevant market is less than 20 percent or where there exist at least four alternative technologies in the relevant market that are accessible at reasonable costs and that are under independent control will not be deemed to be monopolistic under Article 13 of the AML.[11] Likewise, an agreement between noncompetitors whose respective market shares in the relevant markets are less than 30 percent or where there exist at least two alternative technologies in the relevant markets that are accessible at reasonable costs and are under independent control will not be deemed to be monopolistic under Article 14 of the AML.[12] The applicability of the safe harbor, especially the second prong based on the existence of alternative technologies,

[8] Id.
[9] Id.
[10] Id.
[11] Id.
[12] Id.

depends on how the SAIC interprets reasonable costs. Would the SAIC insist on complete cost parity? Or would the SAIC accept cost variations so long as the downstream operator remains viable after taking into account the cost of the technology? These are questions that will be answered only in future enforcement action. Another ambiguity in the Provisions is whether the safe harbor applies to all kinds of conduct, or only "non-hardcore" conduct. In the Interpretations of the Provisions that were subsequently published by the SAIC, it was clarified that the safe harbor only applies to conduct that is not specifically prohibited in Articles 13 and 14 of the AML (Interpretations of Provisions 2015).[13] These include price fixing, output restriction, market allocation, restriction of acquisition of new technology or new facility, restriction of the development of new technology or new product, concerted refusal to deal, and resale price maintenance.

Another complication is the classification of competitors. Are business operators considered competitors if they only become so as a result of a licensing agreement, under which normally the licensor and the licensee would be considered to be in a vertical relationship? In Article 15 of the Provisions, the SAIC sheds some light on this issue. If the licensor and the licensee were originally competitors before the licensing agreement, and both firms utilize the intellectual property right to produce products in the relevant market, they are deemed to be competitors. However, if the licensor and the licensee were originally not competitors, but become competitors after the licensing agreement, they will not be deemed to be competitors unless the original licensing agreement is substantively changed. It is not clear what the SAIC refers to by substantive changes. What sort of changes would alter the nature of the relationship? When does the change have to take place? Does it depend on the nature of the change at issue? These are issues that will have to await future enforcement action. What is clear, though, is that the SAIC generally takes the view that what governs the nature of the relationship between two firms is their pre-agreement relationship. The SAIC will generally ignore how the licensing agreement changes the nature of the relationship. This approach seems rather unusual in that what matters from a competition law perspective generally is the state of competition post-agreement. The competitive relationship between two firms pre-agreement is of little relevance to assessing the competitive impact of an agreement. The analysis is usually focused on how the agreement changes the competitive dynamics.

The remainder of the Provisions is largely devoted to abuse of dominance. Article 6 explains how dominance will be determined as far as intellectual property right owners are concerned.[14] The analysis will be conducted in accordance with the factors delineated in Articles 18 and 19 of the AML. What is particularly helpful is

---

[13] SAIC AML Enforcement Authority Interpretations of "Provisions on Prohibiting the Abuse of Intellectual Property Rights to Exclude and Restrain Competition." Available at www.cicn.com.cn/zggsb/2015-07/28/cms74478article.shtml.

[14] Id.

the clarification that ownership of an intellectual property right does not create a presumption of dominance. It is only one of the factors to be considered. This is consistent with the holding of the US Supreme Court in *Illinois Tool Works v. Independent Ink*.

One of the most controversial parts of the Provisions is probably Article 7, which discusses unilateral refusal to license intellectual property rights as an abuse. This Article justifiably causes unease among US commentators and practitioners as US courts rarely use antitrust law to compel licensing of intellectual property rights. The only notable exception is probably the Ninth Circuit decision in *Image Technical Services v. Eastman Kodak*. Article 7 explicitly discusses the application of the essential facilities doctrine to intellectual property rights, which probably causes yet more discomfort among US commentators and practitioners, as the essential facilities doctrine has never been used to compel the licensing of intellectual property rights in the United States.[15] The scope of the application of the doctrine to intellectual property rights in China remains to be seen. Article 7 states that a dominant business operator shall not, for the purpose of restricting or excluding competition, refuse to license to other business operators, on reasonable terms, intellectual property rights that constitute an essential facility.[16] There are two key terms in this provision. First, what constitute reasonable terms would be crucial to the application of the provision. If the SAIC interprets reasonable terms broadly to allow the licensor more leeway, then Article 7 may be less damaging to intellectual property right owners. Second, how seriously the SAIC takes the proviso "for the purpose of restricting or excluding competition" would have serious implications. Would the intellectual property right owner be able to plead that its purpose is not to restrict competition, but simply to exercise its prerogative as an intellectual property right owner? Would the SAIC interpret restriction of competition as exclusion of a competitor or genuine restriction of competition? In other words, would a licensor be able to justify refusal by arguing that there is already a sufficient number of producers in the relevant market for the product incorporating the intellectual property right, and hence refusal does not harm effective competition? The SAIC's answer to these questions would determine the reach of the essential facilities doctrine with respect to intellectual property rights in China.

Article 7 enumerates three factors for consideration when applying the doctrine to intellectual property rights.[17] First, whether there are no reasonable alternatives to the intellectual property right concerned in the relevant market and whether the intellectual property is required by other business operators to compete in the relevant market. Second, whether the refusal to license would adversely affect competition or innovation in the relevant market, and prejudice consumer or public

---

[15] Genevaz (2004).
[16] State Administration of Trade and Commerce (2015).
[17] Id.

interests. Third, whether the compulsory license will cause unreasonable damage to the owner of the intellectual property right. There are two noteworthy features about these three factors. First, the mention of public interests can potentially expand the scope of application of the Article significantly. If denial of access to a technology to a domestic manufacturer is deemed to be prejudicial to public interests, then the doctrine can be very readily invoked. Second, there is recognition of the need to balance the benefit to competition against the harm to the intellectual property right owner. Here the key is what constitutes unreasonable damage. If the introduction of new competition that threatens the market position of the owner counts as unreasonable damage, then the applicability of the doctrine would be more limited.

There are two noticeable differences between the SAIC's formulation of the essential facilities doctrine and the formulation of the doctrine in the United States, most notably in the Seventh Circuit case *MCI v. AT&T*. First, under the Seventh Circuit's formulation, the plaintiff is required to demonstrate the competitor's inability practically or reasonably to replicate the facility, as opposed to under the SAIC's formulation, under which the SAIC only needs to show that there are no reasonable alternatives in the market. Second, under the Seventh Circuit's formulation, the plaintiff is required to demonstrate the feasibility of supply, while under the SAIC's formulation, the SAIC considers whether there is unreasonable damage to the right owner. The US formulation is stricter on one count but probably more permissive on the other count. Under the US approach, in addition to showing that there are no alternatives in the market, the plaintiff needs to demonstrate that it cannot practically or reasonably replicate the facility. Under this formulation, the doctrine would not be applied if it can be shown that a competitor has the innovative capacity to create comparable intellectual property, even though it does not have it at present. Meanwhile, under the SAIC formulation, it would seem that a competitor's capacity to come up with a similar innovation would not be relevant. In short, the SAIC's view of the market seems to be more static whereas the Seventh Circuit's formulation is more dynamic in application. However, under the US approach, mere feasibility to supply suffices, regardless of the harm suffered by the facility owner, whereas under the SAIC approach, unreasonable harm to the owner may negate application of the doctrine. Feasibility includes economic feasibility. But that is still a more permissive standard in that a facility owner can suffer a lot of harm or damage, perhaps to an unreasonable extent, before it becomes economically unfeasible for it to supply the facility.

Article 8 prohibits exclusive dealing without justifications.[18] Article 9 prohibits tying sales that satisfy the following two criteria: (1) imposing tied sales against transaction practices or consumption habits, and (2) the tie allows the business operator to leverage its dominance in the tying product market to the tied product

---

[18] *Id.*

market.[19] The first criterion opens the possibility of a defense based on consumption habit. If it can be argued that the tie is justified by consumer preference, then it would not be prohibited under Article 9. What is not clear is how prevalent the consumer preference must be to justify the tie. Is it sufficient that only some consumers prefer to buy the two products together? Or a majority of the consumers? Or perhaps most of the consumers? This is again not clarified in the Provisions. If preference on the part of some consumers suffices to justify a tie, it would amount to a mirror image of Justice Sandra Day O'Connor's assertion in *Jefferson Parish Hospital No. 2 v. Hyde* that "[f]or products to be treated as distinct, the tied product must, at a minimum, be one that some consumers might wish to purchase separately *without also purchasing the tying product.*"[20] The SAIC's formulation would represent a more permissive attitude toward tying.

Article 10 prohibits a number of licensing restrictions,[21] including exclusive grantbacks, no-challenge clauses, charging of post-expiration royalties, exclusive dealing, restricting licensees from utilizing competing products or technologies upon expiration of the licensing agreement, or imposing other unreasonable restrictive licensing conditions.[22] The SAIC's stance toward these licensing restrictions is largely uncontroversial. One notable exception is the treatment of no-challenge clauses, which have not been prohibited under US antitrust law[23] even though they have been held to be a patent misuse since *Lear v. Adkins, Inc*. Charging of post-expiration royalties likewise has been largely dealt with under patent law, and has been held to be a misuse since *Brulotte v. Thys Co.*, and very recently reaffirmed in *Kimble v. Marvel Entertainment*. Article 11 prohibits discriminatory treatment by an intellectual property right owner for the purpose of restricting or excluding competition without justifications.

Articles 12 and 13 are concerned with patent pools and standardization respectively.[24] With respect to patent pools, Article 12 prohibits members of a pool from exchanging sensitive information regarding output or market division to reach a monopolistic agreement prohibited by Articles 13 and 14 of the AML. Article 12 further prohibits a patent pool with a dominant position from the following conduct: (1) requiring exclusive licensing by pool members, (2) prohibiting members from developing, independently or jointly with third parties, technologies that may compete with the pool, (3) requiring exclusive grantbacks from pool members, (4) imposing no-challenge clauses on pool members, and (5) adopting discriminatory treatment against pool members. Article 13 broadly prohibits members of standard setting organizations (SSOs) from engaging in activities that

---

[19] *Id.*
[20] *Jefferson Parish* 1984 (emphasis in original).
[21] SAIC Provisions (2015).
[22] *Id.*
[23] Miller and Gal (2015),
[24] State Administration of Trade and Commerce (2015).

restrict competition. In addition, it prohibits a dominant business operator from engaging in the following activities in the course of standard setting: (1) deliberately refusing to disclose information concerning its patent rights or asserting its patent rights after clearly waiving them in the standard setting process, (2) refusing to license, engaging in tying, or imposing other unreasonable licensing conditions in violation of its FRAND obligations.

Of these two articles, Article 13 is more controversial. There are a number of controversial aspects to it. First, what is not clear is whether the FRAND obligations only apply if the SSO imposes them on members, or whether the FRAND obligations come from the AML itself. Article 13 does not specify the source of the FRAND obligations, and can be interpreted as imposing general FRAND obligations in standard setting activities. That would render the SAIC's position on standard setting activities relatively aggressive when compared to other jurisdictions. In the Interpretations of the Provisions issued by the SAIC, the SAIC alludes to the fact that SSOs often impose FRAND obligations (Interpretations of Provisions 2015). This allusion to the origin of FRAND obligations perhaps can be read as suggesting that Article 13 only applies when the SSO imposes FRAND obligations. Second, some SSOs impose FRAND obligations, while others impose RAND obligations. However, Article 13 seems to mandate a FRAND standard, which could limit the room for maneuver for SSOs. Third, most SSOs impose a disclosure obligation. However, in most cases, the disclosure is based on a good faith effort to search the company's patent portfolio. This is important because a major technology company such as Microsoft or Qualcomm will have a massive patent portfolio and it may be impossible for it to undertake an exhaustive search for every patent. To add further complication to this search effort, it is not often immediately clear whether a patent covers a particular technology. This requires interpretation of claims, which is not often straightforward. Article 13 speaks of deliberate refusal to disclose. If this goes beyond the usual requirement of good faith search, it could potentially expose SSO members to expansive liability. What seems clear is that the obligation to disclose goes beyond what was laid down by the D.C. Circuit in the *Rambus v. FTC* case. The prohibition against deliberate refusal to disclose seems to be a flat obligation that is not premised on whether the SSO would have adopted a different technology had a disclosure been made, as in the *Rambus* case.

It is important to note that this is not the first time Chinese authorities have required patent holders to disclose their patents in the standard setting process. The Interim Provisions on the Administration of National Standards Involving Patents, which was issued by the Standardization Administration of China and the State Intellectual Property Office (SIPO) and became effective in January 2014, also stipulated that:

> [A]t any stage of development and revision of a national standard, any organization or individual that participates in the development and revision of the national

standard shall, as soon as possible, disclose the essential patents it, he or she owns or knows to the relevant national standardization technical committee or the entity in charge of standardization, simultaneously provide the relevant patent information and corresponding evidential materials, and be liable for the veracity of the evidential materials provided.

7.3. OTHER RELEVANT LAWS AND REGULATIONS

In addition to the AML and the SAIC IP-AML provisions, there are a few other laws and regulations that are also relevant to the IP–antitrust interface in China. These include the Patent Law, which provides for compulsory licensing under certain circumstances, and the Rules on Regulating the Market Order of Internet Information Services ("MIIT Internet Rules")[25] and an Interpretation of the Internet Rules ("MIIT Interpretation")[26] issued by the powerful Ministry of Industry and Information Technology.

7.3.1. Patent Law

Articles 48, 50, and 51 of the Chinese Patent Law provide for compulsory licensing under limited circumstances[27] (Patent Law 2008). Article 48 allows someone who possesses the capability for exploitation to apply for a compulsory license to an invention patent or a utility model patent (1) if the patentee has failed to exploit the patent without legitimate reasons after three years since the patent was granted, or (2) if the patentee's exercise of the patent has been confirmed as monopolistic conduct and a compulsory license is needed in order to reduce or eliminate the negative impact on competition. Article 50 provides for the grant of a compulsory license on public health grounds. Article 51 allows SIPO, China's patent regulator, to issue a compulsory license to an earlier patent when a subsequent invention represents a major technological advancement of remarkable economic significance and can only be exploited with access to the earlier patent. The provision which has the greatest antitrust implication is Article 48(2). The Patent Law itself does not define what constitutes monopolistic conduct. It is unclear whether this is to be determined in accordance with the AML. SIPO issued Implementing Regulations of the Patent Law to provide further guidance on the interpretation and implementation of the Patent Law.[28] The Implementing Regulations explain the application of Article 48(1) and Article 50, but are silent on Article 48(2). It is unclear whether Article 48(2) provides for an independent basis for the imposition of compulsory licenses on

---

[25] Ministry of Industry and Information Technology (2011a).
[26] Ministry of Industry and Information Technology (2011b).
[27] National People's Congress Standing Committee (2008).
[28] State Intellectual Property Office (2001).

competition grounds, or merely affirms that compulsory licenses are an available remedy under the AML.

The Measures for Compulsory Licensing of Patent Implementation, however, shed some light on the issue.[29] Article 11 of the Measures states that a petition for compulsory license under Article 48(2) should be accompanied by a judgment or determination by a court or an enforcement authority that the patent exploitation practice at issue has been deemed monopolistic according to the law.[30] This seems to suggest that the AML is to be used as a basis for determining monopolistic conduct and therefore the Patent Law does not have an independent definition of monopolistic conduct. Article 11 of the Measures seems to confirm the view that Article 48(2) merely reiterates the availability of compulsory licensing as an available remedy under the AML.

A further ambiguity in Article 48(2) is what constitutes negative impact on competition. Does it refer to actual instances of competition being restricted, such as where a competitor has been excluded causing consumer harm? Or does it simply refer to market prices being higher than the competitive level? If it refers to the latter, Article 48(2) could have broad application. One consoling fact is that, as of 2013, SIPO apparently had never issued a compulsory license under the various provisions of the Patent Law.[31] And the author is aware of no cases in which a court or either of the enforcement authorities has imposed compulsory licensing as a remedy in an AML case.

7.3.2. MIIT's Internet Rules

The MIIT Internet Rules are not framed in explicit antitrust terms. And they only apply to providers of internet information services (PIIS). Nonetheless, some of the provisions have antitrust implications and should be briefly mentioned. The first thing to note is that the MIIT Internet Rules do not define PIIS. However, Internet information services are defined as the provision of information services through the Internet to online subscribers. Article 5 of the MIIT Internet Rules prohibits a PIIS from maliciously introducing incompatibility with a competitor's products or services and maliciously interfering with the downloading, installation, operation, and updating of a competitor's products. Product incompatibility has occasionally been grounds for antitrust violations, as in the EU *Microsoft* case. Product incompatibility involving Internet services would often concern software code, which implicates the interface between copyright law and antitrust. However, the MIIT Internet Rules differ from the AML in that they impose an intent requirement and require malice in the act. In that sense, their scope of application is fairly limited. Article 7 further

---

[29] State Intellectual Property Office (2012).
[30] Id.
[31] Bloch and Milner (2013).

prohibits a PIIS from requiring customers to use or refrain from using PIIS-designated Internet information services without reasonable justification. This provision can be applied to a tying scenario. The MIIT Internet Rules are more expansive than the AML in that there is no requirement of dominance for tying to be proscribed. Article 7 seems to apply to all PIIS regardless of market power.

7.4. CASES

There have been few cases, either in the courts or brought by the enforcement authorities, involving the IP–antitrust interface in China since the AML came into effect in 2008. There was *Qihoo v. Tencent*, which was the first AML case to reach the Supreme People's Court and which involved the Internet. At issue in the case was a tying claim and a product incompatibility claim which did not directly implicate any intellectual property rights. Therefore, it does not count as an IP-antitrust case. The first proper court case implicating the IP-antitrust interface was probably *Huawei v. Interdigital*, which involved a licensing dispute between Huawei, a Chinese telecom equipment and handset manufacturer, and InterDigital, an American technology licensing company. The case originated in the Shenzhen Intermediate People's Court and was appealed to the Guangdong High People's Court. Both courts ruled in favor of Huawei, accusing InterDigital of abusing its dominance through its licensing practices. The highest-profile case involving IP-antitrust issues in China thus far is undoubtedly the NDRC's investigation of the US technological giant Qualcomm. The NDRC accused Qualcomm of abusing its dominance in the telecommunications technology licensing markets and the baseband chips markets through a range of licensing restrictions.[32] The NDRC eventually fined Qualcomm close to US$1 billion, required Qualcomm to reduce its royalty rate by 35 percent, and imposed a range of behavioral remedies.

7.4.1. Huawei v. Interdigital

The case concerned the standard essential patents (SEPs) in the 3G telecommunications technologies of WCDMA, CDMA2000, and TD-SCDMA owned by InterDigital. Huawei sought to license these SEPs and disputed that the royalty rates offered by InterDigital violated the FRAND obligations which the latter had undertaken when its patents were included in the standards formulated by the European Telecommunications Standards Institute (ETSI). In addition, the case also appeared to involve tying claims involving InterDigital's package licenses (one cannot be certain about the precise scope of the tying claims as the judgment was heavily redacted with the precise scope of the tying claims concealed). InterDigital apparently offered package licenses consisting of both SEPs and non-SEPs.

[32] National Development and Reform Commission (2015).

The Shenzhen Intermediate People's Court defined the relevant product market as consisting of each of the SEPs. In other words, each SEP constitutes its own relevant product market. The relevant geographic markets were China and the United States. The Shenzhen court also determined InterDigital to be dominant in the SEP licensing markets because it was the only operator in each of these markets. It further noted that InterDigital wielded unfettered market power because it was a nonpracticing entity. It was merely engaged in the licensing business. Therefore, Huawei could not counter-balance against InterDigital's market power through reciprocal licensing of its patents to InterDigital. The court found that InterDigital had clearly undertaken FRAND obligations with the ETSI. It also mentioned, without specifying the precise source, that Chinese law imposes FRAND obligations on SEPs as well. The court found that InterDigital charged Huawei a much higher royalty rate than it did Samsung and Apple. The court further found that InterDigital violated its FRAND obligations by requiring Huawei to reciprocally license its patents to InterDigital. It found that Huawei's patents were much more numerous and valuable than InterDigital's, and therefore InterDigital's reciprocal licensing demand violated its FRAND obligations. On an issue that has garnered much attention around the globe recently, the Shenzhen court objected to InterDigital's application for an injunction in the US International Trade Commission and Federal District Court for the District of Delaware, despite Huawei's good faith and best effort to reach an agreement with InterDigital. The court noted that the injunction would exclude Huawei's export to the United States and concluded that InterDigital's application for an injunction was a violation of its FRAND obligations. Finally, the court held that InterDigital's package licensing of its SEPs and non-SEPs constituted an illegal tie, but a package license of all the SEPs was not.

The Shenzhen court's decision was appealed to the Guangdong High People's Court. The Guangdong court affirmed the Shenzhen court's decision on all counts. The Guangdong court affirmed the lower court's product market definition. On appeal, InterDigital argued that the relevant geographic market should be global. The Guangdong court rejected the argument on the grounds that patent licensing is highly dependent on the patent law of the local jurisdiction. What InterDigital licenses in China are its patents under Chinese law, not US patents. Therefore, it does not make sense to speak of a global licensing market. The Guangdong court affirmed the lower court's finding of dominance on the same grounds that InterDigital was the sole supplier of SEPs in the relevant market and a nonpracticing entity.

The Guangdong court concurred with the Shenzhen court that InterDigital had violated its FRAND obligations. On appeal, regarding the discriminatory royalty claim, InterDigital argued that the lower court had erroneously compared the royalties of Samsung and Apple, which were one-time flat-fee royalties, and Huawei's, which was a percentage-based royalty. The Guangdong court acknowledged that in general, the two types of royalty are different and should not be compared. However, the court said in light of InterDigital's refusal to disclose its licensing agreements and percentage-based royalty arrangements with other entities to the court, the Shenzhen court had done the best it could and was justified in comparing a flat-fee royalty with a percentage-based royalty. The court further noted that InterDigital's high royalty rate against Huawei was unjustified in light of its own internal forecast that licensing revenue was in decline. The Guangdong court reiterated the unreasonableness of InterDigital's royalties in light of its demand for reciprocal licensing from Huawei and its application for an injunction in the US courts.

Regarding the package license claims, the Guangdong court reaffirmed the lower court's conclusions that packaging licensing of SEPs and non-SEPs was an illegal tie while package licensing of SEPs was justified. The court noted that the package license practice allowed InterDigital to leverage its dominance in the SEP markets into the non-SEP markets. The court rejected InterDigital's defense that it was very difficult to distinguish between SEPs and non-SEPs, pointing out that InterDigital had declared its SEPs to ETSI during the standard setting process. In contrast, the court upheld the package licensing of SEPs on the grounds that the practice is efficient, does not create foreclosure effect, is not coercive, and is a standard practice in the industry. The Guangdong court upheld the lower court's award of RMB20 million (about US$3 million) of damages to Huawei.

7.4.2. *The* NDRC Qualcomm *Case*

The NDRC launched its investigation of Qualcomm in November 2013. The investigation covered a range of Qualcomm's licensing practices. The NDRC eventually announced its decision in February 2015, imposing a fine on Qualcomm to the tune of US$975 million, which is the largest fine ever imposed on a single entity in the history of AML enforcement, and a host of behavioral remedies. In particular, the NDRC took exception to four of Qualcomm's licensing practices: (1) charging of royalties for expired patents, (2) compelling royalty-free grantbacks of licenses from licensees, (3) tying of wireless SEPs and non-wireless SEPs, and (4) imposing no-challenge clauses on licensees that prohibit them from challenging the validity of the patents.

The NDRC defined the relevant product markets as consisting of the licensing of CDMA, WCDMA, and LTE technologies and the baseband chips incorporating these technologies. Like the Shenzhen and Guangdong courts, the NDRC regarded each SEP as a stand-alone relevant product market, which means Qualcomm had a 100 percent market share. The NDRC substantiated its market definition on the grounds of low demand substitutability – licensees will need to license Qualcomm's SEPs if they want to deploy these technologies regardless of the royalty rate, and low supply substitutability – it is practically impossible for substitute patents to be developed and a terminal manufacturer must purchase compliant baseband chips if it desires to produce a terminal under a certain technology standard. The NDRC further concluded that Qualcomm was dominant in the SEP licensing markets because it was the sole supplier in all of them and was dominant in the baseband chips markets in light of its very high market share and high technical barriers to entry.

The NDRC found that Qualcomm committed four abuses in its licensing practices. The crux of the charge of charging royalties for expired patents pertains to Qualcomm's package licensing practice, under which Qualcomm did not provide an itemized list of patents that were licensed out, and charged a flat fee for all the SEPs in its portfolio. As patents expired over time, new patents were added to replenish the package. Qualcomm maintained the same license fee without providing evidence that the new patents added to the portfolio were of equivalent value to the expired patents. Qualcomm did not provide an itemized list of patents which it licensed, and licensees were not given the opportunity to negotiate to avoid paying royalties for expired patents. From the decision, it can be gleaned that the NDRC particularly objected to Qualcomm's failure to provide an itemized list of licensed patents, to assess and indicate whether the newly added patents were needed or valuable to the licensees, and to compare the change in value of the expired patents against newly added patents.

The NDRC alleged that Qualcomm compelled its licensees to grant royalty-free licenses to the licensees' non-wireless SEPs as part of the consideration for Qualcomm's licenses. Non-wireless SEPs pertained to cases, screens, cameras, microphones, amplifiers, batteries, internal storage, and operating systems of wireless communications terminals. The NDRC objected to this behavior because Qualcomm did not consider and assess the value of the licensees' patents and refused to pay reasonable consideration for the patents licensed back by the licensees. In response, Qualcomm offered three defenses: (1) the license grantbacks were necessary to protect Qualcomm's own business and its baseband chip customers from patent infringement concerns, (2) Qualcomm's request for royalty-free grantbacks was part of the consideration in Qualcomm's licensing agreements with the licensees, and (3) many licensees in China did not own patent portfolios of sufficient value for exchange with Qualcomm. The NDRC rejected all of them. In particular, the NDRC took offense to Qualcomm's third defense that Chinese companies did not have patent portfolios of value, arguing that many of them had valuable patents.

The NDRC also accused Qualcomm of tying the licensing of wireless SEPs with the licensing of non-wireless SEPs. A related accusation is that Qualcomm charged unfairly high license fees by including non-wireless SEPs in its license package and by calculating the fees based on the net wholesale prices of the wireless communications terminals. There are two components to this charge. The first is that the license fees were artificially inflated through the inclusion of the non-wireless SEPs. That is in fact another way to characterize the tying claim. The second is an objection to the basis for the calculation of the license fees. In response to the tying claim, Qualcomm offered three defenses. First, Qualcomm argued that it had in fact offered licensees the option of licensing only wireless SEPs, but most licensees nonetheless chose the bundle. Qualcomm effectively argued that there was no tie. The NDRC rejected this defense on the grounds that it was not borne out by the factual record. Second, Qualcomm argued that it was difficult to distinguish between wireless SEPs and non-wireless SEPs, and the licensees would be exposed to litigation risks if they only licensed the wireless SEPs. This amounts to an argument that wireless SEPs and non-wireless SEPs constitute one product on transactional costs grounds. Third, Qualcomm asserted that the tie did not in fact restrict competition and the licensees could still choose alternative technologies. In other words, the tie did not have a foreclosure effect on the market for the licensing of non-wireless SEPs. These defenses were both rejected by the NDRC.

Lastly, Qualcomm allegedly required licensees to undertake not to challenge the validity of the agreement before it would supply baseband chips to the licensees. If a dispute arose between Qualcomm and a licensee over the licensing agreement and litigation ensued, Qualcomm would cease the supply of baseband chips. Although the decision referred to the validity of the licensing agreement, it presumably referred to the validity of the underlying patents. Qualcomm admitted to the imposition of these so-called no-contest clauses or no-challenge clauses in the licensing agreements, but it argued that its conduct was justifiable. The NDRC did not detail what the justifications were but dismissed them as insufficient. According to the NDRC, it is within the licensees' right to challenge or institute litigation with respect to the licensing agreements. However, Qualcomm's imposition of the no-challenge clauses restricted the availability of this right to the licensees, if not depriving them of it outright. Moreover, the NDRC argued that competition was restricted when potential licensees which were unwilling to accept the no-challenge clauses were excluded from the market.

7.5. CONCLUSION

Given the size of the Chinese economy and the importance of its market, antitrust law enforcement is bound to take on added significance in the global antitrust community. And with the involvement of Chinese companies in the manufacturing of many technological products, which makes it necessary for them to obtain licenses to patents, patent licensing disputes are destined to arise and the IP–antitrust interface will be particularly relevant. China has taken a measured approach to the interface, with its enforcement authorities taking rather a long time to issue guidelines and to take on AML cases involving intellectual property rights. However, much remains uncertain about the law on the IP–antitrust interface in China. The SAIC IP-antitrust Provisions are typical of most AML-related guidelines and regulations in their brevity and vagueness. Although the Provisions indicate the

SAIC's attitude toward a range of patent exploitation practices, how the law will be applied in practice remains to be seen. The Guangdong court's *Huawei v. Interdigital* case pits a Chinese licensee against an American nonpracticing licensor. Even though the setting could easily have lent itself to charges of protectionism, this author believes that the decision was largely consistent with sound competition law principles. In particular, the Chinese court's attitude toward a SEP holder obtaining an injunction against putative licensees is consistent with that of the European courts and the emerging thinking of the Japan Fair Trade Commission, as indicated in its draft IP-competition guidelines. The NDRC Qualcomm decision is more controversial and complicated and a detailed evaluation of the decision is beyond the scope of this chapter. Suffice it to note for now that some of the issues dealt with in the decision, such as no-challenge clauses, are rather cutting-edge and have not been examined in detail by more advanced jurisdictions. This shows that despite their relatively short history, the AML enforcement authorities do not shy away from challenges and are not hesitant to chart their own territory.

REFERENCES

Bloch, Gabriel, and Gordon A. Milner. 2013. Compulsory Patent Licensing in China. Available at www.lexology.com/library/detail.aspx?g=3d04e562-16df-49b8-b88c-6cd535adffc9.

Brulotte v. Thys Co., 379 U.S. 29 (1964).

Genevaz, Simon. 2004. Against Immunity for Unilateral Refusals to Deal in Intellectual Property: Why Antitrust Law Should Not Distinguish Between IP and Other Property Rights. *Berkeley Technology Law Journal*, 19, 741–84.

Hovenkamp, Herbert, Mark D. Janis, Mark A. Lemley, and Christopher R. Leslie. 2009. *IP and Antitrust: An Analysis of Antitrust Principles Applied to Intellectual Property Law*. Austin, TX: Aspen.

Huawei v. Interdigital, Westlaw (Guangdong High People's Ct., Oct. 21, 2013).

Illinois Tool Works, Inc. v. Independent Ink, Inc., 547 U.S. 28 (2006).

Image Technical Services v. Eastman Kodak, 125 F.3d 1195 (9th Cir. 1997).

Jamaica Fair Trade Commission. 2006. Response to Unilateral Conduct Working Group Questionnaire. International Competition Network 4. Available at www.internationalcompetitionnetwork.org/uploads/questionnaires/uc%20objectives/jamaica%20response.pdf.

Jefferson Parish Hospital No. 2 v. Hyde, 466 U.S. 2 (1984).

Kaplow, Louis. 1984. The Patent-Antitrust Intersection: A Reappraisal. *Harvard Law Review*, 97, 1813–90.

Kimble v. Marvel Entertainment, LLC, 576 U.S. __, 135 S. Ct. 2401 (2015).

Lear v. Adkins, Inc., 395 U.S. 653 (1969).

MCI Communications v. AT&T Co., 708 F.2d 1081 (7th Cir. 1983).

Miller, Alan S. and Michal S. Gal. 2015. Licensee Patent Challenges. *Yale Journal on Regulation*, 32, 121–60.

Ministry of Industry and Information Technology. 2011a. Rules on Regulating the Market Order of Internet Information Services. Available at www.miit.gov.cn/n11293472/n11293832/n12843926/n13917012/14414975.html.

  2011b. An Interpretation on the Rules on Regulating the Market Order of Internet Information Services. Available at www.miit.gov.cn/n11293472/n11293832/n12771663/14417071.html.

National Development and Reform Commission. 2015. NDRC Price Supervision Bureau Penalty [2015] No.1. Available at http://jjs.ndrc.gov.cn/fjgld/201503/t20150302_666170.html.
National People's Congress Standing Committee. 2007. Anti-Monopoly Law. Available at www.saic.gov.cn/zcfg/fl/200712/t20071206_45811.html.
National People's Congress Standing Committee. 2008. Patent Law. Available at www.sipo.gov.cn/zcfg/flfg/zl/fljxzfg/200812/t20081230_435796.html.
Qihoo v. Tencent, Supreme People's Court, (2013) Min 3 Zhong No. 4.
Rambus, Inc. v. Federal Trade Commission, 522 F.3d 456 (Fed. Cir. 2008).
State Administration of Trade and Commerce. 2015. Provisions on Prohibiting the Abuse of Intellectual Property Rights to Exclude and Restrain Competition. Available at www.saic.gov.cn/zcfg/xzgzjgfxwj/201508/t20150820_160841.html.
State Intellectual Property Office. 2001. Implementing Regulations of the Patent Law. Available at http://english.sipo.gov.cn/laws/lawsregulations/200804/t20080416_380326.html.
  2012. Order of the Director of State Intellectual Property Office No. 64 – Measures for Compulsory Licensing of Patent Implementation. Available at www.wipo.int/edocs/lexdocs/laws/en/cn/cn191en.pdf.

8

Intellectual Property and Antitrust in Japan

Masako Wakui

8.1. INTRODUCTION

It has been a long time since Japan was called "the world's factory." Since the 1990s, the notion that Japan cannot maintain its economic strength and has to change to steer its economy to be an innovation-driven one has been widespread. Accordingly, measures to strengthen intellectual property (IP) protection have been introduced.[1] Where a strong IP system exists, there is a potential for encountering serious antitrust issues. The chapter examines how Japan maintains the balance (or does not) between IP and antitrust.

One of the characteristics of IP-related antitrust enforcement in Japan is the lack of legal precedent. As we shall see in detail below, there are only a few formally filed cases both with the Japan Fair Trade Commission (JFTC) and the courts in Japan. As a result, the JFTC's guidelines pertaining to IP and licensing are an essential source of legal rules. The current and most relevant JFTC guidelines are the Guidelines for the Use of Intellectual Property under the Antimonopoly Act (September 28, 2007, as amended on January 1, 2010, hereinafter referred to as IP Guidelines). In the following analysis, too, I often refer to the IP Guidelines.

8.2. RELEVANT PROVISIONS

The most relevant provisions in the Antimonopoly Law (AML) are Articles 3 and 19. While Article 3 prohibits practices that have significant anticompetitive effects in the relevant market either by way of horizontal agreement or exclusion, Article 19 prohibits miscellaneous conducts that tend to inhibit a free and fair competition order, such as tying, exclusive dealing, and resale price maintenance. The basic outlines of these articles are explained below.

---

[1] Taplin (2009).

8.2.1. Article 3

The first part of Article 3 prohibits private monopolization, which is constituted of three elements, namely, (1) anticompetitive exclusion or control of another person's business activities, (2) substantially restricting the competition in a particular field of trade, and (3) acting against the public interest (Article 2(5)). The second part of Article 3 prohibits unreasonable restraint of trade, covering the following elements: (1) agreement or communication of intent between entrepreneurs (or undertakings), (2) imposition of restriction on a member's business activities, (3) substantially restricting competition in a particular field of trade, and (4) acting against the public interest (Article 2(6)). Under this provision, "entrepreneurs" has been interpreted as competitors only, and therefore, the second part of Article 3 is applicable to horizontal agreement.[2] The first part of Article 3, in contrast, can be applied to horizontal and vertical agreement and an individual firm's unilateral conduct.[3] Price fixing and market allocation among competitors, for example, are dealt with in the first part, while refusal to license and tying are tackled in the second part.

Restricting the competition is a common element of both prohibitions. Under case-law, this means "a state in which there actually appears or at least is going to emerge a situation in which a specific entrepreneur or groups of entrepreneurs can control the market by controlling the price, quality, quantity or other conditions freely at its own will to a certain degree as a result of reducing competition in a market,"[4] and a decided Supreme Court case explains that this implies establishment, maintenance, or enhancement of market power.[5]

Another common element is acting against the public interest. Although the Supreme Court once suggested that parties accused of anticompetitive practices may negate the condemnation by demonstrating the existence of overriding public interest,[6] there has been no case in which unlawfulness was denied solely on this ground[7] and thus, the element is not significant in practice.

Competitive effect on the market is assessed on a case-by-case basis by considering the relevant parties' market shares, market concentration, potential for new entrants, and other competitive constraints from neighboring markets and customers, as well as the procompetitive effect. The efficiency achievable through innovation and licensing activities has been regarded as a classic example of potential procompetitive effect.[8]

---

[2] Takigawa (2013).
[3] Lin and Ohashi (2015).
[4] *Toho/Subaru* Case, Tokyo High Court Judgment of September 19, 1951, Shinketsushu 3:166; *Toho/Shin-Toho* Case, Tokyo High Court Judgment of December 7, 1953, Shinketsushu 5:118.
[5] *NTT-East* Case, Supreme Court Judgment of December 17, 2010, Shinketsushu 57:215.
[6] *Oil Price Fixing (Criminal)* Case, Supreme Court Judgment of February 24, 1984, Keishu 38:4:1287.
[7] Tsuchida (2014).
[8] See, e.g., JFTC, Guidelines Concerning Joint Research and Development under the Antimonopoly Act (1993) (hereinafter referred to as Joint R&D Guidelines); JFTC, Guidelines for Exclusionary Private Monopolization under the Antimonopoly Act (2009).

The relevant market, or "particular field of trade" to use the statutory term, may be a product, service, or technology market. The act of innovation, or the research and development activity, itself is not regarded as a relevant market because it is not tradable.[9] Yet, a strong ability to innovate might be considered when assessing the competitive effect.[10]

The IP Guidelines set the soft safe harbor by stating that generally, the anticompetitive effect is unlikely to occur in cases where the market share of the defendant is lower than 20 percent or in the presence of more than four competitive technologies in the market.[11]

The JFTC would issue a cease and desist order and a surcharge payment order against the infringer (Articles 7 and 7-2). Although the JFTC has wide discretion when crafting cease and desist orders, such an order must be necessary to eliminate unlawful conduct only and not its effects. Consequently, the order to reduce the royalty rate, which has increased because of unlawful conduct, cannot be a part of the cease and desist order.[12]

Although Article 2(6) is applicable to hard-core cartels and cases of collaboration among competitors, no formal case has been filed in relation to the latter.[13] As for violation of Article 2(5), no surcharge order existed until the 2005 AML amendments came into force, and since the amendment, Article 2(5) violation has been found in only one case.[14] This was an atypical controlling activities case in which price fixing among competitors, too, was involved and the surcharge payment order was not issued. This implies that there is no Article 2(5) case where surcharge was imposed.[15]

Criminal penalty may be imposed for an Article 3 violation, but this is the only case in which hard-core cartels were found in practice[16] and thus far, criminal penalty has not been imposed in any IP rights case.

Private enforcement, too, has been rare in intellectual property right (IPR)-related cases. Although the aggrieved party may bring a suit for damage, the chance of obtaining a sufficient amount of damage tends to be limited in general.[17] As far as is known, there has been only one IPR-related case in which a private suit was brought (*Hokkaido Shimbun* Case, see Section 8.4.1).

---

[9] Hirabayashi (1993).
[10] See, e.g., Joint R&D Guidelines I.2(1).
[11] IP Guidelines, 2.1(5).
[12] Negishi (2009).
[13] The informal consultation cases exist. See JFTC, Soudan Jirei Shu [Consultation Cases], available at www.jftc.go.jp/dk/soudanjirei/index.html. As for the JFTC consultation procedure, see JFTC, Prior Consultation System for Activities of Business (2001), available at www.jftc.go.jp/en/legislation_gls/imonopoly_guidelines.html.
[14] *JA Fukui* Case, JFTC Cease and Desist Order of January 16, 2015. English summary is available at www.jftc.go.jp/en/pressreleases/yearly-2015/January/150116.html.
[15] Hayashi (2015).
[16] Kawai and Tanaka (2014).
[17] Walle (2013).

8.2.2. Article 19

Under Article 19, practices that reduce free competition, although not to the extent of establishing a market power, are prohibited. Practices that inhibit well-informed, reasonable decisions from being made are also prohibited as unfair methods of competition. Furthermore, practices that impose unjustified disadvantages on a trading partner, often involving retrospectively breaching the agreement between the parties, are regulated because they undermine the system of free competition.[18]

The prohibited practices are listed under Article 2(9)1–5 (inclusive). Additionally, the JFTC can designate certain practice as such by virtue of Article 2(9)6. The so-called JFTC "General Designation (GD)"[19] is relevant to IPRs. All provisions that prescribe the prohibited practices include the anticompetitive effect element, either reduction of free competition, unfair method of competition, or undermining system of free competition, and a violation is found to have occurred only when a relevant anticompetitive effect arising from one of the listed practices is established.

Among the relevant anticompetitive effects, it is more difficult to define employment of unfair methods of competition and/or undermining the system of free competition, and this blurs the standard of lawfulness. These concepts, however, tend to be of limited importance in practice. For instance, in the IP-related case, the surcharge will be imposed mainly on the kinds of practices that are condemned on the ground of reducing free competition.

The concept of reducing free competition, in contrast, is more closely related to market power. Reduction of free competition may be found where (1) establishing, maintaining, or strengthening of market power is likely; (2) minor degree of market power is established, maintained, or strengthened; or (3) the exercise of such power is facilitated.[20] The concept of reduction of free competition thus aims at regulating market power in its incipiency and appears to fit well with conventional competition policy. However, determining whether free competition is sufficiently reduced to establish a violation is not always straightforward. In practice, the relevant market is generally not rigorously defined under Article 19, which implies there is no rigid market share threshold to assess lawfulness. Violation is often found in the case of exclusion by a company that imposes competitive constraints by way of low price or in the case of exclusion of an innovative product or service in the market where competition is not deemed inactive either due to the existence of a large company or because of an oligopolistic market structure. Fortunately, with regard to IPR-related cases, the same soft safe harbor mentioned above is applicable, which creates certainty to some extent.

---

[18] Dokusenkinshi ho kenkyu kai [The AMA study group], Hukosei na torihiki hoho ni kansuru kihon teki na kangaekata [Report on the Basic Principles in Relation to Unfair Trade Practices] (July 8, 1982). For more details, see, e.g., Wakui (2008) 4.1.

[19] JFTC, Designation of Unfair Trade Practices (1982, revised 2009).

[20] Sensui (2015).

The procompetitive effect is relevant in the assessment.[21] For some categories of practice, which include concerted refusal to deal or license and resale price maintenance, an adverse effect is presumed. Yet, this is a rebuttable presumption and the defendant is allowed to argue that the practice has a redeeming procompetitive effect.

The penalty for violation is minor, reflecting the practice's minor anticompetitive effects and the precautionary nature of Article 19. Essentially, the cease and desist order issued by the JFTC under Article 20 is the only sanction imposed in the majority of the cases. The surcharge payment order has been associated with a few types of Article 19 practices since the 2009 AML amendments, but it is applied when an offense is repeated within ten years (Article 20-2–20-5 (inclusive)). Exceptionally, surcharge payment is ordered in a case where a superior bargaining position has been abused since the first violation (Article 20-6); all abuse cases in which surcharges were ordered involved such abuses and, in all cases, powerful retailers unfairly used their bargaining position against their suppliers.[22] There is no IP-related case in which a surcharge order was issued. Criminal penalty is not imposed for violation of Article 19. Private enforcement is inactive in terms of both damages and injunction orders in general,[23] and there is no reported private enforcement case in which the court accepted such a claim against the IP owners on the ground of abusive use of IP contrary to Article 19.

8.3. GENERAL PRINCIPLE AND IP EXEMPTION

8.3.1. *General Principle Between IP and Antitrust*

There is a widely recognized view that IP and antitrust should work in a complementary manner.[24] The basis of this view is explained in the following. While a reasonably competitive market tends to encourage innovation, R&D activities promote competition through new technology resulting in cost reduction or new products. This means that antitrust is good for innovation, while IP is good for antitrust.

The language of the statutes also supports the view. Particularly, the statutes are clear that both the antitrust and the IP protection systems aim to ensure development of the economy. While Article 1 of the AML provides that the law is aimed at "promot[ing] fair and free competition" and thereby "stimulat[ing] the creative initiative of enterprises, encourag[ing] business activity … and thereby promot[ing] the democratic and wholesome development of the national economy," the

---

[21] IP Guidelines, 2.3 and 4.1(2).
[22] Wakui and Cheng (2015).
[23] Walle (2013).
[24] Negishi (1990).

patent law, for example, provides that "the purpose of this Act is ... to contribute to the development of industry."

The aforementioned complementarity is ensured by the IP laws. The IP laws have various safeguards to prevent them from having a negative effect on competition. Such safeguards include limited period of protection, exhaustion principle, and a compulsory license system under IP law. Furthermore, the Intellectual Property Basic Act, in which the national IP framework policy was set out, provides that the IP system should be crafted considering the requirements to be fulfilled for securing fair and free competition (Article 10).

It is true that IP creates the right to exclude competitors from using certain technology and may be used to drive them out of the market. This, however, occurs rarely. In most cases, alternatives to a patented technology are available. Even if a rival is indeed excluded from the exclusion, it is only for a period, and this is often necessary to secure the incentive to engage in R&D.

Additionally, we can say that the AML is harmonious with IP. There has been no IP-related case where a surcharge or a criminal penalty was imposed. Market power has never been presumed to exist merely because of patent ownership both by the JFTC and by the courts. Although the JFTC appeared to view the licensor as having a stronger bargaining position against the licensee when it issued the International Technology Transfer Guidelines in 1968, that view had changed by the time it published the Guidelines for Patent and Know-how Licensing Agreements under the Antimonopoly Act on July 20, 1999.[25] Furthermore, anticompetitive conduct that affects competition has been distinguished from the existence and abuse of market power, and only the former is prohibited. This implies that merely charging a supracompetitive price is not prohibited. As discussed below, a unilateral refusal to deal is generally lawful, regardless of whether it relates to IP. Indeed, arguably there is no case in which unilateral refusal to license was found contrary to the AML; some might say there is one case, but it is only one anyway. The safe harbor set in the IP Guidelines used to be generous relative to the ones set in the guidelines pertaining to non-IP practices. In the non-IP-related case, the threshold was often 10 percent.[26]

8.3.2. The Exemption: Article 21

The AML specifically refers to IP under Article 21, in which it is provided that "the provisions of [the Antimonopoly Law] do not apply to acts found to constitute an exercise of rights under the Copyright Act, Patent Act, Utility Model Act, Design Act or Trademark Act." Unlike its appearance, the outcome of antitrust analysis is hardly

---

[25] Uesugi (1993).
[26] JFTC, Guidelines Concerning Distribution Systems and Business Practices (as of 1991–2015) (hereinafter referred to as Distribution Systems Guidelines). The guideline was amended in May 2016 and the current threshold is 20 percent.

affected by the provision. Yet, the provision defines the analytical framework and thus has some importance.[27]

The "exercise of rights" implies an IP owner's actions to exclude others from using a technology, or mark, for which exclusive rights were granted under the IP act. This includes the following actions: (1) bringing a law suit against or sending a warning letter to the infringer, (2) refusal of license, or (3) licensing a part of the right, while retaining the remainder, such as licensing with a field restriction.

The exercise of rights is generally exempted under Article 21, yet there is a qualification. The exercise has to be legitimate; in other words, the benefit of exemption is not given to practices deemed to be contrary to the aim and purpose of the IP protection system. Legitimacy is assessed in the light of the stated aim in the relevant laws. For example, for the Patent Act, it is encouraging invention. Furthermore, on the basis of the complementarity principle explained above and Article 10 of the Intellectual Property Basic Act, the impact on competition is considered and using rights to suppress competition might be deemed an illegitimate exercise of rights in the light of the aim of IP laws.

Overall, an examination of the IP-related practice is conducted as shown in Figure 8.1. It would be noteworthy that non-application of Article 21 does not always result in AML violation; the action at issue also needs to satisfy other relevant elements, including the element pertaining to the anticompetitive effect. In that anticompetitive effect, the redeeming procompetitive effect is also taken into account.

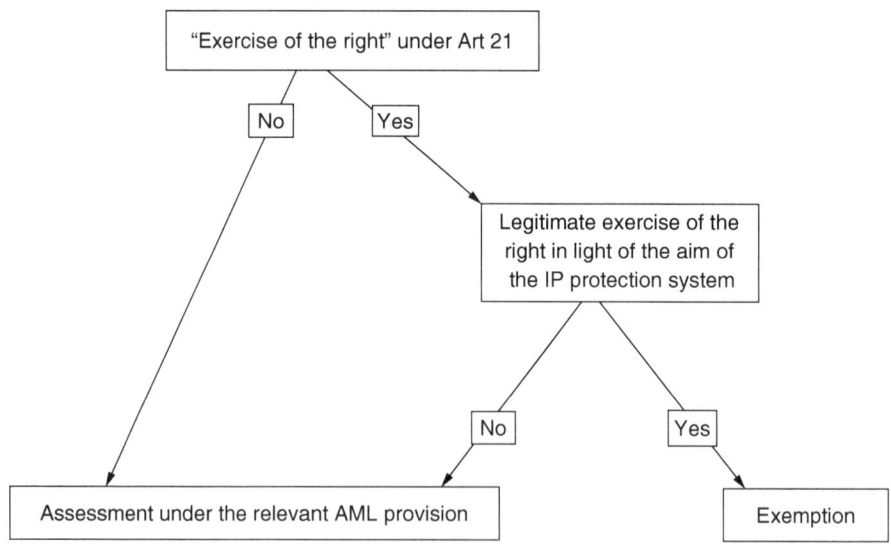

FIGURE 8.1: Examination of IP-related practice under Article 21

[27] The following explanation is in line with the position taken in IP Guidelines, 2.1.

8.4. APPLICATION OF GENERAL PRINCIPLES AND ASSESSMENT

8.4.1. *Acquisition of Rights – Patent and Trademark Application*

Applying for IP rights is of course lawful generally. The practice, however, may constitute an AML violation when the IP system is abused.

In the *Hokkaido Shimbun* case,[28] Hokkaido Shimbun Press, a dominant newspaper publishing company in the region, applied for nine trademarks. This was done only to deter a new entrant and the nine trademarks were selected because they were likely to be chosen by the new entrant. The entrant indeed chose one of the trademarks, and the incumbent continued to harass the new rival using the trademark application as well as through other anticompetitive practices. The incumbent did not have the intention of using any of the trademarks applied for, and the practice was clearly an abuse of the trademark registration system. Together with other exclusive practices, the conduct was found to infringe Article 2(5) and the JFTC issued a cease and desist order. Thereafter, the entrant brought a follow-up suit claiming damages, which resulted in a settlement of JPY 220 million (approximately US$1.8 million) paid by the incumbent.[29] Interestingly, while the JFTC was examining the case, the Japan Patent Office (JPO) rejected the trademark application on the ground of its being contrary to the public interest by virtue of Article 4(7) of the Trademark Act.[30]

8.4.2. *Patent Infringement Lawsuit*

Obviously, bringing a lawsuit against a (likely) infringer is "exercise of the right" under Article 21, and the AML is inapplicable. However, AML violation might be found to occur when the IP system is abused.

Abuse can be found in cases where a patent is obtained by fraud. Another example is the case where the patentee furnishes fraudulent information and misleads the government to adopt his/her patented technology for its standard promulgated for governmental procurement procedures.

The *Paramount Bed* Case[31] is illustrative of the latter scenario. The case relates to the circumstances in which the Tokyo metropolitan government procured beds for its hospitals through public tenders. The beds were required to meet the requirements set by the government. To enhance competiveness in the public tendering process, the Tokyo metropolitan government, as a policy, wrote its specifications in such a way that all major bed manufactures could meet the specifications. There

---

[28] JFTC Consent Decision of February 28, 2000, Shinketsushu 46:144.
[29] Kyodo News, '22000 man yen Shiharai Wakai [JPY 220 million payment to settle]', available at www.47news.jp/CN/200610/CN200610240100466.html.
[30] JPO Decision to dismiss a statement of dissatisfaction with a decision of dismissal on March 10, 1999.
[31] JFTC Recommendation Decision of March 31, 1998, Shinketsushu 44:362.

were three bed manufacturers, and Paramount Bed Co., Ltd was one of them. Despite knowledge of the abovementioned Tokyo metropolitan government public tender policy, Paramount Bed induced the official who was in charge to write the specification to include designs that were covered by Paramount Bed's intellectual property (utility model patent, etc.). Obviously, Paramount Bed did not license the IPs to the competitors, and as a result, the beds manufactured by the competitors were excluded from the public tendering process. The JFTC found that Paramount Bed's conduct was an illegal exclusion under Article 2(5) and issued a cease and desist order.

Standard essential patents (SEPs) have been debated fiercely. Some argue that pursuing infringement action in relation to SEPs may constitute AML violation in cases where the patentee is committed to grant licenses on a fair, reasonable, and non-discriminatory basis and the potential licensee has been willingly and reasonably negotiating for a license. On one occasion, the Tokyo High Court adjudged that in such a situation, the patentee's claim for an injunction and damages was unacceptable under the Japanese patent and civil law because such a claim was against the good faith principle.[32] Some argue that such a lawsuit should also constitute AML infringement. The JFTC, too, amended the IP Guidelines to clarify such conducts would constitute AML violation in 2016.[33]

In any case, mere abuse of the IP system cannot be a basis for AML infringement, and the practice should be assessed further in the light of the relevant AML provisions.

8.4.3. Unilateral Refusal to License

Unilateral refusal to license is "exercise of the right" for the purpose of Article 21. Furthermore, unilateral refusal to deal is generally lawful, regardless of whether it is IP related. Exceptionally, AML violation may be found in cases where the patent was obtained by fraud. Additionally, the argument pertaining to bringing lawsuits relating to the SEPs applies here.

The *Daiichi Kosho (DK)* case[34] is controversial, and some may argue that the case exemplifies the unilateral termination of copyright license as a possible AML violation. Daiichi Kosho is a Kara-OK system provider and owns a few record company subsidiaries. Having been sued by a patentee, DK decided to terminate its subsidiaries' copyright license contracts with the patentee's subsidiary company, which is also a Kara-OK system provider. It was DK's retaliation to the patent lawsuit, and DK explicitly informed the rival's customer of the fact and warned that

---

[32] *Apple v. Samsung*, Tokyo High Court Judgment of May 16, 2014, Hanrei-Jiho 2224:146.
[33] JFTC, Press Release, Partial Amendment of "Guidelines for the Use of Intellectual Property under the Antimonopoly Act," www.jftc.go.jp/en/pressreleases/yearly-2016/January/160121.html.
[34] JFTC Examination Decision of February 16, 2009, Shinketsushu 55:500.

the rival company would be unable to use some music. The JFTC found that DK had employed unfair methods of competition and that the series of practices employed reduced free competition given the significance of the copyright at issue. Based on these findings, the JFTC concluded that DK's practices constituted a GD Item ex 15 (current ex 14) violation, but it did not issue any remedial order because DK had already ceased the conduct.

Some view the *DK* case as not being an ordinary unilateral refusal of license because the refusal was retaliatory in nature, and in addition to the refusal, the acts of warning and informing the rival's customers of the fact were involved. However, one may doubt whether these factors are sufficient to characterize otherwise lawful conducts as unlawful. The condemnation was based partly on unfair method of competition, which is a generally unclear concept that is difficult to define. Furthermore, in the present case, the use of IP as a retaliatory measure was viewed as unfair. Considering the prevalence of IP-related conflicts, the appropriateness of such evaluation is doubtful.[35] In relation to the second ground of condemnation, namely, the reduction of free competition, the relevant market was not rigorously defined and the likelihood of the market power was not thoroughly established.[36] Overall, it is considered that the JFTC assessment was not thorough enough in view of the significant impact the case could have on IP licensing practices.

8.4.4. Restrictive Licenses

Patentee practices that restrict licensee activities are grouped into two categories depending on whether the licensee infringes the patent right after violating the restrictive condition set under the license agreement. When a license is granted only for manufacturing PCs, using the patented technology for manufacturing TVs would amount to patent infringement. In such a case, the license is granted with the so-called "field-of-use" restriction, and a practice of this type amounts to "exercise of the right" under Article 21 because the practice is simply to grant a license to a part of the right while retaining the remainder.[37] Similarly, setting the license time period and territorial scope is "exercise of the right" for the purpose of Article 21 and, thus, generally lawful, unless some abusive use of the IP system is found.[38] For example, abusive use may be found in cases where territorial restriction is used to disguise a market-divining cartel. This is in contrast to the case where such IP infringement would not be found in any case, where Article 21 is inapplicable.

---

[35] Shiraishi (2009).
[36] Sensui (2010).
[37] IP Guidelines, 3.1.2 and 4.3.
[38] *Id.*, 3.1.2 and 4.3.

8.4.4.1. Output and Price Restrictions in Relation to Patented Products

The line between "exercise of the right" and non-exercise is not always clear. As for output restriction, the IP Guidelines generally see it as an exercise of right and consider it lawful,[39] although some argue otherwise.[40] Given the guidelines' position, the AML is generally inapplicable to the patentee's practices to limit the licensee's output of patented products. The exception to the rule may be found in the case of disguised cartels. There are some precedents that are arguably in line with such understanding.[41]

Vertical price restriction in general is unlawful under Article 19, although it may be justifiable on the ground of prevention of free ride. Restricting the licensee's price in relation to licensed products is also taken seriously, and the IP Guidelines say that the practice is generally unlawful.[42] In the *Twentieth Century Fox Japan (TCFJ)* case,[43] the film distributor, who had the right to license copyrights to movie theaters, inhibited theaters from providing discounts on the entry fee. The JFTC found TCFJ to be in violation of Article 19 and issued a cease and desist order.

8.4.4.2. Restriction in Relation to an Expired or Exhausted Patent

The licensee's activity after expiration or exhaustion of a patent cannot constitute IP infringement, and thus, Article 21 is inapplicable.[44] Similarly, once IP is transferred, there is no IP to exercise, and Article 21 is inapplicable. Inapplicability does not imply AML violation, and practices are examined on a case-by-case basis in the light of the relevant AML provisions.

Resale price maintenance (RPM) is not exempted because the relevant IP has been exhausted by first sales. Under the general rule on RPM, the conduct almost always constitutes a violation of Article 2(9)4 (RPM),[45] against which the JFTC issues a cease and desist order, and in the event of repeated violation, a surcharge payment order. There are a not-insignificant number of cases in which illegal RPM was found in relation to exhausted trademarked products.[46]

---

[39] *Id.*, 4.3(2).
[40] Nakayama (2012).
[41] *Hinode-suido (Kita-kyushu)* Case, JFTC Examination Decision of September 10, 1993, Shinketsushu 40:29; *Hinode-suido (Fukuoka)* Case, JFTC Examination Decision of September 10, 1993, Shinketsushu 40:3; *Kaiware* case, JFTC warning of February 17, 1994.
[42] IP Guidelines, 4.4(3).
[43] JFTC Recommendation Decision of November 25, 2003, Shinketsushu 50:389.
[44] IP Guidelines, 2.1.
[45] Distribution Systems Guidelines, 2.1(1).
[46] *Nike Japan* Case, JFTC Recommendation Decision of July 28, 1998, Shinketsushu 45:130; *Sony Computer Entertainment* Case, JFTC Examination Decision of August 1, 2001, Shinketsushu 48:3; *Scubapro Asia* Case, JFTC Recommendation Decision of December 26, 2002, Shinketsushu 49:247. However, note that for certain copyrighted products such as books, newspapers, journals, and music CDs, setting resale price is specifically exempted under Art 23(4) of AML.

Apart from price restriction, rarely are cases of vertical restriction thought to reduce free competition in general. In relation to IP, the JFTC has issued a cease and desist order on only two occasions.[47] The case involved transfer of know-how to a Taiwanese manufacturer, and it was understood that the transferor did not own any IP to exercise after the transfer. Despite this, the transferor prohibited the transferee from manufacturing and sales of the relevant product in Japan. Although the JFTC decision did not contain details of the facts and reasons, it is known that the practice effectively deterred the significantly cheaper products from the market in Japan and, thus, the transferor's actions seemed to have an anticompetitive effect.[48] Yet, the practice may have promoted competition by promoting licensing, and some may argue that the JFTC should have considered such an effect more rigorously.

The procompetitive effect associated with protection of interest of the IP owner is considered in assessing the anticompetitive effect, as in other cases. For instance, in the *Mediplorer* case, the Osaka District Court dismissed the licensee's counterclaim alleging that the licensor violated the AML by prohibiting the licensee from selling certain products during the period of license plus nine months after the period. The court found that the restriction was justifiable for protecting the licensed know-how.[49]

8.4.4.3. Tying and Package License

Clearly, tying the other product or technology, which is distinct from the licensed patent, does not amount to "exercise of right" for the purpose of Article 21.[50] Similarly, a package license under which a licensee is obliged to license more than one distinct patent right does not amount to "exercise of rights." Such a practice may have an exclusionary effect and violate Section 2(5) if the practice is found anticompetitive, particularly in the sense that there is no justifiable reason to do so and market power is established, maintained, or strengthened by the practice.[51] The same conduct may also violate GD Item 10 (tying)[52] if (1) two distinct products, namely, tying product and tied product, exist in the light of demand, standard commercial practice, and any associated efficiency; (2) licensor coerces to take these; and (3) the practice has an adverse effect on competition. The adverse effect may be lessening competition or employing unfair methods of competition to undermine the system of fair competition.[53]

---

[47] *Asahi-denka* Case, JFTC Recommendation Decision of October 13, 1995, Shinketsushu 42:163; *Oxirane Chemical* Case, JFTC Recommendation Decision of October 13, 1995, Shinketsushu 42:166.
[48] Tsukada (1996).
[49] Osaka District Court Judgment of April 27, 2006, Handei-Jiho 1958:155.
[50] To that effect, see *Pot Cutter (Patent)* Case, Osaka High Court Judgment of May 27, 2003.
[51] IP Guidelines, 3.1(3).
[52] *Id.*, 4.4(1) and 4.5(4).
[53] Kanai (2015).

In relation to packaging licenses, investigating the existence of distinguished products tends to be more complicated than doing so for tangible products or services. Under the prescribed IP Guidelines, a packaging license is not unlawful so long as it accompanies a justifiable reason, for instance when implementing several patents is necessary for effectuating a specific function. Although the IP Guidelines specifically do not note this point, it is possible to consider that, in that case, there is no distinct product.[54]

In the *Microsoft (Tying)* case,[55] the computer manufacturers who wished to obtain a license for the spreadsheet software "Excel" were compelled to obtain licenses also for word processing software "Word." When the practice was commenced, Microsoft ranked first in the former market, but not in the latter market. The JFTC found that as a result of the practice, Microsoft came into first position in the latter market, specifically by excluding its rival, which used to rank first. The JFTC subsequently issued a cease and desist order to Microsoft Japan.

8.4.4.4. Assign/Grant-back

Although assign- or grant-back obligations can increase a patentee's incentive to license out and promote wider adoption of the improved technology developed by the licensee, the obligation can also be anticompetitive by discouraging the licensee's R&D activities aimed at further improvement or development of the next generation of the licensed technology. Considering these effects, the IP Guidelines take different positions toward assign-back and grant-back obligations, namely, Article 19 violations may be found in case of (1) imposition of assign-back obligation or exclusive grant-back obligation if the technology developed by the licensee is implementable independently of the patented technology originally licensed to the licensee or (2) imposition of the obligation to share ownership in relation to the improvement. In both cases (1) and (2), the practice is in violation of Article 19 when it has an anticompetitive effect. In contrast, the licensee's obligation to grant a nonexclusive license or to inform the licensor of the improvement does not violate Article 19 generally.[56]

8.4.4.5. Non-challenge Clause

A non-challenge clause is generally valid under Japanese patent law. Furthermore, the argument or claim that the licensed patent is invalid may not be accepted by the court as well as by the JPO, for example when the licensee and licensor are in a close cooperative relationship and such a challenge is considered contrary to the

---

[54] Kawahama (2011).
[55] JFTC Recommendation Decision of December 14, 1998, Shinketsushu 45:153.
[56] IP Guidelines, 4.5(8)–(10)(inclusive).

good faith principle or when such a challenge is expressly restricted under the license agreement.[57]

With regard to treatment under the AML, although the IP Guidelines admit that the practice may constitute Article 19 violation, it takes the position that the obligation is generally procompetitive because it tends to help smooth licensing activities and is unlikely to have an anticompetitive effect.[58] Moreover, the IP Guidelines clarify that stipulating termination of the license agreement once the licensee challenges the validity of the licensed patent does not generally violate Article 19.[59]

8.4.5. Restriction of R&D Activities

The IP Guidelines take the position that restriction of R&D activities on the licensee side generally constitutes infringement of GD Item 12.[60] The exception to this rule exists when the licensee is prevented from engaging in joint R&D with third parties and such restriction is necessary to prevent the licensor's know-how from being disclosed to said party. There has been no formal reported case where such a violation was found.

8.4.6. Royalty Payment and Calculation

In most cases, it is not conceivable that charging royalties and setting out the calculation method has an anticompetitive effect and thus violates the AML. Yet, imposing the royalty payment obligation regardless of the usage of licensed technology or setting a flat rate may have a foreclosure effect by discouraging the effort to find or develop alternative technologies.

Such possibility is acknowledged in the IP Guidelines, and it is stated that imposing the royalty payment obligation, which does not relate to the usage of licensed technology, may violate Article 19 when it has an anticompetitive effect.[61] The guidelines further note that such conduct is not unlawful when the calculation method is reasonable. Examples of such reasonableness include cases in which the technology in question is involved in the manufacturing of products or parts of the products and the amount of the royalty is calculated on the basis of output or the sales amount of the final products.

The JASRAC case exemplifies that flat-rate packaging licensing may violate the AML. The Japanese Society for Rights of Authors, Composers and Publishers (JASRAC) is a collective music copyright management organization and was once

---

[57] For more detail, see Nakayama (2012).
[58] IP Guidelines, 4.4(7).
[59] Id.
[60] Id., 4.5(7).
[61] Id., 4.5(2).

a statutory monopolist. After several years of market liberalization, JASRAC still had a dominant position in the market. The JFTC found its practice of selling packaging licenses at a flat rate, according to which users pay the same amount of royalty regardless of the amount of usage, to be anticompetitive and to have an exclusionary effect against new entrants, in particular those who started the business and had an attractive music copyright portfolio, and issued a cease and desist order under Article 2(5).[62] JASRAC disputed the finding, and the JFTC revoked its original order by an examination decision on June 12, 2012. However, the Tokyo High Court revoked JFTC's decision and determined that the practice had an exclusionary effect, which was upheld by the Supreme Court.[63] The courts did not examine other necessary elements under Article 2(5), namely the effect of establishing, maintaining, or strengthening market power in the relevant market. Following the courts' decisions, JASRAC has withdrawn its appeal against the original JFTC's order.

8.4.7. Agreement Among IP Owners

8.4.7.1. Agreement in Relation to IP Usage

Agreement among IP owners who separately own IPs is clearly not "exercise of the right" for the purpose of Article 21. Such practice is examined on a case-by-case basis.

In the *Concrete Pile* case,[64] the companies who owned patents and utility model patents separately agreed upon their share of sales and, on receipt of the order, the companies coordinated in compliance with the agreed shares and determined who should manage sales. The JFTC found that the companies had violated Article 2(6).

8.4.7.2. Agreement in Relation to Licensing

Fixing the royalty rate among IP owners may amount to a violation of Article 2(6) if the IP owners are in a competitive relationship either in the product or the technology market and the agreement has an anticompetitive effect in the relevant market. An analysis is to be made on a case-by-case basis even in the case of an outright price fixing cartel, and the burden of proving the presence of an anticompetitive effect is eased owing to the current practice of allowing quick determination of the relevant market based on the scope of the agreement.[65] As in other cases, the defendant may argue that the practice has a procompetitive effect and, thus, is lawful. There has not been an IP royalty fixing case in which the JFTC issued a

---

[62] JFTC Cease and Desist Order of February 27, 2009, Shinketsushu 55:712.
[63] Tokyo High Court Decision of November 1, 2013, Shinketsushu 60-2:22; Supreme Court Decision of April 28, 2015.
[64] JFTC Recommendation Decision of August 5, 1970, Shinketsushu 17:86.
[65] Shinagawa (2015).

cease and desist order, although there is a published consultation case that suggests that the JFTC will view such practices negatively.[66]

A "collective decision of royalty" may take place in the context of IP-pooling. The issue is examined in Section 8.4.8.

An IP owner's agreement to refuse the license to a particular potential licensee is outside Article 21 as well[67] and is assessed on a case-by-case basis in the light of the relevant AML provisions. There are a few famous AML violation cases, but they are related to patent pooling or joint licensing arrangements and are discussed in Section 8.4.8. The *Concrete Pile* case mentioned earlier (Section 8.4.7.1) is a rare example in which the no-pooling arrangement was involved. In this case, in addition to the share agreement, the IP owners agreed not to grant a license unless the other cartel members agreed upon it. The JFTC saw this practice as a part of illegal conduct and ordered it be stopped.

Concerted refusal to license is not always unlawful. The anticompetitive effect arising from the practice is assessed on a case-by-case basis and the procompetitive effect is taken into consideration in the assessment. Furthermore, the Joint R&D Guidelines make it clear that agreeing not to license IP that is the outcome of the joint R&D effort to third parties is generally lawful.[68]

8.4.7.3. Cross-licensing and Non-assertion of Patent (NAP) Obligation

Cross-licensing should be deemed lawful in general and is not condemned unless, for example, it is in fact a disguised cartel.[69] It is likely that stipulating that the licensee not assert the licensee's patent against the licensor, which effectively works as cross-licensing, will be considered lawful in general. The IP Guidelines stipulate, however, that the practice may constitute a GD 12 violation where it results in strengthening the market power of the licensor or in diminishing licensees' incentive to innovate.[70]

In the *Microsoft (NAP)* case,[71] the JFTC issued a cease and desist order for imposition of the NAP clause on the ground of violation of GD Item 12. In this case, the NAP obligation was applied not only against the licensor but also against Microsoft licensees, which effectively meant virtually all personal computer manufacturers. The scope of the obligation was unclear because it was defined only as any patent that needs to be implemented in Microsoft's operating system (OS) software and the function incorporated in the OS would be under sole control of Microsoft.

---

[66] JFTC Consultation Case #7 (FY 2010), available at www.jftc.go.jp/dk/soudanjirei/h23/h22nendomokuji/h22nendo07.html.
[67] *Ringtone* Case, Tokyo High Court Judgment of January 29, 2010, Shinketsushu 56:498.
[68] Joint R&D Guidelines, 2.2(2).
[69] IP Guidelines, 3.2(3).
[70] *Id.*, 4.5(6).
[71] JFTC Examination Decision of September 16, 2008, Shinketsushu 55:380.

Furthermore, as the source code was not disclosed to the licensees, the licensees could not tell which patents were and were likely to be relevant. Microsoft was expanding the OS' functionality, which led to broadening of the scope of the NAP obligation. The patents to which NAP was applied included the patents related to audiovisual technology, which had been a major source of the revenue for the licensee as well as the means to differentiate their PC from those of competitors. Under the NAP obligation, the patents were made valueless and ceased to function as differentiating technologies. The JFTC found the Microsoft practice was anticompetitive on the ground of its R&D chilling effect on licensees and for strengthening Microsoft's already dominant position.[72]

8.4.8. Patent Pool

Pooling IPs and joint license arrangements among IP owners tends to have a procompetitive effect by saving transaction cost, lowering royalty level by resolving the double marginalization problem, and facilitating the use of IP. The arrangement, however, may have an anticompetitive effect by suppressing competition among IP owners or excluding outsiders.

In the IP Guidelines, as well as in the guidelines on standard setting and patent pool, the JFTC takes the position that the pooling arrangement and the corresponding agreement should be assessed on a case-by-case basis.[73] The importance, or essentiality, of the pooled patent relationship among the pooled patents (in terms of their coverage of competitive technologies or supplemental technologies), and state of competition in the relevant market are taken into account when assessing the competitive effect. Furthermore, whether the patent pool is operated by an independent party may be taken into consideration.

In the *Pachinko Patent Pool* case,[74] the pachinko manufacturers, who collectively hold the dominant position in the product market, accumulated the patents in the pool and made it their policy not to license their patents to new entrants. The pooled patents included the essential patents for complying with the standard set under the relevant law. Further, the pachinko product market was not competitive due to the collusive practices of existing players. The entry of new players, to whom licenses were not granted, had substantial potential to make the market competitive. The JFTC found a violation of Article 2(5) and issued a cease and desist order.[75]

In the *Ringtone* case, five record companies collectively established a company to operate the ringtone music provider business. In the process, the record companies agreed not to grant the copyrights, which were owned separately by various

---

[72] For further details, see Kameoka (2014).
[73] IP Guidelines, 3.1(1) and 3.2(1); JFTC, Guidelines on Standardization and Patent Pool Arrangements (2005) pt 3.
[74] JFTC Recommendation Decision of August 6, 1997, Shinketsushu 44:238.
[75] For further details, see Kameoka (2014). Other cases are explained in Wakui (2004).

companies, to other ringtone companies. The JFTC issued a cease and desist order on the ground of GD [ex] Item 1(1) violation (currently AML Article 2(9)(i) violation).[76] On appeal, the Tokyo High Court sustained the JFTC decision and order.[77] The defendants could have tried to justify the exclusiveness as being necessary for the smooth operation of their joint venture, but they did not. Accordingly, the issue was examined neither by the JFTC nor by the court.

8.5. CONCLUSION

It has been discussed that IP and antitrust should work in harmony in Japan. The above analysis appears to demonstrate that they are indeed working complementarily and in harmony. In particular, antitrust does not seem to be presenting serious obstacles to IP enforcement and licensing activities. The penalty for AML violations is, or has been, weak in general, and superficially, licensees and alleged IP infringers, who are in a position to find and litigate AML violation on the side of IP owners, do not play any significant role. The JFTC's position toward IP enforcement and licensing seems to be generous, as observed, for example, in the lack of presumption of market power, no-question policy for refusal to license, and lax attitude toward the non-challenge clause. The JFTC has filed cases in relation to IP violation only on a few occasions. Generally, the JFTC enforcement effort tends to be concentrated on cartels and bid riggings, and vertical agreements and exclusive practices tend to be less prioritized.[78] Given that IP-related issues tend to be either of the vertical agreement type or the exclusion type, the JFTC's inactivity in the field is understandable. This analysis casts doubt on whether there is imbalance between IP and antitrust in Japan, and the AML's lack, or shortage, of oversight is the key concern in the IP/antitrust field.

Yet, one may criticize the AML for presenting excessive regulation, and inhibiting efficient licensing activities and undermining the incentive to innovate. This is valid criticism considering the broad coverage of Article 19, the low safe-harbor threshold in the IP Guidelines, and the lack of precedents. The lack of precedents is fatal because businesses and practitioners are left not knowing the underlying principles and concrete legal standards, and the way to put forward claims. The recently broadened surcharge system and introduction of the civil injunction system under the AML do not lend sufficient impetus to the relevant parties to pursue legal procedures, and more measures must be implemented to encourage the development of case-law as well as AML private enforcement in the IP/antitrust area.

---

[76] JFTC Recommendation Decision of April 26, 2005, Shinketsushu 52:348; JFTC Examination Decision of July 24, 2008, Shinketsushu 55:294.
[77] Tokyo High Court Decision of January 29, 2010, Shinketsushu 56:498.
[78] JFTC, Annual Reports (2000–2015).

REFERENCES

Hayashi, Shuya. 2015. *Discretionary Surcharge System Model in the Japanese Cartel Law Enforcement*. In Thomas Cheng, Sandra Marco Colino and Burton Ong (eds.), *Cartels in Asia: Law & Practice*. Hong Kong: Wolters Kluwer Hong Kong, 1–39.

Hirabayashi, Hidekatsu. 1993. *Kyodo Kenkyu Kaihatsu ni Kansuru Dokusenkinshi ho Guidelines [Guidelines Concerning Joint Research and Development under the Antimonopoly Act]*. Tokyo: Shojihomu.

Japan Fair Trade Commission. 2000–2013. *Annual Reports*.

Kameoka, Etsuko. 2014. *Competition Law and Policy in Japan and the EU*. Cheltenham: Edward Elgar.

Kanai, Takaji. 2015. *Futo-na Torihiki Kyosei [Unfair compulsory trade]*. In Takaji Kanai, Noboru Kawahama and Fumio Sensui (eds.), *Dokusenkinshi Ho [Antimonopoly Law]*, 5th edn. Tokyo: Koubundou.

Kawahama, Noboru. 2011. *Package License*. In Kazunori Yamagami and Yoshito Fujikawa (eds.), *Chizai License Keiyaku no Horitsu Sodan [IP Licensing in Counselling]*, rev. edn. Tokyo: Seirin-Shoin.

Kawai, Kozo and Nobuhiro Tanaka. 2014. Article 89. In Masahiro Murakami (ed.), *Jokai Dokusenkinshi Ho [Commentary on Antimonopoly Law]*. Tokyo: Koubundou, 817–22.

Lin, Ping and Hiroshi Ohashi. 2015. *Treatments of Monopolization in Japan and China*. In Roger D. Blair and D. Daniel Sokol (eds.), *International Antitrust Economics 2*. Oxford University Press, 188–233.

Nakayama, Nobuhiro. 2012. *Tokkyo Ho [Patent Law]*, 2nd edn. Tokyo: Koubundou.

Negishi, Akira. 1990. *Chiteki Zaisanken Ho to Dokunsenkinsi Ho no Kihonteki Kankei [Basic Principles of Relationship between Intellectual Property Law and Antimonopoly Law]*. In Akira Negishi (ed.), *Dokusenkinshi Ho no Kihon Mondai [Fundamental Issues under Antimonopoly Law]*. Tokyo: Yuhikaku, 185–98.

  2009. Article 7. In Akira Negishi (ed.), *Chusyaku Dokunsenkinshi Ho [A Commentary on the Antimonopoly Law of Japan]*. Tokyo: Yuhikaku.

Sensui, Fumio. 2010. *[Case analysis in Japanese] Daiichi Kosho Jiken*. NBL, 925, 62–71.

  2015. *Kihon Gainen [Basic Concepts]*. In Takaji Kanai, Noboru Kawahama and Fumio Sensui (eds.), *Dokusenkinshi Ho [Antimonopoly Law]*, 5th edn. Tokyo: Koubundou.

Shinagawa, Takeshi. 2015. *Futona Torihiki Seigen [Unreasonable Restraint of Trade]*. In Shuichi Sugahisa (ed.), *Dokusenkinshi Ho [Antimonopoly Law]*, 2nd edn. Tokyo: Shojihomu.

Shiraishi, Tadashi. 2009. *Daiichi Kosho Shinketsu ni Tsuite [On the Daiichi Kosho Decision]*. Kosei Torihiki, 703, 54–9.

Takigawa, Toshiaki. 2013. *Japan*. In Mark Williams (ed.), *The Political Economy of Competition Law in Asia*. Cheltenham: Edward Elgar, 11–46.

Taplin, Ruth. 2009. *Intellectual Property and the New Global Japanese Economy*. London: Routledge.

Tsuchida, Kazuhiro. 2014. *Jyosho [Introduction]*. In Kazuhiro Tsuchida, Makoto Kurita, Yoshizumi Tojo and Kuninobu Takeda (eds.), *Jobun Kara Manabu Dokusenkinshi Ho [Antimonopoly Law: Text, Outline, and Cases]*. Tokyo: Yuhikaku, 1–29.

Tsukada, Masunori. 1996. *Kokusai Gijutsu Enjo Keiyaku ni kakaru Dokusenkinshi Ho Ihan Jiken [AML Violation Cases in relation to International Technology Transfer]*. Kosei Torihiki, 543, 48–51.

Uesugi, Akinori. 1993. *Toi 17 [Question 17]*. In Akinori Uesugi (ed.), *Q&A Tokkyo/Know-how License Keiyaku to Kyodo Kenkyu Kaihatsu [Patent and Know-how License and Joint R&D: Questions and Answers]*. Tokyo: Shojihomu.

Wakui, Masako. 2004. Standardisation and Patent Pools in Japan. In Ruth Taplin (ed.), *Valuing Intellectual Property in Japan, Britain and the United States*. London: RoutledgeCurzon.

——— 2008. *Antimonopoly Law: Competition Law and Policy in Japan*. Bury St. Edmunds: Arima.

Wakui, Masako and Thomas K. Cheng. 2015. Regulating Abuse of Superior Bargaining Position under the Japanese Competition Law: An Anomaly or a Necessity? *Journal of Antitrust Enforcement*, 3(2), 302–33.

Walle, Simon Vande. 2013. *Private Antitrust Litigation in the European Union and Japan*. Antwerp-Apeldorn: Maklu.

9

Competition Law and Intellectual Property in Korea

Hwang Lee

9.1. IMPORTANCE OF IPR AND COMPETITION LAW IN THE KOREAN ECONOMY

Recent years have seen exploding attention on the interface between competition and intellectual property rights (IPR) laws, most notably on the so-called patent wars between global smartphone giants. In this process, novel issues have arisen, such as injunctions sought by standard essential patent (SEP) holders and patent privateering by nonpracticing entities (NPEs). Along with relatively traditional issues like patent ambush and royalty stacking, these issues have become the new center of debate over how far patents should go and what role competition law should play in fixing perceived flaws in the patent system.

Korean economic development has depended heavily on active IPR policies. The importance of IPR policies has been increasing due to needs of increased productivity for continuous development after the depressed growth during the past decades. In fact, ever since the 2008 global financial crisis, Korea has experienced significant economic difficulties. Experts are concerned about the limitations of existing growth strategies while the current government attempts to resolve the issue by emphasizing the so-called "creative economy" as a key economic policy. Along these lines, promotion of IPR is a major initiative while effective control of potential abuse of IPRs is also drawing attention. These issues are more critical in Korea because the

---

This chapter is an updated and edited version of the following book chapter; Hwang Lee, 2014, *Enhancing Enforcement of Competition Law to Prevent Abuse of IPR in IT Industry*, 2013 Knowledge Sharing Program with China: Sharing Experience of the KFTC on Enforcing Competition Laws and Policies, 23–64. The author thanks the Korea Development Institute for allowing the use. The author also thanks Ms. Yee Wah Chin, a Counsel at Ingram Yuzek Gainen Carroll & Bertolotti, LLP, and Mr. Jipil Choi, a researcher at the ICR Law Center, for helpful comments and research. Details and updates of Korean IP and competition law enforcement are expected to be found annually at MENG Yan Bei and LEE Hwang ed., China-Korea IP & Competition Law Annual Report 2014, China-Korea Market & Regulation Law Center, 2015.

IT industry, which is heavily dependent on intellectual property (IP) makes up a large portion of the Korean economy against the backdrop of rampant patent disputes and alleged abuse. As a result, with an advanced IT industry and aggressive competition law enforcement by the Korea Fair Trade Commission (KFTC), novel cases continue to emerge.

9.1.1. Policy Direction

It is largely accepted that current Korean policy agrees on several important issues with other major jurisdictions.[1] First, the goals of competition law and intellectual property law are considered to be along the same lines. Also, major legal regimes share a common goal of promoting innovation and consumer welfare. Second, determining the illegality of IPR licensing requires balancing of the procompetitive effects (such as innovation) and the suspected anticompetitive effects. Third, the mere existence of IPR does not in itself confer market power. Fourth, the purpose of safe harbors is to promote technology transfer.

Korean competition law enforcement is regarded as strict, especially due to its unfairness-based legal precedents and broad language of relevant guidelines that encompass a wider regulatory scope than traditional antitrust enforcement focused on economic efficiency. While regulation in the field of IPR developed later than in some other jurisdictions, the degree of enforcement is in line with them. The KFTC is known for its active enforcement and does not hesitate to navigate novel areas such as FRAND (fair, reasonable and non-discriminatory) and NPEs-related issues.

Yet the evolution of legal and economic thinking strongly supports deference to IPR in most cases, with only a small number of exceptions. In the meantime, judicial decisions and the enforcement activities of KFTC seem consistent with global trends at large. And in a significant move, in 2014, the KFTC issued amendments to the KFTC Guideline on the Unfair Exercise of Intellectual Property Rights ("IPR Guidelines"), which prioritized the application of the abuse of market dominance clause to pursue economic efficiency rather than unfair business practices, as illustrated below.

9.2. RELEVANT PROVISIONS AND INSTITUTIONS

9.2.1. Korea Fair Trade Commission

As an independent regulatory body, the KFTC enforces the Monopoly Regulation and Fair Trade Act (MRFTA) and related laws while also functioning as a quasi-legislative and quasi-judicial body.

---

[1] Noh (2011).

The KFTC has nine commissioners: the chairman, vice chairman, three standing commissioners, and four non-standing commissioners. The Secretariat, with a staff of 528 (as of September 2015), supports the Commission as a working body. Under the Secretariat are multiple bureaus and offices; among them, the Anti-Monopoly Bureau mainly monitors IPR abuse. The KFTC recently hired two patent attorneys to join its corps of experts to cultivate deeper understanding of IPR.

9.2.2. *Cooperation with KIPO and Other Agencies*

The principal authority in charge of administering IP policy is the Korea Intellectual Property Office (KIPO), an executive agency under the Ministry of Trade, Industry and Energy. KIPO is responsible for the examination and registration of patents, utility models, designs, trademarks, and layout designs of semiconductor chips. Also, a tribunal established within KIPO reviews and decides trial cases involving disputes over eligibility of IPR registration and the scope and validity of registered IPR.

The Framework Act on Intellectual Property was enacted in 2011 to ensure uniformity and consistency of policies on IP within the government. According to this law, the Presidential Council on Intellectual Property (PCIP) was established to devise, review, and administer major policies and plans for IP and review and evaluate the progress of such policies and plans. The KFTC chairman is a member of the PCIP.

The KFTC has closely cooperated with KIPO and the PCIP in developing its IPR policies and institutions, but no official cooperation or consultation channel exists for the investigation or review of a case to guarantee independence.

9.2.3. *Statutory Scheme*

Article 59 of the MRFTA provides an exemption clause stating: "The provisions of this Act do not apply to a legitimate exercise of rights under the Copyright Act, the Patent Act, the Utility Models Act, the Design Act or the Trademark Act." The KFTC espouses the view that "illegitimate" exercise of IPR may be subject to sanctions under the MRFTA (IPR Guideline II.2.A).

A debate has lingered over whether the legitimacy of an exercise of IPR should be reviewed from the perspective of IPR law or the MRFTA. In the landmark *GSK* case decided on February 27, 2014, the Supreme Court held that "conduct deemed not to be a legitimate exercise of the patent right" included conduct that ostensibly seems to be an exercise of patent rights, but in substance departs from the fundamental tenet of the patent system and goes against the institution's essential purpose.[2] The illegitimate exercise of a patent right, the ruling said, should be determined in consideration of the objectives and tenet of the Patent Act, the substance of patent

---

[2] Supreme Court 2012du24498, delivered on February 27, 2014 (on the complaint of GSK).

rights, and circumstances including the effects the conduct in question could exert on fair and free competition.

The GSK decision tried to strike a balance between conflicting views. The decision that the MRFTA can apply if the exercise of the patent obviously tends to reduce consumer welfare beyond a tolerable degree from the perspective of competition law is considered to be in line with the KFTC's existing views.[3]

Substantive violations of the MRFTA are classified into four major categories: abuse of market-dominating positions; anticompetitive business combinations; unjust concerted practices; and unfair trade practices. Traditionally the enforcement of the prohibitions of unfair trade practices has been strong, but in recent years, the KFTC has tried to reduce such regulatory scope due to an increasing need to advocate efficiency rather than fairness. It is noteworthy that efficiency is considered increasingly critical, especially related to IPR regulation and foreign firms.

The KFTC adopted the IPR Guidelines to ensure consistency and predictability in enforcing the MRFTA. It was first adopted in 2000 and has been amended four times since then. The IPR Guidelines are widely considered to share similar views to those of the United States or European Union.[4]

9.3. ANALYTIC FRAMEWORK FOR IPR ABUSE UNDER THE KFTC IPR GUIDELINES

9.3.1. Introduction

Although the IPR Guidelines were adopted in 2000, for various reasons, including lack of enforcement experience, they were rarely utilized. To strengthen the basis for enforcement and cover new issues, major revisions were made in 2010 and 2014.[5] A 2015 revision was proposed and under public consultation as of this writing in January 2016.[6]

The IPR Guidelines explicitly recognize that the IPR system and MRFTA share common goals of encouraging creative business activities and promoting sound development of the economy (IPR Guidelines Sec. II.1). It suggests that IPR should be legitimately exercised to the extent that they do not distort market competition, while providing an incentive for new technological innovation.

---

[3] Lee (2014). The current version of the IPR Guideline still uses the term "going beyond the legitimate scope of patent rights" on its face. However, KFTC enforcement is expected to follow the GSK decision, which is similar to the test of the US *Actavis* decision.
[4] Some foreign commentators raise concerns about the IPR Guideline (Kobayashi *et al.* 2015).
[5] An English-language version of the text can be found at the English-language website of the KFTC at http://eng.ftc.go.kr/bbs.do?command=getList&type_cd=62&pageId=0401.
[6] See Section 9.3.12.

9.3.2. *Scope of Application (Sec. I.2)*

The IPR Guidelines apply to the exercise of IPR such as patent rights, utility models, designs, trademarks, and copyrights. Though the Guideline focuses on patent rights that are representative of IPR for the convenience of description, it may be applied by analogy to the exercise of IPR other than patent rights, accounting for the peculiarity of IPR in each case.

The IPR Guidelines shall apply to the conduct of foreign enterprises established outside Korea if such activity affects the Korean market. The MRFTA expressly adopts the "effects doctrine" regarding extraterritorial application (Article 2-2).

The Guideline also warns that even if a certain exercise of IPR is not specified in the Guideline, it shall not be excluded from application of the MRFTA and may be prosecuted under general provisions of the MRFTA.

9.3.3. *Basic Principle*

As described earlier, under Article 59 of the MRFTA, illegitimate exercise of IPR can be subject to the MRFTA. Finding illegitimate exercise of rights based on IPR mainly depends on whether the conduct that appears to be an exercise of patent rights on the surface departs from the fundamental tenet of the patent system and goes against the institution's essential purpose in substance. The relevant inquiry would involve the objectives and tenets of the Patent Act, the substance of IPR, and other circumstances, including the effects the conduct in question could have on competition in the relevant market (Sec. II.2.A of the IPR Guidelines).

One of the most notable features in the 2014 amendment is the declaration of the principle that the IPR Guidelines shall be applied only when a firm possesses market dominance with regard to unilateral conduct, in addition to unjust concerted practices (Sec. II.2.B). The KFTC can still apply general provisions in the MRFTA or the Review Guideline for Unfair Trade Practices, but this revision may significantly reduce the risk of the form-based application of the MRFTA. As described in the recent *Dolby* case[7] (Section 9.4.6), the unfair trade practices provision still has teeth, but this clause signals a major turn in policy direction. The IPR Guidelines also clarify that the theory of harm or unreasonableness within the Guideline means restriction of competition instead of "impeding fair trade" (which is considered to encompass unfairness as well as anticompetitiveness).

The IPR Guidelines provide that the mere possession of an IPR does not confer the presumption of the market power, and consideration should be given not only to the existence of the IPR, but also to the influence of the technology, existence of alternative technologies, and the competitive landscape of the relevant market

---

[7] KFTC Decision No. 2015-125, 2015.8.3.

(Sec. II.2.C). Even a SEP holder is not presumed to have market power, although the IPR Guidelines say the SEP holder is "more likely" to have market power.

If the exercise of IPR results in anticompetitive effects and efficiency-enhancing effects at the same time, finding whether the exercise violates the MRFTA shall be determined, in principle, after balancing both effects. If the exercise of an IPR results in increased efficiency exceeding its anticompetitive effects, such exercise could be considered not to be in violation of the MRFTA (Sec. II.2.D). But clear evidence (as opposed to vague speculation) must show that such effects are highly likely to occur (Sec. III.C).

The notion of restricting competition includes product competition, technology competition and R&D competition between actual and potential market participants (Sec II.2.E).

9.3.4. Approach for Determining Illegality

The IPR Guidelines adopted a four-stage approach to find illegality. In the first stage, the rationale for and concerns of competition in regulating problematic conduct and basic considerations in determining illegality are set forth. The second stage prescribes the applicable provisions of the MRFTA. In the third stage, the type of conduct that can be assessed to constitute abuse is stipulated. This is not a categorical prohibition and each conduct must be subject to the rule of reason approach to find liability under relevant MRFTA provisions in addition to the IPR Guidelines. And in the fourth stage, detailed examples of potentially problematic conduct based on domestic and foreign cases are provided (when deemed appropriate) with modifications to help comprehension of the Guideline.

The 2014 amendment introduces the "innovation market" as a possible relevant market, similar to the US Antitrust Guidelines for the Licensing of Intellectual Property (1995)(Sec. II.3.A.(3)). In the proposed Applied Materials/Tokyo Electron merger, which was abandoned after the KFTC challenged it in cooperation with US and Chinese agencies, the competition in innovation was a major concern,[8] although it was not clear whether the KFTC defined a separate innovation market.

9.3.5. IPR Guidelines' Approach to the Acquisition of the Patent Rights (Sec. III.1).

The IPR Guidelines acknowledge that an agreement on the assignment of patent rights that constitute a main part of a business or an exclusive license that generates

---

[8] See Brent Kendall and Don Clark, "Applied Materials, Tokyo Electron Cancel Merger Plan," *Wall Street Journal*, April 27, 2015, available at www.wsj.com/articles/applied-materials-tokyo-electron-scrap-merger-plan-1430117758.

the same effect will undergo review according to relevant merger review provisions (Sec. III.1.A).

9.3.6. *Abuse of Patent Actions (Sec. III.2)*

According to the IPR Guidelines, legal procedures such as patent infringement actions are important means to enforce patent holders' rights. But such actions incur considerable costs of time and expense, and could also interfere with the parties' business activities by influencing their reputations in the relevant market. Hence, abuse of legal/administrative procedures/actions, including illegitimate patent infringement actions, could be deemed to go beyond the legitimate scope of patent rights. If an infringement action is brought when the patent holder is aware of noninfringement or when it is objectively clear that no infringement has occurred, it is highly likely to constitute IPR abuse. Yet if the patent holder's expectation of legal action is deemed reasonable and justified, the sole fact that the patent holder failed to prevail in legal action does not automatically prove abuse of a patent action.

9.3.7. *IPR Guidelines' Approach to Licensing Terms*

9.3.7.1. Royalty for License (Sec. III.3.A)

The IPR Guidelines recognize that obtaining a patent through innovative development of technology requires a considerable amount of time, money, and investment risk. Considering the need for just compensation for patent holders' technological accomplishments and inducing new technological innovation, the IPR Guidelines generally consider imposition of royalties as a legitimate exercise of patent rights.

On the other hand, practices likely to restrict competition in the relevant market, such as making unreasonable demands for payment in exchange for licensing, can be considered to go beyond the legitimate scope of patent rights:

(1) an act of unduly deciding, maintaining, or changing royalties in collaboration with another enterprise;
(2) an act of discriminatorily charging royalties depending on factors such as the licensee's identity;
(3) an act of unduly charging royalties for parts that have not used the technology licensed;[9]

---

[9] The act of including the use of a competitor's technology in calculating royalties for licensed technology can increase the cost of using the competitor's technology; moreover, it can decrease demand for the technology, which is most likely to be considered an anticompetitive act. Exceptions can be made when such methods of calculating royalties are acknowledged as inevitable (e.g., due to limitations in measuring the volume of licensed technology use) (Sec. III.1.A.(3) note).

(4) an act of unduly charging royalties for a term extending beyond the expiration of patent rights; and
(5) an act of allowing the patent holder to decide or change royalty calculation methods without clearly stating them in the contract.

One of the notable changes in 2014 was the deletion of the sub-clause which prohibited "an act of charging an excessively unreasonable level of royalty in light of customary commercial practice" although the clause reappeared with regard to SEPs (see Section 9.3.9) and in the new chapter on NPEs (see Section 9.3.11). This does not necessarily mean that the KFTC excluded exploitation as a theory of harm for IPR abuse, but it is considered meaningful progress.

9.3.7.2. Refusal to License (Sec. III.3.B)

To provide just compensation for a patent holder's technological accomplishments and promote new technological innovation, the patent system grants a patent holder exclusive rights over the use of the relevant invention. Hence refusal to grant a license by a patent holder can be considered a legitimate exercise of patent rights.

But practices such as unduly refusing to license can be considered to go beyond the legitimate scope of patent rights:

(1) an act of refusing to license a certain enterprise together with a competitor without a legitimate reason;
(2) an act of unduly refusing to license a certain enterprise;[10] and
(3) an act of refusing to license to secure the effectiveness of other unfair acts, such as refusing to grant a license to another party if the latter rejected unreasonable conditions imposed by the patent holder.

9.3.7.3. Limitations on License Scope (Sec. III.3.C)

A patent holder may not only refuse to license within a reasonable scope, but may also set boundaries on the use of patented inventions, allowing only partial implementation. This limitation can result in procompetitive effects of promoting trade of a patent holder's technology when the owner would otherwise refuse to grant the license. Hence a patent holder's granting of a license with reasonable limitations (e.g., number of licenses, territory, or period for protecting the patent holder's rights) can generally be considered a legitimate exercise of patent rights. But practices such

---

[10] In cases where (1) the purpose of the refusal to trade relates to restraining competition, (2) the refused technology is an essential element of business activities, (3) the technology is extremely influential in the relevant market (e.g., technology standard), (4) use of the technology has been excessively impeded through a refusal to grant licenses; though the patent holder itself did not have the intent to use the technology, the act will probably be considered unreasonable (Sec. III.3.B.(2) note).

as unfair restriction on the scope of the license can be considered to go beyond the legitimate scope of patent rights:

(1) an act in which the patent holder and licensee makes an agreement with unreasonable conditions, such as restricting the number of licenses, territory, and period related to either products concerned with the license ("contracted products") or technology ("contracted technology");
(2) an act of discriminatorily restricting the number of licenses, location, and duration related to the contracted products or technology depending on the other party to the transaction.

9.3.7.4. Imposition of Conditions when Licensing (Sec. III.3.D)

In addition to being able to restrict the scope of use of patented inventions by granting a partial license, a patent holder may impose conditions not directly related to specifying the scope of the license. In general, acts of a patent holder imposing terms within a reasonable scope to effectively implement the relevant patented invention, enhance the safety of contracted products, or prevent the technology's misappropriation can be considered a legitimate exercise of patent rights.

But imposition of unreasonable conditions when licensing (as listed below) can be considered to go beyond the legitimate scope of patent rights:

(1) restrictions on prices of contracted products;
(2) restrictions on suppliers of raw materials;
(3) restrictions on purchasers of contracted products;
(4) restrictions on trade in competing goods or technology;
(5) tie-in sales;[11]
(6) imposition of no-challenge obligation;
(7) restrictions on improving technology and research activities;
(8) restrictions on use after expiration of patent rights; and
(9) provisions on termination of contract or disputes.

When determining whether the patent holder's addition of terms in licensing a grant is anticompetitive, the following factors shall be considered:

(1) whether the added conditions are essential in implementing the relevant patented invention (i.e., relationship between patented invention and imposed conditions),

---

[11] A note, newly added in 2014, distinguishes procompetitive package licensing from tie-in sales that are coerced purchases of unnecessary patents, and provides that it is likely to be deemed unreasonable to require the licensee to accept unnecessary non-SEPs on the condition of licensing SEPs.

(2) whether the conditions contribute to promoting implementation of the relevant technology, and
(3) whether the patent rights relating to the conditions have been exhausted.

9.3.8. Treatment of Patent Pools and Cross-Licensing (Sec. III.4)

A patent pool refers to an arrangement of multiple patent holders putting together individual patents for cross-licensing or collective licensing to third parties. The IPR Guidelines recognize that patent pools can have the procompetitive effects of boosting efficiency in relevant markets and promoting technology use by cutting the search cost in the relevant technical field or expenses needed for negotiating with multiple patent holders, while also reducing the risk in technology use caused by infringement through integrated operation of complementary technologies

But certain practices of patent pools (as listed below) can be considered to go beyond the legitimate scope of patent rights:

(1) an act of unduly agreeing on certain related terms such as price, quantity, territory, licensees, and restriction on technology improvement during the operation of the patent pool;
(2) an act of unduly refusing to grant licenses or executing a license agreement containing discriminatory terms with other enterprises that did not participate in the patent pool;
(3) an act of unduly requiring the other enterprise to share its independently acquired knowledge, experience, and technical achievements during the operation of the patent pool;
(4) an act of coercing package licensing by unduly including invalid or non-essential patents in the patent pool;
(5) an act of causing excessive disadvantage to the licensee by imposing a lump-sum royalty that is excessively higher than that for each patent included in the patent pool.

Specifically, determination of illegality in the exercise of IPR in a patent pool is to be based mainly on technologies that comprise the pool, the form in which the technologies are licensed, and method of operation.

The exercise of rights of a patent pool will most likely be deemed unreasonable if the technologies that comprise such a pool can be mutually substituted or the patent pool includes non-essential or invalid patents.

The exercise of rights of the patent pool will most likely also be deemed unreasonable if package licensing of the relevant technologies is the only method offered and independent licensing of each technology is prohibited.

On the method of operation, the exercise of rights of the relevant patent pool will less likely be deemed unreasonable if the patent pool is independently operated by a specialist group separate from the patent holders.

Cross-licensing is an arrangement among multiple patent holders to grant mutual licenses for patents held by each patent holder and is often used as a settlement tool, especially in patent disputes. Cross-licensing is similar to patent pools in that it can restrict competition through enabling collaborative acts between enterprises and excluding third-party competitors, notwithstanding its procompetitive effects (such as promoting technology use and cutting transaction costs). Accordingly, the provisions applicable to patent pools under Sec. 4.A.(1), (2) and (3) of the Guideline may be applied with proper modifications to cross-licenses.

9.3.9. Special Issues in the Exercise of Standard-related Patent Rights (Sec. III.5)

9.3.9.1. In General (Sec. III.5.A)

Once a technology is adopted as a standard, it can have enormous and lasting influence in the relevant market if the standard is widely adopted, partly because considerable switching costs may be necessary to replace it. If a technical standard is protected as a patent with exclusive/monopolistic characteristics, this could seriously restrict competition in the relevant market. To address these concerns, numerous standardization organizations usually require relevant patent data to be disclosed in advance before adopting technical standards. Prior consultation is also needed to ensure FRAND licensing terms if the technology is protected by patents. The disclosure of and prior consultation on licensing terms are necessary to prevent abuse of SEPs, and compliance therewith is an important consideration when judging legitimacy in the exercise of the patent rights related to the technical standard.

Generally, consultation on selection of technical standards and exercise of patent rights related thereto create procompetitive effects in that they promote use of the relevant technology and contribute to consumer welfare by creating efficiency.

But abuse of the standardization procedure or demanding unreasonable terms after adoption of the technical standard can be considered to go beyond the legitimate scope of patent rights:

(1) an act of unduly agreeing on certain terms such as price, quantity, territory, licensees and restriction on technology improvement during consultation on technical standard selection;
(2) an act of unduly refusing to disclose information about the related patent application or registered patents owned by oneself to increase the possibility of being selected as the technical standard or avoid prior consultation on licensing terms;

(3) an act of avoiding or circumventing licensing on FRAND terms to strengthen market dominance or to exclude competitors;[12]
(4) an act of unduly refusing to grant licenses for SEPs;
(5) an act of unduly discriminatory licensing terms for a SEPs or imposing an excessive royalty;
(6) an act of imposing conditions unduly restricting the exercise of the patent rights held by licensees or unduly requiring the licensees to cross-license their non-SEPs.[13]

9.3.9.2. Injunctions Sought by SEP Holders (Sec. III.5.B)

The KFTC became the first competition authority to adopt an official guideline for injunctions sought by SEP holders, while Japan and Canada are still undergoing a consultation process. The basic principle is that, when a SEP holder who voluntarily offered a FRAND commitment to standard setting organizations (SSOs) seeks to enjoin a willing licensee from practicing the patent, this can be considered to go beyond the legitimate scope of patent rights. Beside the willingness on the licensee's side, the IPR Guidelines also require good faith negotiation efforts by SEP holders. In recognition of the possibility of a reverse holdup, the Guideline notes that an injunction against an unwilling licensee is less likely to violate the MRFTA. The IPR Guidelines also state that when an injunction is recognized as the only viable remedy for certain reasons (such as impending bankruptcy of the potential licensee), it is less likely to be found illegal.

9.3.10. Anticompetitive Agreements during Patent Disputes (Sec. III.6)

Settlements covering a patent's validity and infringement can be effective for resolving a dispute between the patent holder and interested parties. This is because it can reduce the cost of the lawsuit and risk in technology use.

But settlement with the effect of unduly sustaining the exclusive authority of an invalid patent and unduly preventing the entry of competing enterprises into the market can be considered to go beyond the legitimate scope of patent rights.

In cases where (1) the parties to the settlement agreement are competitors, (2) the purpose of the settlement is related to restriction of competition in the relevant market, (3) market entry of the relevant enterprises is delayed until after the patent rights expire, (4) the entry of related enterprises into the market not directly related to the patent is delayed, or (5) the parties to the settlement are aware that the

---

[12] This sub-clause was introduced in 2014 to clarify the applicability of the MRFTA to cases such as that of N-Data, as seen in *In re Negotiated Data Solutions LLC*, No. 0510094 (2008).
[13] This sub-clause was added in 2014. While the meaning of the first part is unclear, it seems to target covenants not to assert patents against a licensor's other licensees.

disputed patent is invalid or it is objectively obvious that the patent is invalid, such a settlement agreement for a patent dispute is likely to be deemed unreasonable.

9.3.11. *Exercise of Patent Rights by Nonpracticing Entities (Sec. III.7)*

The KFTC took the leading role of navigating *terra incognita* regarding NPEs in the 2014 amendment. The IPR Guidelines adopt a broad definition of NPEs as "entities that generate profit by practicing patents while not manufacturing or selling goods or providing services using the patent" (Sec. I.3.A.(7)). But the description of NPEs' business model under Sec. III.7 is somewhat narrower in that it presents the NPEs as having a business model that combines buying and licensing of patents. The IPR Guidelines clearly recognize procompetitive aspects of NPEs in commercialization, rewarding innovators, and acting as an intermediary. But it also espouses the view that NPEs may have greater incentive to abuse patents since they do not undertake manufacturing and thus place no value on cross-licensing or are not vulnerable to counterattack. Abuse of patent rights by NPEs can be considered to go beyond the legitimate scope of patent rights:

(1) an act of imposing a significantly unreasonable level of royalty compared to normal trade practices;
(2) when patents were acquired from a third party, an act of demanding an unreasonable level of royalties in light of the normal trade practices, while refusing to honor the FRAND obligation applied to the former patent holders;
(3) when the NPE was established through a consortium of enterprises, an act of unduly refusing to license or discriminating against non-participants with licensing terms;
(4) an act of filing a patent lawsuit or sending a warning by concealing, omitting, or misleading material information necessary for a counterparty to respond properly to the NPE's claim.
(5) an act of transferring patents to NPEs, then having the NPEs engage in conduct described in Sec III.7.A and B.

9.3.12. *Proposed Amendments to the 2014 IPR Guidelines*

On December 16, 2015, the KFTC announced draft amendments to its 2014 IPR Guideline and started public comment procedures.[14] The most important feature of this proposal is the clarification of the definition of SEPs to differentiate the illegality standards applied to FRAND-encumbered SEPs from those applied to patents that are recognized under the de facto standard.

---

[14] Press release, KFTC, "KFTC initiates public comment period on the amendment to its IP guidelines" (December 16, 2015).

The Microsoft/Nokia merger case (see Section 9.4.7) raised concerns regarding the treatment of key non-SEPs or de facto SEPs because the current definition of "standard technologies" and "standard essential patents" under IPR guidelines and substantive provisions encompass both de jure SEPs which are adopted by SSOs and patents that are recognized under the de facto standard. Although issues involving de facto standard patents remain clearly within the reach of the MRFTA, the KFTC recognizes that it is not proper to apply the same criteria as FRAND-encumbered SEPs since the de facto standard is usually the result of market competition, and can be better regulated by general provisions that apply to ordinary patent rights.

Minor changes include:

(1) clarifying that the objective of the IPR guidelines is to promote fair and free competition rather than to promote fair trade practices;
(2) deleting a provision prohibiting unfair contract terms regarding arbitration rule, arbitration tribunal, and governing law in the event of termination;
(3) refining the illegality standard regarding the refusal to license more closely to that of the abuse of market-dominant position.[15]

The proposed amendment seems to represent a step forward, in line with the 2014 IPR Guidelines, that focuses on anticompetitive behavior and shows that the KFTC is making continuous efforts to refine its IPR policy.[16]

9.4. MAJOR CASES OF IPR ABUSE

Analysis of cases of IPR abuse reveals that the focus of liability findings has shifted from fairness in conduct to anticompetitive effects on the market. This reflects the development of competition law enforcement from a historical perspective.

When determining anticompetitive effects to find illegality, the main focus of the KFTC has been on the exclusionary effects by market-dominant players based on a rule of reason standard as opposed to a focus on simple pricing issues. In most cases, the high licensing fee of an IPR alone was not considered a major interest of the KFTC. This is because the applicable regulatory scope for excessive pricing by a market-dominant business is extremely narrow under the MRFTA, and subtle legal and economic grounds exist for price regulation.

---

[15] Relevant considerations for finding illegitimate refusal to license include the anticompetitive intent or purpose of such refusal, exclusive control of the technology, practical, legal or economic viability of substituting the technology, and impossibility to enter the market or being placed at an unavoidable competitive disadvantage without the technology (proposed IPR Guidelines Sec. III.B.(2). Note). See Section 9.3.7.2 for current provision.

[16] Some commentators, however, expressed a disappointment with the limited scope of revision (Wright et al. 2016).

9.4.1. Earlier Cases

Earlier cases in the mid-2000s mostly involved a licensor requiring a licensee to sign a contract irrelevant to the patent, or denying access to the new technology in the context of public construction bidding that incorporated a patented technology.

In *Rainbow Scape*[17] decided on July 26, 2005, the KFTC ruled against a licensor who forced licensees to grant subcontracts to the licensor for construction work to which the patented technology did not apply. Rainbow Scape owned a patent for a method of installing fountains. Certain municipal governments required use of Rainbow Scape's patented technology as a qualification for participating in fountain construction bids. Taking advantage of its superior bargaining position, which stemmed from the necessity of the patent to win the bid, Rainbow Scape demanded that prospective licensees sign subcontracts with the company covering not only portions of work involving the patented technology, but also portions of work irrelevant to the patent, consisting of 74 percent of the combined construction work. The KFTC found that (1) the patent did not confer the power to condition licensing on the subcontracting of unrelated works and that (2) this practice was at odds with the patent protection's purpose to encourage innovation and technological development.

Kobec[18] involved an issue similar to *Rainbow Scape*. In exchange for the licensing of a patented technology incorporated in bridge improvement work, Kobec was found to have required subcontracting of excessive portions of work to Kobec or a company it designated, and required prepayment of a part of the proceeds in cash. Although the KFTC left open the possibility that requiring licensees to subcontract construction covered by a patent to a capable company can be justified on the grounds of preventing nonuse of the patented technology and safety, it also held that designating the party to whom the work should be subcontracted, the high proportion of subcontracting, and excessive amounts of prepaid money went beyond the legitimate scope of the patent and would thus be deemed an impediment to fair trade.

In *Insung Industry*[19] decided on September 9, 2005, at issue was a refusal to license a technology to the successful bidder for a public construction work. To pass the qualification examination, the successful bidder was required to submit a technology use agreement with Insung Industry. At first, Insung refused to license based on inappropriate timing of the construction, but even after the defect was remedied, it refused to license on other unreasonable grounds. The KFTC held that the company's conduct was an illegal refusal to deal. Such conduct was deemed to go beyond the limits of freedom to choose trading partners since the respondent

---

[17] KFTC Decision No. 2005-151, 2005.7.26.
[18] KFTC Decision No. 2006-266, 2006.11.23.
[19] KFTC Decision No. 2005-174, 2005.9.9.

could have been compensated for its technology by incorporating its capital investment and technology costs in calculating royalties. Similar issues to the Insung case also emerged in *Space*.[20]

The four defendants above were deemed in violation of Article 23 of the MRFTA, which prohibits unfair trading practices. But in one case, the KFTC applied the provision on abuse of dominant position as prescribed in Article 3(2) of the MRFTA. In *Royal Information Technology Corporation (RITCO)*,[21] Hyundai Engineering & Construction, which was awarded a land improvement contract from a local government, gave a public bid notice for installing automatic fire detectors. The project was designed to use a specific model of fire detectors manufactured by a Swiss company, which was exclusively distributed by RITCO. Three companies bid and RITCO was the third-lowest bidder. The lowest bidder contacted RITCO to order supplies of the fire detector, but got no response and eventually had to abandon the contract. The second-lowest bidder also gave up the contract after receiving an estimate from RITCO that was deemed too expensive. Hyundai had no choice but to sign the contract with RITCO. The KFTC found that RITCO's conduct constituted refusal to deal under Article 3(2) of the MRFTA.

9.4.2. Qualcomm: *Abuse of SEP by Breaching FRAND Commitment*

The first significant enforcement action against IPR abuse by the KFTC was sanctions against Qualcomm for its CDMA technology patent licensing in 2009.[22]

The KFTC imposed a corrective order and fine of US$210 million on Qualcomm for (1) charging discriminatorily high royalties for using non-Qualcomm modem chips rather than its modem chips, (2) offering loyalty rebates on the condition that licensees meet a great portion of their demand with Qualcomm modem chips and RF (radio frequency) chips, and (3) requiring royalties even after the expiration or invalidation of the relevant patents.[23] The first and third issues were related to IPR licensing.[24]

The KFTC found that Qualcomm was a dominant player in the Korean market for code division multiple access ("CDMA") modem chips and that Qualcomm charged discriminatively high royalties for export models when it licensed its CDMA technology to mobile handset makers. For example, the company charged a 5 percent royalty when a handset maker used Qualcomm modem chips and a 5.75 percent royalty when non-Qualcomm modem chips were used. In addition,

---

[20] KFTC Decision No. 2006-166, 2006.7.26.
[21] KFTC Decision No. 2006-221, 2006.10.10.
[22] KFTC Decision No. 2009-281, 2009.12.30.
[23] Press Release, KFTC, "Qualcomm's Abuse of Market Dominance" (July 23, 2009).
[24] Failure to disclose the application digital signal processor interface was also raised, but the KFTC closed the investigation on that issue as Qualcomm voluntarily pledged to disclose the interface to third parties.

Qualcomm discriminatorily set a royalty ceiling of $20 for handset makers using Qualcomm modem chips and $30 for handset makers that incorporated competing modem chips. The company also netted the price of its modem chip when calculating royalties applied to the handset for domestic use. In this case, the KFTC referred to Qualcomm making a written FRAND commitment to the Telecommunications Technology Association, a standard setting body for CDMA.

On the third issue, a provision in the license agreement in question provided under the title of "Invalid or Expired Patents" that "The obligation for LICENSEE to pay royalties to QUALCOMM under Section 5.2 shall be reduced by XX percent (XX%) of the applicable royalty rate in the event all patents that are licensed hereunder by QUALCOMM and used by LICENSEE have expired or are invalid." The KFTC found that the provision constituted abuse of superior bargaining position under Article 23, Paragraph 1, Item 4 of the MRFTA ("unduly takes advantage of its bargaining position to harm others").

At the time of writing, the final review of this case remains pending with the Supreme Court.

9.4.3. GSK *Decision: Anticompetitive Patent Settlement*

Inspired by the European Commission's inquiry into the pharmaceutical sector in 2009, the KFTC conducted a market-wide survey to gather information about patent licensing agreements and legal disputes among pharmaceutical companies in 2010. The inquiry involved thirty multinational and eighteen domestic companies, and eventually included all major pharmaceutical companies doing business in Korea.[25] After a lengthy analysis of the results, the KFTC concluded that GlaxoSmithKline Korea (GSK), one of the world's largest pharmaceutical multinationals, and Dong-A Pharmaceutical Co. (Dong-A), Korea's industry leader, had violated Article 19, Paragraph 1, Item 4 (Limiting the Territory of Trade) of the MRFTA by agreeing not to sell generic drugs, something which resulted in high prices and harm to consumer interests.

GSK developed Ondansetron, an ingredient that fights nausea, and subsequently produced and sold Zofran, which uses Ondansetron as an active pharmaceutical ingredient (API). Subsequently Dong-A developed Ondaron, a medication that used Ondansetron API that was claimed to be developed by a different method. In a subsequent patent dispute, both parties settled on the condition that in exchange for the withdrawal of the patent lawsuit, GSK would grant Dong-A (1) exclusive rights to sell Zofran to national and public hospitals in Korea and (2) exclusive domestic rights to sell Valtrex, a separate medication. The duration of the settlement agreement went beyond the expiration date of the GSK patent. Dong-A also agreed not to

---

[25] See KFTC Press Release, "Beginning of IPR Licensing Market Inquiry of Pharmaceutical Industry" (June 2010).

develop, produce, or sell any medication with identical or similar ingredients to Zofran or Valtrex.

The KFTC imposed fines of US$2.9 million on GSK and US$2 million on Dong-A, along with a cease and desist order in November 2011.[26] Both companies filed appeals to overturn the KFTC decision with the Seoul High Court, which had exclusive jurisdiction for cases dealing with KFTC actions.

The Supreme Court largely affirmed the High Court's decision, which mostly agreed with KFTC decision and held that (1) the agreement at issue was that GSK provided Dong-A, a company that disputed the validity of the patent GSK held and put on sale a competing product, with a significantly larger amount of economic benefit compared to the cost of the patent lawsuit in exchange for withdrawing the competing product from the market and restricting its sale beyond the term of the patent; (2) the agreement affected fair and free trade by providing a portion of the monopoly profit to the other party in return for maintaining its monopoly power; (3) the agreement was not deemed a legitimate exercise of patent rights; (4) thus, the MRFTA was applicable to the conduct; and (5) harm to consumers exceeded efficiency.[27] The Supreme Court also ruled that the KFTC found an anticompetitive agreement for the other new drug (Valtrex), which bore no relationship to the patent for Ondansetron without properly defining the market or assessing the anticompetitive effect, and repealed part of the KFTC decision.

9.4.4. SK Telecom: *Unfair Licensing Practices for SMEs*

On November 11, 2011, the KFTC issued a corrective order against SK Telecom, Korea's largest telecommunication carrier, for abusing its superior bargaining position to conclude "unfair" technology transfer agreements with SMEs that needed SK's technology to deliver gap-fillers ordered from the telecom giant.[28] This action followed a sweeping factfinding survey on IPR abuse in the IT industry.

The licensing contract at issue included an unfair clause that the validity of SK patents would not affect the contract, and thus meant that the licensee should continue to pay royalties regardless of the patents' validity. But if SK used the patented technology of its contracting party, a clause effectively exempted SK from paying royalties after the patent expired or was invalidated. The KFTC found that SK abused its superior bargaining position to unilaterally transfer the risk caused by the uncertainty around a patent's invalidity and unreasonably restrain use of the technology after invalidation, which the other party would otherwise be free to use.

---

[26] KFTC Decision 2011-300, 2011.12.13.
[27] Supreme Court 2012du24498, delivered on February 27, 2014 (on the complaint of GSK).
[28] KFTC Decision No. 2011-120, 2011.11.30.

9.4.5. Samsung Electronics: No Abuse of SEP

On February 25, 2014, the KFTC announced that the injunction obtained by Samsung Electronics against Apple based on the SEP did not constitute abuse of market dominant position or unfair trading practices under the MRFTA.

Samsung was deemed to have a monopoly in each of the four essential patent technology markets and thus would be considered a market-dominant operator in the mobile telecommunication device market. But the KFTC rejected all of Apple's arguments on the illegality of Samsung's conduct.

The KFTC made it clear that FRAND commitments did not automatically exclude the right of patent holders to seek an injunction against potential licensees. The critical issue was whether the potential licensee negotiated in good faith (adopting the *willing licensee* standard). But Apple had filed a patent infringement lawsuit during negotiations and proposed unfavorable license terms that underestimated patent value. Apple also showed an unwillingness to pay any royalty before the conclusion of the suit, which the KFTC considered a typical example of a reverse holdup.

The KFTC concluded after review that the SEP held by Samsung did not amount to an "essential factor." Samsung lacked exclusive control over SEPs because it was obliged to license to any potential licensee on FRAND terms and more than fifty companies held 15,000 SEPs for 3G mobile communication technologies. In addition, the KFTC found that Samsung did not fail to disclose its patent properly in the standardization process by intentionally withholding patent information.

In *Samsung v. Apple*, a parallel civil lawsuit preceding the KFTC investigation, the Seoul Central District Court addressed the same issues in reviewing Apple's defenses to Samsung's injunction and damages claim.[29] The court rejected Apple's claim that Samsung demanded excessive or discriminatory royalties in breach of FRAND terms. Although the court found that SEPs constituted essential facilities (a different conclusion from the KFTC decision), it held that Samsung's conduct did not constitute unreasonable refusal to deal because when considering relevant facts (which were similar to the facts findings in the KFTC above) Apple failed to prove anticompetitive effect or an intent to influence the market order, which is required by the *POSCO* judgment[30] to support a MRFTA violation. The court ruled that even if the royalty rates suggested by Samsung were unreasonably high it remained a specific harm to Apple rather than a harm to market competition. The court also rejected Apple's claim that Samsung was engaged in royalty discrimination in breach of a FRAND commitment or Samsung was engaged in deception by failing to disclose its SEPS or make a FRAND commitment in a timely manner.

---

[29] Seoul Central District Court Decision No. 2011Gahap39552, rendered on August 24, 2012.
[30] Supreme Court Decision No. 2008Du16322, rendered on June 10, 2011.

9.4.6. Dolby: *Unfair Licensing Practices Case after 2014 IPR Guidelines*

On August 3, 2015, the KFTC found that Dolby, a global company that held SEPs for audio compression format AC-3 and E-AC-3, unfairly abused its superior bargaining position by (1) imposing a no-challenge obligation with a terminate-on-challenge clause, (2) including a termination clause for any threatened infringement or misuse of Dolby's IPRs, (3) setting out excessive liquidated damages or inspection fee clauses for insignificant violations of reporting obligations, and (4) limiting the transfer or exercise of the licensee's proprietary technology acquired using Dolby's technology, e.g. prohibiting the assertion of the licensee's patent against Dolby or its other licensees.[31]

The KFTC only imposed corrective measures without surcharges and the case was not appealed to the court. The case remains somewhat controversial because the KFTC defined Dolby's conduct as an unfair business practice based upon unfairness after the 2014 IPR Guidelines had been issued, by which the application of an abuse of dominance clause was expressly prioritized.

9.4.7. *Microsoft/Nokia Merger: First Consent Decision Addressing IP Deal with Holdup Concern*

On August 24, 2015, the KFTC issued a consent decision accepting a remedy plan offered by Microsoft, which had acquired Nokia's handset business.[32] Nokia retained its extensive patent portfolios but granted licenses to Microsoft. The concerns were that Microsoft might vertically foreclose its handset competitors by raising royalties for its Android OS patents or filing for injunctions, or that business collaboration agreements between Microsoft and handset competitors might have coordinated anticompetitive effects. The remedy addressed potential holdup by reaffirming the commitment to comply with the FRAND obligation, prohibiting injunctions for SEPs (regardless of the willingness of the licensee) and requests for cross-license of non-SEPs in return for licensing SEPs, and requiring encumbrance of FRAND obligation in case of patent transfer. The remedy plan also addressed non-SEP matters in the absence of the FRAND protection and appeared more severe compared with SEP-related remedies. For example, Microsoft must continue to license its non-SEPs, cannot raise royalties or change conditions unfavorably for seven years, cannot seek injunctions against willing licensees, and cannot transfer non-SEPs for five years. The remedy plan was more extensive than that of the parallel case of Mofcom in China in that the patents at issue included foreign patents and injunction cannot be sought even abroad. But it should be noted that

---

[31] KFTC Decision No. 2015-125, 2015.8.3.
[32] KFTC Decision No. 2015-316, 2015.8.24.

the remedy plan was the voluntary result of negotiation, and it is not necessarily limited to the contours of the MRFTA.

The KFTC also sent an examiner's report to Nokia regarding potential dangers from increased incentive to abuse SEPs after becoming an NPE after transferring its handset business, but decided not to issue orders due to the competence issues in line with US and EU competition authorities.

9.5. THE FUTURE OF KOREAN REGULATION OF IPR ABUSE

Regulations on the interface between IPR and competition law are constantly evolving in Korea to meet new challenges and reflect recent developments around the globe. As discussed above, this is especially true in Korea because the IT industry, with several global firms, makes up a significant portion of the economy.

The KFTC recently announced that it would take a closer look at potentially abusive practices of NPEs and included relevant provisions in the IPR Guidelines, as described earlier. This is due to the perception that Korean firms are becoming the target of attacks by notorious NPEs.[33] The KFTC also pledged to closely monitor patent abuse in the pharmaceutical sector and improve relevant institutions.[34] Along with the introduction of the patent regulatory linkage system modeled after the Hatch-Waxman Act of the United States, patent disputes between brand-name drug giants and generic companies are rapidly increasing.[35] As a result, illegal settlement payments may take place, causing harm to consumer welfare. Based on the recent Supreme Court GSK decision, the KFTC is expected to monitor reverse payment and develop standards for determining illegality in such practices. The KFTC is also expected to strengthen its focus on the abuse of SEPs.[36]

Over the long run, the KFTC can be expected to strengthen competition law enforcement for both domestic and foreign businesses. One reminder is that regulators should take a prudent approach amidst rapidly changing industry structures and global trends. As the Korean economy requires the contributions of IPRs for further development, achieving the right balance between the interests of IPR holders and potential licensees has become crucial to improving creativity and efficiency.

REFERENCES

Jin, Hyunsuk. 2015, Ensuring Effectiveness of Marketing Prevention and First Generic Exclusivity, in ICR Law Center, Current issues of the Drug Approval-Patent Linkage System. Available at http://media.wix.com/ugd/7496b6_7ecee32e982a4839a11eb4beboe8ea51.pdf.
KFTC, 2014. 2014 Annual Work Plan.

---

[33] KFTC (2014).
[34] Id.
[35] Jin (2015).
[36] KFTC (2015); Yi (2015).

2015. 2015 Annual Work Plan.

Kobayashi, Bruce H., Douglas H. Ginsburg, Koren W. Wong-Ervin, and Joshua D. Wright. 2015. Comment on the Korea Fair Trade Commission's Revised Review Guidelines on Unfair Exercise of Intellectual Property Rights, George Mason University Legal Studies Research Paper Series LS 15–35.

Lee, Hwang. 2014. Actavis Decision in U.S on Reverse Payment and Its Implication. *Judicial Practice Review*, 11(2), 551–92 (in Korean).

Noh, Sangsup. 2011. The Current Status of the Operation of IPR Guideline and Future Tasks. In *ICR Law Center, Regulatory Framework for Licensing of Intellectual Property Rights*, 2011 ICR Law Center Seminar Series: 519–39 (in Korean).

Wright, Joshua D., Koren W. Wong-Ervin, Douglas H. Ginsburg, and Bruce H. Kobayashi. 2016. Comment of the Global Antitrust Institute, George Mason University School of Law, on the Korea Fair Trade Commission's Amendment to Its Review Guidelines on Unfair Exercise of Intellectual Property Rights.

Yi, Sangseung. 2015. Responsibilities of FRAND-Encumbered SEP Owners under Competition Law. In *Legal and Economic Analysis on SEP Holder's Abuse of Patent Rights*, Sogang University Conference, 96–139.

PART III

Monopolization

10

Is Pepsi Really a Substitute for Coke? Market Definition in Antitrust and IP

Mark A. Lemley and Mark P. McKenna

Does Pepsi compete with Coke? It seems a straightforward question; perhaps you think it has a straightforward answer. But in fact the answer is far from clear. For a significant number of consumers, it would take a pretty dramatic change in price to get them to give up their preferred brand. And for antitrust, this insensitivity to price has a simple – and shocking – implication: Coke and Pepsi don't compete in the same market. Antitrust defines markets by asking, in part, whether a small but significant nontransitory increase in price (or SSNIP) above marginal cost would cause customers to switch from one good to another. And by small, the regulators mean something like 5 percent.[1] So if a bottle of Coke costs $1.60, unless a price increase of eight cents would send so many consumers running to buy Pepsi that Coke would lose money, the two don't compete. And while supply substitution can also constrain market power, that won't work here either; Pepsi-Cola can't (legally) start making Coke rather than Pepsi just because the price went up. By our classic antitrust definition, then, Coke and Pepsi are not in the same market. And assuming Coke's price change doesn't send customers off to any other drink either, then Coke is in a market by itself, one in which it has market power.

Nor is this an isolated result. In market after market, intellectual property (IP)-based product differentiation and brand loyalty mean that small price changes have little or no effect on purchasing behavior. Customers who want a Harry Potter novel won't be satisfied with a somewhat cheaper Stephen King novel. New York Yankees fans won't accept a Boston Red Sox hat instead of one supporting their team. And, perhaps most remarkably, brand-loyal drug customers will happily choose Advil even though a substitute guaranteed by the Food and Drug

---

This chapter is adapted from Mark A. Lemley and Mark P. McKenna. 2012. Is Pepsi Really a Substitute for Coke? Market Definition in Antitrust and IP. *Georgetown Law Journal*, 100, 2055.

[1] U.S. DOJ and FTC (1992).

Administration (FDA) to be functionally identical is sitting right next to it on the drugstore shelf at as little as half the price. To antitrust law, all these products have market power in their own, individualized markets.

If we took market definition seriously, antitrust law would look very different as applied to IP. Monopolists are subject to a variety of constraints that ordinary competitors don't face. To take just one example, exclusive dealing arrangements are subject to the rule of reason: they are unlawful if they foreclose a substantial share of the market. If any well-known brand has power in its own, individually defined market, exclusive deals or licenses signed by those brands would probably be illegal. That result would require a major change in business practice.

IP law also depends to some extent on market definition. A variety of IP doctrines also require courts to define markets, though they often do so in a more ad hoc way than does antitrust. Here too, actual practice seems at odds with what antitrust's approach to market definition requires. If generic drugs don't compete with their branded counterparts, then it makes little sense to award damages in patent cases on the assumption that sales of the infringing product cause the patentee to lose sales. In short, a world that recognized the market power apparently conferred by IP rights would impose many more restrictions on IP owners than our cases do today, as a matter both of antitrust doctrine and IP law. And, we would be a lot more worried about the anticompetitive effects of IP rights generally.

Perhaps the problem is that antitrust market definition has become too ossified, dependent on assumptions about competition that are static and presuppose homogeneous products that compete solely on price and quality. This may be why courts have not been able to figure out how to define markets appropriately in the many IP settings where market definition is required: they don't really know what it means for IP products to compete. So perhaps courts in IP cases are right when they ignore antitrust's traditional approach to market definition, even if they haven't been doing so consciously.

In our view, antitrust law must recognize that the point of IP rights generally is to create some form of differentiation, and therefore some (usually constrained) power over price. Treating any product that does so as a monopoly can be counterproductive, both because it will put antitrust law into regular conflict with IP policy and because it may discourage *inter*brand competition more than it promotes *intra*brand competition. At the same time, both IP and antitrust law are wrong to ignore the implications of product differentiation and the power it gives IP owners over customers who strongly prefer one brand or type of product. The law needs to acknowledge that power in setting IP policies, in awarding remedies, and in evaluating the competitive effects of IP licensing. And both IP and antitrust law need to focus less on artificial market boundaries and more on the actual effects of IP rules and IP owner conduct.

10.1. MARKET DEFINITION IN IP DOCTRINE

A wide variety of IP doctrines require courts to engage, explicitly or implicitly, in market definition, or at least something that closely resembles it. Yet courts have developed no apparent methodology for defining markets. Indeed, courts frequently don't even acknowledge that they are drawing difficult lines between markets.

10.1.1. *Trademark*

The question of competitive need for access to particular features arises in a variety of contexts in trademark law, and courts necessarily engage in something quite similar to market definition in a number of areas not conventionally thought of in these terms.

To take one example, market definition is necessarily implicated to the extent courts view functionality in terms of competitive need, because one cannot determine what is "essential to effective competition" without understanding the market in which the parties participate. And despite *TrafFix Devices, Inc. v. Marketing Displays, Inc.*,[2] which seemed to downplay that conception of functionality, competitive need remains relevant in functionality doctrine. For one thing, several courts have maintained that the availability of alternative designs, a measure of competitive need, is relevant to the question of whether a feature is *"essential* to the use or purpose of the device."[3] Competitive need also remains important in cases of aesthetic functionality, where, according to the Court "[i]t is proper to inquire into a significant non-reputation-related disadvantage."[4]

Importantly, where market definition is relevant, in many cases it is essentially dispositive: the more broadly a court defines the relevant market, the less any particular feature appears necessary; the more narrowly it describes the market, the more granting exclusive rights to particular features seems to impose competitive harm. Yet courts appear to define markets in these cases entirely by their own intuition, without any apparent methodology.

10.1.2. *Patent*

Market definition plays a role in several patent contexts too. The clearest example is in the context of lost profit damages. Patent law awards damages "adequate to compensate for the infringement."[5] The traditional measure of those damages is the patentee's lost profits: the amount of money the patentee would have made selling the invention had the defendant not infringed. To determine that, we need

---

[2] 532 U.S. 23 (2001).
[3] *Id.* at 33 (emphasis added).
[4] *Id.*
[5] 35 U.S.C. §284 (2006).

to know which other competitors in the market might take the sale away from the patentee, what noninfringing alternatives might serve as substitutes for the patented good, how the presence of infringement affected the price the patentee could charge, and how an increase in price absent that infringement would have affected total demand. In some circumstances, we may also need to know how the patented product interacts with other, unpatented complements, and even how it interacts with unpatented non-complementary products sold by the patentee. The analysis of whether a patentee lost profits because of an infringement is accordingly quite sophisticated, and requires the estimation of supply and demand curves and cross-elasticities of demand. These are the classic hallmarks of antitrust market definition and power analysis.

Market definition is also relevant in reasonable royalty cases. In those cases, courts distinguish among prior comparable licenses in part by asking which are the most relevant to the current patent, and that requires some assessment of competitive proximity. Further, reasonable royalties, like lost profits, require consideration of available noninfringing alternatives, which means that courts must identify other technologies that are sufficiently different from the patent to avoid infringement but still close enough that they serve as functional substitutes for the patented product in the minds of actual consumers of that product. Notably, in contrast to the sophisticated market definition process associated with lost profits cases, market definition in reasonable royalty cases remains rudimentary and largely implicit, subsumed in the overarching question of what people in "this industry" would have paid for a license.

10.1.3. *Copyright*

Just as in the trademark and patent contexts, courts in copyright cases sometimes delineate the boundaries of copyright rights by considering market conditions. That means that courts sometimes engage in market definition here as well.

It is black-letter copyright law that protection attaches only to expression and not to the ideas embodied therein. Courts must therefore distinguish ideas from expression both to determine what the plaintiff in a copyright case actually owns and to determine whether a particular defendant's work infringes. To take a modern example, when Mattel sued the makers of the Bratz line of dolls, claiming the dolls too closely resembled those depicted in preliminary sketches to which Mattel owned rights, the Ninth Circuit held that the district court erred because it failed to filter out the unprotected idea of "fashion dolls with a bratty look or attitude, or dolls sporting trendy clothing."[6] For the Ninth Circuit, without those elements filtered out, Mattel would gain monopoly control over a certain type of doll.

But of course the court's description of the dolls at that level of specificity depended on an implicit judgment about what sorts of things compete in a relevant

---

[6] *Mattel, Inc. v. MGA Entm't, Inc.*, 616 F.3d 904, 916 (9th Cir. 2010).

market. And *Mattel* is fairly representative in this respect: the idea–expression dichotomy relies on an implicit market definition, and application of that rule tends to imply fairly narrow markets. In common parlance, we might speak broadly of markets for "books" or "romance novels," but copyright law parses the markets much more narrowly, defining markets as something more akin to the one identified by Learned Hand: stories in which two children of feuding families fall in love, marry in secret, and meet a tragic end, *and* which are designed to appeal to a certain ethnic or religious group.[7]

Market definition is also necessary in fair use cases, given the heavy weight most courts give to the fourth fair use factor – the effect on the market for the copyrighted work. Here, in contrast to the idea–expression dichotomy, courts tend to conceive of the market for a copyrighted work very broadly – so that not only do courts consider the extent of market harm caused by the particular actions of the alleged infringer, but also whether unrestricted and widespread conduct of the sort engaged in by the defendant would result in a substantially adverse impact on the potential market for the original. Moreover, according to the Supreme Court, the enquiry must take account not only of harm to the original, but also of harm to the market for derivative works. Notably, however, courts' determination of the relevant market tends not to be very sophisticated; they rarely attempt to parse the market beyond genre, nor do they offer any reasons to think genres correspond to markets.

10.1.4. *The Problem with Implicit Market Definition*

Lack of clarity about market definition in IP cases is to some extent a consequence of inconsistency in defining the terms of the debate. Courts and scholars alike sometimes talk about relevant markets in purely functional terms, so that markets are defined by the functional characteristics of the products at issue. Other times, however, markets are defined by consumer demand, which may or may not track functional considerations.

This framing choice, however, tends to track normative views about the scope of the rights in question. When making the case for strong rights, advocates emphasize the need for incentives and the important role IP rights play in enabling creators to appropriate economic value. They also cite enormous economic losses attributable to infringement, relying heavily on the assumption that most infringements represent lost sales. This is reflected in the damages phase of litigation, where parties define markets narrowly and dismiss possible substitutes in order to emphasize the harm they have suffered.

---

[7] *Nichols v. Universal Pictures Corp.*, 45 F.2d 119 (2d Cir. 1930) (L. Hand, J.) (distinguishing between two plays about feuding Jewish and Irish families whose children secretly fall in love, marry, and have children in part because they appeal to separate markets).

Where the issue is concern about the scope of IP rights and their general effects on competitors or on the ability of others to speak, by contrast, parties and advocates of broad IP rights define markets broadly so as to downplay the consequences of IP rights. So, for example, advocates of broad trademark rights dismiss concerns about the effects of those rights by claiming that trademark protection has few competitive costs since others can simply choose another mark at little cost. In copyright, concerns about the costs of protection are brushed aside on the ground that copyright doesn't create any market power since it allows others to create competing, functionally equivalent works. And in patent law, we have seen a consensus develop that patents do not confer market power – a view that depends on the belief that patented inventions compete with other products in more or less competitive markets.

Perhaps the problem is the lack of clear standards for market definition in IP cases. If we had clear rules for defining markets, parties couldn't take inconsistent positions when it suited them to do so. Unlike IP law, antitrust law has a long-standing, well-developed methodology for defining markets. Accordingly, we turn in Section 10.2 to consider antitrust market definition and whether it offers a way to rationalize the conflicting understandings of market scope and market power in IP cases.

10.2. MARKET DEFINITION AND PRODUCT DIFFERENTIATION

10.2.1. *Market Definition in Antitrust Law*

The first step in virtually any antitrust case is to define the market in which the competitive harm is alleged.[8] Mergers, joint ventures, and some agreements among competitors are harmless in competitive markets but can impose serious competitive threats in highly concentrated markets. Before those agreements can be characterized as unlawful, the fact finder must establish either the existence of market power or the likelihood that the conduct at issue will create such power or facilitate its exercise. And even in the shrinking set of cases involving conduct antitrust law treats as illegal per se – where market *power* is irrelevant – market *definition* is still relevant. In order to know whether an agreement is illegal per se, we need to know whether it is an agreement between competitors, which means we need to determine whether the agreeing parties would have competed absent the agreement.

Antitrust determines market power circumstantially by defining a relevant market and then computing the defendant's share of this market. A relevant market is a collection of goods or services that can profitably be sold at a monopoly price.[9] For

---

[8] Areeda and Hovenkamp (2006) vol. 2B, ch. 5.
[9] See, e.g., *H.J. Inc. v. Int'l Tel. & Tel. Corp.*, 867 F.2d 1531, 1537 (8th Cir. 1989) (explaining that a relevant market is "any grouping of sales whose sellers, if unified by a hypothetical cartel or

that to be true, the goods in the collection must be effective substitutes for one another, and goods outside the group must not be effective substitutes for goods inside the group. A substitute is effective when the cross-elasticity of demand between the two goods is high – a change in the price of one good will affect the demand for the other good. When cross-elasticity of demand is low – modest price increases will not cause consumers to switch – then goods are not considered to be in the same market.

Antitrust market definition also considers supply substitution and barriers to entry. The ease with which suppliers of related goods can move into the market in response to an increase in price determines the "elasticity of supply" or "supply substitutability." A barrier to entry is something that deters or delays the entry of new rivals even when the firm or firms in the market are already charging supracompetitive prices. If entry is so easy that new entrants can flood a market any time incumbents raise prices to supracompetitive levels, the products those incumbents sell cannot be effectively monopolized or cartelized, and it doesn't make sense to talk about those products as a separate market.

A relevant antitrust market is a grouping of sales for which *both* the cross-elasticity of demand with other products and the cross-elasticity of supply with other products are sufficiently low to warrant the conclusion that a significant price increase of that grouping to a supracompetitive level would be profitable and relatively durable. This typically means not merely that the increase will be immediately profitable, but that the monopoly price increase will be profitable for a nontransitory period of time.

Finally, the size of an antitrust market necessitates a policy judgment about how large a price increase above the competitive level should be tolerated. Antitrust market definition generally includes in the same market firms that, while not perfect competitors, are able to hold one another's prices to within 5 or 10 percent above the competitive price.[10] The federal Merger Guidelines refer to this test as a "small but significant nontransitory increase in price." If a SSNIP is profitable because the price increase more than compensates the producer for any lost sales, the substitutes are not effective and the product whose price is increased defines a separate market.

10.2.2. *The Role of IP Rights in Defining Markets*

It is an article of faith in antitrust law that particular companies do not define a relevant market, because for most users different brands compete with one another. But even when products clearly vary in quality – cars, say, or laptop computers – the basic assumption in antitrust law is that the products produced by individual

---

merger, could raise prices significantly above the competitive level") (citation omitted) (see also Areeda and Hovenkamp 2006, vol. 2B, para. 533b).

[10] U.S. DOJ and FTC (1992) Sec. 1.11, 551–55; U.S. DOJ and FTC (2010) Sec. 4.1.2.

companies compete in a larger market of (mostly) like goods. To an antitrust lawyer, brands aren't markets.

Yet unlike the commodity products antitrust had in mind when it conceived of market definition, products today often are far more significantly differentiated, and sometimes brands are the differentiators. Neuroscience research shows that brands convey emotional content as well as information about product characteristics, and indeed that people react to their favorite brands in ways that mirror their reaction to religious icons.[11] When preferences created by that information or those attachments are substantial and rivals cannot readily attain the same status, then it is simply wrong to say that the brand does not constitute its own relevant market. Demand is not price-elastic between the two. And the trademark itself prevents supply substitution.

The point is more general. By definition, IP rights attach to things that are differentiated from their putative competitors, and the existence of IP rights means that the things that differentiate my product from yours cannot be copied. An IP right is a barrier to entry that prevents the sort of quick and easy supply substitution that might undermine market power.

This IP-induced product differentiation can sometimes have the effect of excluding other goods in the same functional category from an antitrust-defined market. Suppose, for example, that three different pain pills – Bayer Aspirin, Tylenol, and Advil – all sell for 10 cents per pill. Though the three pills perform many similar functions and are considered roughly interchangeable for certain pains, they are also quite different and have different side effects. Acetaminophen, a generic version of Tylenol, then enters the market at a price of 3 cents per pill. In response, the price of branded Tylenol drops to 4 cents, while the prices of branded Bayer Aspirin and Advil show no effect. On these facts, Bayer Aspirin and Advil do not compete with Tylenol in a traditionally defined antitrust market. If they did, their higher price would cause so many customers to switch to the now-lower-priced Tylenol that they would be forced to lower their own prices in response.

And the effect of IP rights on market definition can be even more dramatic. Tylenol and Advil are brand names for acetaminophen and ibuprofen, respectively. But neither is the subject of a current patent. So anyone is free to copy the chemical formula of those drugs exactly and to produce an acetaminophen or ibuprofen product that competes with Tylenol or Advil. And they do. Curiously, however, those generic alternatives sell at a significant discount compared to the price of brand-name Tylenol or Advil, often as much as 50 percent less per dose.[12] Even though generic companies can and do advertise that they have the same ingredients as Tylenol or Advil, and even though they are often placed next to each other on

---

[11] See Lindstrom (2008) 123–6.
[12] See Anderson *et al.* (2012) tbl. 1. On the same dynamic in prescription drugs, see Ching (2010) fig. 1, showing that on average, generic prices gradually drop from about 60% of brand price to 50% as their market share rises to about 60%.

pharmacy shelves, the brand name is sufficient to cause many consumers to pay a higher price for a functionally equivalent good.

These brand-driven price differences have an interesting antitrust consequence: just as Advil and aspirin might not compete with each other in a relevant antitrust market, Advil and generic ibuprofen might not compete with each other on traditional antitrust analysis. It might seem odd to say that two functionally identical products are in separate markets. But that is because markets aren't always about function; consumers' beliefs about products can drive purchasing behavior even if those beliefs are not based in reality.

The implications are not limited to pharmaceuticals. IP rights can create their own antitrust market in any product in which consumer loyalty to the brand is strong enough, or in which IP rights make the products different enough. We see brand-based price differentiation among goods of undifferentiated quality in a wide variety of consumer markets. Consumers willingly pay substantially more for Clorox-brand bleach than for generic bleach,[13] even though the essence of bleach is a simple chemical formula. And sports fans won't be willing to replace a hat with their team's logo with a hat from a competing sports franchise. Under traditional antitrust market definition, each of these things is in a separate market.

Similarly, a wide variety of copyrighted works also command significant willingness to pay. As previously noted, if you want to read the final book in the Harry Potter series, you are unlikely to be satisfied by the latest Stephen King novel. And indeed people line up in advance to get access to the newest Harry Potter novel, or iPhone, or movie, while ignoring a variety of other creative products of the same type available with no wait and frequently at the same or a similar price. Whether a copyrighted work or group of works constitutes a relevant product market for antitrust purposes depends on the consumer response to price increases, not simply whether there are other works in the same genre. The copyrighted works of a particular musical artist are not a relevant product market so long as enough consumers, in response to a supracompetitive price, would instead purchase the works of other artists that the price increase would be unprofitable.[14] But if consumers wouldn't switch in response to a modest increase in price, the works are not in the same market.

In short, the evidence does not support the conclusion that IP rights automatically confer market power. But it does suggest that a surprisingly large number of IP rights do in fact give their owners power over price. Under traditional antitrust principles, those products are not in the same market as the goods that would seem to be their closest competitors.

---

[13] Clorox has a 65% market share (Datamonitor 2011).
[14] See *Rock River Commc'ns, Inc. v. Universal Music Grp., Inc.*, No. 08-cv-635, 2011 WL 1598916, at *15 (C.D. Cal. April 27, 2011) (finding that Bob Marley albums do not constitute a relevant product market unto themselves). But see *Polygram Holding, Inc. v. FTC*, 416 F.3d 29, 37 (D.C. Cir. 2005) (treating *Three Tenors* albums as a discrete market).

10.2.3. Implications of Market Definition in IP Cases

It is difficult to overstate just how shocking this conclusion ought to be to antitrust lawyers. Consider the implications of concluding that Coke (or Harry Potter books, or Shell gasoline, or Clorox bleach, or brand-name pharmaceuticals) really is in an antitrust market by itself. Antitrust law draws sharp distinctions between unilateral conduct, agreements with competitors, and agreements with noncompetitors. Agreements between horizontal competitors to create cartels that restrict price or competition get the harshest condemnation; they are illegal per se.[15] Agreements between parties in a buyer–seller relationship (vertical agreements) are treated much more leniently.[16] And agreements between companies that are not in a competitive relationship at all generally get no antitrust scrutiny at all. Unilateral conduct is not condemned at all unless the actor is a monopolist. If it is, a number of otherwise legal acts become antitrust problems if they have the purpose or effect of acquiring or maintaining that market power.[17]

A world in which individual books and individual brands define their own markets and hence confer market power turns that world upside down. To oversimplify, if IP rights define their own markets, everyone is monopolizing and no one is cartelizing. On the one hand, a number of accepted business practices become illegal if the firm that engages in them has market power. For instance, exclusive dealing arrangements, in which a party agrees to sell to or buy from only one party, are subject to antitrust's rule of reason. They are illegal if entered into by firms with market power and if they foreclose a "substantial share" of the market.[18] But if the market is defined not as "novels" or even "fantasy novels" but "Harry Potter books," J.K. Rowling entered into an illegal exclusive dealing arrangement when she allowed only Scholastic to produce and sell her books. Grants of exclusive franchise territories by McDonald's look like a market allocation scheme if we don't think Burger King competes in the same market. And so on. Antitrust spent the last thirty-five years undoing its rules restricting vertical restraints,[19] in large part on the theory that it was worth sacrificing "intrabrand competition" among dealers or franchisees in the service of promoting more "interbrand competition." But if interbrand competition doesn't work – if the brands aren't really competing in the same market at all – the last forty years of antitrust history have been a colossal mistake.

Concluding that many IP rights define markets turns a wide array of companies into monopolists, and so in many ways it would dramatically expand the role of

---

[15] Areeda and Hovenkamp (2006) vol. 3A, para. 980.
[16] Nominally vertical agreements are subject to the rule of reason, but in practice they are virtually always held legal (see, e.g., Reeves and Stucke 2011; Schwartz 2011, applying the rule of reason to sports leagues).
[17] See generally Hovenkamp et al. (2010) vol. 1, ch. 10.
[18] See *Tampa Elec. Co. v. Nashville Coal Co.*, 365 U.S. 320, 327 (1961).
[19] Reeves and Stucke (2011) 1549–50.

antitrust law. But in other respects it would shrink it. Suppose Pepsi and Coke were to enter into a horizontal market division agreement, in which Pepsi agreed to sell only north of the Mason–Dixon Line, and Coke agreed to sell only south of that line. If Pepsi and Coke compete in the same product market, that agreement would be illegal per se, just as it would be if they both agreed on the price they would charge.[20] But if they aren't competitors, traditional antitrust analysis doesn't have much to say about agreements they enter into, any more than it would object to Coke's agreeing with the two of us on the price it might charge.[21] Similarly, if brand-name drugs are not in the same market as their generic equivalents because the branded drugs command different prices than the generics, an agreement to fix the prices of the branded and generic drugs, or to pay the generic drug company to leave the market, would not seem to be an antitrust problem. And companies that make one of only two drugs for a particular disease could acquire the other, avoiding competition concerns by arguing that the drugs are differentiated and are prescribed by different doctors in different circumstances.[22]

In still other markets the effects may be more ambiguous. In sports leagues, the effect of brand-specific market definition may point in the opposite direction. If an entire sports league is treated as a single entity, it cannot be held liable for conspiring with itself. But if individual teams are separate actors, their agreement to league rules implicates Section 1 of the Sherman Act. That is what the Court held in *American Needle*.[23] But if the teams are not only separate actors, but also ones that don't compete with each other in a market for merchandise at all, those agreements presumably aren't horizontal restraints on trade. Similarly, antitrust analysis of the Google book search settlement has proceeded on the assumption that "books" (or maybe "out-of-print books") constitute the relevant market.[24] But if each of the ten million books is its own market, Google presumably isn't creating a horizontal aggregation, though it might face heightened concerns about monopoly, particularly over out-of-print books for which other sources are not available.

If we live in a world in which IP rights regularly confer market power, antitrust law has been doing pretty much everything wrong – finding illegal practices we shouldn't care about, and ignoring things that should be quite troubling. But before we reach that startling conclusion, perhaps it's worth revisiting the concept of

---

[20] See *Palmer v. BRG of Ga., Inc.*, 498 U.S. 46, 49–50 (1990) (per curiam) (finding "unlawful on its face" a territory-allocation and revenue-sharing agreement between two bar-review companies).
[21] It is true that product differentiation does make cartelization harder, (see Leslie 2004, 579 n453), but it surely doesn't make it categorically impossible.
[22] See *FTC v. Lundbeck, Inc.*, 650 F.3d 1236, 1239 (8th Cir. 2011) (upholding that finding by the trial court). Cf. *Delano Farms Co. v. Cal. Table Grape Comm'n*, 655 F.3d 1337, 1351–2 (Fed. Cir. 2011) (holding that plaintiff failed to sufficiently allege that Sweet Scarlet variety table grapes were in a separate market from Autumn King table grapes).
[23] Schwartz (2011) 314–15.
[24] See *Author's Guild v. Google Inc.*, 770 F. Supp. 2d 666, 682–3 (S.D.N.Y. 2011). Full disclosure: one of the authors (Lemley) represents Google in this matter.

market definition itself. For maybe the problem is less that courts have drawn all the wrong conclusions in antitrust cases and more that their market definition tool is broken.

10.3. RETHINKING MARKET DEFINITION IN ANTITRUST

10.3.1. Is Antitrust Miscalculating Markets in IP Cases?

One possible explanation for the phenomenon we have identified is that IP markets are different than other goods markets in a way traditional antitrust market definition fails to consider. Market power is a firm's ability to profit by raising price above the competitive level, with the competitive level generally defined as marginal cost. But with IP, the most significant costs are often fixed and fully "sunk" at the time the IP right is developed. The cost to invent a new device, or make a new movie, doesn't change with the number of copies of the product or movie the creator has produced, and that cost must be paid whether or not the creator produces anything at all. The *marginal* cost of selling the invention or creative work is limited to the cost of educating individual buyers and negotiating license agreements, plus whatever it costs to make each incremental version of the product or copy of the work. These costs are often, but not always, quite small in comparison to the costs of initial development.

Perhaps marginal cost is the wrong measure of market power in IP cases. At a price of $2.00 per DVD, *Titanic* might have lost money, and a firm that had contemplated such a price would never have made *Titanic* in the first place. A price that merely covers marginal cost in this situation would be insufficient to maintain investment in the market for making movies. To justify being in the movie business, the moviemaker must be compensated not only for the marginal cost of placing a copy of the movie on a DVD, but also for the much greater cost of producing and promoting the movie in the first place. So maybe total cost including initial investment, not marginal cost, should be the measure of market power.

10.3.2. Is Market Definition Itself a Flawed Enterprise?

It is also possible that the problem lies not with how market definition is applied to IP cases, but with the enterprise of market definition itself. Antitrust traditionally defines markets in a binary sense – we put a product (and a geographic space) in or out of the market. Then we use mathematical metrics to decide whether a market is too concentrated to allow a merger, or sufficiently concentrated that the defendant is a monopolist. One problem is that we are making yes–no decisions in cases that are analog, not digital. It is not the case that a video store three miles away is a potential competitor while another one four miles away is not. Rather, some slightly larger number of consumers will travel three miles than will travel four miles. And clearly some – but not all – customers will switch from brand-name drugs to lower-priced generics when they become available. The current approach to market definition draws an arbitrary line when what we need is a continuum that reflects the partial differentiation of products and differences in the cost and convenience of those products. That is especially true in IP markets, where the existence of the IP right all but guarantees some amount of product differentiation, making the yes–no framework of classical market definition particularly problematic.

The problem with market definition may be more intractable than just how it is assessed, however. In a provocative series of articles, Louis Kaplow has argued that the entire enterprise of market definition is fundamentally flawed because defining markets properly requires the kind of information that would allow us to assess power over price directly.[25] We think Kaplow has a point. Antitrust needs some way to assess the competitive effects of behavior, and that requires some inquiry into who is likely to be affected by that behavior and how. But it probably doesn't need (or want) the kind of formalistic, mathematical framework of the old 1992 Merger Guidelines. The 2010 Merger Guidelines move away from some of the line drawing of the traditional approach, placing much more emphasis on observed competitive effects – looking, for example, at upward pricing pressure from mergers and companies' responses to actual or threatened entry by a differentiated product.[26] As a general matter, this is a good thing, because it is more realistic. And it may lead the way toward a good compromise for IP cases, focusing not on artificial constructs nor on special rules designed to make IP market power go away but on the observed reality of how consumers interact with IP rights.

But abandoning market definition, as Kaplow suggests, doesn't mean we can abandon the effort to decide when IP rights confer market power. We may well conclude that branded pharmaceuticals and their generic equivalents aren't in the same market because of the large difference in their prices, but the presence of the generic on a pharmacy shelf next to the branded product surely has *some* influence on the price and output decisions of the brand owner – more than does, say, the fact that the pharmacy also sells greeting cards. The equilibrium after generic entry may separate into products that are priced differently, but that doesn't mean the introduction of new generics or the prices they set have no effect on branded pharmaceutical pricing or profits. In evaluating efforts by pharmaceutical firms to exclude generic entry, antitrust must take those effects into account, whether or not it calls the process of doing so "market definition."[27]

Similarly, deciding where to look for effects on market outcomes requires some implicit judgments about the relatedness of goods that traditional market definition might class as in separate markets. If we're looking at the effects on customers of

---

[25] Kaplow (2011; 2010, 440). For earlier discussion of the topic, see Markovits (1978).
[26] U.S. DOJ and FTC (2010) 2–6: s. 2.
[27] Hovenkamp (2012); see also Coate and Simmons (2012).

regular ice cream in *Dippin' Dots, Inc. v. Frosty Bites Distribution, Inc.*,[28] it doesn't look like there are any; but if you look for effects on buyers of flash-frozen ice cream, you get a different result. Classical market definition requires us to choose one of those markets and only look for effects there. A more realistic analysis looks for competitive effects in a number of possible places and makes the presence or absence of those effects the critical factor in the analysis.

Eliminating market definition, then, doesn't mean we stop trying to decide how differentiated products relate to each other. That assessment is necessary to any sort of antitrust analysis. And the evidence suggests that IP rights quite often confer significant power over price. That shouldn't mean that different IP-protected products are necessarily segregated into separate markets. But it does mean that we can no longer assume that IP rights have little or no effect on price and market power. Consistent with the economic literature on IP and product differentiation, IP rights will often confer significant power over price and purchasing decisions, which means those rights will often tread on territory we think of as reserved for antitrust law. Efforts by courts and scholars to assume away any conflict between IP and antitrust are accordingly misguided. The reality of this conflict doesn't mean antitrust law must always triumph; there are good reasons for IP law to create at least some modest product differentiation. But it does mean that the conflict cannot be wished away by invoking the mantra that IP and antitrust law serve the same goals.

10.4. LESSONS FOR IP

In our view, the difficulty courts have experienced with market definition in IP cases stems at least in part from their (perhaps intuitive) resistance to relying on the antitrust understanding of markets. The antitrust approach derives from a classical conception of competition involving undifferentiated goods where producers compete solely on price and quality. But markets for goods encumbered by IP rights are different: these goods are, by definition, differentiated, either in terms of some product feature or because of the product's branding. This product differentiation isn't unequivocally bad, of course. Indeed, markets characterized by differentiated products often give consumers choices, and that can enhance consumer welfare. And in some cases we need to confer some power over price on IP owners as part of a government policy to drive investment in innovation. But these benefits come at a cost to the extent differentiation insulates some IP owners from competition.

This is not to say, however, that all IP rights create market power that antitrust should care about, or even that all IP rights have the same effect on substitution. Indeed, we suspect that the effects of an IP right are likely to be quite different in a variety of respects. They may vary by IP right (patent vs. copyright vs. trademark)

[28] 369 F.3d 1197 (11th Cir. 2004).

because each of those rights will attach, generally speaking, to different features. They may vary by the type of product or service at issue. And they may vary by industry, even perhaps with respect to the same type of product or service. The general point is that the competitive costs of IP rights vary with the extent to which they enable the owners of those rights to exclude close substitutes, and courts should therefore conceive of competition in the IP space in terms of the quality of available substitutes. Rather than considering whether a particular product is or is not in the relevant market, courts must at a minimum think of competition as existing along a spectrum. The further away one product is from another on that spectrum (that is, the more imperfect it is as a substitute), the less we can see the product as effectively constraining the price of the first product or offering a realistic alternative to many consumers. Indeed, there may be multiple dimensions to the differentiation, including price, quality, and (for IP goods) brand association.

Conceiving of competition in this way does not obviate the need for a policy judgment, of course, any more than avoiding explicit market definition in antitrust allowed us to escape a focus on competitive effects. For even if courts take the approach we suggest and focus on the relative closeness of available substitutes, determining when the substitutes are sufficiently close that we can count them as competing goods entails a judgment about how much competition we want in this space.

At the level of specific cases, we think that IP law can learn something from a more flexible antitrust market definition, despite the many flaws in antitrust's approach. Because courts in antitrust cases use market definition for the ultimate purpose of assessing market power, and because those courts treat market power differently in different contexts, market definition itself tends to look different in different antitrust contexts. Market definition, in other words, has many masters in antitrust, just as it does in IP. Hence, while IP courts should think more coherently about the nature of competition between IP-protected products, they might well make different policy judgments in different circumstances about when products are sufficiently close that they raise competitive concerns. Some of the contexts we identified in which market definition arises in IP cases reflect courts' attempts to limit the competitive consequences of IP rights. And the extent to which courts ought to be concerned about those competitive costs depends on the importance of the type of protection the plaintiff in that case seeks relative to the purposes of the IP doctrine at issue.

At the more general level of the appropriate scope of IP rights, the question is when an IP right creates or enables so much differentiation that the costs of IP protection are too high to justify their benefits. Resolving that question requires policymakers to ask how much market power we want an IP owner to have, and that determination is bound up with the policy justifications of IP rights in the first place. We don't purport to answer the question of "how much market power is sufficient" here. Our point is that IP rights are not free.

If we wish to restrict the extent to which IP rights confer market power, and therefore minimize the range of conflict between the doctrines, we should focus on IP doctrines through which differentiation could be reduced. Those are the most efficient ways to reduce the harm caused by IP while still protecting IP rights. Put differently, to reduce the competitive costs of IP rights we can adjust doctrinal rules for the purpose of enabling closer substitutes. In this section, we identify some doctrinal contexts we see as especially conducive to tailoring for the purpose of enabling closer substitution.

10.4.1. Copyright

In the copyright area, courts could use a number of doctrinal levers to lessen the distance between the copyrighted work and available substitutes. First, courts could require much greater similarity between a plaintiff's and defendant's works in order for the defendant's work to be considered "substantially similar" and therefore infringing. Whereas the defendant's work once would have needed to be virtually identical to the plaintiff's in order to be considered an infringement, courts now find infringement based on fragmented literal similarity and non-literal or pattern similarity, including similarities in plot and/or characters. As a consequence, a copyright owner now can block works that are less similar to their own, meaning they can force third parties to produce less perfect substitutes. Courts could reduce the competitive costs of copyright, then, by requiring greater similarity to prove infringement.

Likewise, the derivative work right currently enables copyright owners to create further space around their own work by excluding related or follow-on works that might be closer substitutes for the copyright owner's own work than would be unrelated third-party works. The right therefore forces third parties to market more imperfect substitutes than they might if they were able to build more directly on the copyrighted work. We could decrease copyright owners' power here by significantly curtailing or eliminating the derivative work right, thereby allowing better substitutes in the market.

The economics of derivative rights are more complex, however, than the economics of substantial similarity. While substantial similarity affects how close a substitute a defendant can make, in many cases derivative works are complements rather than substitutes for the original work, as the derivative works are frequently directed at different markets. A Harry Potter encyclopedia doesn't substitute for sales of the series; it is likely useful only to those who have already read the series. By contrast, other derivative works might serve as both complements (or other non-competing goods) and substitutes. Movie versions of books are complements for those who want to read the book and then see the movie, but substitutes for those who opt for one but not the other. Finally, even when a derivative work is clearly a complement for some consumers (those who read the Harry Potter books and then also want to see the movie), it may be that this kind of complement is also a better substitute for the original than some other type of work. Some consumers, for example, might rather watch the Harry Potter movies in place of the Harry Potter books than switch to a Stephen King novel. This doesn't mean Harry Potter movies are particularly good substitutes for the book, only that Stephen King novels, or whatever else is the next best substitute for Harry Potter books, are comparatively worse substitutes.

Depending on how frequently the last situation arises, where a derivative work is both a complement and a better substitute than the next best substitute in the same medium as the original, our analysis might have a surprising implication: if we are concerned only with market power and substitution, we might want to focus more on limiting the scope of the copying right than on restricting the derivative works right. Among derivative works, we might care more about controlling power over that subset of works likely to serve as a substitute for the existing work, rather than derivatives that expand into a new market or act as complements to the original work.

10.4.2. *Trademark*

A number of scholars, including us, have argued recently that trademark rights should be narrower, and more certain, in scope. In large measure, these calls have focused on the requisite similarity of goods – the range of goods or services against which a mark owner could assert its rights. But if our goal is to reduce the differentiation afforded by a trademark and make other products better substitutes, then we should want at least as much to make more available the differentiating features, which in the trademark context means the marks themselves. We should, in other words, want to allow other parties to get closer to the claimed trademark. One means of accomplishing this in the trade dress context is a robust functionality doctrine, which would prevent parties from claiming certain features as trademarks at all. But in the broad range of cases, the best way to accomplish this would be for courts to put significant weight on the similarity of the marks factor of the likelihood of confusion test, requiring as a necessary condition of liability that the defendant's mark be extremely similar (if not identical) to the plaintiff's claimed mark. We think this may well be particularly important in the trade dress context, both because trade dress is often less clearly defined and because in many cases product packaging or design elements may be the most effective way to communicate that one's product competes with another known product.

To the extent non-trademark characteristics of products or services make them good substitutes for the plaintiff's products, comparative advertising can be an effective way of constraining market power by bringing those potential substitutes to the attention of consumers. And trademark law has long allowed parties to use the trademarks of their competitors in order to enable them to draw attention to the substitutability of their products. The difficulty is that courts have articulated the comparative advertising defense in ways that make it both less sure and less predictable because they have always attached a caveat that the defendant's use cannot cause confusion.

The uncertainty of the defense is compounded by the fact that plaintiffs of late have used the advent of new technology that enables comparative advertising as a reason to try to get courts to restrict the ability of competitors to engage in comparative ads. Particularly concerning are cases in which plaintiffs assert trademark claims against intermediaries like search engines, or even against advertisers themselves – despite the fact that the actual advertisement at issue is not even alleged to be false – claiming that the simple fact of using the plaintiff's mark as a keyword or as part of the text (which might allow it to turn up in response to an Internet search) is itself enough to cause confusion, thereby disqualifying it from being considered comparative advertising. Even though advertisers and search engines ultimately have prevailed in most of these cases, they generally have done so only after litigation on the likelihood of confusion issue.

Our analysis provides a strong defense for comparative advertising and suggests that recent potential narrowing of the defense is a mistake from the standpoint of concerns about substitution. To the extent other parties' goods or services can serve as substitutes, trademark law should make it easy for those parties to bring their goods to consumers' attention.

10.4.3. *The Scope of Patent Rights*

The scope of patent rights is a function of the language of the patent claims. Accordingly, any effort to modulate the market power conferred by a patent needs to start with the doctrines of claim construction and infringement. As a general matter, application of those doctrines is case-specific: much turns on the particular words chosen by the patentee and what the court construes those words to mean. But precisely because the focus is on the words chosen by the patent lawyers, rather than on what the patentee actually built, patentees have been able to obtain patents that cover broad swaths of conceptual space.

One reason patents are often read broadly is that patent lawyers frequently claim their inventions in functional terms. Rather than limiting the patent to a particular device, patentees claim any device that performs a particular function. This will often, though not always, coincide with market power, since control over a particular function may make the development of an effective substitute impossible. One way to limit the reach of patent claims, especially in software, is to take seriously the old idea that functional claiming is not permitted, and to limit the patentee to what she actually invented or equivalents thereof.

The control that patentees have over markets is also a function of a power that patents confer that copyrights and trade secrets don't: the power to prevent independent development. Indeed, patents today are overwhelmingly asserted not against accused copyists but against independent developers. Some have suggested that patent law, like copyright, should limit its reach to those who obtain the idea from the patentee. That may or may not be a good idea. But if adopted, it would substantially reduce the number of cases in which patents conferred market power by allowing an important source of substitutes.

10.4.4. Exhaustion and the First-Sale Doctrine

Finally, one set of doctrines crosses all doctrinal boundaries. Going by the name "exhaustion" or "first sale," copyright, patent, and trademark all impose limits on the ability of IP rights holders to control what happens to particular copies of their goods once they have sold those goods into the open market.

The first-sale or exhaustion doctrine is a powerful tool for reducing the market power of an IP owner by enabling better substitutes. Used copies of the original work are likely in many cases to be much better substitutes than other works even in the same genre would be. Antitrust law has long acknowledged this fact, and a number of antitrust cases depend on the ability of fringe competitors to offer used or reconditioned goods originally manufactured by a company that would otherwise hold unrestrained market power. That is true even if the goods are protected by IP rights, because all IP doctrines have some sort of first-sale or exhaustion principle limiting control of resales.

The first-sale doctrine is under sustained attack in each area of law, however. In copyright, that is in part a function of the Internet; the principle that one owns a particular copy and can pass it along to others is harder to apply when virtually anything one does involves the making of new copies. But in larger part it results from the growth of shrinkwrap, clickwrap, and browsewrap licenses and judicial deference to those "contracts." IP owners have increasingly sought either to prevent the resale of their goods altogether or to condition the price at which those goods can be resold. Trademark courts too are nearly always equivocal about this "right" to resell branded products, in that they universally express that right as subject to the requirement that the use not cause confusion.

Efforts to undermine the first-sale doctrine threaten to confer substantial additional market power on IP owners by constraining the closest source of substitutes. If the goal is to reduce differentiation, we should want few limitations on parties' ability to resell (or rent or otherwise recirculate) both patented and branded goods.

10.5. CONCLUSION

Courts in a variety of IP contexts engage more or less explicitly in market definition as a predicate to some doctrinal determination. They do this, however, without any apparent methodology, and that has the pernicious effect of making market definition simply a tool for reaching desired results. Antitrust law has a well-developed methodology for defining markets, but as we argued, that methodology does not work well in the context of IP goods. Specifically, antitrust's traditional approach is based on a classical conception of competition in which parties sell largely undifferentiated goods and compete only on price or quality. IP goods, however, differ precisely because they are differentiated; indeed that is largely the point of IP rights – to enable owners to prevent others from copying too closely so that the IP owners can reap a profit sufficient to provide incentives to invent or distribute (or, in the case of trademarks, to enable consumers to differentiate between goods of different source).

Traditional market definition is too rigid to accommodate this reality, because it asks whether particular goods are in or out of the market rather than focusing, as we would, on the closeness of available substitutes, taking into account the inevitability of some differentiation. In other words, in a market full of imperfect substitutes, it makes more sense to ask just how imperfect the substitutes are. To the extent we want to reduce IP owners' market power, there are doctrinal levers within IP that can be used to enable more and better substitutes.

REFERENCES

Anderson, Simon P. et al. 2012. "Push-Me Pull-You: Comparative Advertising in the OTC Analgesics Industry." Unpublished manuscript, last updated June 18, 2015. doi: 10.2139/2047106.

Areeda, Phillip E. and Herbert Hovenkamp. 2006. *Antitrust Law: An Analysis of Antitrust Principles and Their Application*, 3rd edn. Vols. 2B, 3A New York: Aspen Publishers.

Ching, Andrew T. 2010. Consumer Learning and Heterogeneity: Dynamics of Demand for Prescription Drugs after Patent Expiration. *International Journal of Industrial Organization*, 28, 619–38.

Coate, Malcolm B. and Joseph J. Simons. 2012. In Defense of Market Definition. *The Antitrust Bulletin*, 57, 667.

Datamonitor. 2011. Industry Profile: Bleach in the United States. 2 Ref. No. 0072–0013.

Hovenkamp, Herbert. 2012. Markets in IP & Antitrust. *Georgetown Law Journal*, 100, 2133–58.

Hovenkamp, Herbert, Mark Janis, Mark Lemley, and Christopher Leslie. 2010. *IP and Antitrust*, 2nd edn. Vol. 1. New York: Aspen Publishers.

Kaplow, Louis. 2010. Why (Ever) Define Markets?. *Harvard Law Review*, 124, 437–517.
  2011. "Market Share Thresholds: On the Conflation of Empirical Assessments and Legal Policy Judgments." Harvard Law and Economics Discussion Paper No. 692. doi: 10.2139/1873431.

Leslie, Christopher R. 2004. Trust, Distrust, and Antitrust, *Texas Law Review*, 82, 515–680.

Lindstrom, Martin. 2008. *Buyology: Truth and Lies about Why We Buy*. New York: Doubleday.

Markovits, Richard S. 1978. Predicting the Competitive Impact of Horizontal Mergers in a Monopolistically Competitive World: A Non-Market-Oriented Proposal and Critique of the Market Definition-Market Share-Market Concentration Approach. *Texas Law Review*, 56, 587–731.

Reeves, Amanda P. and Maurice E. Stucke. 2011. Behavioral Antitrust. *Indiana Law Journal*, 86, 1527–86.
Schwartz, Daniel A. 2011. Shutting the Back Door: Using American Needle to Cure the Problem of Improper Market Definition. *Michigan Law Review*, 110, 295–318.
U.S. DOJ and FTC (United States Department of Justice and Federal Trade Commission). 1992. Horizontal Merger Guidelines. 57 Fed. Reg. 41,552 (Sept. 10).
  2010. Horizontal Merger Guidelines.

11

Monopoly Power and Intellectual Property

Roger D. Blair and Wenche Wang

11.1. INTRODUCTION

There is an undeniable tension between antitrust law and intellectual property (IP) law. The former seeks to promote and protect competition by limiting the formation of monopolies and cartels. The latter rewards some inventors and creators by conferring legal monopolies. In spite of this conflict, the ultimate aim of both antitrust and IP law is the protection and promotion of consumer welfare. Protecting competition promotes consumer welfare by reducing deadweight social welfare losses that result from monopolistic restrictions on output and the accompanying allocative inefficiency. Protecting monopoly that results from the production of information that qualifies for IP protection enhances social welfare by encouraging the allocation of resources to invention. The economic rationale for IP protection flows from the fact that investments in inventive endeavors yield substantial future benefits in the form of new or improved products, more efficient production processes, literary and musical creations, and the like. Thus, the antitrust–IP tension involves the sacrifice of some consumer welfare in the present for enhanced consumer welfare in the future.[1]

In this chapter, we begin in Section 11.2 with a brief overview of the economic rationale for IP protection. In Section 11.3, we examine the nature of IP rights involved in patents, trade secrets, copyrights, and trademarks. In Section 11.4, we turn our attention to the economic consequences of the legal monopoly conferred on inventors by IP law. We also examine the measurement of monopoly power. In Section 11.5, we address several economic issues at the interface of IP law and antitrust policy. Those include dramatic monopoly markups, compulsory licensing,

---

Nearly everything that we know about intellectual property was learned from Thomas Cotter. If there are errors in what follows, it is because we are poor students and not because he is a poor teacher. We also received many useful suggestions from our colleague, D. Daniel Sokol.

[1] Nordhaus (1969).

patent pools, cross-licensing, and standard setting organizations. In Section 11.6, we close with some concluding remarks.

11.2. THE ECONOMIC RATIONALE FOR INTELLECTUAL PROPERTY RIGHTS

Inventors of new products or new production processes must invest their time, money, and other resources before – sometimes *long* before – they realize any return.[2] Similarly, creators of literary works, music, film, and choreography incur the costs of creation before they enjoy any income from their endeavor. The sequential nature of investment and return produces incentive problems. There are, however, other difficulties for inventors and creators, which we will explore below.

Investing in the production of information is a highly risky venture. There is a substantial stochastic element in the production process and, therefore, the output will be a random variable. Many innovative efforts will produce no output at all. Others will produce something, but it might turn out to be of little economic value. But some projects will produce positive profits to one degree or another. The decision to invest in such a risky venture is the result of weighing the expected costs and benefits. If the latter outweigh the former, the investment should be made.

We can formalize this decision process by examining the expected net present value E[NPV] of the project.

$$NPV = p \sum_{t=1}^{T} \frac{E[\pi_t]}{(1+i)^t} - I_0$$

where p is the probability that the investment yields successful and profitable results, $E[\pi_t]$ is the expected profit earned in year t, i is the interest rate, I_0 is the initial investment,[3] T is the economic life of the creation, and $\sum$ is the summation operator. In this formulation, the costs of producing the information (I_0) are incurred before any operating profits can be earned. The operating profits, if any, are calculated as the difference between the revenue realized and the costs of production and distribution.

Notice that the demand for the innovation is correlated with the success of the project (probability p) as well as the quality of the innovation, which is positively related to the initial investment I_0. Future improvement or innovation by other investors could alter consumer choice and therefore affect demand; consequently, the investors must take into account the stochastic nature of demand. The costs of

---

[2] This is particularly troublesome for pharmaceuticals since the prolonged Food and Drug Administration approval process must be tacked on to the time that the inventive process requires.

[3] I_0 is the capitalized value of the expenditure necessary for the invention and creation. I_0 is assumed to be known with certainty.

production are unknown and hard to estimate before the innovation process begins. The prospective inventor must focus on expected profits, which will be earned over the time horizon T.

If the expected net present value is positive, the investment should be made. Note, however, that the expected profits will be earned in the future, i.e. after the information is produced.

Consequently, there must be some assurance that the flow of future profits will not be undermined by rivals who simply copy the results of the production without incurring any costs. In other words, there is a need for some mechanism to permit the information to be appropriable.[4] If creative efforts are to be encouraged, the fruits of those efforts must be protected from free riders who would make few, if any, investments of their own. Preserving the incentive to invest resources in a risky venture demands a means of appropriating the economic value of the new product, new song, or new sculpture. This is precisely what our system of intellectual property intends to do.

Our system of IP rights shields an inventor from opportunistic competition, i.e., simple copying or imitation. But it cannot insulate the invention from all forms of competition nor should it. If a rival invents a new product that is superior to the original, that rival is entitled to IP protection of his or her own. That, however, is not the only danger. Future operating profits may be undermined by the discovery of unfortunate side-effects, changes in consumer preferences, government regulations, or financial conditions among other things. Consequently, the future returns to inventive effort face various degrees of uncertainty which reduce the incentive to innovate. All that our IP system can do (or should do) is to limit uncertainty by punishing unethical or unlawful conduct, which in turn would reduce some forms of competition.

11.3. INTELLECTUAL PROPERTY

There are four basic forms of intellectual property: patents, trade secrets, copyrights, and trademarks. They cover different types of invention or creation, but serve to make the creations appropriable for at least some period of time. These forms of IP vary in the scope and duration of the protection afforded. They also vary in the sanctions imposed for infringement.[5]

11.3.1. *Patents*

To qualify for patent protection, an invention must be novel, useful, and not obvious.[6] The novelty requirement normally is satisfied as long as the patent

---

[4] Arrow (1962) explored various incentive problems surrounding investments in the production of information. A central focus of his analysis was appropriability.
[5] This section relies on the more extensive treatment provided by Blair and Cotter (2005).
[6] These are statutory requirements for an invention to qualify for patent protection.

applicant was the first to invent the claimed innovation. If several research teams are pursuing the same objective simultaneously, this all-or-nothing feature creates an incentive to accelerate the inventive process in ways that are not socially optimal![7] For an innovation to be useful, it is only necessary that the invention work as intended and that it serves some minimal human need. Since patent protection is aimed at providing an incentive to invest resources in innovative activities, there is no need to award patents to improvements that are obvious. Consequently, the nonobviousness requirement denies patentability if the differences between the claimed invention and other earlier inventions are such that the claimed invention would have been obvious at the time the invention was made to a person having ordinary skill in the art to which said subject matter pertains. This requirement denies patentability for insubstantial improvements over existing technology.

Once the patent application is granted, the patentee may exclude others from making, using, or selling the invention in the United States for a term ending twenty years from the date on which the application was filed.

In most instances, the police power of the state does not pursue those who make, sell, or use a patented innovation without the patentee's permission. Instead, the patentee must defend its IP rights through patent infringement suits. In such suits, damages awards are intended to restore the patentee to the financial position that he or she would have occupied if the infringement had not occurred.[8]

In a suit for patent infringement, the court may award the prevailing plaintiff injunctive relief, as well as damages adequate to compensate for the infringement, but in no event less than a reasonable royalty for the use made of the invention by the infringer, together with interest and costs as fixed by the court. Damages adequate to compensate for the infringement may include an award of the plaintiff's lost profits attributable to the infringement, the amount of an established royalty, or a *reasonable* royalty.

11.3.2. Trade Secrets

A trade secret is a piece of economically valuable information that firms want to conceal from their competitors. Trade secrets are usually in the form of a business plan, design, commercial method, or manufacturing process that provides its owner with a competitive advantage. The best known example of a trade secret is the recipe for the Coca-Cola syrup. Trade secrets are only protected by law if they are disclosed through a breach of contract. No legal remedy is given if the secrets are revealed by accident or through reverse engineering by a competitor.

---

[7] For example, Kultti, Takalo, and Toikka (2006) argue that firms seek patents for defensive purpose and Kwon (2012) shows that a strong patent policy increases the number of patent applications but weakens the incentive to innovate.

[8] See Blair and Cotter (1998) for a more complete examination of damages for patent infringement. For a comparative analysis of remedies, see Cotter (2013).

Trade secret law differs from the law of patents in several ways. First, trade secret protection is based primarily on common law and state statutory law, and requires no act of recognition on the part of the government as a precondition of enforceability. Second, any information that provides a person with a competitive advantage as long as it remains secret is potentially protectable as a trade secret. Something as simple as a customer list, for example, may qualify for trade secret protection. Thus, the stringent nonobviousness condition of patent law does not apply. Third, trade secret law affirmatively discourages the owner from making any public disclosure because any such disclosure of trade secret information may result in the information losing its protectable status. Fourth, trade secret protection is of uncertain duration and can be forfeited much more easily than can patent protection. While the protection lasts, the owner of a trade secret may exclude another from, among other things, acquiring the secret by "improper means" such as theft or espionage. He or she may also prevent another from using or disclosing the secret if the other knew, or had reason to know, at the time of disclosure or use that the secret was derived from a person that: (1) had used improper means to acquire it; (2) had acquired it under circumstances giving rise to a duty to maintain secrecy; or (3) owed a duty of secrecy to another. Unlike a patentee, however, the owner of a trade secret cannot exclude one who independently invents or discovers the subject matter of the secret from making use of it, nor can he or she prevent another from attempting to discover, and subsequently exploiting, the secret through reverse engineering. In this respect, trade secret protection is more tenuous and less valuable than patent protection. As long as the owner takes steps to maintain secrecy and there is no independent discovery, the protected status can last forever.

If the plaintiff proves the actual or threatened misappropriation of a trade secret, the court may award injunctive relief. The plaintiff also may recover damages, which may include "both the actual loss caused by misappropriation and the unjust enrichment caused by misappropriation that is not taken into account in computing actual loss." Alternatively, the court may award a reasonable royalty for a misappropriator's unauthorized disclosure or use of the secret.

In either event, the remedy for the victim of a breach is intended to make the victim whole.[9] Due to the consideration of unjust enrichment, it is clear that the remedy is also intended to make infringement unprofitable for the infringer. Obviously the law provides more protection to patentees than to the trade secret possessors.

11.3.3. *Copyrights*

Unlike patent and trade secret law, copyright law forbids certain uses only of the expression of a given idea and not of the idea itself. Specifically, copyright inheres in

---

[9] This, of course, is necessary to protect the incentive to invest.

"works of authorship" that are "original" and "fixed" in a "tangible medium of expression." Works of authorship potentially subject to copyright protection include literary, musical, dramatic, and choreographic works; pictorial, graphic, and sculptural works; motion pictures and other audiovisual works; sound recordings; and architectural works. Unlike patents, copyright is an unregistered right. The *originality* condition requires only independent creation and some minimal degree of creativity, either in the expression of underlying facts or ideas, or in the selection or arrangement of those facts. The *fixation* condition requires that the work be embodied in either a copy or phonorecord, by or under the authority of the author of the work, in sufficiently permanent or stable condition "to permit it to be perceived, reproduced, or otherwise communicated for a period of more than transitory duration." Copyright exists from the moment of creation and initially vests in the author. The copyright term for such works consists of the life of the author plus eighty years.

Copyright ownership entails the five exclusive rights of reproduction, adaptation, distribution, public performance, and public display. Of these, the most important is usually the reproduction right, which entitles the copyright owner to make copies of his own work and to authorize others to do so. Thus, one who copies protectable expression from another's copyrighted work without permission infringes the owner's copyright, absent some overriding defense such as fair use.[10] Note, however, that independent discovery is not actionable in copyright; absent a copying of another's work there can be no liability.

In addition to injunctive relief, the prevailing plaintiff will recover the actual damages suffered by him or her as a result of the infringement, and any profits of the infringer that are attributable to the infringement and are not taken into account in computing the actual damages. Normally, this means that the plaintiff may recover the larger of either his or her lost profits or the defendant's profits attributable to the infringement. Thus, copyright damages are intended to render infringement unprofitable in addition to compensating the copyright holder. In many instances, determining the actual damages or the profits realized through the infringement is decidedly problematic.

Consider the case of Lebbeus Woods, who is the author of several copyrighted books of fantasy architecture. One of Woods' drawings is of a room containing a high-backed chair attached to a wall, a sphere hanging from the ceiling, and other paraphernalia. The director Terry Gilliam saw a copy of this drawing and used it as his inspiration for a handful of scenes in the movie 12 *Monkeys*. In these scenes, the Bruce Willis character is strapped to a similar chair attached to a wall, in a room resembling the one depicted in Woods' drawing. After Universal Studios released the movie, Woods sued for copyright infringement, and the court entered a

---

[10] For a formal examination of the "fair use" doctrine, which allows the use of copyrighted material without prior permission, see Miceli and Adelstein (2006).

preliminary injunction against the public performance or distribution of the film. The defendants, after all, had copied Woods' drawing.

It is obvious that it would be difficult to tease out the portion of the film's profits that flowed from the alleged copyright infringement. Woods' injury may well have been minimal if it were confined to a reasonable royalty. In any event, following injunctive relief the case settled.[11]

The incentive theory of IP suggests that, in the absence of copyright protection, the number of works created and published will be suboptimal, due to the ability of others to free ride upon the efforts of creators and publishers and thereby prevent them from recouping their investments in creation and dissemination. Copyright protection, however, provides copyright holders with monopoly power which could create distortions in the market.

11.3.4. Trademarks

A trademark is any word, name, device, or symbol that identifies a unique source of a product or service. Familiar examples include McDonald's golden arches, Coca-Cola's use of Coke for its signature soft drink, and Nike's swoosh. Trademarks convey information about product quality and affect consumer preference and purchasing decisions. Trademarks therefore reduce search costs for consumers to acquire product information. Trademarks also create incentives for producers to maintain product and service quality that benefit consumers and markets.[12]

In general, the first person to make a lawful, commercial use of a mark to identify his or her product or service acquires the right to exclude others from using the same or a confusingly similar mark for the same or related product or service. The more famous a mark is, however, the greater the likelihood that a court will enjoin others from using a similar designation even on products or services that are not closely related to the owner's business. For example, a motel was precluded from using the name "McSleep" when McDonald's complained that consumers might believe that McDonald's had a business interest in the motel. Presumably, Rolls Royce could preclude the owner of Roger's Ribs from using its logo to protect unwary consumers who might think that Rolls Royce had branched out into the barbeque business. Trademark rights persist for as long as consumers continue to identify the mark with a unique source.

The prevailing plaintiff in a federal trademark infringement action is entitled to injunctive relief and also may recover the defendant's profit attributable to the infringement and any damages sustained by the plaintiff, as long as the court avoids double counting. With respect to actual damages, the court may award; (1) the

---

[11] The injunction prevented Universal from licensing the film for exhibition and thereby gave Woods considerable leverage in negotiating a settlement.
[12] See Shapiro (1982).

plaintiff's lost profits attributable to the infringement; (2) the amount necessary to undertake a corrective advertising campaign; or (3) a reasonable royalty for use of the mark. These options reflect some recognition that defining and calculating damages for trademark infringement can be quite difficult.

11.3.5. Impact of IP Duration

The duration of IP protection varies across the IP areas that we have examined. To the extent that IP rights do not last forever, they provide incomplete protection and, therefore, provide a smaller incentive to invent than would be the case with complete protection. Limited duration of IP protection is intended to balance the incentive for innovation and the distortion from monopoly power.

Since patent protection is of limited duration the patentee cannot extract all of the monopoly profit. A simple numerical example will illustrate this observation. Suppose that an invention has received a patent on filing day and the resulting product is brought to market immediately. Let the demand for the product be

$$P = 100 - 0.001Q$$

and assume further that it remains the same over time. If the marginal (and average) cost of production and distribution is constant and equal to 50, the per year profit will be

$$\Pi = (100 - 0.001Q)Q - 50Q$$

The profit-maximizing quantity will be found where $d\pi/dQ$ equals zero: $Q = 25,000$, which means that $P = 75$. Profit per year would then be \$625,000.

The present value of \$625,000 per year for 20 years is $\sum_{t=1}^{20} \frac{625,000}{(1+i)^t}$, where i is the constant interest rate. This, of course, is the economic value of the patent. If the patent life were infinite; then the value would be $\frac{625,000}{i}$. If, for example, interest rate $i = 0.08$, then the value of the twenty year patent would be \$6,136,340. If the patent life were infinite, the patent owner would be able to get \$7,812,500. As one can readily see, the patentee is able to extract 78.55 percent of the maximum value due to patent protection.

Similarly for copyrights, if the per year profit were, say, \$10,000 and the copyright holder had twenty years of natural life left, then the value of the copyright would be

$$v = \sum_{t=1}^{100} \frac{10,000}{(1+i)^t}$$

which is equal to \$124,940. If the copyright protection lasted forever, the value would be \$125,000. The actual fraction would be 99.95 percent.

The fraction of the value of a trade secret or a trademark that the IP holder can extract is difficult to measure because there is no set limit to their protected status.

11.3.6. Remedies and Investment Incentives

The IP laws provide property rights to inventors and creators, which allow them to profit from their inventions. These profits provide the incentive to invest in an inventive or a creative venture. But these profits also provide an incentive for infringement, which would undermine the incentive to invest in inventive activities. Accordingly, as we have seen, the IP laws contain remedial provisions that are intended to allow the IP holder to recover for injuries that can be traced to infringement.[13] As a general proposition, the remedies are inadequate.

In an ideal world, the remedies for IP infringement would restore the victim to the financial position that he or she would have occupied if the infringement had not occurred. Put differently, the remedies should leave the IP holder indifferent between the two states of the world. In the real world, however, it is unlikely that the victim of infringement will be made whole.

Damage suits for patent infringement are not apt to provide full protection for the injured patentee for several reasons. First, the patentee is responsible for its own costs of litigation. Second, the damage award may be insufficient to cover all past and future losses caused by the infringement. Third, nearly all IP damage suits are settled and any settlement will involve some sacrifices by the patentee. Fourth, the infringer may file a counter-suit, which will have to be defended. This, of course, will impose costs on the patentee.

11.4. MONOPOLY AND MONOPOLY POWER

By definition, a monopolist is the sole supplier of a well-specified product for which there are no close substitutes.[14] Being a monopolist, however, does not confer unlimited power over the market. In typical circumstances, a monopolist cannot *coerce* consumers to pay more for its wares than the amount that they are voluntarily willing to pay. Put differently, the monopolist's ability to raise price is limited by the demand for the monopolized product. The power that a monopolist has derives from the simple fact that it is the sole source of supply. Due to the absence of competition, the monopolist can control the total quantity of its product that is offered for sale and, thereby, control the price. In maximizing its profit, the monopolist restricts output below the competitive level and thereby raises price above the competitive level. Thus, the economic concept of monopoly power is the ability *profitably* to charge prices above the competitive level and to maintain those prices for some period of time. The latter requirement necessarily implies that some

---

[13] The remedies also deter infringement, which is a socially unproductive activity.
[14] If an invention qualifies for IP protection, the inventor will be the sole source of that specific product, but that does not necessarily mean that there are no close substitutes. For example, Apple has a patent monopoly on the production and sale of its iPhones. Nonetheless, Samsung's Galaxy is a somewhat imperfect substitute.

barriers to entry must be present if the monopoly is to survive for very long. This is where intellectual property rights play a vital role.

Although the IP holder must be the guardian of his or her IP rights, IP prohibits the obvious and easiest avenues into the market. Simply copying the product, the production process, or the musical score is made unlawful by the IP laws. Consequently, an entrant must discover another way of entering. In some cases, this may be relatively easy through reverse engineering, but it can be quite difficult. When it is hard to invent around the IP protection, noncompetitive prices can persist as long as the IP protection lasts.

11.4.1. The Lerner Index of Monopoly

IP rights confer a legal monopoly on the successful invention. It does not follow, however, that the owner has much – if any – economic power.[15] That, it turns out, depends upon consumer demand, the existence of substitutes, and production costs. The most commonly used measure of monopoly power can be traced to the seminal work of Abba Lerner.[16] Lerner defined monopoly power as the percentage deviation of price from the competitive outcome. He measured this as

$$\lambda = \frac{P_M - P_C}{P_M}$$

Thus, the Lerner Index is given as the difference between the monopolist's price (P_M) and the competitive price (P_C) divided by the monopolist's price.

Since the competitive price is equal to marginal cost, the Lerner Index can be written as:

$$\lambda = \frac{P_M - MC}{P_M}$$

A profit-maximizing monopolist optimizes where marginal revenue equals marginal costs. Since marginal revenue is equal to $P_M(1 + 1/\eta)$, where η is the price elasticity of demand. Substitution and algebraic rearrangement yields the following expression for the Lerner Index:

$$\lambda = -1/\eta$$

In other words, the Lerner Index equals the absolute value of the reciprocal of the elasticity of demand.

This result makes sense intuitively. As the firm's demand becomes more elastic, the ability to deviate from the competitive price falls. This can be seen in Figure 11.1.

---

[15] The Supreme Court recognized this explicitly in the *Illinois Tool Works v. Independent Ink* case. For an analysis, see Kobayashi (2008).
[16] Lerner (1934).

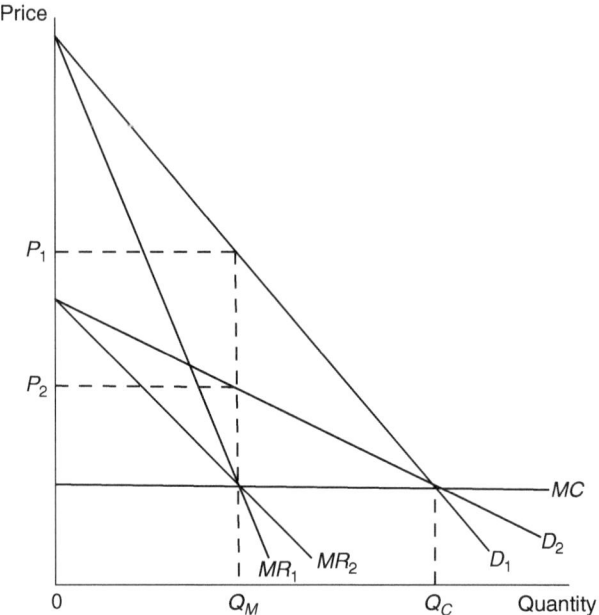

FIGURE 11.1: An illustration of the Lerner Index to measure market power

Consider a special example with two demand curves, D_1 and D_2 in Figure 11.1, which can be written as

$$P = 100 - 0.1Q$$

and

$$P = 70 - 0.05Q, \text{respectively}$$

The marginal cost of production and average cost are assumed to be 40. The two demand functions are constructed such that they intersect the marginal cost curve at the same output. If the marginal (and average) cost is constant, then the competitive output is the same for both demands as is the monopoly output. In this example, the competitive price is $40, the monopoly price for good 1 is $70, and the monopoly price for good 2 is 55.

For D_1, the Lerner Index is

$$\lambda_1 = \frac{70 - 40}{70} = \frac{3}{7}$$

while the Lerner Index for D_2 is

$$\lambda_2 = \frac{55 - 40}{55} = \frac{3}{11}$$

Thus $\lambda_1 > \lambda_2$.

11.4.2. Monopoly Power or Market Power

If a firm can unilaterally and profitably raise its price above the competitive level, it is said to have *monopoly power*. When the firm is a textbook monopoly or a truly dominant firm, the term causes no confusion. In cases where the firm is not dominant, but can still deviate profitably from competitive outcomes, some commentators prefer *market* power to *monopoly* power. In other words, market power and monopoly power differ only in degree, but not in principle. This distinction is not useful[17] from an economic perspective.

To the extent that IP law confers monopoly rights it does so narrowly. Unless permission is granted to others, a patentee may be the sole producer of a product that implements the patent. This does not mean, however, that there are no reasonably close substitutes for the patented product. Consequently, a patentee may enjoy precious little monopoly power in spite of its legal monopoly.

11.5. COMPETITIVE ISSUES

Our system of intellectual property confers a legal monopoly on inventors and creators. This, of course, limits some forms of competition, but it does not eliminate all competition. Nor does it remove complaints about the exercise of the legal monopoly power resulting from IP rights. In this section, we begin by examining the monopoly markups that have been linked to patent protection. Following that, we examine the rationale for the wisdom of compulsory licensing. The practical realities of doing business in some circumstances have resulted in patent pools, cross-licensing agreements, and standard setting organizations. They have also resulted in opportunistic behavior, with the emergence of patent trolls and holdup problems. These raise competitive issues that we will review briefly.

11.5.1. Monopoly Markups and Demand Elasticity

Once a patentee gets to market with its innovation, it will settle on a price and output that will maximize its profit. To do so, it will operate where marginal cost equals marginal revenue, which can be written as:

$$P\left(1 + \frac{1}{\eta}\right) = MC$$

where P represents price, MC represents the marginal cost of production and distribution, and η is the price elasticity of demand. Rearranging the profit-maximizing condition yields:

---

[17] For a more complete discussion, see Krattenmaker, Lande, and Salop (1987).

$$-\eta = \frac{P}{P-MC}$$

Obviously, the larger the markup, the closer $-\eta$ will be to one.

Very large markups result from profit maximization by a patentee in the presence of decidedly inelastic demand. Pharmaceutical products provide excellent examples of substantial markups due to demand inelasticity. Consider Gilead Sciences Inc.'s hepatitis C drug, Sovaldi. Compared to rival treatments, it is far more effective and has far fewer disagreeable side-effects. The treatment requires taking one pill per day for seven weeks. The price per pill was $1,000 while the cost of production may have been about $2. Thus, the markup was $998 above the marginal cost. This markup will be consistent with profit maximization if the demand is sufficiently inelastic.

We can find the price elasticity of demand by substituting for P and MC:

$$\frac{1000}{998} = 1.002$$

Thus, Gilead's dramatic markup, which has gained a good deal of criticism, is consistent with profit maximization. When demand is close to unitary elastic at the quantity where marginal revenue equals marginal cost. After substituting for P and MC, we find $\eta = -1.002$. Thus, the patent monopolist operates in the elastic region of the demand, but just barely.

Whenever we observe substantial markups and the corresponding economic profits that result, questions regarding the wisdom of IP protection are raised. In the Sovaldi example, the price per pill is 500 times the marginal cost of production.[18] This extreme markup may seem excessive, but the substantial profits enjoyed by Gilead provide an incentive for others to invest in creating something better and thereby replace Gilead. In fact, this is precisely what happened. Gilead's Sovaldi displaced an inferior treatment that Merck & Co. sold. In response to Gilead's success, Merck & Co. has now developed Elbasvir and Grazoprevir, which Merck is selling for much less than Gilead's price for Sovaldi. Thus, market forces are doing their work. Nonetheless, many express outrage at dramatic markups and demand some remedy such as compulsory patent licensing.

11.5.2. Compulsory Licensing

Under most circumstances, a patentee may grant permission to others to use its IP or it may refuse to grant such permission. There are occasions, however, where that choice is taken out of the patentee's hands and it is compelled to issue licenses to others. The rationale for compulsory licensing in a misguided effort to lower prices

---

[18] The Massachusetts Attorney General strongly urged Gilead to reduce its price and suggested that its pricing may constitute an unfair business practice, see the letter from Maura Healy (2016) for a more extensive discussion.

is the patentee's reluctance or even outright refusal to license its patent. Under a compulsory license, a patented innovation can be used without the consent of the patent owner but with compensation in the form of a license fee.

If a patentee employs the intellectual property in producing a product that no one else can produce due to the patent protection, it will be able to extract whatever monopoly profit is feasible given the market demand and its cost conditions. Assuming that the patentee maximizes its profit, the monopoly markup represents a reasonable royalty for permission to employ the patent. For example, if demand can be written as:

$$P = 100 - 0.01Q$$

and marginal cost is constant and equal to 20, the patentee's profit-maximizing quantity would be 4,000 and the corresponding price would be 60. The value of the patent per unit of output would then be:

$$P - MC = 40$$

This is then a reasonable royalty in the sense that it is not confiscatory. If the patentee's royalty is limited to some amount below 40, the compulsory licensing will, in effect, confiscate some of the return to the patentee's inventive effort. This possibility will reduce the incentives of others to make similar investments as there is at least some probability that returns will be depressed.

For example, the Agreement on Trade Related Aspects of Intellectual Property Rights (TRIPS) allows member countries to issue a compulsory license to produce foreign patented drugs when there is a national medical emergency. If a reasonable royalty is paid to the patentee, it is unclear why compulsory licensing is necessary. If the patentee is not capacity constrained, the increased demand for its product should cause the patentee to expand its supply of the needed drug. If it is capacity constrained, the pursuit of profits should lead the patentee to contract voluntarily with a competent pharmaceutical manufacturer for the extra production. If compulsory licensing is not going to be confiscatory, however, the gains are unclear. The license fee would have to make the patentee whole. If so, the patentee should be indifferent between licensing and not licensing. To the extent that compulsory licensing makes the patentee worse off, this system will weaken the incentives to invest in inventive activity.

Another aspect of compulsory licensing lies in the impact on countries or firms that use the innovation. Compulsory licensing may discourage domestic innovation if access to foreign inventions at below-market rates provides a disincentive to invent alternative technologies. On the other hand, compulsory licensing may provide incentives to the domestic firms to invest complementary research and skills and create opportunities for learning by doing.[19] Recent theoretical analyses have suggested that the overall welfare effects of compulsory licensing can be positive.[20]

---

[19] Arrow (1962); Stokey (1988); Irwin and Klenow (1994).
[20] Bond and Saggi (2014); Seifert (2015); Stavropoulou and Valletti (2015).

Empirical evidence also supports the positive effect of compulsory licenses on licensing countries' innovation.[21] The bottom line is unclear. Would-be inventors may fear confiscation through compulsory licensing and alter their investment decisions in ways that are predictable but difficult to measure. This will serve to retard progress. On the other hand, there appear to be some benefits for third parties that offset at least some of the adverse effects.

11.5.3. Patent Pools[22]

When patentees commit their patents to a single licensor, they form a patent pool. The patentees then pay a license fee for the privilege of implementing the patents in the pool. In many instances, the patents cover a specific technology. The basic agreement under which this patent pool operates can take many forms and may contain provisions that have both procompetitive and anticompetitive consequences.

Lerner and Tirole suggested a modified patent policy that would create patent pools.[23] Their objective was to maximize the social value of inventive activity. Inefficiency is hard to avoid in our current patent system. The determination of a patent's scope is complicated by difficulties in defining the boundaries of technology space, especially when there are overlapping innovations. Awarding overlapping patents increases the costs of downstream production and may lead to inadvertent patent infringement and consequent litigation, which eventually discourages new innovation. Patent pools are aimed at improving patent system efficiency by issuing combined patents that are shared by a group of firms. This would eliminate patent infringement suits involving patents in the pool. It would also eliminate the need to conduct extensive (and expensive) patent searches. Although patent pools may reduce competition among the patentees, the cost reduction would lead to improved social welfare.[24] The effects of patent pools on the incentive for innovation are not conclusive. The patent pools reduce the risk of litigation among the inventors, which will provide a better environment for research and development. On the other hand, the patent pools also provide free riding possibilities, which would lead to smaller investment in research and development. The empirical literature has found different effects of patents pools on social welfare. Lampe and Moser suggest that patent pools can encourage innovation in substitute technologies[25] while Delcamp found that patent pools lead to more litigation, which tends to dampen incentives to invest in inventive activity.[26]

---

[21] Moser and Venoa (2012).
[22] For a more extensive treatment, see Hovenkamp and Hovenkamp (2015).
[23] Lerner and Tirole (2004).
[24] A complete rigorous analysis can be found in Lerner and Tirole (2004).
[25] Lampe and Moser (2013).
[26] Delcamp (2015).

The jury is still out on the desirability of patent pools. Economic theory does not provide clear guidance and the empirical results are also mixed. Thus, there is no firm foundation for an aggressive antitrust response to the emergence of patent pools, but there is no support for benign neglect.

11.5.4. Cross-Licensing Agreements

Cross-licensing agreements are usually bilateral agreements between rivals that hold patents on similar technology. The purpose of these agreements is to reduce the threat of patent infringement suits and thereby enhance each firm's design freedom. If the two firms have similarly valuable patent portfolios, then they may simply swap IP for IP and no money will change hands. In cases where one firm's patent portfolio is clearly more valuable than the other firm's portfolio, the exchange may well involve a balancing payment, which would take the form of a lump-sum initial payment. In most instances, the agreement would not involve a running royalty.

The competitive consequences of cross-licensing agreements are similar to those of patent pools. Since each firm has access to the IP of its rival, there is some danger that the firms will not invest as much in invention. This, of course, would slow the pace of technological change to the detriment of consumers. At the same time, the reduced danger of patent infringement suits reduces the expected costs of design and production. This would lead to benefits for consumers. On a priori grounds, it is difficult to predict which tendency would predominate. Shapiro (2000) recounts some anecdotal evidence from the microprocessor industry that suggests a positive effect of cross-licensing agreements.[27]

11.5.5. Standard Setting Organizations

Complex products often implement a substantial number of patents in their design and manufacture. In most instances, there are several alternatives and it is desirable that the best one be chosen. Standard setting Organizations (SSOs) coordinate the selection of industry standards to facilitate interoperability across rival brands. This is necessary to guarantee that the iPhone caller can get through to the owner of a Galaxy. The benefits of interoperability to owners are obvious.

In specifying an industry standard, the SSO confers substantial economic power upon the patentees whose patents then become essential. To blunt that power to some extent, the patentees ordinarily limit themselves to "fair, reasonable, and non-discriminatory" license fees. In some instances, however, one or more patentees may behave opportunistically by demanding substantial license fees after a manufacturer has made a commitment to the standard. This is the holdup problem. [28]

---

[27] Shapiro (2000).
[28] For an extensive survey of the hold-up problem, see Lim (2015).

A patentee may deny permission to make, sell, or use a product that implements his or her patent. In most circumstances, there is no obligation to license others to implement a patent. If a patentee issues exclusive or nonexclusive patent licenses, the patentee is free to set the license fee or royalty rate at whatever level maximizes the patentee's profit. In some instances, however, a patentee may behave opportunistically and thereby extract a disproportionate share of the surplus that results from implementing the patent. This is a patent holdup. Those who have made technological commitments to standard essential patents are particularly vulnerable to such opportunism.

11.6. CONCLUDING REMARKS

Our system of intellectual property law creates legal monopolies, but not necessarily economic monopolies. When the demand for the product or process is considerable, IP rights are substantial, and the economic monopoly results in supracompetitive prices and supracompetitive profits returns for the IP owners. These noncompetitive outcomes have adverse consequences for social welfare in a static sense. At the same time, the lure of economic profits provides an incentive to invest in the innovation or creative efforts that provide products, processes, and other creations that would not otherwise exist.

REFERENCES

Arrow, Kenneth J. 1962. Economic Welfare and the Allocation of Resources for Invention. In *The Rate and Direction of Inventive Activity: Economic and Social Factors*. Princeton, N.J.: Princeton University Press.

Blair, Roger D. and Thomas Cotter. 1998. An Economic Analysis of Damages Rules in Intellectual Property Law. *William & Mary Law Review*, 39, 1585–694.

　　2005. *Intellectual Property: Economic and Legal Dimensions of Rights and Remedies*. New York: Cambridge University Press.

Bond, Eric W. and Kamal Saggi. 2014. Compulsory Licensing, Price Controls, and Access to Patented Foreign Products. *Journal of Development Economics*, 109, 217–28.

Cotter, Thomas. 2013. *Comparative Patent Remedies: A Legal and Economic Analysis*. New York: Oxford University Press.

Delcamp, Henry. 2015. Are Patent Pools a Way to Help Patent Owners Enforce their Rights? *International Review of Law and Economics*, 41: 68–76.

Healy, Maura. 2016. Letter to Dr. John C. Martin, Chairman and CEO, of Gilead Sciences, Inc.. dated January 22, 2016; available at http://src.bna.com/cfb.

Hovenkamp, Herbert and Eric Hovenkamp. 2015. *Patent Pools and Related Technology Sharing*. Antitrust and Intellectual Property, Edition 1, Cambridge University Press.

Irwin, Douglas A. and Peter J. Klenow. 1994. Learning-by-Doing Spillovers in the Semiconductor Industry. *Journal of Political Economy*, 102(6), 1200–27.

Kobayashi, Bruce. 2008. Spilled Ink or Economic Progress? The Supreme Court's Decision in Illinois Tool Works v. Independent Ink. *Antitrust Bulletin*, 53, 5–33.

Krattenmaker, Thomas G., Robert H. Lande, and Steven C. Salop. 1987. Monopoly Power and Market Power in Antitrust Law. *Georgetown Law Journal*, 76(2), 241–69.

Kultti, Klaus, Tuomas Takalo, and Juuso Toikka. 2006. Simultaneous Model of Innovation, Secrecy, and Patent Policy. *American Economic Review*, 96(2), 82–6.

Kwon, Illoong. 2012. Patent Races with Secrecy. *Journal of Industrial Economics*, 60(3), 499–516.

Lampe, Ryan and Petra Moser. 2013. Patent Pools and Innovation in Substitute Technologies – Evidence from the 19th-Century Sewing Machine Industry. *RAND Journal of Economics*, 44(4), 757–78.

Lerner, Abba. 1934. The Concept of Monopoly and the Measurement of Monopoly Power. *Review of Economic Studies*, 1, 157–75.

Lerner, Josh and Jean Tirole. 2004. Efficient Patent Pools. *American Economic Review*, 94(3), 691–711.

Lim, Daryl. 2015. *Patent Holdups*. Antitrust and Intellectual Property, Edition 1, Cambridge University Press.

Miceli, Thomas J. and Richard P. Adelstein. 2006. An Economic Model of Fair Use. *Information Economics and Policy*, 18, 359–73.

Moser, Petra and Alessandra Voena. 2012. Compulsory Licensing: Evidence from the Trading with the Enemy Act. *American Economic Review*, 102(1), 396–427.

Nordhaus, William D. 1969. An Economic Theory of Technological Change. *American Economic Review*, 59(2), 18–28.

Seifert, Jacob. 2015. Welfare Effects of Compulsory Licensing. *Journal of Regulatory Economics*, 48, 317–50.

Shapiro, Carl. 1982. Consumer Information, Product Quality, and Seller Reputation. *Bell Journal of Economics*, 13, 20–35.

2000. Navigating the Patent Thicket: Cross Licenses, Patent Pools, and Standard Setting. *Innovation Policy and the Economy*, 1, 119–50.

Stavropoulou, Charitini and Tommaso Valletti. 2015. Compulsory Licensing and Access to Drugs. *European Journal of Health Economics*, 16(1), 83–94.

Stokey, Nancy L. 1988. Learning by Doing and the Introduction of New Goods. *Journal of Political Economy*, 96(4), 701–17.

12

Exploitative Abuses of Intellectual Property Rights

Harry First

12.1. INTRODUCTION

It is the standard view in the United States that US antitrust law does not reach acts of exploitation by a monopolist. The focus in monopolization cases is on exclusionary conduct – conduct that excludes competitors on some basis other than efficiency and thereby allows a firm either to gain or to maintain monopoly. Courts do not pay attention to a monopolist's conduct that is just unfair to its rivals, or even to conduct that is flat-out deceptive. Section 2 of the Sherman Act is concerned with harm to competition, the courts remind us, not harm to competitors.[1] Indeed, and perhaps surprisingly, courts in Section 2 cases are not even concerned with higher prices in themselves – "rent extraction." As the D.C. Circuit Court of Appeals wrote in *Rambus* (a case to which we will return), "[e]ven if deception raises the price secured by a seller, ... it is beyond the antitrust laws' reach."[2]

When it comes to the use of intellectual property rights, this unwillingness to look at exploitation would appear to be even stronger. Early on the Supreme Court affirmed the right of a monopoly patent holder to exploit its rights to the fullest, constrained only by market demand. In *United States v. General Electric*, decided in 1926, the Court allowed GE to license its light bulb patents to a competing light bulb manufacturer and to set the price at which the competitor could sell its bulbs. Chief Justice Taft wrote:

> A research grant from the Filomen D'Agostino and Max E. Greenberg Research Fund at New York University School of Law provided financial assistance for this chapter. I thank Michael Casaburi, Anna Park, Maximilian Riege, and Adam Shamah for their excellent research assistance. I was a consultant with regard to a proceeding related to the Japan Fair Trade Commission's investigation of Qualcomm, but the views expressed in this chapter are mine alone.
>
> [1] See, e.g., *United States v. Microsoft Corp.*, 253 F.3d 34, 58 (D.C. Cir. 2001) ("harm to one or more competitors will not suffice").
>
> [2] *Rambus, Inc. v. FTC*, 522 F.3d 456, 464 (D.C. Cir. 2008). See also *Berkey Photo, Inc. v Eastman Kodak Co.*, 603 F.2d 263, 297 (2d Cir. 1979) ("[a] pristine monopolist ... may charge as high a rate as the market will bear").

[T]he patentee may grant a license ... under the specifications of his patent for any royalty or upon any condition the performance of which is reasonably within the reward which the patentee by the grant of the patent is entitled to secure ... One of the valuable elements of the exclusive right of a patentee is to acquire profit by the price at which the article is sold. The higher the price, the greater the profit, unless it is prohibitory.[3]

Outside the United States, however, the law seems to be otherwise. Article 102 of the TFEU prohibits "abuse" of a dominant position, with a specific clause to catch the imposition of "unfair" selling prices or trading practices. Many countries follow the European Union's approach. China's condemnation of abuse of dominance includes selling at "unfairly high prices," or "other abusive practices" as determined by the enforcement authority (Art. 17). South Africa specifically condemns, as an abuse of dominance, the charging of an "excessive price" (Sec. 8). India prohibits a dominant firm from imposing "unfair" prices in the purchase or sale of goods or services (Sec. 4(2)(a)(ii)). Korea prohibits a dominant firm from pricing "unreasonably" or "unreasonably interfering" with the business activities of other enterprises (Art. 3-2). Japan prohibits "private monopolization" à la United States (Art. 3), but also condemns unfair trade practices, which include "dealing at unjust prices" and dealing with another party on terms that "restrict unjustly" the other party's business (Arts. 19, 2(9)).

Without denying this substantial divergence in general between the United States and the rest of the world, it turns out that there may be fewer differences between the United States and other jurisdictions when it comes to judging exploitative behavior by intellectual property rights holders with market power. For despite the oft-stated unwillingness to condemn exploitation under US antitrust laws, and even despite the broad license given to intellectual property rights holders in *General Electric*, legal doctrine and enforcement policy in the United States is much more willing to rein in exploitative behavior by intellectual property rights holders than might otherwise be supposed.

The purpose of this chapter is to describe the areas in which antitrust law (or competition law, as it is generally referred to outside the United States) constrains intellectual property rights holders from unduly exploiting their monopoly power when licensing or using their intellectual property rights. "Exploitation" is here used in the sense of "taking advantage" of downstream purchasers by extracting rents, either through higher prices or though the imposition of nonprice terms.[4] Some forms of exploitation might lead to exclusion, for example where exploitative behavior raises rivals' costs or increases entry barriers, and thus the line between exploitation and exclusion is not always perfectly clear. Nevertheless, a critical distinction is the focus on harm to the immediate buyer (or licensee) without any

---

[3] *United States v. General Electric Co.*, 272 U.S. 476, 489-90 (1926).
[4] Akman (2009); O'Donoghue and Padilla (2006).

necessary concern for ultimate effects on consumer welfare or deadweight welfare loss.

From a normative standpoint, this chapter argues that intervention to prevent this type of exploitation is consistent with sound competition policy. Preventing intellectual property rights holders from undue exploitation of their rights is an important aspect of economizing on the reward that we give intellectual property rights holders to incentivize innovation. The argument over how much short-term monopoly loss we are willing to incur so as to get long-term innovation is a familiar one in the intellectual property literature.[5] Although some argue, in effect, that "too much is not enough,"[6] antitrust tradition is on the side of placing some limits on monopoly profits and placing greater reliance on the incentives that competitive markets provide. Many of the cases this chapter discusses are in that tradition.

This chapter focuses on three areas in which antitrust enforcers have intervened to prevent exploitation. The first involves patents subject to FRAND (fair, reasonable, and non-discriminatory) licensing obligations (FRAND-committed patents[7]), a major area in which courts and agencies have been willing to prevent excessive pricing. The second area involves disclosure requirements that can be imposed on patent holders to prevent the exploitation of licensees or potential licensees. The third is post-expiration royalties. The chapter concludes with some observations about the emerging policy consensus regarding abusive licensing by patent holders with market power.

12.2. EXCESSIVE PRICING: FRAND OBLIGATIONS

12.2.1. *The General Problem*

FRAND licensing obligations have arisen in the context of the establishment of industry standards that allow interoperability among diverse products. These standards have been adopted through the efforts of private industry standard setting organizations (SSOs) and have been particularly important in high-technology industries, such as electronics and communications equipment, where common platforms are necessary if firms are to manufacture compatible but competing products.

SSOs have adopted FRAND obligations to solve a particular problem. Industry participants know that once a standard is chosen, firms will design products that incorporate the standard. If there are patents that are essential for utilizing the standard ("standard essential patents," or SEPs), firms that use the standard will be

---

[5] First (2007).
[6] Scotchmer (1991).
[7] Commentators generally call such patents "FRAND-encumbered." This chapter will use the term "FRAND-committed," reflecting the commitment a patent holder makes to get its patent adopted as part of a standard, a commitment from which the patent holder benefits substantially.

liable for royalties. When an SSO chooses a standard, it will, of course, give consideration to issues of technical superiority, but industry participants will also want to minimize royalty rates. The SSO could try to negotiate royalties in advance of adopting a standard, but the SSO is just a collective of buyers and its negotiations could be viewed as a buyer's cartel, raising the potential for antitrust liability. On the other hand, if implementers of the standard wait until they design products that need to use SEPs, they might be subject to opportunistic behavior from patent holders. Implementers will likely have made substantial investments in standards-compatible products and become effectively locked into the standard. Patent holders would then be able to hold up licensees for high royalties, not because of the intrinsic innovative value of the patent but because of the value of the investments that the potential licensee has made.

To solve this holdup problem SSOs began requiring firms that owned patents reading on a proposed standard to agree to license those patents on RAND (or FRAND) terms (FRAND and RAND are used interchangeably).[8] What might be "reasonable," however, was left to later bargaining between licensees and licensors.

Besides leaving royalty rates unspecified, the FRAND approach did not take account of two particularly difficult problems in a number of industries heavily reliant on multiple SEPs. Products in the electronics and communications industries use a large number of patents (in contrast to pharmaceutical drugs, for example), and product innovations tend to be incrementally made within a broad portfolio of patents. Under FRAND, how would one assess, ex post, the value of any particular patent to the production of, say, a smartphone that uses 250,000 patents?[9] And even if any individual patent holder asked for a "reasonable" royalty, the stack of patents necessary to produce a complicated electronics product might lead to extremely high royalties in total, with potentially adverse effects on downstream pricing and innovation.[10]

12.2.2. United States: Private Litigation and Government Enforcement

Not surprisingly, FRAND obligations have led to litigation brought by implementing firms that believe that holders of FRAND-committed patents are demanding excessively high royalty rates. The major example of this type of litigation has been the "smartphone wars."[11] Some of this litigation has been framed as breach of contract, with the implementer arguing that the patent holder had effectively agreed to bargain in good faith, but had breached that agreement.[12] Other cases, however,

---

[8] Crane (2010). For a broader discussion of the reasons for using FRAND, see Tsai and Wright (2015).
[9] Contreras (2015a).
[10] The potential for royalty stacking, derived from the theory of double marginalization by monopolists of complementary products, was developed by Lemley and Shapiro (2007). See also Shapiro (2001).
[11] Maldonado (2014).
[12] On the problematics of contract enforcement, see Contreras (2015b).

have been framed as Sherman Act Section 2 cases. No matter how framed, however, the key competition concern has been the exploitation of the patent licensee through rent extraction.

For example, in *Microsoft Corp. v. Motorola*, Microsoft alleged that Motorola had breached its RAND obligations when it offered to license two SEP portfolios based on 2.25 percent of the price of the end product in which they were incorporated (which was all computers running Windows, no matter the manufacturer, and all Xbox video game consoles). The district court found that the RAND royalty rate should have been much lower – $.00555 per unit for one portfolio and $.0371 for the other.[13] Even though this was a contract action, the Ninth Circuit Court of Appeals spoke the language of competition policy, supporting its liability decision by referring to competition and rent extraction:

> The development of standards ... creates an opportunity for companies to engage in anticompetitive behavior ... Using that standard development leverage, the SEP holders are in a position to demand more for a license than the patented technology, had it not been adopted by the SSO, would be worth ... [and] extract more than the fair value of its patented technology.[14]

A direct antitrust attack on high FRAND rates came in *Broadcom v. Qualcomm*. Broadcom was a manufacturer of chipsets for mobile phones that employed the Wide Band Code Division Multiple Access standard (WCDMA).[15] Qualcomm's patents were essential for that standard. Broadcom alleged that Qualcomm had monopolized the WCDMA technology market by making a "false promise" to license its WCDMA essential patents on FRAND terms, a promise on which the relevant standards development organizations had relied when adopting the WCDMA standard, and then had refused to license its technology on FRAND terms. The district court dismissed the complaint, but the Court of Appeals reversed.

The court of appeals put weight on Broadcom's bare allegation of deception, but the court's policy concerns focused on rent extraction: "In [its] unique position of bargaining power, the patent holder may be able to extract supracompetitive royalties from the industry participants." Qualcomm's alleged deception "obscure[ed] the costs of including proprietary technology in a standard."[16]

---

[13] *Microsoft Corp. v. Motorola, Inc.*, 795 F.3d 1024, 1033 (9th Cir. 2015). The patent portfolios were for Wi-Fi and for playing back high-definition video. *Id.* at 1046.

[14] *Id.* at 1030–1.

[15] Code Division Multiple Access is one of the two mobile telephony "paths," or families of standards, used in countries around the world. The other "path" is the Universal Mobile Telecommunications System (UMTS), also known as the Global System for Mobility (GSM), created by the European Telecommunications Standards Institution (ETSI) and its standards organization counterparts in the United States and elsewhere. Wideband CDMA is a component of the technologies for the UMTS standard.

[16] *Broadcom Corp. v. Qualcomm Inc.*, 501 F.3d 297. 310, 314 (3d Cir. 2007). For a more recent antitrust case making similar allegations, see *Microsoft Mobile Inc. v. InterDigital, Inc.*, Case 1:15-cv-00723-UNA, ¶¶ 4, 6 (D. Del., filed August 20, 2015) (alleging monopolization of

United States antitrust enforcement agencies have also been concerned that patent owners have been "imposing excessive royalty obligations on licensees."[17] The Federal Trade Commission (FTC), which has been the more active of the two federal US agencies in this area, has recommended that courts, when awarding reasonable royalty damages in patent litigation in general, should base the award on the "incremental value of the patented invention" as of the time the alleged infringer makes its design choice so as not to "overcompensate" the patentee. Similarly, the Antitrust Division and the Patent and Trademark Office have argued that owners of FRAND-committed SEPs can "hold up" the patent holder to "obtain a higher price" for the use of patented technology "than would have been possible before the standard was set, when alternative technologies could have been chosen."[18]

This concern for excessively high royalty rates has been reflected in more than government reports. In 2013, for example, the FTC issued a complaint against Google, alleging a violation of Section 5 of the Federal Trade Commission (FTC) Act arising out of Google's breach of commitments to license certain SEPs on FRAND terms, commitments to which its newly acquired subsidiary, Motorola Mobility, had previously agreed. The breach was alleged to be the likely result of Google's prosecution of claims for infringement of its SEP patents before the International Trade Commission and the courts, seeking, respectively, exclusion orders and injunctions. Of course, patent holders are generally thought to be able to seek this sort of relief when their patents are infringed, so why are such relief requests an "unfair method of competition" when FRAND-committed SEPs are involved? The Commission explained: threats to use injunctions to deprive implementers of future sales allowed Google to "demand licensing terms that tended to exceed the FRAND range." The "anticompetitive effect" of this conduct was "increase[ed] costs," which the Commission termed a "substantial consumer injury." The Commission alleged that "[i]f Google's practices are allowed to continue, many consumer electronics manufacturers will agree to pay unreasonable royalties simply to avoid an injunction or exclusion order. Manufacturers will likely pass on some portion of these costs to end consumers."[19]

The FTC had made an even more direct attack on high licensing rates in *Negotiated Data Solutions* (*N-Data*), decided five years before the Google case. *N-Data* involved the standard setting process, but not a commitment to license on undefined FRAND terms. Rather, N-Data's predecessor, in the course of an SSO's

---

technology markets by falsely promising to license its SEPs on FRAND terms and then demanding "excessive and discriminatory royalties").

[17] US Federal Trade Commission (2011).
[18] US Department of Justice and US Patent & Trademark Office (2013).
[19] *In the Matter of Motorola Mobility LLC and Google, Inc.*, Docket No. C-4410, Complaint ¶¶ 25–28, 30 (2013). Available at www.ftc.gov/sites/default/files/documents/cases/2013/07/130724goo glemotorolacmpt.pdf.

adopting an Ethernet standard that allowed backward compatibility, had promised to license the patents covering the technology to any requesting party for a one-time fee of $1,000. The relevant patents were later assigned to another company, and eventually to N-Data. The later assignees, although aware of the commitment, decided that the patents were worth more and set out to collect the higher royalties from a group of target companies that included many large computer hardware manufacturers. The royalties demanded represented a "substantial increase" over the original $1,000 fee.[20]

The FTC's complaint charged that N-Data's conduct was an "unfair method of competition" in violation of Section 5 of the FTC Act. The "threatened or actual anticompetitive effects," the FTC asserted, included "increased royalties" for the manufacture or sale of products that implement the standard.[21] As the Commission explained, even if N-Data's conduct did not violate the Sherman Act, the conduct "threatened to raise prices for an entire industry."[22]

The Commission's concerns were not exclusively focused on price raising. These cases arose in the context of the standard setting process. The Commission emphasized that protecting the integrity of that process by bolstering assurances against holdup was critical to innovation as an overall matter. It wasn't just that buyers (licensees and downstream consumers) should be free of excessively high prices. The Commission felt that industries that innovate around standards will be more innovative if royalties for implementing the standards are ex ante fairer and if participants keep their promises.[23]

In neither of these cases, however, did the Commission focus on whether the patent holder had monopoly power in some well-defined market or whether it exercised that power in an exclusionary manner. No real attention was even paid to the extent that the practices restricted output or raised entry barriers. Rather, the Commission talked about exploitation. Patent holders, the Commission wrote, should not be allowed to "exploit the power [they] enjoy" over firms that "lack[] any practical alternatives" because they are locked into a standard.[24]

What of the legal argument that the antitrust laws don't reach pure exploitation? The legal answer is that the Commission felt that it was operating under a "stand-alone" approach to Section 5 of the FTC Act. Section 5 gives the Commission

---

[20] *In the Matter of Negotiated Data Solutions, LLC.*, Docket No. C-4234, Complaint, ¶ 28 (2008). Available at www.ftc.gov/sites/default/files/documents/cases/2008/09/080923ndscomplaint.pdf.

[21] *Id.* ¶ 37a.

[22] N-Data, Analysis of Proposed Consent Order to Aid Public Comment, at 4. Available at www.ftc.gov/sites/default/files/documents/cases/2008/01/080122analysis.pdf.

[23] E.g., *id.* at 6; Motorola Mobility/Google, Analysis of Proposed Consent Order to Aid Public Comment, at 2–3. Available at www.ftc.gov/sites/default/files/documents/cases/2013/01/130103googlemotorolaanalysis.pdf.

[24] N-Data, Analysis to Aid Public Comment, at 5. See Motorola Mobility/Google, Analysis to Aid Public Comment at 5 ("opportunistic breach of its licensing commitment had the tendency of leading to higher prices for consumers and undermining the standard-setting process").

authority to stop "unfair methods of competition," but Congress chose not to define that vague term when enacting the FTC Act. All agree that the inner bounds of the term include anything that violates the Sherman Act, but the outer bounds are highly contested. Both *Google* and *N-Data* represent significant efforts by the Commission to mark an outer bound that would be sensitive to the exploitative behavior of patent holders, albeit exploitative behavior that still had an adverse (if quantitatively undefined) effect on competition in downstream markets and on innovation generally.

Whether the courts would accept the Commission's legal approach to this type of conduct, however, is uncertain. Both cases were settled by consent and no court has recently been called on to review the Commission's approach to a stand-alone theory of Section 5. Indeed, the Commission continues to struggle with its own interpretation of its authority under Section 5, subsequently releasing a one-page statement that attempts to set out a general approach to the issue but that does not specifically address the question of exploitative pricing of FRAND-committed patents (or patents more generally).[25] Whatever the legal uncertainties, however, the Commission continues to investigate abuses in licensing FRAND-committed patents.[26]

12.2.3. Enforcement and Litigation Outside the United States

Competition enforcement authorities outside the United States do not face the same legal constraint as US enforcers do with regard to reprehending high prices, leaving these agencies (and private litigants) freer to pursue cases of excessive pricing by patent holders. Four jurisdictions have been particularly active: China, the European Union, Japan, and Korea.[27]

12.2.3.1. China

Huawei v. InterDigital was a suit for damages brought under the Antimonopoly Law by the Chinese mobile phone producer, Huawei Technologies, against InterDigital, a US licensor of patents related to mobile communications standards. Among Huawei's claims were that InterDigital abused its dominance when it charged unreasonably high and discriminatory patent royalty rates, in violation of its FRAND commitments.[28] In 2013 the Guangdong High People's Court, affirming the court of

---

[25] U.S. Federal Trade Comm'n, Statement of Enforcement Principles Regarding "Unfair Methods of Competition" Under Section 5 of the Federal Trade Commission Act, 80 Fed. Reg. 57056 (September 21, 2015). Commissioner Olhausen, in dissent, noted the lack of specificity with regard to breach of standard-setting commitments, *id.* at 57058.

[26] Qualcomm (2015), reporting that the FTC notified it in 2014 that it was investigating Qualcomm's licensing practices, "including potential breach of FRAND commitments."

[27] For similar developments in India, see Gandhi, Metanios, and Dadwal (2015).

[28] Chin (2015).

first instance, found that each of InterDigital's SEPs constituted a relevant market and that InterDigital had a dominant position in each market because Huawei had no substitutes if it wanted to continue to produce mobile phones. Dominance thus established, the court found that InterDigital's rates were unreasonable – at least seven times higher than the rates it charged other licensees – a difference that was not cost-justified.[29] The court imposed $3.2 million in damages and, in a companion case, set the royalty rate for InterDigital's SEPs at 0.019 percent of the sales price of Huawei's products, which was not only lower than the original rate, but apparently lower than what Huawei itself was charging licensees for its own handset patents.[30]

On the government side, the National Development and Reform Commission (NDRC), which is responsible for conduct involving price-related anticompetitive conduct, also investigated InterDigital's licensing practices. Following the High Court's 2013 decision in *Huawei*, however, the NDRC suspended its investigation, receiving commitments from InterDigital to lower its royalty rates for its portfolio of patents for 2G, 3G, and 4G wireless mobile technology.[31]

More high profile was the NDRC's investigation of Qualcomm and the settlement of that investigation in 2015. The NDRC found that the licensing of each of Qualcomm's SEPs constituted a relevant market and that the relevant market for the case was a "collection" of those markets."[32] As in the High Court decision in *Huawei*, the NDRC concluded that Qualcomm had a dominant position in its SEP markets. Each SEP is "indispensable and irreplaceable" for terminal equipment manufacturers, making these manufacturers "highly reliant on Qualcomm's patent portfolios," and firms with competing technologies are not able to enter the market with standards-compliant technology.[33]

Having found dominance, the Commission then decided that Qualcomm abused its dominant position by charging "unfairly high royalties" in violation of Article 17 of the Antimonopoly Law, which forbids unfair high prices. Unlike the High Court's approach in *Huawei v. InterDigital*, however, the NDRC did not compare Qualcomm's rates to those it gave other licensees. Instead, the NDRC based its decision on three of Qualcomm's practices that it felt resulted in improperly high rates: not offering its licensees a list of the patents they were licensing (Qualcomm might require payment for patents that had already expired), requiring licensees to give it royalty-free grant-back licenses (for which presumably Qualcomm would otherwise have had to pay), and requiring licensees to take a patent portfolio that included non-SEPs as well as SEPs.[34] "The

---

[29] Hou (2015a).
[30] Chin (2015); InterDigital (2016). The parties have settled their disputes in every jurisdiction other than China, agreeing to binding arbitration (Chin 2015). The Chinese court's royalties order is under review in a proceeding before the Supreme People's Court (InterDigital 2016).
[31] Chin (2015).
[32] See *Qualcomm*, NDRC [2015] Nr. 1, Feb. 9, 2015, Sec. I.1 (MLex translation).
[33] *Id.* Sec. I.1c. The case is also usefully discussed in Hou (2015b).
[34] *Id.*, Secs. II.1.1-.2. Hou (2015b).

combination of these factors," the Commission explained, "[led] to high licensing fee[s]."[35]

The NDRC not only ordered Qualcomm to stop its abusive licensing practices, but also required Qualcomm to pay a 6.088 billion yuan fine (nearly $1 billion), which was 8 percent of Qualcomm's revenue in China for 2013.[36] Qualcomm accepted both orders, further agreeing to reduce the royalty base for Chinese licensees from 100 percent of the net wholesale price for devices sold for use within China to 65 percent (a "sharp discount" from what Qualcomm charged elsewhere)[37] and to "continue to invest" in China.[38] In doing so, it avoided the possibility of even more serious penalties, such as being required to provide royalty-free licenses or to pay the maximum penalty under the Antimonopoly Act of 10 percent of sales.[39]

12.2.3.2. Europe

High royalty rates by SEP holders have also been a concern in Europe. In 2007, the European Commission, acting on complaints lodged by six of Qualcomm's competitors and customers (Ericsson, Nokia, Texas Instruments, Broadcom, NEC, and Panasonic), opened an investigation into Qualcomm's licensing practices for WCDMA. The "economic principle" on which the complaints were based, the Commission wrote, "is that essential patent holders should not be able to exploit the extra power they have gained as a result of having technology based on their patent incorporated in the standard."[40] This "extra power," one might argue, could be viewed as the legitimate return to innovation that comes from the grant of patents that are essential for the mobile telephone standard. Apparently, though, the Commission felt that the exercise of that power might lead to rates that were not "fair," in contravention of Qualcomm's commitment to FRAND licensing.

Two years later the complaints were withdrawn and the Commission abandoned its investigation into Qualcomm's licensing practices, recognizing that pricing assessments "may be very complex, and any antitrust enforcer has to be careful

---

[35] Press Release, NDRC, National Development and Reform Commission ordered rectification Qualcomm monopolistic behavior and fined six billion yuan (August 30, 2015) at 3. Available at http://jjs.ndrc.gov.cn/gzdt/201502/t20150210_663872.html (Google translation).
[36] *Qualcomm* Sec. III.2. Note that the conduct for which the fine was imposed also included a finding of abuse of dominance in baseband chips by imposing "unfair" trading conditions. *Id.* Sec. II.3.
[37] Mozur and Hardy (2015).
[38] NDRC Press Release at 2.
[39] Dou (2015).
[40] Press Release, European Comm'n, Antitrust: Commission initiates formal proceedings against Qualcomm, MEMO/07/389 at 1 (October 1, 2007). Available at http://europa.eu/rapid/press-release_MEMO-07-389_en.htm?locale=en.

about overturning commercial agreements."[41] Nevertheless, the Commission did not abandon its concern over the ability of SEP holders to extract high licensing fees.

In 2011 the Commission issued Horizontal Cooperation Guidelines, which discussed, among other topics, standards agreements. The Commission pointed out that the holder of an essential intellectual property right could hold up users ex post "by extracting excess rents by way of excessive royalty fees."[42] Although the Commission stated that it viewed this behavior as "anti-competitive," it eschewed a firm conclusion on whether such behavior would be an abuse of dominance under Article 102, leaving for a case-by-case determination the question whether a SEP holder had "market power" and whether the rates it was charging would be considered excessive under Article 102 and European case-law.[43]

Shortly thereafter the Commission opened two proceedings under Article 102, one against Samsung, the other against Motorola/Google, dealing with their efforts to seek injunctions against Apple in various European courts to enforce their standard essential patents relating to smartphone technology. In 2014 the Commission concluded that Motorola/Google had abused its dominant position under Article 102 because Apple was "not unwilling" to accept a license on FRAND terms, as Motorola had promised the standard setting organization it would offer.[44] At the same time Samsung entered into a commitment (the Commission's form of a binding settlement) that specified detailed procedures that Samsung would be required to follow to ensure that its licensing terms were fair and reasonable.[45] Both the decision and the commitment were bottomed on the notion expressed in the Horizontal Cooperation Guidelines that the holder of a FRAND-committed SEP is entitled only to "appropriate remuneration." Seeking an injunction for infringement against a SEP licensee that would be willing to accept FRAND terms constitutes an

---

[41] Press Release, European Comm'n, Antitrust: Commission closes formal proceedings against Qualcomm, MEMO/09/516 (November 24, 2009). Available at http://europa.eu/rapid/press-release_MEMO-09-516_en.doc.

[42] Guidelines on the applicability of Article 101 of the Treaty on the Functioning of the European Union to horizontal co-operation agreements, para. 269 [2011] OJ C11/01.

[43] Id. With regard to the case-law, the Commission referred to *United Brands* [1978] ECR 207, the leading EU case on the standard for judging when a price is excessive.

[44] Case AT.39985 *Motorola - Enforcement of GPRS standard essential patents*, C(2014) 2892 (April 29, 2014). Available at http://ec.europa.eu/competition/antitrust/cases/dec_docs/39985/39985_928_16.pdf.

[45] See Case AT.39939 *Samsung - Enforcement of UMTS standard essential patents*, C(2014) 2891, paras. 75-118 (April 29, 2014). Available at http://ec.europa.eu/competition/antitrust/cases/dec_docs/39939/39939_1501_5.pdf. Although the Commission in both cases adhered to its view that a SEP holder does not automatically have dominance, the Commission rested a finding of dominance on the parties' 100% market share in the licensing of their SEP technologies necessary for practicing the relevant standard, the indispensability of the standard for manufacturers of standards-compliant products, and industry lock-in to that standard as a result of heavy investment in standards-compliant products and assets. See *Motorola*, paras. 221–26; *Samsung*, paras. 41–51.

abuse of dominance because the result of a threat of an injunction would result in improper rent extraction.[46]

12.2.3.3. Japan

Japan has also been interested in excessive prices charged by the holders of FRAND-committed SEPs. In 2006 the Japan Fair Trade Commission (JFTC) began a long-running proceeding against Qualcomm focused on two of its practices regarding the licensing of FRAND-committed SEPs for CDMA technology. One required Japanese handset manufacturers to grant Qualcomm royalty-free licenses to their patents; the other forbade Japanese manufacturers from asserting their patents against Qualcomm or Qualcomm's customers (a non-assertion of patents clause).[47] Both practices prevented licensees from taking actions that could have lowered the net royalties due Qualcomm.

In 2009 the JFTC found Qualcomm's practices to be a violation of the Antimonopoly Act, not as monopolization but as a designated "Unfair Trade Practice," specifically, "trading on restrictive terms."[48] Unlike China's similar case against Qualcomm, the JFTC did not focus its attention on high prices. Instead, the Commission argued that the inability of Japanese manufacturers to get royalties from Qualcomm "impeded their incentive to engage in research and development" (presumably because they were denied royalties on their innovations). In this sense the JFTC's policy concerns were more like the US FTC's concerns in *Google/Motorola* and *N-Data* regarding the effect of a SEP holder's practices on innovation. The difference is that the US FTC was concerned about the effect on innovation in a standard setting environment if royalties are ex ante fair whereas the JFTC was more explicitly focused on downstream innovation. The Commission thus made an implicit tradeoff between upstream innovation (Qualcomm's) and downstream innovation (Japanese manufacturers'). Figuring this tradeoff is difficult, though, and the Commission's result is, at best, speculative (something one could say about the US FTC's conclusions about innovation in *Google/Motorola* and *N-Data* as well). The more predictable result of the JFTC's decision – whether intended or not – is easier to see. Qualcomm's licensees will pay lower net royalties to Qualcomm for Qualcomm's FRAND-committed SEPs.[49]

---

[46] E.g., *Motorola*, para. 76; *Samsung*, para. 58. The European Court of Justice has also accepted the idea that the effort of a holder of a FRAND-committed SEP to enjoin a willing licensee can be an abuse of dominance under Article 102, although the Court did not base its decision on a desire to prevent improper rent extraction. See *Huawei Technologies Co. v. ZTE Corp.*, C-170/13, EU:C:2015:477.

[47] Non-assertion of patents clauses were also involved in the NDRC proceeding against Qualcomm. See *Qualcomm* Sec. II.1.b.

[48] See Cease and Desist Order against QUALCOMM Incorporated (English version), September 30, 2009. Available at www.jftc.go.jp/en/pressreleases/yearly-2009/sep/individual-000038.files/2009-Sep-30.pdf.

[49] The JFTC began hearing proceedings against Qualcomm in 2010, see Decision to Commence Hearing Procedures against QUALCOMM Incorporated (English) (January 7, 2010). Available at www.jftc.go.jp/en/pressreleases/yearly-2010/jan/individual-000033.files/2010-Jan-7.pdf, but at the time of writing the proceedings had not yet been concluded.

The JFTC's willingness to side with downstream licensees over upstream SEP holders was carried forward in its subsequent decision to amend its intellectual property guidelines to cover the case of a FRAND-committed SEP holder attempting to enforce its patent against a willing licensee by seeking to enjoin the licensee from practicing the patent rather than entering into a licensing agreement.[50] Consistently with the views of the US Department of Justice, the US FTC, and the European Commission (to which the JFTC referred), the Commission announced that such efforts would be treated as violations of the Antimonopoly Act.[51] The Commission indicated that it viewed such conduct as the equivalent of a refusal to trade, violating two designated unfair trade practices ("refusal to trade" and, depending on the businesses of the licensor and licensee, "interference with a competitor's transactions") as well as possibly violating the Antimonopoly Act's prohibition on private monopolization.[52] Consistently with its approach in its case against Qualcomm, the Commission did not focus on rent extraction. In fact, the Commission specifically noted that US and EU competition authorities believe that the charging of high royalties "is not considered as a [competition policy] problem."[53] Instead, the JFTC continued to emphasize the impact of seeking an injunction on the ability of downstream firms to "research & develop, produce and sell the products adopting the standards."[54] Of course, this effect does not come from actually withholding licenses or excluding downstream firms from markets – SEP holders are in the business of licensing downstream users, after all – but from the impact of "prohibitively expensive" royalties on downstream producers that succumb to the threat of an injunction.[55] In other words, the problem really is rent extraction.

12.2.3.4. Korea

In 2006 the Korea Fair Trade Commission (KFTC) began a three-year investigation of Qualcomm that culminated in a 2009 finding that Qualcomm's licensing of its FRAND-committed SEPs violated the Monopoly Regulation and Fair Trade Act (MRFTA). Korea fined Qualcomm more than $200 million, the largest fine it had ever imposed on a single company.[56]

---

[50] See Japan Fair Trade Commission, Guidelines for the Use of Intellectual Property under the Antimonopoly Act, Sec. 3(1)(i)(e). Available at www.jftc.go.jp/en/pressreleases/yearly-2016/January/160121.files/IPGL_Frand_attachment.pdf.
[51] See Japan Fair Trade Commission, Survey Report on Issues Related to Essential Patent at 6 (July 8, 2015). Available at www.jftc.go.jp/en/pressreleases/yearly-2015/July/150708.files/Attachment2.pdf.
[52] Id. at 11. The legal grounds for the Commission's conclusion are different than the ones it asserted against Qualcomm, supra. For critical comment on the draft of the JFTC Guidelines, see Wright and Ginsburg (2015).
[53] Survey Report, supra, at 7.
[54] Id. at 12.
[55] Id. at 14.
[56] Lee (2012).

The KFTC's case is a rare example of a competition enforcement agency focused on the "non-discriminatory" part of FRAND. The KFTC found that Qualcomm had charged discriminatory rates in three ways: (1) for domestically sold mobile phones, Qualcomm deducted from royalties the value of chips and components that the manufacturer purchased from Qualcomm but not from other providers; (2) for mobile phones sold for export, Qualcomm similarly charged lower royalties for phones that included Qualcomm modem chips; and (3) Qualcomm capped its per phone royalties at a lower amount if the manufacturer used Qualcomm modem chips and not a competitor's.[57] After finding that the rates were discriminatory, the Commission then did a very thorough analysis of the competitive impact of that discrimination on competitors that Qualcomm faced in downstream markets for modem and other types of chips. In its 188-page opinion, the Commission analyzed Qualcomm's successful effort to use its position in licensing CDMA technology to exclude competitors (such as Samsung, EoNex, VIA Telecom, and Texas Instruments) in the downstream CDMA chip market. This market-impact analysis is probably the most thorough that any competition enforcement agency has provided in FRAND cases.

The Commission found that Qualcomm's discriminatory rates were an abuse of dominance under Section 3-2 of the MRFTA, for "unreasonably interfering" with the business activities of other enterprises. The Commission also found that the rates constituted an unfair trade practice under Article 23(1), which forbids "unjustly ... treating a trading partner in a discriminatory manner." Although the latter does not require proof of dominance, and is not necessarily grounded in proof of competitive effects, the Commission's unfairness analysis nevertheless focused on the effect of Qualcomm's discriminatory rates on competition from chip competitors it was attempting to exclude in downstream markets.[58]

The Commission's case against Qualcomm's discrimination was thus bottomed on exclusionary effect in other markets in which Qualcomm was competing, and not on unfair pricing in the licensing of its patents. Nevertheless, the Commission also hinted that it might take the view that the decision of a SEP holder to renege on a commitment to FRAND rates might in itself raise competitive concerns.[59] Indeed, the KFTC continues to investigate Qualcomm's licensing of its SEPs, apparently now focusing on whether it is "unfair" for Qualcomm to base royalties on the price of a handset rather than using some other measure.[60]

---

[57] Korea Fair Trade Commission, Qualcomm, Decision and Order No. 2009-281 (2009) (Korean) (translation, author's files).

[58] Kim and Yang (2015). In 2013 the Seoul High Court affirmed the KFTC for the most part; the case is on appeal to the South Korea Supreme Court. The Seoul High Court did not pass on the Article 23 theory, perhaps because that theory was unnecessary for its decision (*id.*).

[59] *Id.*

[60] Mu-hyun (2015). Qualcomm reports a KFTC notification on March 17, 2015, with regard to an investigation of its "licensing business" (Qualcomm 2015).

12.3. DISCLOSURE REQUIREMENTS AND DECEPTION

A second area in which an intellectual property right holder can unfairly exploit its licensees involves the refusal of a right holder to disclose information to its licensees (or potential licensees). These cases generally do not have as clear a price-raising effect as the FRAND-committed SEP cases have and they tend to be closer to the border between exploitation and exclusion. Nevertheless, decisions in these cases show more of an immediate concern for the effect of this behavior on customers, with weaker proof of a diminution of competition in downstream markets.

An early case to examine a patent holder's refusal to disclose information is the US FTC's 1995 complaint against the computer manufacturer Dell.[61] Involved in that case was an effort by an SSO to set a design standard for technology to improve the transmission of video technology between a computer's CPU and peripheral devices. The Commission alleged that Dell, as a member of the SSO, had failed to disclose a pre-existing patent that the standard would infringe. After the standard was adopted, and after 1.4 million computers were sold with this technology, Dell informed computer manufacturers that their implementation of the standard infringed its patent. Dell threatened enforcement unless royalties were paid.

Dell eventually settled with the FTC, agreeing not to enforce its patent for ten years. The Commission's analysis of the effects of Dell's actions was brief. The Commission noted that Dell's collection of royalties "would likely" have increased prices to end-user consumers and that Dell violated Section 5 of the FTC Act because it was trying to "take advantage" of market power resulting from the standard rather than simply getting royalties based on the "inherent value of the patent."[62] Put otherwise, Dell was unfairly trying to exploit the computer-maker licensees.

Three years after settling the *Dell* case the FTC filed a complaint against Intel for withholding proprietary technical information that it had previously furnished to three computer manufacturers, done in retaliation for their assertion of patent rights adverse to Intel and to force these customers to license their technology to Intel.[63] The FTC alleged that Intel had monopoly power in the general-purpose microprocessor market, that all three customers were highly dependent on Intel microprocessors, and that the proprietary information Intel withheld was important for enabling these customers to make computers compatible with Intel chips. Although the Commission framed its case as one involving exclusionary conduct, "entrenching" Intel in its dominant position in the microprocessor market, none of these computer manufacturers competed with Intel in that market nor was it clear how Intel's

---

[61] In the Matter of Dell Computer Corp., 121 FTC 616 (1996).
[62] See *id.* at 624 n. 2 (Statement of the Commission); 60 Fed. Reg. 57872 (November 22, 1995) (Analysis of Proposed Consent Order To Aid Public Comment).
[63] In the Matter of Intel Corp.,128 FTC 213, ¶ 26 (1998). Available at www.ftc.gov/sites/default/files/documents/cases/1998/06/intelcmp_o.pdf.

behavior might have made it more difficult for competing microprocessor chip makers to succeed in the market.

What was actually critical to the Commission's complaint was that the three original equipment manufacturers (OEMs) were highly dependent customers, subject to Intel's "coercive business tactics."[64] The "[c]ontinued denial of advance technical information to an OEM by a dominant supplier can make a customer's very existence as an OEM untenable."[65] The only way to avoid this potential exclusionary impact was to accede to Intel's demands to give up their patent rights. In this way Intel had effectively raised the prices these customers were paying to Intel, extracting rents through the imposition of nonprice terms. The Commission was never put to the test of proving that Intel's coercive tactics violated Section 5, however, because the parties settled the case just before the start of the FTC's administrative hearing.[66]

Intel was followed by two cases that returned to the problem of patent-holder deception of a standard setting body. One, *Rambus*, brought in 2002, involved the alleged deception of a private SSO that developed technical standards for a form of computer memory known as SDRAM.[67] The other, *Unocal*, brought in 2003, involved the alleged deception of the California Air Resources Board, a state agency that set standards for low-emissions reformulated gasoline to be sold in California.[68]

The complaints in the two cases were quite similar. In both the Commission alleged that the patent holder had failed to disclose the existence of its patents prior to the decision adopting the standard. Both cases also had similar allegations as to competitive harm. In each the Commission alleged that the respondents' conduct led to increased prices to licensees and, eventually, to increased prices to the purchasers of the licensees' products.[69] Finally, in both cases the Commission charged the respondents not only with monopolization and attempted monopolization of the technology markets involved (that is, the licensing of their patents), but also with "unreasonably restrain[ing] trade" in those technology markets.[70] The implication of the third charge was that improper price-raising conduct by a single

---

[64] *Id.* at ¶ 14.
[65] In the Matter of Intel Corp., Analysis of Proposed Consent Order to Aid Public Comment (1999). Available at www.ftc.gov/sites/default/files/documents/cases/1999/03/d09288intelanalysis_o.htm.
[66] Intel agreed not to withhold certain advanced technical information for reasons related to an intellectual property dispute. See *In the Matter of Intel Corp.*, Docket No. 9288, Decision and Order (1999). Available at www.ftc.gov/sites/default/files/documents/cases/1999/08/intel.do__o.htm.
[67] *In the Matter of Rambus, Inc.*, Docket No. 9302 (2002). Available at www.ftc.gov/sites/default/files/documents/cases/2002/06/020618admincmp.pdf.
[68] *In the Matter of Union Oil Co. of Cal.*, Docket No. 9305 (2003). Available at www.ftc.gov/sites/default/files/documents/cases/2003/03/030304unocaladmincmplt.pdf.
[69] Rambus Complaint ¶ 120; Unocal Complaint ¶ 98 (alleging 90 percent pass-through to consumers of higher royalty costs).
[70] Rambus Complaint ¶ 124; Unocal Complaint ¶ 102. The Complaint in Unocal added a similar violation with regard to the downstream market for gasoline, ¶ 103; Rambus did not include a similar allegation with regard to prices in the downstream DRAM market.

firm harms competition even if it does not result in monopoly, a position not recognized under the Sherman Act but potentially recognized as an "unfair method of competition" under Section 5 of the FTC Act.

Only Rambus litigated its case.[71] The Commission decided that Rambus's deceptive conduct violated the FTC Act under a monopolization theory, finding that the SSO either would have chosen a different standard or would have required Rambus to license on FRAND terms. The FTC dropped its potentially more expansive theory that Rambus's conduct violated the FTC Act simply because Rambus's deception raised royalty rates and increased the prices of DRAM chips. Still, as part of its remedy the Commission ordered Rambus to license its patents at lower "reasonable" royalty rates, otherwise unspecified.[72]

On appeal to the D.C. Circuit, Rambus argued that the Commission erred in not deciding whether a truthful disclosure would have caused the SSO to choose a different standard that would not have infringed Rambus's patents, thus improperly excluding Rambus's competitors, or would have resulted only in lower FRAND prices. The Court of Appeals agreed that the failure to decide between the two outcomes was an error. Preventing the FRAND outcome was not "anticompetitive" because (as noted at the beginning of this chapter) the only harm from Rambus's deception would then have been higher prices. The Commission had therefore not met its burden of proving that Rambus's conduct caused harm to competition.

The decision in *Rambus* has been criticized on its facts,[73] even by those who generally view rent extraction as not subject to antitrust scrutiny.[74] Although in a broad sense the D.C. Circuit's decision was consistent with the general view that US antitrust law does not protect against rent extraction *simpliciter*, the court's decision appears to be an outlier in the intellectual property context. Neither subsequent litigation nor administrative practice has been constrained by *Rambus*'s holding. As our earlier discussion indicates, the FTC and the courts continue to be concerned about the effects of excessive pricing by intellectual property rights holders, particularly in the FRAND context.

12.4. POST-EXPIRATION ROYALTIES

Royalties imposed in a patent license that continue past the expiration of the patent would seem to be an obvious target for competition enforcers seeking to restrict the

---

[71] The Unocal complaint was settled as part of an agreement allowing Chevron to acquire Unocal, with the parties agreeing that Unocal would neither enforce the patents in question nor attempt to collect royalties for their use. *In the Matter of Union Oil Co. of Cal.*, Docket No. 9305, Decision and Order (2005). Available at www.ftc.gov/sites/default/files/documents/cases/2005/08/050802do.pdf.

[72] See *In the Matter of Rambus Inc.*, Docket No. 9302, Final Order at 2–4 (2007). Available at www.ftc.gov/sites/default/files/documents/cases/2007/02/070205finalorder.pdf; *In the Matter of Rambus*, Docket No. 9302, Opinion of the Commission on Remedy at 22–23.

[73] Besen and Levinson (2009).

[74] Cotter (2009).

ability of patent holders to extract rents from licensees in excess of what is necessary to incentivize innovation. In the United States, however, such clauses have not been the subject of antitrust attack. Rather, they have been litigated by private parties as a matter of patent law rather than antitrust.

The US Supreme Court confronted the issue in a 1964 decision, *Brulotte v. Thys Co.*[75] That case involved the sale of a patented hop-picking machine for a flat sum plus a license to use the machine subject to annual royalties (for a fixed number of years) based on the amount of hops the machine harvested. When some of the patents on the machine expired, the purchaser refused to pay further royalties, claiming that the patent holder had misused the patent by extending its terms beyond its expiration. The lower courts found the license agreement enforceable, but the Supreme Court reversed. The Court acknowledged that a patentee can use the "leverage" of the patent to "exact royalties as high as he can negotiate," but held that the patentee cannot "enlarge" the power of the patent by "project[ing] those royalty payments beyond the life of the patent."[76]

Brulotte's condemnation of post-sale royalties has been much criticized for its misperception of the economics of the licensing transaction, its potential adverse effect on innovation from denying licensors and licensees the ability to structure their transactions optimally, and its indifference to competition issues.[77] From a competition viewpoint, critics have argued that a patent loses its exclusionary effect once it expires, because then others are free to enter the market and use the technology without paying royalties to the inventor. Even if the former licensee is saddled with higher costs, consumers could have lower-cost alternatives to choose from, at least if entry barriers are low.[78] Consumer welfare would not necessarily be harmed.

Despite these criticisms, however, the US Supreme Court turned away a challenge to *Brulotte's* bright-line rule in *Kimble v. Marvel Entertainment*, decided in 2015. Although the Court basically agreed with the criticism that *Brulotte* got its economics wrong in terms of competitive effects, the Court also pointed out that *Brulotte* "did not undertake to assess that practice's likely competitive effects." Instead, the Court chose to follow the judgment that it believed Congress made in the Patent Act – "the day after a patent lapses, the formerly protected invention must be available to all for free."[79] The Court concluded that if *Brulotte's* rule is to be changed, it is up to Congress to do the changing.

---

[75] 379 U.S. 29 (1964).
[76] *Id.* at 33.
[77] Dreyfuss (1986).
[78] See, e.g., *Scheiber v. Dolby Labs., Inc.*, 293 F.3d 1014 (7th Cir. 2002) (pointing out, inter alia, that "[a]fter the patent expires, anyone can make the patented process or product without being guilty of patent infringement. The patent can no longer be used to exclude anybody from such production" (Posner J.)).
[79] *Kimble v. Marvel Entertainment, LLC*, 135 S.Ct. 2401, 2413 (2015).

In two cases outside of the United States, however, competition authorities have attacked post-expiration royalties as a violation of competition law. One case was brought by China's NDRC, the other by the KFTC. Both involved Qualcomm's licensing of its SEPs for wireless communications technology and both were part of the broader challenges to Qualcomm's licensing described earlier.

In the China case, Qualcomm argued that the royalties collected on expired patents were made up for by royalties not collected on new patents that were added to the portfolio during the term of the license. The Commission found, however, that Qualcomm did not prove that the value of the added patents was equivalent to the expired ones and did not provide licensees with a list of what patents were being licensed in the portfolio at any given time.[80] "Qualcomm's practice," the Commission concluded, "obscured the specific objects being licensed, and the licensees had to keep paying for the expired SEPs."[81] This was an abuse of dominance in violation of Article 17 of the Antimonopoly Law, which prohibits sales at "unfairly high prices." Qualcomm agreed to discontinue the practice.[82]

In the Korea case, Qualcomm's SEP licenses required post-expiration royalties at 50 percent of the level charged during the life of the patent. Qualcomm argued that the provision should not be found violative of the MRFTA because the provision had not been implemented (none of the patents had yet expired) and was unlikely to be implemented. The KFTC disagreed. Qualcomm had a dominant position in the market and its post-expiration royalties would have "unreasonably increased the financial burden of local handset manufacturers."[83] Despite the Commission's finding of dominance, however, the Commission did not charge Qualcomm under the Article dealing with abuse of dominance, perhaps because post-expiration royalties had not yet been imposed, which meant that anticompetitive effects in the handset market could not be shown. Instead, the Commission found that Qualcomm had engaged in the "unfair trade practice" of unreasonably taking advantage of its superior position in violation of Article 23 of the Act.[84] The possibility that a dominant firm could raise rates in the future on customers that lacked alternative supplies was enough to constitute unfair exploitation.[85]

The difference between how post-expiration royalties are handled in the United States and how they are handled in China and Korea offers three useful insights. First, handling this as a competition law matter rather than a matter of patent law interpretation allows for a more careful assessment of the economic impact of the

---

[80] Harris (2015).
[81] NDRC, *Qualcomm, supra*, at II.1.a.
[82] NDRC, Press Release, *supra*, at 2.
[83] Korea Fair Trade Commission, 2010 Annual Report, at 51. Available at http://eng.ftc.go.kr/bbs.do?command=getList&type_cd=53&pageId=0301.
[84] Lee (2012).
[85] Qualcomm did not appeal this part of the Commission's decision.

practice. Taking the patent law approach, it does not matter whether the patent holder has monopoly power, either when the license is negotiated or when the patent expires; under a competition law approach, however, the market power of the licensor is critical because a violation depends on whether the licensor has a dominant position. Second, assuming that the dominant position continues into the post-expiration period (as is likely in the *Qualcomm* cases in China and Korea), such that the licensee lacks competitive alternatives, the result will be continued high prices, that is, continued rent extraction beyond the legislative determination of what is sufficient to incentivize innovation. Third, the FRAND setting of the China and Korea cases offers an example of a possible exception to the standard criticism of the ban on post-expiration royalties. Commentators have argued that higher prices post-expiration simply make up for lower prices during the patent term, with the parties choosing to spread the payments as a way to allocate risk. In other words, the net will be the same.[86] But in a FRAND setting, that will not be the case. The SEP holder's pricing is constrained in the patent period, for it has committed to pricing below a monopoly level. This means that there are unrealized monopoly profits to be had. Getting them in the post-expiration period is one way to do it, effectively raising the net above FRAND rates and extracting rents from licensees beyond what the parties had considered fair.

12.5. CONCLUSION

The goal of antitrust law is generally considered to be to promote competition (or at least to remove restraints that hinder competition). The goal is not generally thought to be to prevent exploitation. Yet the examples reviewed in this chapter show that courts and enforcement agencies, in the United States and around the world, have taken a different view when it comes to intellectual property rights exploitation. Contrary to conventional wisdom, antitrust law is being used today to control the ability of intellectual property rights holders to exploit their licensees through excessively high prices or the imposition of particular nonprice terms.

These decisions are most apparent in the area of FRAND licensing – by its terms, a restriction on the prices that the patent holder can charge – but they also occur outside the area of FRAND-committed patents when intellectual property rights holders use their monopoly power (or dominant position) to impose onerous terms on their licensees, whether through direct price raising or through other terms that will effectively raise price. The critical insight from competition policy is that the ability to raise prices is not just a reflection of the existence of a patent – which does not, in itself, grant an economic monopoly – but of market power, assessed through

---

[86] See *Schreiber*, 293 F.3d at 1017 ("The duration of the patent fixes the limit of the patentee's power to extract royalties; it is a detail whether he extracts them at a higher rate over a shorter period of time or a lower rate over a longer period of time").

conventional market analysis that considers substitutes for the technology involved and barriers that restrict the ability of competitors to enter and compete in that market. Outside of monopoly, patent holders are free to charge what the market will bear for their technology, just like other sellers.

The examples reviewed in this chapter also show that enforcers concerned about monopoly exploitation by intellectual property right holders have tended to look for some conduct in addition to the imposition of high prices. Enforcers have focused on the effort of SEP holders to bring infringement cases seeking injunctions or exclusion orders, or the deception of SSOs or regulators that leads to high prices. This effort to look for specific conduct is true both for enforcers in the United States, constrained by the current view that Section 2 of the Sherman Act does not make monopoly prices illegal, and for enforcers outside the United States that operate without this legal constraint. The reason for this focus is not that these enforcers are concerned with deceit (the allegations of deceit are often quite thin) or think that litigation is a bad thing. The reason is that it is administratively challenging to judge whether prices are sufficiently high as to be "excessive."

Competition agencies and courts may fear the administrability problem too much in the patents area. Litigants are often forced into court to determine damages in patent cases, which may involve determining what constitutes a reasonable royalty; FRAND rates are a particular subcategory of this type of litigation. In this patent litigation part of the effort in assessing a reasonable rate is to avoid overcompensating the patent holder so as to not discourage follow-on innovation.[87] If reasonable rates can be determined in the context of patent infringement litigation, in which overcompensation is a concern, why is it so much harder to figure out reasonable rates in the context of antitrust litigation?

Intellectual property rights that give their holders long-term monopoly power are particularly costly for society. Competition law intervention that can reduce this cost is warranted. This intervention is consistent with the rough judgment made through patent law's limited term that returns to innovation can be limited without destroying the "progress of Science and useful Arts." Preventing exploitation through high prices in the area of intellectual property rights can thus be a welcome departure from the general view that such conduct is not the province of competition law.

REFERENCES

Akman, Pinar. 2009. The Role of Exploitation in Abuse under Article 82 EC. In *Cambridge Yearbook of European Legal Studies* 11, 165–88.
Besen, Stanley M. and Robert J. Levinson. 2009. Standards, Intellectual Property Disclosure, and Patent Royalties after Rambus. *North Carolina Journal of Law & Technology*, 10, 233, 249–53.

---

[87] Lee and Melamed (2016).

Chin, Yee Wah. 2015. Intellectual Property Rights and Intellectual Property in China. In Donna Suchy (ed.), *IP Protection in China*. ABA, pp. 313, 314.

Contreras, Jorge L. 2015a. Standards, Royalty Stacking, and Collective Action. *CPI Antitrust Chronicle*, March (1). Available at www.competitionpolicyinternational.com/file/view/7353.

2015b. A Market Reliance Theory for FRAND Commitments and Other Patent Pledges. *Utah Law Review*, 501–17.

Cotter, Thomas F. 2009. Patent Holdup, Patent Remedies, and Antitrust Responses. *Iowa Journal of Corporate Law*, 34, 1151, 1194–1200.

Crane, Daniel A., 2010. Patent Pools, RAND Commitments, and the Problematics of Price Discrimination. In Rochelle C. Dreyfuss, Harry First, and Diane L. Zimmerman (eds.), *Working Within the Boundaries of Intellectual Property*. New York: Oxford University Press, 372–8.

Dou, Eva. 2015. Qualcomm Spared Some of the Worst Penalties by Chinese Authorities, *Wall Street Journal*, February 10.

Dreyfuss, Rochelle C. 1986. Dethroning Lear: Licensee Estoppel and the Incentive to Innovate. *Virginia Law Review*, 72, 709–12.

First, Harry. 2007. Controlling the Intellectual Property Grab: Protect Innovation, Not Innovators. *Rutgers Law Review*, 38, 365–98.

Gandhi, Samir R., Fadi Metanios, and Hemangini Dadwal. 2015. Competition Law and FRAND: Developments and Challenges in India. *CPI Antitrust Chronicle*, October (1). Available at www.competitionpolicyinternational.com/assets/Uploads/GandhietalOct-151.pdf.

Harris, H. Stephen, Jr. 2015. An Overview of the NDRC Decision in the Qualcomm Investigation. *CPI Antitrust Chronicle*, July, 2, 4. Available at www.competitionpolicyinternational.com/file/view/7409.

Hou, Liyang. 2015a. Antitrust Regulation of Intellectual Property Rights in China, 6–7. Available at http://ssrn.com/abstract=2648736.

2015b. The Qualcomm Decision: Protectionism? And for Whom? Available at http://ssrn.com/abstract=2648741.

InterDigital, Inc., Form 10-K Annual Report for 2015 (2016). Available at http://files.shareholder.com/downloads/IDCC/1618089504x0xS1405495-16-47/1405495/filing.pdf.

Kim, Yoonhee and Hui-Jin Yang. 2015. A Brief Overview of Qualcomm v. Korea Fair Trade Commission. *CPI Antitrust Chronicle*, 2, 6, 7. Available at www.competitionpolicyinternational.com/file/view/7352.

Lee, In Ho, 2012. Qualcomm's Abuse of Dominance, at 22–23. Available at http://eng.ftc.go.kr/bbs.do?command=getList&type_cd=54&pageId=0302.

Lee, William F. and A. Douglas Melamed. 2016. Breaking the Vicious Cycle of Patent Damages. *Cornell Law Review*, 101, 385–466.

Lemley, Mark A. and Carl Shapiro. 2007. Patent Holdup and Royalty Stacking. *Texas Law Review*, 85, 1991–2049.

Maldonado, Kassandra. 2014. Breaching RAND and Reaching for Reasonable: Microsoft v. Motorola and Standard-Essential Patent Litigation. *Berkeley Technology Law Journal*, 29, 419–20.

Mozur, Paul and Quentin Hardy. 2015. China Hits Qualcomm with Fine, *New York Times*, February 9.

Mu-hyun, Cho. 2015. Qualcomm Facing Penalty from South Korean Antitrust Regulator. ZD Net. Available at www.zdnet.com/article/qualcomm-facing-penalty-from-south-korean-antitrust-regulator/.

O'Donoghue, Robert and Jorge Padilla. 2006. *The Law and Economics of Article 82 EC* 174. Oxford: Hart Publishing.

Qualcomm, Inc. 2015. 10-Q for period ending 12/28/14, at 13. Available at http://files.shareholder.com/downloads/QCOM/343973291x0xS1234452-15-24/804328/filing.pdf.

Scotchmer, Suzanne. 1991. Standing on the Shoulders of Giants: Cumulative Research and the Patent Law. *Journal of Economic Perspectives*, 5, 31.

Shapiro, Carl. 2001. Navigating the Patent Thicket: Cross-Licenses, Patent Pools, and Standard Setting. In Adam B. Jaffe *et al.* (eds.), *Innovation Policy and the Economy.* Cambridge, MA: MIT Press.

Tsai, Joanna and Joshua D. Wright. 2015. Standard Setting, Intellectual Property Rights, and the Role of Antitrust in Regulating Incomplete Contracts. *Antitrust Law Journal*, 80, 157, 162–65.

US Department of Justice and U.S. Patent & Trademark Office. 2013. Policy Statement on Remedies for Standards-Essential Patents Subject to Voluntary F/Rand Commitments, 4.

US Federal Trade Commission 2011. The Evolving IP Marketplace: Aligning Patent Notice and Remedies with Competition, 189, 190, 192.

Wright, Joshua D. and Douglas H. Ginsburg. 2015. Comment Regarding the Japan Fair Trade Commission's Draft Partial Amendment to the Guidelines for the Use of Intellectual Property Under the Antimonopoly Act. Available at www.ftc.gov/system/files/documents/public_statements/693631/150803japantradecomments.pdf.

13

Patent Holdups

Daryl Lim

13.1. INTRODUCTION

This chapter explores holdups in the standard setting organization (SSO) and non-SSO setting. It observes that patentees are not always the aggressors. On the other hand, holdups are harmful and need to be deterred before they occur and decisively quashed when they do. Much progress has been made, but the results of those efforts, particularly with patent assertion entities (PAEs) remains to be seen.

Patent holdups happen when patentees demand more than they deserve from implementers who made technology-specific investments.[1] In a world still reeling from the aftershocks of the greatest economic crisis since the Great Depression, greed ranks high in the order of cardinal sins. While extortionists have swiftly found themselves vilified,[2] some patentees may simply be ferreting out uncooperative infringers attempting a holdout (reverse holdup) aimed at depressing royalties owed to patentees below what the technology is worth.[3]

Holdups have become more prominent due to a convergence of technologies and a divergence of interests. Companies like Motorola, Nokia, and Ericsson once shared a common interest in promoting open or free access to each other's technology as rivals and partners with a shared business model and corporate culture.[4] Today, consumer electronics devices are a mere portal to an "Internet of Things," shattering traditional boundaries between industries such as software and wireless telecommunications, drawing unlikely rivals such as Apple, Google, Huawei,

---

I am grateful to Danny Sokol for his helpful comments. Many thanks to Associate Dean Julie Spanbauer for the summer research grant that supported the writing of this book chapter, as well as to Amy Taylor and Qi Qiong for their editorial assistance.

[1] US Federal Trade Commission (2011).
[2] Lim (2014).
[3] Sidak (2015).
[4] Hollman (2015).

Microsoft, Qualcomm, Samsung, and ZTE into a collision course.[5] They exist at different points on the value chain, and have different incentives. For instance, a handset maker views patent royalties as a cost, while the owner of an operating system would view it as a source of revenue.

Systems prone to holdups consist of complementary hardware and software components spanning different industries. Each smartphone embodies about 250,000 patents covering thirty different standards for wireless communications, video display, Internet access and other services.[6] With each new patentee, the holdup risk rises.[7] Just like land owned by many different people, it becomes costly to identify and negotiate for rights to use the technology.[8] The sheer number of patents of indeterminate scope and validity cause the transaction costs to preclude bargaining.[9] It would take 2 million attorneys working full time to compare every software patent issued each year to a firm's products, and cost $400 billion, twice the value of the software industry.[10] As one commentator observed, "[w]hen you're paying lawyers between $10,000 to $15,000 per patent to drill down and research each patent, it's usually less expensive to cut a licensing deal. Anything over 200, nobody's [sic] talking about merits."[11]

Fragmentation also heightens the risk of royalty stacking, which happens when patentees cumulatively maximize profits, "resulting in excessive royalty payments such that (1) the cumulative royalties paid for patents incorporated into a standard exceed the value of the feature implementing the standard, and (2) the aggregate royalties obtained for the various features of a product exceed the value of the product itself."[12]

Holdups generally happen in two ways. First, they can infect collaboratively set standards administered by SSOs. Standards allow devices to interoperate, increasing competition by lowering entry barriers and adding value to manufacturers' products by encouraging production of complementary products.[13] Since patented technologies represent best-in-breed solutions to technical problems, broad adoption is socially beneficial. For standards to succeed, implementers need attractive terms of access to incentivize investment in product development while patentees need sufficient rewards to contribute their technology in the first place.

Once a standard is set, implementers make technology-specific investments to comply with that technology. The common denominator providing consumers with the convenience of interoperability across different devices also "locks-in"

---

[5] New York Times (2012).
[6] Blind *et al.* (2011).
[7] Cass (2015).
[8] Santore, McKee, and Bjornstad (2010).
[9] Chien (2010).
[10] Mulligan and Lee (2012).
[11] Hollman (2015).
[12] *Microsoft Corp. v. Motorola, Inc.*, 795 F.3d 1024, 1031 (9th Cir. 2015).
[13] *Id.* at 1030.

consumers and manufacturers. This confers significant leverage on patentees to demand high royalties since implementers cannot avoid infringing standard essential patents (SEPs) embedded within them.[14]

Patentees may choose to strategically ambush implementers by failing to disclose their patents or by inducing adoption with false representations. Courts and antitrust agencies have been largely successful in clamping down on these cases of opportunistic subterfuge, and patentees today would be ill-advised to follow in the footsteps of companies such as Dell, N-Data, and Rambus.

Patentees and implementers may also battle over the nature and extent of their commitments to SSOs. Fair, reasonable and non-discriminatory (FRAND) commitments on SEPs encourages potential implementers to adopt the technology on the basis that they will have access to the technology on FRAND terms.[15] The commitments may form part of the bylaws or may be the object of a letter of assurance.[16] Today, it is settled law that FRAND-encumbered patents may enjoy injunctive relief against unwilling licensees but not otherwise. In assessing a "fair" royalty, courts focus on the patentee's technological contribution to the standard. On the other hand, implementers complaining of holdups and stacking must prove them.

The second way holdups occur is outside of the SSO context. Patent assertion entities, more popularly known as patent trolls, are the antagonists. PAEs subsist mostly or entirely from extracting royalties or settlements from those who implement technology that their patents read on. They acquire patents or inherit them from unsuccessful operating companies.[17] The worst sort of PAEs harass small businesses for nuisance settlements. Others amass swarms of patents and strategically use shell companies to pummel their victims with successive lawsuits. A number of PAEs are "privateers," servicing operating companies seeking to monetize their patents that might be left fallow, or who simply prefer to keep their image wholesome.[18]

Injunctions feature prominently in holdups.[19] Sometimes, patentees want a rival eliminated from the market so that they can appropriate their rivals' sales.[20] While noninfringing alternatives may be technologically available to alleged infringers, they may not be commercially feasible. For instance, a noninfringing iPhone model that does not work on a 4G network is commercially useless. Trapped in the interlaced web of 4G technologies, patentees with SEPs essential to 4G interoperability can threaten even a giant like Apple with catastrophic losses. At other times, patentees realize that while a dead hostage makes a poor bargaining chip for ransom demands, the *threat* of extermination is more useful in loosening purse strings.[21]

---

[14] *Ericsson, Inc. v. D-Link Sys., Inc.*, 773 F.3d 1201, 1209 (Fed. Cir. 2014).
[15] *Microsoft Corp.*, 795 F.3d at 1031.
[16] Contreras (2015).
[17] Morton and Shapiro (2015).
[18] Sokol (in press).
[19] Morton and Shapiro (2015).
[20] Cass (2015).
[21] *Microsoft Corp.*, 795 F.3d at 1046.

Unjustified wealth transfers discourage industry and innovation and harm consumers. No one seeing looting and pillaging taking place around them will be motivated to open shop. No one wants to be next. Implementers, hesitant to be another holdup statistic, fear to venture into the minefield of patents. Follow-on innovation is chilled, intra-standard competition suppressed. The result is "delayed hiring, business pivots, and the killing of products and product lines."[22]

Patents are constitutionally mandated to promote scientific progress. When patents end up overcompensating patentees at the expense of that progress, patent rules run contrary to patent policy, and must be fixed.[23] Injunctions must be kept on a tight leash since "the creation of a right is distinct from the provision of remedies for violations of that right."[24] Similarly, damages must reflect the patentee's technical contribution and not the threat value of the holdup.[25] The former is its economic value, the latter is a naked attempt at extortion masquerading as the stark but simple costs of patent litigation.[26]

Today, software patents are harder to come by. Courts find it easier to saddle vexatious patentees with attorney's fees penalties. State and federal legislatures have successfully taken steps to curb abusive PAE litigation.[27] The focus has been to cauterize the worst features in patent litigation that facilitate holdups and make adjudication more efficient so that courts can sieve and quickly dismiss baseless allegations when they arise.

The proper role of antitrust law in addressing holdups is controversial.[28] The most likely avenue for an antitrust attack is a claim under § 2 of the Sherman Act. A monopolist violates antitrust law if they acquire or maintain market power by means other than "offering a superior product, business acumen or historic accident."[29] Patentees must wield power over prices or exclude competition.[30] This may be shown by high market share and entry barriers that allow them to exercise market power over an appreciable amount of time.[31] To be unlawful, market power must be coupled with anticompetitive conduct that harms the competitive process rather than just competitors.[32] Section 7 of the Clayton Act also prohibits PAE acquisitions of patents from another firm where it may substantially lessen competition in a relevant market.[33] That may happen if the transaction is likely to enable the merging

---

[22] Chien (2014).
[23] Harris (2014).
[24] *eBay Inc. v. MercExchange, L.L.C.*, 547 U.S. 388, 392 (2006).
[25] *Microsoft Corp. v. Motorola, Inc.* 864 F. Supp. 2d 1023, 1030–33 (W.D. Wash. 2012).
[26] Harris (2014).
[27] Whitfield (2015).
[28] Merges and Kuhn (2009).
[29] *United States v. Grinnell Corp.*, 384 U.S. 563, 571, 86 S. Ct. 1698 (1966).
[30] *United States v. E.I. du Pont de Nemours & Co.*, 351 U.S. 377, 391 (1956).
[31] *W. Parcel Express v. UPS*, 190 F.3d 974, 975 (9th Cir. 1999).
[32] *United States v. Microsoft Corp.*, 253 F.3d 34, 58 (D.C. Cir. 2001).
[33] 15 U.S.C. § 18 (2012).

entity to profitably increase prices post-merger or increase the likelihood of price coordination.[34]

13.2. HOLDUPS IN THE STANDARD SETTING CONTEXT

13.2.1. *Patent Ambush*

Patent ambush is a species of holdup. It arises in three circumstances. First, patentees remain silent and spring their patents on locked-in defendants. Second, patentees induce adoption by representing that the invention can be practiced for free or at reasonable cost. The central enquiry in both cases is whether the patentee breached its duty to disclose its interests in the prospective standard, and in doing so misled SSO members into adopting it. In the third instance, a patent acquirer attempts to renege on a promise to license given by the original owner.

Dell Computer Corp illustrates the first instance.[35] Dell failed to disclose its patent and twice affirmed that it had no relevant patents before suing SSO members.[36] The Federal Trade Commission (FTC) intervened, resulting in a consent decree where Dell agreed not to sue implementers.[37] Similarly in *Rambus, Inc. v. Infineon Techs. AG*, Rambus filed patent applications intending to sue implementers once the standard became widely adopted.[38] The Court of Appeals for the Federal Circuit held that the SSO rules did not expressly impose a duty on Rambus to disclose its patents, and instead faulted the SSO for failing to draft a policy to address attempts by Rambus to "mine a disclosed specification for broader undisclosed claims."[39]

In a related case, the Court of Appeals for the D.C. Circuit held that the antitrust plaintiff had to show that the SSO would have adopted a different standard if it had been aware of Rambus' patents.[40] Rambus' deceit "simply to obtain higher prices normally has no particular tendency to exclude rivals and thus to diminish competition."[41] The cases indicate that patentees who gain market power by inducing adoption by concealment can be penalized. However, the absence of clear SSO rules requiring disclosure and failure to show that SSO members would have adopted the standard but for that disclosure are fatal to an antitrust claim.

Qualcomm Inc. v. Broadcom Corp illustrates the second instance. Here, Qualcomm's FRAND commitment was material in the SSO adopting the standard.[42] The Court of Appeals for the Third Circuit held that "in a consensus-oriented

---

[34] US DOJ-FTC (2010).
[35] In re Dell Computer Corp., No. 93-10097 (F.T.C. 1995).
[36] *Id.*
[37] *Id.*
[38] *Rambus, Inc. v. Infineon Techs. AG*, 318 F.3d 1081 (Fed. Cir. 2003).
[39] *Id.*
[40] *Rambus Inc. v. FTC*, 522 F.3d 456 (D.C. Cir. 2008).
[41] *Id.* at 464.
[42] *Broadcom v. Qualcomm*, 501 F.3d 297 (3d Cir. 2007).

private standard-setting environment, a patent holder's intentionally false promise to license essential proprietary technology on [FRAND] terms, coupled with an [SSO]'s reliance on that promise when including the technology in a standard, and the patent holder's subsequent breach of that promise, is actionable anticompetitive conduct."[43]

In a related case, the Federal Circuit found that Qualcomm agreed to disclose patents that "reasonably might be necessary to practice" the standard but deliberately hid them from holdup implementers by alleging infringement primarily based on compliance with the standard.[44] The court construed an implied waiver by Qualcomm to enforce its rights based on a duty to disclose, rendering the patents unenforceable against the implementers.

In re Negotiated Data Solutions LLC illustrates the third instance.[45] The FTC entered into a consent decree that prevented a patent transferor from reneging on FRAND commitments made by its predecessor.[46] N-Data knew of its predecessor's commitment prior to purchasing the patents and exploited industry lock-in by waiting until widespread adoption before attempting its holdup.

Having more rigorous disclosure requirements seem like an obvious solution, but they are also fraught with inherent difficulty. Standards evolve dynamically, and patents may become essential, optional, or only preferable, upon adoption.[47] Owners of large portfolios will also find it costly and difficult to determine whether their patents will read on a standard as it changes.[48] It can also be unclear whether technologies fall within the scope of a patent claim until they have been judicially construed.[49] Finally, innovators may be reluctant to disclose unpublished patent applications when a standard is still under development for fear that they will also be divulging their innovation trajectories and product features that give them a competitive advantage. SSOs therefore require or encourage disclosure only of issued patents and published patent applications, which exacerbates the likelihood of holdups post-standardization.[50]

13.2.2. FRAND Disputes

FRAND is a binding contractual commitment proposed by SSOs on the one hand and accepted by patentees on the other.[51] It is meant to benefit SSO members and

---

[43] *Id.* at 314.
[44] *Qualcomm Inc. v. Broadcom Corp.*, 548 F.3d 1004 (Fed. Cir. 2008).
[45] *In re Negotiated Data Solutions LLC*, No. 051-0094 (F.T.C. Jan. 23, 2008).
[46] *Id.*
[47] Morrow, Levin, and Hill (2014).
[48] Lim (2011).
[49] Sadler (2013).
[50] Lim (2014).
[51] *Microsoft Corp. v. Motorola, Inc.*, 795 F.3d 1024, 1051 (9th Cir. 2015).

third parties who implement the standard.[52] Despite the participation of sophisticated patentees and implementers, those terms are mere guidelines, if that. As with disclosure requirements, some SSOs allow SEP owners to couch FRAND commitments on their own terms or explicitly disclaim responsibility for identifying SEPs, their validity, or any role in policing compliance with FRAND obligations.[53] Courts and agencies stepped in to help delineate those boundaries, and they have largely succeeded in shepherding parties toward negotiated settlements or court adjudications.

13.2.2.1. Injunctions

Since the English Chancery Court devised the equitable remedy of injunctions, courts have had to consider whether it is in the public's interest to grant them.[54] With the patentee's reward ancillary to incentivizing technological progress, the focus is on the public interest rather than on the preservation of the patentee's property rights. At the same time, there is no per se rule against SEP owners seeking an injunction against an implementer.[55]

This is good policy, since a per se ban would discourage participation in SSOs.[56] When confronted by the patentee, rouge implementers may also quibble that the patentee's offer is not "fair" nor "non-discriminatory," forcing the patentee to sue.[57] Implementers may also attempt a holdout,[58] or not respond or accept offers to license.[59] They may also insist on licensing terms that may put the patentee in breach of its "non-discriminatory" obligations to other patentees.[60]

At the same time, patentees today will not likely get injunctions on FRAND-encumbered patents. A willingness to license on FRAND terms means that patentees will not likely suffer the prerequisite irreparable injury if compelled to license, particularly where they licensed others, including rivals.[61] Courts have imposed a duty of good faith on both sides and are usually astute in picking out bad actors no matter the side. For instance, in *Apple Inc. v. Motorola*, the court found that negotiations were ongoing and the licensee, Apple, had not refused to agree to a deal, contrary to Motorola's allegations of Apple holding out by refusing to accept its initial licensing offer and stalling negotiations.[62] The court denied Motorola's

---

[52] Lim (2014).
[53] *Id.*
[54] Raack (1986); Larouche and Zingales (2014).
[55] *Apple Inc. v. Motorola, Inc.*, 757 F.3d 1286, 1330 (Fed. Cir. 2014).
[56] Ginsburg, Owings, and Wright (2014).
[57] US International Trade Commission (2013).
[58] *Microsoft Corp.*, 795 F.3d at 1049 n. 19.
[59] Chien (2014).
[60] Israel (2015).
[61] *Apple Inc. v. Samsung Elecs. Co.*, 695 F.3d 1370, 1374 (Fed. Cir. 2012).
[62] *Apple Inc.*, 757 F.3d 1286.

injunction request. Conversely, in *Ericsson v. D-Link Systems*, the district court found that because D-Link "never meaningfully engaged Ericsson in RAND licensing negotiations after the initial offer," they were not "willing licensees" and vulnerable to an injunction.[63]

At a more granular level, the patentee's initial offer is a starting point and does not constitute a violation of FRAND commitments merely because it is greater than what the implementer deems reasonable even if the rate is ultimately litigated.[64] Similarly, the implementer does not violate its obligations by refusing that offer as long as it showed it was willing to "negotiate in good faith and earnestly seek an amicable royalty rate."[65] As one court put it, "[i]f two parties negotiating a RAND license are unable to agree to the financial terms of an agreement, it is entirely appropriate to resolve their dispute in court."[66] Implementers must prove holdup with evidence that goes beyond "theoretical" arguments. With respect to royalty stacking, the Federal Circuit has held that "[t]he mere fact that thousands of patents are declared to be essential to a standard does not mean that a standards-compliant company will necessarily have to pay a royalty to each [patent] holder."[67] According to commentators, "[t]his balance should make most antitrust claims based on breach of a [F]RAND commitment unnecessary."[68]

In contrast, Europe has opted for an antitrust response to the injunction issue. The Court of Justice of the European Union (CJEU) in *Huawei v. ZTE* held that patentees seeking injunctions on SEPs against "willing" licensees would violate competition law, as it would harm competition in a market that has already been weakened by the patentee's SEPs.[69] The FRAND commitment gives rise to implementers' "legitimate expectations" of continued access to the technology in order to make standards-compliant products.[70] Wielding market power obtained by virtue of its FRAND undertaking would be anticompetitive.

The CJEU laid out a protocol for both sides. Patentees must present "a specific written offer for a license" alerting implementers "of the infringement complained about by designating that SEP and specifying the way in which it has been infringed."[71] The large number of SEPs means that implementers will usually be unaware of them.[72] If the implementer is willing to conclude a FRAND license, patentees must present "a specific, written offer" on FRAND terms "specifying, in

---

[63] *Ericsson Inc. v. D-Link Sys., Inc.*, No. 6:10-CV-473, 2013 WL 4046225, at *25 (E.D. Tex. Aug. 6, 2013).
[64] *Id.*
[65] *Id.*
[66] *Id.*
[67] *Ericsson, Inc. v. D-Link Sys., Inc.*, 773 F.3d 1201, 1234 (Fed. Cir. 2014).
[68] Hovenkamp, Janis, and Lemley (2012).
[69] Case C-170/13, *Huawei Tech. Co. Ltd v. ZTE Corp.*, 2015 ECR 477, ¶ 45.
[70] *Id.* at ¶ 53.
[71] *Id.* at ¶ 61.
[72] *Id.* at ¶ 62.

particular, the amount of the royalty and the way in which that royalty is to be calculated."[73]

In turn, the implementer must "diligently respond to that offer, in accordance with recognised commercial practices in the field and in good faith, a point which must be established on the basis of objective factors and which implies, in particular, that there are no delaying tactics."[74] During negotiations, the implementer must also "provide appropriate security, in accordance with recognised commercial practices in the field, for example by providing a bank guarantee or by placing the amounts necessary on deposit."[75] However, the implementer retains its right to challenge both validity and infringement without running afoul of its FRAND commitments, since SSOs do not check whether patents declared essential truly are essential to the standard or valid.[76] Where no agreement is reached, the parties may request that the amount of the royalty be determined by an independent third party, by decision without delay.[77]

The CJEU did not address whether ownership of SEPs necessarily indicates market dominance. The better view is that ownership does not, since claim construction may determine that the technology was not technologically essential. US law also presumes no market power from ownership of SEPs.[78] The CJEU also leaves open two other issues for future courts: first, what constitutes either a FRAND royalty rate or a FRAND royalty base? Second, in noting the danger of patentees leveraging SEP ownership to reserve the downstream market for themselves, what would happen if the patentee was a nonpracticing entity?[79] This chapter offers some thoughts on both of these issues in Sections 13.2.2 and 13.3.

As early as 2007, the antitrust agencies recognized the risk of SSO-related holdups.[80] While the FTC preferred enforcement, the Department of Justice (DOJ) has focused on advocacy.[81] In *Bosch* and *Motorola*, the FTC found that the patentees violated their FRAND commitments and were liable under § 5 of the FTC Act.[82] The consent decrees required the patentees to resolve FRAND disputes via adjudication without recourse to injunctions.[83] Licensees also could neutralize the threat of an injunction by indicating that they are willing licensees.[84]

---

[73] *Id.* at ¶ 63.
[74] *Id.* at ¶ 65.
[75] *Id.* at ¶ 67.
[76] *Id.* at ¶ 69.
[77] *Id.* at ¶ 68.
[78] *Chrimar Systems v. Cisco Systems*, 72 F. Supp. 3d 1012 (N.D. Cal. 2014).
[79] Soames and Rato (2015).
[80] US DOJ-FTC (2007).
[81] Hesse (2012a; 2012b; 2013; 2014a; 2014b).
[82] *In re Robert Bosch GmbH*, No. C-4377, 2013 WL 1911293 (F.T.C. Apr. 24, 2013); *In re Motorola Mobility LLC*, No. C-4410, 2013 WL 3944149 (F.T.C. July 23, 2013).
[83] *In re Robert Bosch*, 2013 WL 1911293; *In re Motorola Mobility*, 2013 WL 3944149.
[84] *In re Robert Bosch*, 2013 WL 1911293; *In re Motorola Mobility*, 2013 WL 3944149.

In contrast, the DOJ's advocacy efforts underscore the importance of SSOs in mitigating the risk of holdups. It has found some measure of success with the Institute of Electrical and Electronics Engineers (IEEE)'s 2015 policy update.[85] Under the policy update, injunctive relief is limited to implementers who fail to participate in, or comply with an adjudicated outcome.[86] Implementers may challenge validity or infringement, and may require reciprocal cross-licensing SEPs, adjusted for the relative value of the technologies licensed regardless of the level of production.[87] This prevents patentees from only licensing manufacturers of the end product rather than component or sub-assembly intermediaries.[88] Reasonable royalties are confined to the patentee's technical contribution, and should consider the smallest saleable unit that practices said invention.[89] Comparable licenses can be used for comparison if they were obtained without threat of injunction and were negotiated under similar circumstances.[90]

The DOJ determined that the IEEE's update "has the potential to benefit competition and consumers by facilitating licensing negotiations, mitigating [holdup] and royalty stacking, and promoting competition among technologies for inclusion in standards."[91] It noted that "additional clarity ... could help speed licensing negotiations, limit patent infringement litigations, enable parties to reach mutually beneficial bargains that appropriately value the patented technology, and lead to increased competition among technologies for inclusion in IEEE standards."[92] A former DOJ chief economist and member of the President's Council of Economic Advisors noted that "[i]n our experience, this is about as enthusiastic as the Department of Justice gets in its business review letters."[93]

Qualcomm has been vocal in its criticism of the IEEE's policy changes.[94] Other SSOs have not made similar changes, with internal resistance as the chief reason.[95] The concern is that royalties will be depressed by opportunistic implementers. Since royalty rates are subject to FRAND, the worst outcome for the implementer is that they are subject to a FRAND rate, giving them little incentive to accept the SEP owner's initial offer unless it is lower than the FRAND rate.[96]

That concern is exaggerated. Patentees have little to gain by withdrawing from SSOs. Current laws allow patentees to enjoin implementers who act in bad faith. Even if the royalty rate is depressed, it may still be "reasonable" and sufficiently

---

[85] IEEE-SA (2015).
[86] Id.
[87] IEEE-SA (2015).
[88] Sundararaman (2015).
[89] IEEE-SA (2015).
[90] Id.
[91] US Department of Justice (2015).
[92] Id.
[93] Morton and Shapiro (2015).
[94] Decker and King (2015).
[95] Morton and Shapiro (2015).
[96] Chemtob (2015).

supracompetitive to induce innovation while also giving patentees the benefits of network effects. Further, unless and until the holdup defense is proven, infringement analysis proceeds in the usual way. In any case, the IEEE policy provides a "natural experiment" to observe the effects of SSO policy changes on licensing negotiations and SSO participation.

13.2.2.2. Damages

Damages can contribute to holdups if they are disproportionately large compared to the patentee's technological contribution. The "reasonable" prong in FRAND is intended to prevent this by mapping quantum to contribution.[97] The "non-discrimination" prong requires the patentee to license on similar terms as similarly situated licensees, and for implementers to accept a higher rate when they are not thus situated.[98] Despite the different methods in calculating a reasonable FRAND royalty, the prevailing view is that the correct measure of royalties is one that reflects only the value of the patented invention, and not holdup or switching costs.[99]

Courts start with the legal fiction of a negotiation that takes place between the parties just before the standard is set to determine what the royalty would be if the patentee had licensed the invention to a willing licensee in the position of the defendant.[100] This negotiation attempts to re-create what would have happened but for the infringement.[101] A notable method is the "top-down," approach which takes the price of the smallest unit and multiplies it by the profit margin of the unit.[102] Where the patented technology relates to the chip in a cell phone, the smallest saleable unit (SSU) would be the chip and not the mobile phone. That figure is then divided by the number of essential patents and adjusted for relative value. An evidentiary rule, the SSU approach attempts to avoid prejudicing a jury with large royalty base figures that result from the sales of the downstream product containing the component part. It also simplifies apportionment of reasonable royalty, since it is simpler to apportion the value of the patented technology in a Wi-Fi chip alone than when it is enmeshed as part of the value of a smartphone that contains the Wi-Fi chip.

There are several limitations to this approach. First, it deviates from the longstanding practice of licensing SEPs on a portfolio basis using the value of the finished product as the royalty base.[103] The Federal Circuit in *D-Link* used the entire product as the royalty base, holding it was appropriate where the entire market

---

[97] *Ericsson, Inc. v. D-Link Sys., Inc.*, 773 F.3d 1201, 1233 (Fed. Cir. 2014).
[98] *In re Innovatio IP Ventures, LLC Patent Litig.*, No. 11 C 9308, 2013 WL 5593609, at *38 (N.D. Ill. Oct. 3, 2013).
[99] *Ericsson, Inc.*, 773 F.3d at 1233.
[100] *Microsoft Corp. v. Motorola, Inc.*, 795 F.3d 1024, 1040 (9th Cir. 2015).
[101] Siebrasse and Cotter (2015).
[102] *In re Innovatio IP Ventures*, 2013 WL 5593609, at *37–9.
[103] Hoffinger (2015).

value of a device was attributable to the patented feature.[104] The flexibility to use the entire product as a base is a sensible one. It is only a matter of metrics; ounces or kilograms do not ultimately change the weight.[105] It is readily ascertainable and auditable by the licensor. One may also look to the precise patents that are at issue, the established royalty rate for those patents, or whether the rate for the unbundled portfolio of SEPs in suit exceeds the FRAND royalties for the bundled portfolio. Second, the component's value may depend on other inputs. Valuing a chip alone separates the value of the invention from its consumer benefit. Third, proportionality assumes there is a quantifiable total contribution that can be neatly divided among patentees proportionally.

China's National Development and Reform Commission (NDRC) issued a record penalty of $975 million against Qualcomm for holding up rivals.[106] In its settlement, Qualcomm agreed to lower royalties by 35 percent.[107] Similar to the United States, Qualcomm was entitled to use the net wholesale price of the end product as its royalty base, and could refuse to sell chips to unwilling licensees.[108] Like the CJEU's decision, the NRDC decision did not explain how to calculate FRAND royalties.

Despite being bound by the non-discriminatory requirement, royalties can differ based on the nature of and extent to which the technology is used in the marketplace and its importance in the standard and the product.[109] Years ahead, the non-discriminatory prong may become more controversial, as foreign competition authorities or courts impose their own requirements on patentees.[110] For instance, now that Qualcomm has been required to license its SEPs in China on specific terms, as have Samsung and Motorola in the European Union, can those help benchmark what constitutes non-discriminatory licenses elsewhere?

The NDRC's *Qualcomm* decision may be juxtaposed against the Shenzhen Intermediate People's Court's decision in *Huawei v. InterDigital*.[111] The patentee was accused of failing to negotiate and license SEPs on FRAND terms. The court held that the patents were essential.[112] The court determined that the royalty rate should not exceed 0.019 percent of the actual sales price of each product manufactured.[113] The court's decision was affirmed by the Guangdong Higher People's Court.[114] In a separate ruling, the Shenzhen court also held that the patentee abused

---

[104] *Ericsson, Inc.*, 773 F.3d at 1227.
[105] Hollman (2015).
[106] Moorhead (2015).
[107] *Id.*
[108] *Id.*
[109] Morrow, Levin, and Hill (2014).
[110] Prywes and Bell (2015).
[111] Ye *et al.* (2013).
[112] Mehra and Meng (2015).
[113] Han and Lin (2013).
[114] *Id.*

its dominant position in violation of China's Antimonopoly Laws in attempting to get an injunction against the defendant in the United States, whom it deemed a willing licensee, while in the midst of negotiations in China.[115]

In the Chinese Ministry of Commerce (Mofcom)'s Nokia/Alcatel merger review, the combined entity was required to adhere to FRAND obligations as a condition of the merger.[116] Mofcom required the undertaking because most Chinese wireless network equipment and handset makers did not have the patents required to effectively negotiate cross-licensing deals with Nokia, putting them at risk of holdup.[117]

With the crystallization of the law on injunctions and damages, lawyers are now advising against asserting SEPs, and instead advising patentees to assert non-FRAND-encumbered patents, some going as far as saying the alternative "would almost certainly be committing malpractice."[118] In either case, licensing disputes between manufacturing entities should eventually end in negotiations based on a rough estimate of the number and significance of each party's portfolio. This outcome is driven by the fear of paralyzing and expensive litigation if the parties fail to forge an agreement.[119]

When patentees transfer their SEPs to PAEs, the powerful impetus to reach a negotiated settlement is lost. Microsoft, Apple and other operating companies have systematically transferred patents to PAEs with the express purpose of monetizing them.[120] Operating companies disaggregate patent portfolios and transfer portions at a discounted rate to multiple PAEs while the operating companies retain a financial interest.[121] A variation of this arrangement involves the use of privateers. Microsoft acquired Nokia's business but left Nokia with its patents, shielding Microsoft from counter-suit while it aggressively asserts its patents against handset makers.[122] Privateering has become sufficiently prevalent and detrimental that companies have jointly made public commitments to prevent it from occurring. For example, Google, Canon, SAP and other companies have jointly agreed to grant licenses to each other in the event their patents are sold.[123] Cisco and Yahoo have undertaken not to sell their patents to PAEs.[124]

PAEs have been at the center of a maelstrom of controversy. Tech executives labeled them "terrorists," and newspapers highlighted their aggressiveness in

---

[115] Mehra and Meng (2015).
[116] Tung (2015).
[117] Schindler (2015).
[118] Hoffinger (2015).
[119] Prywes and Bell (2015).
[120] Faas (2012).
[121] Jurata and Patel (2014).
[122] Popofsky and Laufert (2013).
[123] Google (n.d.).
[124] Non-SDO Patent Statements and Commitments (2014).

extracting settlements from small businesses.[125] United States President Barack Obama characterized their tactics as "trying to essentially leverage and hijack somebody else's idea and see if they can extort money out of them."[126]

Yet, is it possible that PAEs are simply misunderstood? Airbnb is the world's largest accommodation provider, but owns no real estate. Alibaba is the most valuable retailer in the world, but has no inventory. Facebook is the world's most popular content media owner, but creates no content. Uber is the world's largest taxi company but owns no vehicles. By aggregating and licensing the patents needed to practice the technology, PAEs arguably serve a useful middleman function. With products that contain thousands of patented inventions, the technology market fills an invention gap. Like a supermarket, they provide the shopfront for individual inventors to monetize the value of their patents.[127]

Acacia Technologies describes itself as an intermediary that monetizes assets and connects upstream inventions to downstream users.[128] Small businesses lack the $5 to $8 million it costs to litigate against large, well-funded corporate infringers.[129] Financially constrained inventors sell to PAEs because they reduce litigation and liquidity risks.[130] To large tech companies, PAEs are a "garage sale" service, rummaging through the owner's old patents to monetize them for a quick profit.[131] Similarly, privateers may simply be a well-structured risk avoidance strategy. PAEs thus free operating companies to focus on its new trajectory of research and development while obtaining a stream of licensing income.

13.3. HOLDUPS BY PATENT ASSERTION ENTITIES

13.3.1. "As American as Apple Pie"

Supporters of PAEs argue that "patent licensing is as American as apple pie"[132] and estimate that "more than two-thirds of the great inventors of the nineteenth century Industrial Revolution licensed their inventions directly to manufacturers or employed licensing agents."[133] Further, new disruptive technologies have always spurred vigorous enforcement of patents.[134] To condemn increased incentives would bring innovation onto a slippery slope of regulatory overreaching and stifling efficient resource allocation.[135]

---

[125] Economist (2013).
[126] Sperling (2013).
[127] Lim (2010).
[128] Acacia Technologies (n.d.).
[129] Quinn (2015a).
[130] Id.
[131] Hollman (2015).
[132] Pridham (2015).
[133] Id.
[134] Jurata and Patel (2014).
[135] Wright (2008).

However, the reality on the ground is less rosy. More than a hundred thousand firms have received threats from PAEs, and up to a third of startup companies have received them.[136] The "overwhelming majority" of PAE-initiated infringement suits are being brought not against copyists, but rather against those who developed an invention independently.[137] PAE litigation has been on the rise, accounting for more than 60 percent of infringement suits filed in 2013, twice the figure in 2011.[138] Yet they lose nine out of ten times, compared to six out of ten times when practicing entities sue.[139]

PAEs also tax innovation when they bottleneck rather than channel wealth to the original patentees. One study shows that only 10 percent of implementers receive licenses from PAEs with a know-how transfer.[140] The authors surmise that "even the potential for future innovation were almost entirely absent."[141] From a dynamic efficiency standpoint, wealth transfers from implementers to PAEs do not transfer technology.[142] Instead, the tax downstream developers must pay discourages them from building on it.

PAEs have mastered the art of amassing an arsenal of patents to assert against businesses that they know would quail at the cost and uncertainty of litigation and the barrage of follow-on lawsuits that could be filed.[143] These portfolios usually have overlapping patent claims covering and surrounding a product or service.[144] The suits have gained a bad reputation of being a holdup attempt by patentees suing to extract a nuisance settlement from their victims.[145] Only one in ten patentees wins at trial, and most cases do not get there as nine in ten settle before a merits resolution.[146]

As the Federal Circuit observed, one PAE had "a history of filing nearly identical patent infringement complaints against a plethora of diverse defendants, where [it] followed each filing with a demand for a quick settlement at a price far lower than the cost to defend the litigation."[147] Defending a lawsuit can cost upwards of half a million dollars, so in most cases it makes more sense to settle.[148] These costs are incurred during discovery, before the court has defined the claims.[149]

PAEs target small companies because large manufacturers, the true targets, are "more likely to be more patent sophisticated, have better access to defensive prior

---

[136] Bessen (2014); Feldman (2013).
[137] Lemley (2007).
[138] Executive Office of the President (2013).
[139] Allison, Lemley, and Walker (2011).
[140] Feldman and Lemley (2015).
[141] Id.
[142] Chan (2012).
[143] Jurata and Patel (2014).
[144] Hollman (2015).
[145] Morton and Shapiro (2015).
[146] Allison, Lemley, and Walker (2011).
[147] *Eon-Net LP v. Flagstar Bancorp*, 653 F.3d 1314, 1326 (Fed. Cir. 2011).
[148] Divine *et al.* (2013).
[149] US Government Accountability Office (2013).

art, and [are] more invested in establishing a reputation for toughness lest they be targeted by other patent asserters."[150] As a result, the two parties best placed to negotiate or litigate to an efficient outcome do not meet.

For instance, MPHJ Technology Investments accused more than 16,000 small businesses of infringing patents covering network scanning technology.[151] MPHJ's letters demanded thousands of dollars from each owner. The FTC opened an investigation against MPHJ, which resulted in a settlement that would require it to pay $16,000 per demand letter.[152] Commentators note that while the case "establishes a valuable precedent and deterrent to the most egregious and widespread demand-letter campaigns, it seems unlikely to have much impact on PAE activity involving larger individual damages claims directed at specific, selected targets."[153] It is worth noting that the FTC investigated MPHJ under § 5 of the FTC Act as an exercise of its consumer protection, rather than antitrust powers. While the dynamics may be different, it represented the first real attempt by an antitrust agency to deal with the PAE issue.

PAEs' asymmetric litigation incentives compared to operating companies encourages unbridled assertion, risking negative externalities much in the same way that factories without carbon emissions limits tend to overproduce. They offer few or no products and services of their own, insulating them from infringement counterclaims.[154] PAEs also have the incentive to build a reputation for toughness by instilling fear, not goodwill. Since they do not cross-license or manufacture, alienating the firms they target will do them no reputational harm. PAEs exploit vulnerabilities in the legal framework to extort a nuisance settlement. These loopholes include the ease of obtaining software patents, the lack of pre-suit, defendant-specific investigation required of patentees,[155] the ability to assert broadly worded claims against many defendants at once,[156] and the difficulty in punishing PAEs who engage in vexatious litigation.

From the foregoing discussion, PAEs, broadly speaking, fall into three categories. First are "lottery-ticket" PAEs, so called because they bank on having patents that read on a significant area of technology.[157] These PAEs seek big damages awards against entrenched players with a few patents.[158] These come close to the usual sort of patent litigation. If there are excesses to their activity, the solution is found in patent reform, as discussed in Section 13.3.2. Second, "bottom-feeder" PAEs like

---

[150] Chien (2014).
[151] Wolfe (2014).
[152] US Federal Trade Commission (2014).
[153] Morton and Shapiro (2015).
[154] Cass (2015).
[155] Chien (2012).
[156] Chien and Reines (2014).
[157] Lemley and Melamed (2013).
[158] Id.

MPHJ seek quick, low-value settlements with low-quality patents.[159] They avoid trials and induce settlement by threatening costly litigation.[160] Thus far, these have attracted responses under consumer protection law rather than antitrust law. Patent reform of the sort discussed below has also helped curb the worst antagonists. Third, "patent aggregators" amass patents in order to demand royalties in return for a license on its portfolio. They rely more on the quantity of their patents rather than their quality to mount a credible threat against their victims.[161] As seen in Section 13.3.3, "patent aggregators" have the most potential to attract antitrust liability due to their appreciable market power.

13.3.2. Patent Responses

One cause of holdups is poor-quality patents. Examiners are overwhelmed by the volume of patent applications in fast-moving and complex technologies, leading them to grant patents without adequate review.[162] Patentees understand that "in a portfolio, volume makes up for weakness."[163] The probabilistic nature of patents and the sheer number of them that cover a standard can make it very difficult for implementers to avoid ensnaring themselves within a multitude of purportedly infringed claims. Rivals risk expensive and time-consuming litigation if they refuse to take a license. On the other hand, if they take a license over a questionable patent, that also raises costs by unnecessary licensing.

To address this, the Patent and Trademark Office (PTO) ramped up efforts to improve patent quality through the "Enhanced Patent Quality Initiative."[164] The Initiative includes identifying challenges earlier in the examination process, an enhanced automated pre-examination search, sharing results with applicants, and making claim constructions explicit to enhance clarity and completeness of the prosecution record.[165] The PTO's "Patent Litigation Toolkit" provides answers to common questions on patent assertions and also allows recipients of demand letters free access to databases that contain basic information on patent litigation.[166]

Separately, the FTC launched a study focusing on the impact of PAEs on innovation and competition.[167] It issued compulsory orders to PAEs in the wireless communications industry about the cost and benefits of PAE activities.[168] The results would be used to inform FTC policy advocacy and enforcement efforts. It

---

[159] Id.
[160] Id.
[161] Id.
[162] Hollman (2015).
[163] Id.
[164] Lee (2015).
[165] Id.
[166] Id.
[167] US Federal Trade Commission (2013).
[168] Id.

is worth noting that a 2002 FTC study regarding generic drug entry culminated in the FTC's recent victory in *FTC v. Actavis*, where the Supreme Court held that reverse payments were vulnerable to antitrust scrutiny under the rule of reason.[169]

The courts have also been active in stemming PAE litigation. PAEs almost never succeed in getting an injunction.[170] As early as the *eBay* decision in 2008, the Supreme Court recognized that injunctions could be "employed as a bargaining tool to charge exorbitant fees" by firms that use patents "not as a basis for producing and selling goods but, instead, primarily for obtaining licensing fees."[171] The Supreme Court narrowed down the types of business methods and software that are patent-eligible.[172] Courts have since routinely found software innovations ineligible for patent protection.[173] One commentator noted that "[a]s patents have become harder to get and even harder to keep, there is no longer a viable business strategy for those who were simply buying a patent or portfolio and rushing to the nearest patent litigation firm as a monetization strategy."[174] A notable example is Intellectual Venture (IV)'s patents that Capital One successfully challenged and invalidated.[175] The antitrust dimension of the dispute is discussed below in Section 13.3.3. The Court also made it easier for courts to penalize bad-faith litigation by awarding the prevailing party its attorney fees.[176]

Third, Congress passed the America Invents Act of 2011, the Patent Law Treaties Implementation Act of 2012, and the Patent Quality Improvement Act of 2013. Each aims to make it more difficult to obtain patents, easier to challenge them, and harder to enforce them.[177]

One unintended consequence of the American Invents Act is the spawning of a new breed of trolls that use *inter partes* review, or IPR, to extort patentees. The IPR process allows the public to have the PTO reconsider its decision to grant a patent. The high success enjoyed by challengers has caused them to be dubbed "patent terminators." Challengers inform patentees of their intent to institute an IPR unless the owner agrees to pay them. Challengers may or may not provide the evidence. Some challengers identify a publicly traded company that relies heavily on its patented technology. Those challengers then take a short position in the company's stock and file an IPR to invalidate the company's patents and publicize the IPR to spook investors.[178]

---

[169] Breed and Dickinson (2013). *F.T.C. v. Actavis, Inc.*, 133 S. Ct. 2223 (2013).
[170] Seaman (2016).
[171] *eBay, Inc. v. MercExchange, LLC*, 547 U.S. 388, 396 (2006) (Kennedy J. concurring).
[172] *Alice Corp. v. CLS Bank International*, 134 S. Ct. 2347 (2014).
[173] Quinn (2015b).
[174] *Id.*
[175] *Intellectual Ventures I, LLC v. Capital One Fin. Corp.*, 792 F.3d 1363 (Fed. Cir. 2015).
[176] *Highmark Inc. v. Allcare Health Mgmt. Sys.*, 134 S. Ct. 1744, 1747 (2014).
[177] Landau and Neu (2015).
[178] *Id.*

Looking ahead, PAEs may turn to design patents, which allow a design patent's owner to disgorge the defendant of its "total profit" – and when multiple patents are at issue, the patent owner can in theory be awarded the defendant's "total profit" multiple times. Companies may invest substantial amounts into a design only to face the threat of holdup once a design patent emerges.[179]

13.3.3. Antitrust Responses

It is legal for patentees to own a large portfolio, and to transfer ownership from one firm to another.[180] Whom the royalties flow toward is irrelevant. The acquisition becomes vulnerable to merger law only if it harms competition or changes enforcement incentives.[181] PAEs can also "aggregate supplementary or substitute patents in a market and prevent entry by cutting off a necessary input."[182]

Some have defended higher prices resulting from the transfer on the basis that the market power was "underutilized."[183] PAEs simply acquire from their predecessors in title a right to exclude that "neither alters the nature of competition in the relevant market nor changes the competitive process."[184] Those higher prices do not arise from changes in competition since their transfer is legal. That argument puts the cart before the horse. The legality of the transaction is determined by its nature and effect. It is a conclusion that flows from those determinations and should not be their starting point.

One way of pre-empting challenges when PAEs acquire patents is to assure implementers that they can rely on licensing commitments made by predecessors in interest. PAEs must continue to be bound by FRAND, regardless of whether SSOs have included the obligation in its governing documents and that obligation should bind successors in interest through bankruptcy proceedings.[185]

Post-acquisition, PAEs may be vulnerable to § 1 and/or § 2 violations.[186] One notable example is *Intellectual Ventures I LLC v. Capital One Fin. Corp.*[187] When IV sued Capital One for infringement, Capital One counterclaimed accusing IV of abusing its monopoly power and unlawfully acquiring 3,500 patents and asserting them through 2,000 shell companies to holdup Capital One and other banks.[188] The District Court for the District of Maryland found that Capital One had sufficiently

---

[179] Lim (2015).
[180] *Illinois Tool Works Inc. v. Independent Ink, Inc.*, 547 U.S. 28, 43 n. 4 (2006).
[181] Popofsky and Laufert (2013).
[182] Gotts and Sher (2012).
[183] Jurata and Patel (2014).
[184] *Id.*
[185] Israel (2015).
[186] 15 U.S.C. §§ 1–2.
[187] *Intellectual Ventures I, LLC v. Capital One Fin. Corp.*, No. PWG-14-111, 2015 WL 898146 (D. Md. Mar. 2, 2015).
[188] *Id.* at *2. *Intellectual Ventures I LLC v. Capital One Fin. Corp.*, No. 1:13-CV-00740 AJT, 2013 WL 6682981, at *2 (E.D. Va. Dec. 18, 2013).

alleged that IV demanded a license for its portfolio on the pain of "ceaseless litigation" through "a carefully orchestrated campaign of patent aggregation, concealment and sham litigation."[189] In particular, Capital One complained that IV intentionally acquired "a massive patent portfolio," encompassing 3,500 patents, "that [IV] alleges reads, vaguely, on existing products in [the financial-services] industry, regardless of how those products are designed, so that it could hold up banks that have substantially invested in those existing product designs."[190]

On the issue of monopoly power, the court agreed that the banks sufficiently alleged that "they have no viable alternative other than to license the patents" from IV because "an alternative license would not eliminate the threat of IV's threat of ceaseless litigation"[191] for them "to continue to provide the online services they already offer without paying the cost-prohibitive licensing fees."[192] The court found that IV plausibly had monopoly power – a 100 percent market share for "an indispensable body of patents."[193] This was achieved through patent acquisitions which had the effect of "eliminat[ing] all competition between patentees that would otherwise compete with each other for financial-services licensing opportunities."[194] Further, the court found high entry barriers based on IV's control over essential patents and rivals' "lack the ability to expand their output to challenge [IV's] high prices."[195]

The court also found that the banks sufficiently alleged anticompetitive conduct based on IV's patent acquisition. First, it noted that unlike operating companies that acquire patents for defensive purposes or design freedom, IV sought to "assert repeated claims of infringement to tax productive commercial use of existing technology."[196] Second, IV "concealed and obfuscated [their] patent holdings" so that it is "practically impossible for targets like Capital One to assess the portfolio and take steps to avoid IV's claims of patent infringement."[197] Third, the court noted that at the time of the patent acquisitions, IV had no presence in the banking market "when they began their patent acquisition, and the products already were in place and employed by the banking industry."[198] This indicates that the exercise of IV's monopoly power was not benign and customary market behavior, but behavior motivated by an intent to holdup the banks by disrupting their normal business activities.[199]

---

[189] *Intellectual Ventures I*, 2015 WL 898146 at *11.
[190] *Id.* at *13.
[191] *Id.* at *11 (quoting Third Amended Counterclaim, Intellectual Ventures I LLC v. Capital One Fin. Corp., ¶ 156).
[192] *Id.* at *11
[193] *Id.* at *12.
[194] *Id.* at *10.
[195] *Intellectual Ventures I LLC v. Capital One Fin. Corp.*, No. 1:13-CV-00740 AJT, 2013 WL 6682981, at *12 (D. Md. Mar. 2, 2015).
[196] *Id.* at *13.
[197] *Id.* at *14.
[198] *Id.*
[199] *Id.* at *11.

Echoing the earlier discussion, the royalties IV demanded were not based on the value of the patents, but rather on the costs of defending against serial litigation.[200] IV acquired individually weak patents in order to corral defendants like Capital One into settlement at rates that exceeded the technological value of the patents.[201] Specifically, the court noted that "IV reverses the normal from-patent-to-product process by using the designs of existing products as targets for custom-built patent portfolios" and "capitalizes on the fact that 'companies like Capital One have substantial sunk investments in their existing product designs'."[202] Leveraging on sunk costs in bad faith has been a recurrent theme in the holdup narrative, and patentees will find themselves vilified regardless of the form the holdup attempt takes.

Intellectual Ventures indicates that courts are willing to apply the sledgehammer of antitrust law to patentees if targets can present a well-pleaded counterclaim. The case also shows that courts take a dim view of PAEs monetizing patents through assertion in markets they did not help build up, and particularly where their only role is to disrupt the business of operating companies through a blitzkrieg of lawsuits.

13.4. CONCLUSION

Holdups have gained infamy from the image of knuckled-under implementers forced to pay patentees a premium because they are locked in. Like shark attacks, holdups are real but their actual occurrence is sporadic enough to be treated as aberrations rather than a systematic failure in the patent system.[203] A higher price may also reflect the premium associated with calculated convenience and certainty, a premium anyone who chooses Uber's taxi service rather than the public bus system knows. At the same time, while holdups, like crimes, are not widespread, laws must still exist to deter them before they occur and address them when they do.

Patent ambushes are rare, less because of disclosure obligations, and more due to the sting of possible antitrust enforcement. With FRAND disputes, patentees know injunctions are hard to come by and courts will likely map reward to technical contribution. Evidence showing intent can be useful here, as is direct or circumstantial evidence of harm. In the event negotiations fail, agreeing to submit the dispute to third-party adjudication would be the clearest evidence of good faith.[204] Harmful PAE conduct stems from features of the patent system, so the solution is to raise the bar of software patents, make it harder to initiate a suit, and punish those who bring frivolous suits.

---

[200] *Id.*
[201] *Intellectual Ventures I*, 2013 WL 6682981, at *11.
[202] *Id.* at *13.
[203] Larouche and Zingales (2014).
[204] Carlton and Shampine (2014).

Protagonists and antagonists exist on a spectrum that can appear flipped, depending on one ideological point of view, much as one man's terrorist is another man's freedom fighter. As with many things in life, the truth can be complicated and each choice comes with its own set of tradeoffs. Deterring brinkmanship could deter innovators who would otherwise invest more heavily in new technologies or participate as intermediaries in facilitating licensing. Despite this complexity, the law must set down the ground rules for engagement. Those whose everyday lives depend on the law finding a proper balance between competing interests deserve no less.

REFERENCES

Acacia Technologies. n.d. About Us. Available at http://acaciaresearch.com/about-us/.

Allison, John, Mark A. Lemley, and Joshua Walker. 2011. Patent Quality and Settlement among Repeat Patent Litigants. *Georgetown Law Journal*, 99, 677–712.

Bessen, James. 2014. All the Facts: PAEs Are Suing Many More Companies. Patently-O, January. Available at http://patentlyo.com/patent/2014/01/facts-suing-companies.html.

Blind, Knut, Rudi Bekkers, Yann Dietrich, *et al.* 2011. Study on the Interplay between Standards and Intellectual Property Rights (IPRs). Tender No ENTR/09/015 (OJEU S136 of 18/07/2009). Final Report, April.

Breed, Logan M. and Charles E. Dickinson. 2013. FTC Formally Proposes to Launch Section 6(b) Study on Activities of Patent Assertion Entities. Lexology (September). Available at www.lexology.com/library/detail.aspx?g=f0d6333f-ef37-484d-8d09-5428e52b7e06.

Carlton, Dennis W. and Allan L. Shampine. 2014. Identifying Benchmarks for Applying Non-Discrimination in FRAND. *CPI Antitrust Chronicle*, 8(1), 3–6.

Cass, Ronald A. 2015. Lessons from the Smartphone Wars: Patent Litigants, Patent Quality, and Software. *Minnesota Journal of Law, Science, and Technology*, 16, 1–61.

Chan, Nicholas P. 2012. Balancing Judicial Misvaluation and Patent Hold-Up: Some Principles for Considering Injunctive Relief after eBay. *University of California Law Review*, 59, 746–87.

Chemtob, Stuart M. 2015. Carte Blanche for SSOs? The Antitrust Division's Business Review Letter on the IEEE's Patent Policy Update. *CPI Antitrust Chronicle*.

Chien, Colleen V. 2010. From Arms Race to Marketplace: The Complex Patent Ecosystem and Its Implications for the Patent System. *Hastings Law Journal*, 62, 297–355.

  2012. Reforming Software Patents. *Houston Law Review*, 50, 325–90.

  2014. Holding Up and Holding Out. *Michigan Telecommunication and Technology Law Review*, 21, 1–42.

Chien, Colleen V. and Ed Reines. 2014. Why Technology Customers Are Being Sued En Masse for Patent Infringement and What Can Be Done. *Wake Forest Law Review*, 49, 235–57.

Contreras, Jorge L. 2015. A Market Reliance Theory for FRAND Commitments and Other Patent Pledges. *Utah Law Review*, 479–558.

Decker, Susan and Ian King. 2015. Qualcomm says it Won't Follow New Wi-Fi Rules on Patents. Bloomberg Business, February. Available at www.bloomberg.com/news/articles/2015-02-11/qualcomm-says-new-wi-fi-standard-rules-unfair-may-not-take-part.

Divine, David A., Richard W. Goldstein *et al.* 2013. Report of the Economic Survey 2013. American Intellectual Property Law Association. Available at www.patentinsurance.com/custdocs/2013AIPLA%20Survey.pdf.

Economist. 2013, Trolls on the Hill: Congress Takes Aim at Patent Abusers. December.
Executive Office of the President. 2013. Patent Assertion and U.S. Innovation. Available at www.whitehouse.gov/sites/default/files/docs/patent_report.pdf.
Faas, Ryan. 2012. Apple's Secret Weapon in the Patent Wars is a Nuclear NORAD. Cult of Mac, May. Available at www.cultofmac.com/ 168696/apples-secret-weaponin-the-patent-wars-is-a-nuclear-norad/.
Feldman, Robin. 2013. Patent Demands & Startup Companies: The View from the Venture Capital Community. University of California Hastings College of Law Research Paper No. 75. Available at http://papers.ssrn.com/sol3/papers.cfm?abstract_id=2346338.
Feldman, Robin and Mark Lemley. 2015. Does Patent Licensing Mean Innovation?. Working paper. Available at http://papers.ssrn.com/sol3/papers.cfm?abstract_id=2565292.
Ginsburg, Douglas H., Taylor M. Owings, and Joshua D. Wright. 2014. Enjoining Injunctions: The Case against Antitrust Liability for Standard Essential Patents. *Antitrust Source*, October, 1–7.
Google. n.d. The LOT Agreement. Available at www.google.com/patents/licensing/lot/.
Gotts, Ilene Knable and Scott Sher. 2012. The Particular Antitrust Concerns with Patent Acquisitions. *Competition Law International*, August.
Han, Michael and Kexin Lin. 2013. Huawei v. InterDigital: China at the Crossroads of Antitrust and Intellectual Property, Competition and Innovation. Competition Pol'y Int'l, November. Available at www.competitionpolicyinternational.com/huawei-v-interdigital-china-at-the-crossroads-of-antitrust-and-intellectual-property-competition-and-innovation/.
Harris, Robert G. 2014. Patent Assertion Entities and Privateers: Economic Harms to Innovation and Competition. *The Antitrust Bulletin*, 59(2), 281–325.
Hesse, Renata. 2012a. Six "Small" Proposals for SSOs before Lunch. Department of Justice Antitrust Division. Presentation, October 10, 2012.
  2012b. The Antitrust Division and SSOs: Continuing the Dialogue. Department of Justice Antitrust Division. Presentation, November 8, 2012.
  2013. The Art of Persuasion: Competition Advocacy at the Intersection of Antitrust and Intellectual Property. Department of Justice Antitrust Division. Presentation, November 8, 2013.
  2014a. At the Intersection of Antitrust and Hi-Tech: Opportunities for Constructive Engagement. Department of Justice Antitrust Division. Presentation, January 22, 2014.
  2014b. A Year in the Life of the Joint DO-PTO Policy Statement on Remedies for F/RAND Encumbered Standards-Essential Patents. Department of Justice Antitrust Division. Presentation, March 25, 2014.
Hoffinger, Roy E. 2015. The 2015 DOJ IEEE Business Review Letter: The Triumph of Industrial Policy Preferences over Law and Evidence. *CPI Antitrust Chronicle*, March.
Hollman, Hugh M. 2015. IEEE Business Review Letter: The DOJ Reveals Its Hand. *CPI Antitrust Chronicle*, March.
Hovenkamp, Herbert, Mark D. Janis, and Mark A. Lemley. 2012. *IP and Antitrust: An Analysis of Antitrust Principles Applied to Intellectual Property Law*. Austin, TX: Aspen, § 35.5.
IEEE-SA Standard Board Bylaws. 2015. Available at http://standards.ieee.org/develop/policies/bylaws/approved-changes.pdf.
Israel, Sharon A. 2015. Draft AIPLA Comments to Japan FTC IP Guidelines. Drafted July 21, 2015, updated August 6, 2015.
Jurata, Jr., John "Jay" and Amisha R. Patel. 2014. Taming the Trolls: Why Antitrust Is Not a Viable Solution for Stopping Patent Assertion Entities. *George Mason Law Review*, 21, 1251–85.

Landau, Nicholas J. and Jake Neu. 2015. Innovators Beware! Patent Reform Creates the New "Anti-Patent" Troll. Bradley Arant Boult Cummings LLP, July.

Larouche, Pierre and Nicolo Zingales. 2014. Injunctive Relief in Disputes Related to Standard-Essential Patents: Time for the CJEU to Set Fair and Reasonable Presumptions. TILEC Discussion Paper DP, December, 2014-048.

Lee, Michelle K. 2015. Remarks at the IPO Education Foundation PTO IPO Day Luncheon. Keynote speech, March 10, 2015. Available at www.uspto.gov/about-us/news-updates/remarks-michelle-k-lee-ipo-education-foundation-pto-ipo-day-luncheon

Lemley, Mark A. 2007. Should Patent Infringement Require Proof of Copying?. *Michigan Law Review*, 105, 1525–36.

Lemley, Mark A. and, Douglas A. Melamed. 2013. Missing the Forest for the Trolls. *Columbia Law Review*, 113, 2117–89.

Lim, Daryl. 2010. Post-eBay: A Brave New World?. *European Intellectual Property Review*, 10, 483–5.

 2011. Misconduct in Standard Setting: The Case for Patent Misuse. *IDEA*, 51, 559–604.

 2014. Standard Essential Patents, Trolls, and the Smartphone Wars: Triangulating the End Game. *Penn State Law Review*, 119, 1–91.

 2015. Why Samsung Owes Apple $930 Million. RealClearPolicy, June. Available at www.realclearpolicy.com/blog/2015/06/08/why_samsung_owes_apple_930_million_1318.html.

Mehra, Salil and Yanbei Meng. 2015. Essential Facilities with Chinese Characteristics: A Different Perspective on the Conditional Compulsory Licensing of Intellectual Property. *Journal of Antitrust Enforcement*, 3, 1–13.

Merges, Robert and Jeffrey Kuhn. 2009. An Estoppel Doctrine for Patented Standards. *California Law Review*, 97, 1–50.

Moorhead, Patrick. 2015. Qualcomm Settlement With China's NDRC Removes Major Speedbump. Forbes, February. Available at www.forbes.com/sites/patrickmoorhead/2015/02/10/qualcomm-settlement-with-chinas-ndrc-removes-major-speedbump/.

Morrow, Charlene M., Adam M. Levin, and Tammi L. Hill. 2014. To Join or Not to Join: When Membership in a Standard-Setting Organization Is the Question. Fenwick & West LLP, December.

Morton, Fiona Scott and Carl Shapiro. 2015. Patent Assertions: Are We Any Closer to Aligning Reward to Contribution?. National Bureau of Economic Research, April. Available at www.nber.org/chapters/c13587.pdf.

Mulligan, Christina and Timothy B. Lee. 2012. Scaling the Patent System. *New York University Annual Survey of American Law*, 68, 289–317.

New York Times. 2012. Fighters in a Patent War. October. Available at www.nytimes.com/interactive/2012/10/08/business/Fighters-in-a-Patent-War.html?_r=0.

Non-SDO Patent Statements and Commitments. 2014. Program on Information Justice and Intellectual Property. Updated June 21, 2014. Available at www.pijip.org/non-sdo-patent-commitments/.

Popofsky, Mark S. and Michael D. Laufert. 2013. Patent Assertion Entities and Antitrust: Operating Company Patent Transfers. Antitrust Source, April, 1–13. Available at www.ropesgray.com/~/media/Files/articles/2013/04/Antitrust-Attacks-on-Patent-Assertion-Entities.pdf.

Pridham, David. 2015. Patent Licensing Is as American as Apple Pie. IP Watchdog, March.

Prywes, Daniel I. and Robert S. K. Bell. 2015. Patent Hold-Up: Down But Not Out. *Antitrust Source*, 29(3), 25–30.

Quinn, Gene. 2015a. Understanding the valuable role played by Patent Trolls. IP Watchdog, March. Available at www.ipwatchdog.com/2015/03/17/understanding-the-valuable-role-played-by-patent-trolls/id=55787/.

2015b. Tactics for Coping with New Realities of Monetizing Innovation. IP Watchdog, July. Available at www.ipwatchdog.com/2015/07/09/tactics-for-coping-with-new-realities-of-monetizing-innovation/id=59680/.
Raack, D. W. 1986. A History of Injunctions in England Before 1700. *Indiana Law Journal*, 61, 539–92.
Sadler, Rodger. 2013. Reconsidering Claim Construction Standard of Review. Law360, April. Available at www.law360.com/articles/433763/reconsidering-claim-construction-standard-of-review.
Santore, Rudy, Michael McKee, and David Bjornstad. 2010. Patent Pools as a Solution to Efficient Licensing of Complementary Patents? Some Experimental Evidence. *Journal of Law and Economics*, 53, 167–83.
Schindler, Jacob. 2015. Nokia/Alcatel Tie-up Cleared in China following Patent Promises and a Timely Joint Venture. *Intellectual Assert Magazine*. Available at www.iam-media.com/blog/Detail.aspx?g=8f11696a-3c3b-4ff6-a690-09c1439e188c
Seaman, Christopher B. 2016. Permanent Injunctions in Patent Litigation after eBay: An Empirical Study. *Iowa Law Review*, 101, 1949.
Sidak, J. Gregory. 2015. The Antitrust Division's Devaluation of Standard-Essential Patents. *Georgetown Law Journal Online*, 104, 48–73.
Siebrasse, Norman V. and Thomas F. Cotter. 2015. A New Framework for Determining Reasonable Royalties in Patent Litigation. Available at https://papers.ssrn.com/sol3/papers.cfm?abstract_id=2528616.
Soames, Trevor, and Miguel Rato. 2015. The Court of Justice's Preliminary Ruling in Huawei v. ZTE: The Final Word?. Shearman & Sterling LLP, July.
Sokol, D. Daniel. 2017. *"Patent Privateers" in Patent Assertion Entities and Competition Policy*. Cambridge University Press.
Sperling, Gene. 2013. Taking on Patent Trolls to Protect American Innovation. The White House Blog, June. Available at www.whitehouse.gov/blog/2013/06/04/taking-patent-trolls-protect-american-innovation.
Sundararaman, Deepa. 2015. Inside The IEEE's Important Changes To Patent Policy. Law360, April.
Tung, Liam. 2015. China Gives Nod to Nokia-Alcatel Deal After Extracting Patent Promise, ZDNET (October). Available at www.zdnet.com/article/china-gives-nod-to-nokia-alcatel-deal-after-extracting-patent-promise/.
US Department of Justice. 2015. Institute of Electrical and Electronics Engineers, Incorporated. Business Review Letter, February. 2015 WL 557991.
US Department of Justice and the Federal Trade Commission (US DOJ-FTC). 2007. Antitrust Enforcement and Intellectual Property Rights: Promoting Innovation and Competition. April.
   2010. Horizontal Merger Guidelines §§ 6–7.
US Federal Trade Commission. 2011. The Evolving IP Marketplace: Aligning Patent Notice and Remedies with Competition.
   2013. FTC Seeks to Examine Patent Assertion Entities and Their Impact on Innovation, Competition. FTC Press Releases, September 27. Available at www.ftc.gov/news-events/press-releases/2013/09/ftc-seeks-examine-patent-assertion-entities-their-impact.
   2014. Settlement Bars Patent Assertion Entity from using Deceptive Tactics.
US Government Accountability Office. 2013. Intellectual Property: Assessing Factors that Affect Patent Infringement Litigation Could Help Improve Patent Quality. Gao-13-465.

US International Trade Commission. 2013. In re Certain Elec. Devices, Including Wireless Commc'n Devices, Portable Music & Data Processing Devices, & Tablet Computers. Inv. No. 337-TA-794. Commission Opinion, July.

Whitfield, Angelina M. 2015. Blocking Eco-Patent Trolls: Using Federalism to Foster Innovation in Environmental Technology. *Journal of Environmental and Sustainability Law*, 21, 307–29.

Wolfe, Jan. 2014. Accused Patent Troll Takes Aim at FTC, Settles With N.Y. AG. *The Litigation Daily*, January. Available at www.americanlawyer.com/id=1202638423763?slreturn=20140023185327.

Wright, Joshua D. 2008. No Ovation for FTC's Latest Enforcement Theory. Truth on the Market Blog, December. Available at http://truthonthemarket.com/2008/12/17/no-ovation-for-ftcs-latest-enforcement-theory/.

Ye Ruosi *et al*. 2013. Determination of Whether Abuse of Dominance by SEP Owners Constitutes Monopoly: Comments on the Antitrust Lawsuit Huawei v. InterDigital. *Digital Intellectual Property*, 3.

14

Walker Process and Sham Litigation

Leon Greenfield and Mark Ford

14.1. INTRODUCTION

By its very nature, a valid and enforceable patent gives its owner a state-sanctioned right to exclude others from making, using, or selling the patented invention throughout the United States during the term of the patent. There is, moreover, a strong public interest – of such significance that it is reflected in Article I of the US Constitution – in encouraging patenting of useful inventions and facilitating effective enforcement of the exclusionary right that a patent brings. In addition, the right to petition the government, including by filing lawsuits, is enshrined in the First Amendment to the Constitution. Enforcing a patent against a potential infringer is therefore rarely grounds for antitrust intervention. Where, however, a patent is either fraudulently acquired or asserted in bad faith, the general immunity from antitrust attack may no longer apply. The precise application of this principle presents both theoretical and practical challenges.

Courts have developed two complementary doctrines to attempt to strike the right balance between encouraging obtaining and enforcement of patents and a more general First Amendment right to pursue government process (including litigation), on the one hand, and the public interest in avoiding harm to competition based on *fraudulent* or *bad-faith* patent-related conduct on the other. First, under the *Walker Process* doctrine, asserting a patent knowingly procured by fraud will violate the Sherman Act (usually Section 2) where the other elements of a Sherman Act claim are met.[1] Second, under the Supreme Court's decision in *Professional Real Estate Investors*[2] (*PRE*), which expounded on the Ninth Circuit's decision in *Handgards*,[3] "sham" litigation to enforce a patent – that is, objectively baseless litigation intended solely to harm a rival through the litigation *process* itself rather than through the

---

[1] *Walker Process Equipment, Inc. v. Food Machinery & Chemical Corp.*, 382 U.S. 172 (1965).
[2] *Professional Real Estate Investors Inc. v. Columbia Pictures Indus., Inc.*, 508 U.S. 49, 56–57 (1993).
[3] *Handgards, Inc. v. Ethicon, Inc.*, 743 F.2d 1282 (9th Cir. 1984).

outcome of that process – also violates the Sherman Act, again assuming that the other elements of the antitrust claim are met). (In the patent context, sham litigation claims are often called "*Handgards*" claims.)

Debate among commentators and in the courts has focused on a core underlying question: Do the *Walker Process* and *Handgards* doctrines and applications thereof correctly modulate between the dangers of over-deterring desirable patentee-related conduct and under-deterring abuses? Although the courts have generally tended to constrict the availability of *Walker Process* and *Handgards* claims over time, an active group of commentators has advocated for expansion.

This chapter discusses the history and development of the *Walker Process* and sham litigation doctrines, some of the key unresolved issues that courts struggle with today, and scholarship relating to these doctrines. The chapter ends by addressing briefly an issue that has been emerging in the context of suits to enforce declared standard essential patents: Does the general antitrust immunity for non-sham litigation apply when an alleged anticompetitive scheme culminates with patent litigation that is not – itself – alleged to be a sham?

14.2. LEGAL FRAMEWORK

Antitrust actions grounded in patent enforcement overwhelmingly are brought as monopolization or attempted monopolization claims under Section 2 of the Sherman Act, which require proof of (1) monopoly power (or, for attempted monopolization, a dangerous probability of acquiring monopoly power), (2) willful acquisition and maintenance of that power, and (3) antitrust injury, that is, harm flowing from a reduction in competition.[4] When the antitrust claim is based on the defendant's enforcement of a fraudulently obtained patent, or on the baseless assertion of a patent, the plaintiff contends that the lawsuit itself constitutes exclusionary conduct sufficient to show that the patentee's monopoly power was "willfully acquired or maintained."[5]

---

[4] See *Datagate, Inc. v. Hewlett-Packard Co.*, 941 F.2d 864, 868 (9th Cir. 1991); *Spectrum Sports, Inc. v. McQuillan*, 506 U.S. 447, 456, 113 S. Ct. 884, 890 (1993). In antitrust cases with *Walker Process* or sham litigation claims, one recurring question is whether incurring litigation costs to defend against a wrongful patent suit constitutes an *antitrust* injury, i.e., a private loss that arises from an "injury to the market or to competition in general, not merely injury to individuals or individual firms." *McGlinchy v. Shell Chem. Co.*, 845 F.2d 802, 812 (9th Cir. 1988). Some courts have held that legal fees, standing alone, cannot constitute antitrust injury. *See Chip-Mender, Inc. v. Sherwin-Williams Co.*, No. C 05-3465 PJH, 2006 WL 13058, at *5–6 (N.D. Cal. Jan. 3, 2006). The prevailing trend, however, is that legal defense costs can constitute antitrust injury where they are incurred as a result of conduct that has caused harm to competition. See *Apple, Inc. v. Samsung Electronics Co.*, No. 11-CV-01846-LHK, 2012 WL 2571719, at *27 (N.D. Cal. June 30, 2012) ("Litigation costs have been recognized as appropriate antitrust damages in the context of anticompetitive sham litigation."); see also *Handgards*, 601 F.2d at 997 ("In a suit alleging antitrust injury based upon a bad faith prosecution theory it is obvious that the costs incurred in defense of the prior patent infringement suit are an injury which 'flows' from the antitrust wrong").

[5] Sham litigation (but not *Walker Process*) issues can also arise in the copyright or trademark contexts. See, e.g., *PRE*, 508 U.S. at 56–7 (1993) (considering whether copyright infringement

As discussed below, a defendant's enforcement of patent rights through litigation is generally immune from antitrust liability under the *Noerr-Pennington* doctrine. To overcome this threshold hurdle, the plaintiff must prove that the challenged conduct falls within the *Walker Process* or sham litigation exceptions to *Noerr-Pennington* immunity.

14.2.1. Walker Process *Doctrine*

Walker Process claims, which derive from the Supreme Court's decision in *Walker Process Equipment, Inc. v. Food Machinery & Chemical Corp.*, 382 U.S. 172 (1965), are grounded in contentions that the patentee failed to act with candor before the US Patent and Trademark Office (PTO) in applying for a patent. It is useful to ground discussion of *Walker Process* claims in some basic background regarding the PTO's procedure for granting patents, especially the extent to which that process relies on the good faith of the patent applicant. As Hovenkamp et al. explain:

> The PTO receives nearly 400,000 applications and grants nearly 200,000 patents per year. Patent applications are filed *ex parte*, and are kept confidential through much of the application process. Third parties have no opportunity to participate in the decision to grant a patent ... Patent examiners spend very little time with each application – only 18 hours per patent on average over the course of three years. As a result, the patent system of necessity depends heavily on the candor of the patent application. Applications are required to ... disclose any prior art or other information of which they are actually aware that might prevent them from getting a patent.[6]

14.2.1.1. Walker Process *Decision*

In *Walker Process*, the patentee, Food Machinery, brought a patent infringement suit against Walker, a competing suppler of aeration equipment components.[7] In response, Walker argued that the asserted patent was unenforceable because Food Machinery knowingly made a material misrepresentation in its patent application. Walker also filed a Section 2 counterclaim alleging that by asserting its fraudulently obtained patent, Food Machinery "had deprived Walker of business that it otherwise would have enjoyed."[8] The District Court and the Court of Appeals agreed that the alleged fraud on the PTO, if proven, would prevent Food Machinery from

---

action was a sham). Antitrust issues more frequently arise from patent assertions, however, in part because patents are more likely to result in market dominance than copyrights or trademarks. Thus, this chapter will focus on patent assertions.

[6] Hovenkamp et al. (2014).
[7] *Walker Process*, 382 U.S. at 173.
[8] *Id.* at 174.

recovering damages for any infringement. Both courts went on to hold, however, that Walker was not entitled to any *antitrust* remedy for the alleged anticompetitive effect of Food Machinery's patent suit. At the time, lower courts were split over whether an antitrust remedy should be available in these circumstances.[9]

Siding with Walker, the Supreme Court reversed, holding that "the enforcement of a patent procured by fraud on the Patent Office may be violative of Section 2 of the Sherman Act provided the other elements necessary to a section 2 case are present."[10] The Court explained that acquiring a patent by fraud would strip the patentee of "the limited exceptions to the prohibitions of the Sherman Act" normally granted to those asserting patents.[11] According to the Court, the public interest required this outcome:

> A patent by its very nature is affected with a public interest ... [It] is an exception to the general rule against monopolies and to the right to access to a free and open market. The far-reaching social and economic consequences of a patent, therefore, give the public a paramount interest in seeing that patent monopolies spring from backgrounds free from fraud or other inequitable conduct and that such monopolies are kept within their legitimate scope.[12]

The majority made clear, however, that "knowing and willfully misrepresenting facts to the Patent Office" was required, and that the patentee's "good faith would furnish a complete defense" to an antitrust claim arising from the enforcement of the patent, even if the PTO relied on "an honest mistake" by the patentee when issuing the patent.[13] In his oft-cited concurrence, Justice Harlan emphasized this point and stressed the delicate balance between patent law and antitrust in this context. He warned that permitting antitrust lawsuits based merely on the assertion of invalid patents – "patents that for one reason or another may turn out to be voidable under one or more of the numerous technicalities attending the issuance of a patent" – would likely "chill the disclosure of inventions through the obtaining of a patent because of fear of the vexations or punitive consequences of treble-damage suits."[14] With the understanding that antitrust claims would be permitted based only on assertions of patents that had been *fraudulently obtained*, however, Justice Harlan joined the Court's opinion.

14.2.1.2. Current *Walker Process* Standards

In applying *Walker Process*, lower courts have generally been solicitous of Justice Harlan's concerns and limited the availability of claims alleging fraudulent patent

---

[9] *Id.* at 175.
[10] *Id.* at 174.
[11] *Id.* at 176.
[12] *Id.* at 177 (internal quotation marks and citation omitted).
[13] *Id.* at 177.
[14] *Id.* at 180 (Harlan J., concurring).

procurement. For example, courts have held that to prove the "fraud" element of the claim, the plaintiff must present proof that satisfies the standard for common law fraud, including that there is clear and convincing evidence of "deceptive intent together with a clear showing of reliance."[15] In addition, the Federal Circuit has held that fraudulent intent must be "the single most reasonable inference to be drawn from the evidence," rather than just one of several reasonable inferences.[16] Finally, although the Supreme Court did not specify what constituted "enforcement" of a patent, courts have typically held that either an infringement lawsuit or a threatened infringement lawsuit is required, and it is not enough to allege merely that the patentee advertised its patent rights widely or did something else short of actually suing or threatening suit.[17]

Under the Federal Circuit standard that governs today, to bring a *Walker Process* claim, a plaintiff must prove by clear and convincing evidence four robust elements:

- the patent applicant made a fraudulent misrepresentation or omission to the PTO;
- the misrepresentation or omission was made with the specific intent to deceive the PTO;
- but for the misrepresentation or omission the PTO would not have issued the patent; and
- the patent was enforced with knowledge that it had been fraudulently obtained.[18]

14.2.1.3. Knowledge and Intent Requirement in *Walker Process* Claims

In a footnote in *Walker Process*, the Supreme Court observed that its "conclusion [that assertion of a fraudulently obtained patent can give rise to antitrust liability] applies with equal force to an assignee who maintains and enforces the patent with knowledge of the patent's infirmity," i.e., "with knowledge of the fraudulent manner in which it was obtained."[19] In *Nobelpharma*, the Federal Circuit applied this principle to hold that "[t]he plaintiff in the patent infringement suit must also have been aware of the fraud when bringing suit."[20]

Neither the Supreme Court nor the Federal Circuit, however, addressed the state of mind requirement when the *same entity* that initially obtained the patent seeks to

---

[15] *Nobelpharma AB v. Implant Innovations, Inc.*, 141 F.3d 1059, 1071 (Fed. Cir. 1998); Daniel (2009), discussing *Herman & MacLean v. Huddleston*, 459 U.S. 375, 389 (1983) (holding allegations of securities fraud need only be proved by a preponderance of the evidence).
[16] *Ariad Pharm., Inc. v. Eli Lilly & Co.*, 560 F.3d 1366, 1379 (Fed. Cir. 2009).
[17] Daniel (2009).
[18] See *Nobelpharma*, 141 F.3d at 1070; *Dippin' Dots, Inc. v. Mosey*, 476 F.3d 1337, 1346 (Fed. Cir. 2007); *C.R. Bard, Inc. v. M3 Sys., Inc.*, 157 F.3d 1340, 1364 (Fed. Cir. 1998).
[19] *Walker Process*, 382 U.S. at 177 n. 5, 179.
[20] *Nobelpharma*, 141 F.3d at 1068, 1073.

enforce it. When the *same individual* both obtains and enforces the patent, that question is irrelevant because the person will necessarily have knowledge of his or her own specific intent to commit fraud before the PTO. In many instances, however, certain employees obtain a patent and then, many years later, other employees enforce the patent, often without first-hand knowledge regarding the patent prosecution. When this is the case, a critical and unsettled question arises: Precisely what type of unlawful intent or bad faith must accompany the patent enforcement before courts deprive the enforcer of antitrust immunity for bringing suit?

The general principle that knowledge of one corporate employee is imputed to the corporation is not controversial. In *Acme Precision Products, Inc. v. Am. Alloys Corp.*, 422 F.2d 1395, 1398 (8th Cir. 1970), for instance, the Eighth Circuit held that "knowledge of officers and key employees of a corporation, obtained while acting in the course of their employment and within the scope of their authority, is imputed to the corporation itself," even after the officer or key employee has left the corporation.

But that the *knowledge* of one employee can be imputed to the corporation does not resolve the question of whether the *specific intent* to commit fraud of one employee that prosecuted the patent can be imputed to the corporation when another employee seeks to enforce the patent against an alleged infringer. Indeed, several courts have held outside the *Walker Process* context that in determining whether a corporation had the necessary state of mind for liability, "it [is] appropriate to look to the state of mind of the individual corporate official or officials who make or issue the statement ... rather than generally to the collective knowledge of all the corporation's officers and employees acquired in the course of their employment."[21] This is because "[a] corporation can be held to have a particular state of mind only when that state of mind is possessed by a single individual."[22] In the context of a *Walker Process* claim, a defendant might argue that because "specific intent" "cannot be aggregated," the knowledge of one employee who fraudulently obtained a patent cannot mean that another employee who enforced the patent is charged with doing so with the specific intent of enforcing a patent that had been obtained by fraud.[23]

The Federal Circuit's decisions in this area have never explicitly addressed this issue, though they seemingly have relied on contradictory assumptions. In *Nobelpharma*, for instance, the Federal Circuit carefully analyzed whether specific corporate officers who made the decision to enforce the patent had actual knowledge that it was obtained by fraud.[24] In *Unitherm*, however, the Federal Circuit

---

[21] *City of Roseville Employees' Ret. Sys. v. Horizon Lines, Inc.*, 442 F. App'x 672, 676 (3d. Cir. 2011).
[22] *First Equity Corp. of Fla. v. Standard & Poor's Corp.*, 690 F. Supp. 256, 260 (S.D.N.Y. 1988), aff'd, 869 F.2d 175 (2d Cir. 1989).
[23] *Lind v. Jones, Lang LaSalle Americas*, Inc., 135 F. Supp. 2d 616, 622 n.6 (E.D. Pa. 2001).
[24] 141 F.3d at 1072–3.

focused on whether the named inventor had the specific intent to deceive the PTO, and apparently assumed that such intent would be imputed to different corporate employees who decided to enforce the patent (though the court never explicitly addressed that issue).[25]

In the only decision precisely to address the type of corporate intent required under *Walker Process*, the court in *FTC v. Cephalon, Inc.*, 36 F. Supp. 3d 527, 535 (E.D. Pa. 2014), imputed knowledge of the allegedly fraudulent activities of employees who applied for a patent to employees of the same corporation who later enforced it, holding that under "basic principles of agency law ... a corporation is charged with knowledge of acts done by its agents acting within the scope of their employment." The court dismissed as irrelevant evidence Cephalon presented that the corporate officers responsible for bringing suit were unaware of any fraud on the PTO: "Cephalon ... must be held to have had knowledge of its own misconduct, notwithstanding the alleged lack of specific knowledge on the part" of the officials who brought suit.[26] Thus, the court found that the corporation had specific intent to enforce a patent that had been obtained by fraud by imputing to the enforcing employees the fraudulent intent of the other employees who had applied for the patent.

At bottom, the case-law regarding the necessary knowledge or intent on the part of those enforcing patents in the *Walker Process* context is sparse and cuts in different directions. This is an area where thoughtful normative commentary could help move the law in a sensible direction that strikes the right balance between under-deterring exclusion through enforcement of fraudulently obtained patents, while avoiding the chilling of bona fide patent assertions based on unbounded exposure premised on procurement fraud of which the individuals asserting the patent could not reasonably have been expected to be aware.

14.2.1.4. Some Other Recurring Issues Under the *Walker Process* Doctrine

Historically, *Walker Process* claims have only very rarely been successful. Indeed, a 2002 study found that only two claims had succeeded in the previous two decades.[27] Several commentators have argued that *Walker Process* claims could be an important tool in deterring and detecting fraud or material omissions before the PTO, and that such claims should therefore be available in a broader range of circumstances or burdens of proof should be relaxed.[28]

One focus of commentary has been whether direct purchasers, as opposed to actual or potential rivals, should have standing to bring *Walker Process* claims.[29]

---

[25] *Unitherm Food Systems v. Swift-Eckrich*, 375 F.3d 1341, 1360–1 (Fed. Cir. 2004).
[26] 36 F. Supp. 3d at 536.
[27] Steinman and Fitzpatrick (2002).
[28] See, e.g., Daniel (2009); see also Leslie (2008).
[29] Tokic (2012); Himes (2009); Leslie (2007); Leslie (2006); see also *In re DDAVP Direct Purchaser Antitrust Litig.*, 585 F.3d 677, 695 (2d Cir. 2009).

Several courts had assumed, without reference to the text of *Walker Process*, that because fraud on the PTO is "fraud aimed at competing manufacturers," not direct purchasers, only competitors should be permitted to bring *Walker Process* claims.[30] One rationale for this view is that direct purchasers generally cannot bring declaratory judgment actions to challenge the validity of a patent directly, and granting direct purchasers standing to bring *Walker Process* claims would allow them "to challenge a patent's validity ... simply by dressing their patent challenge with a *Walker Process* claim."[31] Courts and commentators have also expressed a policy concern that allowing direct purchasers to bring *Walker Process* claims would lead to a flood of unmeritorious claims that would discourage legitimate patent assertions.[32] Several commentators, however, have supported consumer standing, particularly given their concerns about under-deterrence of fraud before the PTO.[33]

In 2009, the Second Circuit addressed this standing issue in *In re DDAVP Direct Purchaser Antitrust Litig.*, 585 F.3d 677, 695 (2d Cir. 2009), which involved claims by direct purchasers of pharmaceuticals who argued that the patentee had used fraudulently obtained patents to exclude generic entrants from the market, resulting in inflated prices to the direct purchasers.[34] The court first found that under general antitrust standing principles, the direct purchasers would be suitable plaintiffs to bring a *Walker Process* claim.[35] But then, based on policy concerns that granting unfettered *Walker Process* standing to direct purchasers could open the floodgates to litigation and chill legitimate patent assertions, the Second Circuit declined "to decide whether purchaser plaintiffs *per se* have standing to raise *Walker Process* claims."[36] Instead, limiting its decision to the facts of the case before it (where the patent had already been held unenforceable), the court held "only that purchaser plaintiffs have standing to raise *Walker Process* claims for patents that are already unenforceable due to inequitable conduct."[37]

In 2012, the Federal Circuit addressed this issue for the first time, holding in *Ritz Camera & Image, LLC v. SanDisk Corp.*, 700 F.3d 503, 504 (Fed. Cir. 2012), that direct purchasers may bring *Walker Process* claims *de novo*. The court noted that direct purchasers are generally "preferred antitrust plaintiffs," and saw no reason to treat *Walker Process* plaintiffs differently.[38] The court declined to limit *Walker Process* standing to claims involving already "tarnished" patents, as the Second Circuit

---

[30] *In re Ciprofloxacin Hydrochloride Antitrust Litig.*, 363 F. Supp. 2d 514, 547 (E.D.N.Y. 2005), aff'd in part, 544 F.3d 1323 (Fed. Cir. 2008) and aff'd in part sub nom. *Arkansas Carpenters Health & Welfare Fund v. Bayer AG*, 604 F.3d 98 (2d Cir. 2010), as corrected (June 17, 2010) (quoting *Asahi Glass Co. v. Pentech Pharms.*, 289 F.Supp.2d 986, 995 (N.D. Ill. 2003)) (collecting cases).
[31] *In re DDAVP*, 585 F.3d at 690.
[32] See *supra* note 28.
[33] *Id*.
[34] *In re DDAVP*, 585 F.3d at 685.
[35] *Id*. at 688–90; see also Leslie (2006).
[36] *In re DDAVP*, 585 F.3d at 691.
[37] *Id*. at 691-2.
[38] *Ritz Camera.*, 700 F.3d at 506 (internal quotation marks and citation omitted).

had, holding that *Walker Process* contained "no such limitation" and that such a rule would only "generate unproductive wrangling over what counts as a sufficiently 'tarnished' patent to support a *Walker Process*" claim.[39] The court was similarly brief in disposing of the argument that allowing direct purchaser standing would allow too many claims, observing that the Supreme Court had rejected a similar argument in *Walker Process* itself.[40] In May 2015, the district court on remand certified a class of direct purchasers bringing a *Walker Process* claim.[41]

Another issue that has received attention in the literature is how courts should define the types of "enforcement" that can make out a Section 2 violation. Although the case-law is not settled, courts have typically required either an infringement lawsuit or the threat of an infringement lawsuit to find "enforcement."[42] In *Unitherm*, the Federal Circuit held that the standards required for finding jurisdiction in a Declaratory Judgment Act case would also "define the minimum level of 'enforcement' necessary to expose the patentee to a *Walker Process* claim ..."[43] Because the Supreme Court has since abrogated the former "reasonable apprehension of suit" test for Declaratory Judgment Act (DJA) jurisdiction, it is not clear that *Unitherm* is still good law. The few courts to address this issue have construed *Unitherm* to require equivalence between the DJA and *Walker Process* standards and held that the new *MedImmune* test for DJA jurisdiction – whether a "substantial controversy" exists – now defines the "minimum level of 'enforcement' necessary" for *Walker Process*.[44] This conclusion is debatable, however, since the *MedImmune* test is arguably more relaxed than the former "reasonable apprehension" test, and the Federal Circuit has not explicitly addressed whether the equivalence principle it adopted in *Unitherm* is still appropriate in light of the particular legal and policy issues that *Walker Process* claims implicate.

---

[39] *Id.* at 507 and n. 2.
[40] *Id.* at 508.
[41] *Giuliano v. SanDisk Corp.*, No. C 10-02787 SBA, 2014 WL 4685012, at *1 (N.D. Cal. Sept. 19, 2014). One post-*Ritz Camera* district court observed that *Ritz Camera* was "persuasive" but not binding on it, and wrote that it was "not entirely convinced there should be no limits on purchaser standing" (though it ultimately determined it need not decide the standing issue because the sham allegations were implausible on their face). *In re Lipitor Antitrust Litig.*, No. 3:12-CV-2389 PGS, 2013 WL 4780496, at *17 (D.N.J. Sept. 5, 2013). The debate over direct purchaser standing is far from over.
[42] *See, e.g.*, *Cygnus Therapeutics Sys. v. ALZA Corp.*, 92 F.3d 1153, 1159 (Fed. Cir. 1996), overruled on other grounds by *Nobelpharma AB v. Implant Innovations, Inc.*, 141 F.3d 1059, 1068 (Fed. Cir. 1998) (refusing to find enforcement based on various letters and speeches touting patent, including a statement that the company took a "very strong proprietary position" regarding its patents); *K-Lath, Div. of Tree Island Wire (USA), Inc. v. Davis Wire Corp.*, 15 F. Supp. 2d 952, 964 (C.D. Cal. 1998) (refusing to find enforcement based on correspondence that included, among other things, a reservation of the right to sue); *Aguirre v. Powerchute Sports, LLC*, No. SA-10-CV-702-XR, 2011 3359554 (W.D.Tex. Aug. 4, 2011) (heavy advertising was not enforcement).
[43] 375 F.3d at 1358.
[44] *Xitronix Corp. v. KLA-Tencor Corp.*, No. A-14-CA-1113-SS, 2015 WL 5037387, at *1 (W.D. Tex. Aug. 24, 2015) (analyzing issues and collecting cases).

Professor Leslie has argued that whatever the precise contours of the standard for "enforcement" under the evolving case-law, it will not capture enough anticompetitive behavior.[45] He contends that an invalid patent can deter competitive entry even when no infringement lawsuit has been brought or explicitly threatened, if the patent holder "touts" its patent through patent protection markings, letters to the accused infringer, and the like.[46] In that circumstance, entry may be deterred if it would not be economical for the accused infringer to defend infringement litigation, or, alternatively, the accused infringer's costs could be raised if it must "attempt to design around the patent" or is forced to "pay (unnecessary) licensing fees."[47] So long as a *Walker Process* claim can rest only on litigation or a threat thereof, Professor Leslie argues, "the enforcement requirement creates an effective safe harbor for firms that commit fraud against the PTO," and thus:

> The enforcement requirement makes maintenance of an invalid patent cost-beneficial. Under the current legal regime, the holder of an invalid patent can secure monopoly rents for years without ever worrying about antitrust liability. If knowingly maintaining an invalid patent cannot itself give rise to legal liability, then it is perfectly rational for a monopolist to sit tight – neither initiating infringement litigation nor informing the market of the patent's invalidity. Once a patentee learns that its patent is invalid, it costs little, if anything, to remain silent. As a result, maintaining an invalid patent can provide a relatively cost-effective mechanism to deter others from entering one's market ... The current antitrust rule creates an incentive for firms to acquire and maintain invalid patents and then to bide their time.[48]

Notwithstanding the potentially significant competitive harm from imposing a high barrier to showing "enforcement," it is not a simple matter to draw a more expansive, but clear, line delineating what constitutes "enforcement" without chilling legitimate patent-related conduct. This is especially true given First Amendment and patent law considerations that favor allowing a patentee to publicize its patents widely.

Professor Leslie proposes that where a patent holder knows that a patent was obtained fraudulently, it will be deemed to have "enforced" that patent unless and until it publicly disavows enforcement.[49] He acknowledges, however, several potential objections to this regime. The first is that punishing mere *possession* of an invalid patent is "essentially converting monopolization into an inchoate violation." He observes, however, that possession is a punishable offense in many areas of law, and argues "the refusal of a monopolist to disavow a patent that it knows to be invalid is as much conduct as a monopolist's refusal to share access to an essential facility."[50]

---

[45] Leslie (2008).
[46] Leslie (2008); Leslie (2006).
[47] Leslie (2006).
[48] Id.
[49] Id.
[50] Id.

A second argument against a looser "enforcement" standard is that it will encourage nuisance suits – i.e., by creating a plaintiffs' bar dedicated to filing *Walker Process* claims whenever any patent is invalidated. Leslie argues, however, that *Walker Process* claims will remain difficult to prove as a substantive matter, which may discourage filing of true nuisance suits. Moreover, the court in the original *Walker Process* case rejected arguments that allowing fraud-based claims would subject "patentees [to] innumerable vexatious suits,"[51] although it was not specifically addressing the standard for "enforcement"; Professor Leslie contends that the fear of nuisance suits is no more reason to deny an antitrust remedy now than it was when *Walker Process* was decided in 1965.[52]

Professor Leslie has made a provocative proposal, though (as he acknowledges) there is considerable room for debate about how well it would work in application and whether it leans too far in the direction of encouraging *Walker Process* actions. This is a crucial area that would benefit from more scholarly attention and further development in the courts.

14.2.2. *Sham Litigation*

The sham litigation doctrine provides an independent, "alternative legal ground[] on which a patentee may be stripped of its immunity to antitrust laws."[53] Importantly, a sham litigation claim may arise from the enforcement of a patent even where there is no allegation that the patent was fraudulently procured.

As a general rule, petitioning the government, including through the filing of a lawsuit, is constitutionally protected conduct, immune from antitrust liability under the *Noerr-Pennington* doctrine. The Supreme Court set forth the general principle in *E. R. R. Presidents Conference v. Noerr Motor Freight, Inc.*, 365 U.S. 127 (1961). There, a group of truckers brought a Sherman Act claim against railroads that had engaged in deceptive anti-trucking lobbying campaigns. The Supreme Court found that the campaigns "f[ell] far short of the ethical standards generally approved in this country."[54] The Court, however, drew a sharp distinction between those political activities, which it deemed protected by the First Amendment, and conduct within the reach of the antitrust laws:

> Insofar as that [Sherman] Act sets up a code of ethics at all, it is a code that condemns trade restraints, not political activity, and, as we have already pointed out, a publicity campaign to influence governmental action falls clearly into the category of political activity. The proscriptions of the Act, tailored as they are for the

---

[51] *Walker Process*, 382 U.S. at 176 (internal quotation marks and citation omitted).
[52] Leslie (2006).
[53] *Nobelpharma*, 141 F.3d at 1071. A *Walker Process* claim and a sham litigation claim can apply to the same conduct, i.e., when a patent holder pursues litigation that it knows is baseless because the asserted patent was fraudulently obtained. *Id.*
[54] *Noerr*, 365 U.S. at 140.

business world, are not at all appropriate for application in the political arena ... Congress has traditionally exercised extreme caution in legislating with respect to problems relating to the conduct of political activities, a caution which has been reflected in the decisions of this Court interpreting such legislation. All of this caution would go for naught if we permitted an extension of the Sherman Act to regulate activities of that nature simply because those activities have a commercial impact and involve conduct that can be termed unethical.[55]

The same principle was extended to filing lawsuits in *California Motor Transport Co. v. Trucking Unlimited*, 404 U.S. 508, 510 (1972). As discussed in more detail below, however, courts have found that *Noerr-Pennington*[56] immunity applies only to petitions for *discretionary* government action; in other words, it does not protect activities, like certain regulatory filings, that receive only a "ministerial" government response. See FTC Staff Report (2006).

Noerr noted in passing that there would be an exception to the general immunity for petitioning activity if the activities were a sham.[57] As the Supreme Court put it in a subsequent case, sham petitioning activity is that which is designed to make "use [of] the governmental process –as opposed to the outcome of that process – as an anticompetitive weapon."[58] A few years after *Noerr*, in *Handgards, Inc. v. Ethicon, Inc.*, 601 F.2d 986, 994 (9th Cir. 1979), the Ninth Circuit took up the question of sham litigation in the patent enforcement context. There, after a patent that Ethicon asserted against it had been declared invalid at trial, Handgards brought an antitrust claim arguing that although the patent was not fraudulently obtained, Ethicon knew it was invalid for other reasons at the time it was pursuing enforcement. The Ninth Circuit acknowledged that "[p]atentees must be permitted to test the validity of their patents in court," but also observed that "infringement actions initiated and conducted in bad faith contribute nothing to the furtherance of the policies in either patent law or antitrust law."[59] In fashioning a rule to identify bad-faith patent enforcement actions that are not immunized under *Noerr-Pennington*, the court concluded that it should place a heavy thumb on the patentee's side of the scale to avoid chilling legitimate exercises of patent rights and formulated the following standard: an infringement action is presumptively brought in good faith, and the presumption can only be rebutted by clear and convincing evidence.[60]

---

[55] Id. at 140–1.
[56] The other decision that lent its name to the *Noerr-Pennington* doctrine was decided four years after *Noerr*. There, the Supreme Court held that a labor union's campaign to urge the Secretary of Labor and TVA to impose a minimum wage on competitors was exempted from antitrust liability, because "[j]oint efforts to influence public officials do not violate the antitrust laws even though intended to eliminate competition." *United Mine Works of America v. Pennington*, 381 U.S. 657, 669–70 (1965).
[57] *Noerr*, 365 U.S. at 144.
[58] *Columbia v. Omni Outdoor Advertising, Inc.*, 499 U.S. 365, 380 (1990).
[59] 601 F.2d at 983.
[60] Id. at 996.

The Supreme Court largely adopted the *Handgards* approach in *PRE*, although it made the standard more stringent. In *PRE*, Columbia brought a copyright infringement case against PRE. PRE brought a Section 2 counterclaim alleging that the suit was a sham designed to exclude PRE from the market for in-room entertainment services at hotels. PRE asserted that Columbia lacked a subjective belief that the suit was meritorious, even though the district court had found that Columbia had probable cause to sue.[61] The Supreme Court established a two-part framework for antitrust claims grounded in sham litigation. First, the antitrust plaintiff must show that the suit is *objectively* baseless; that is, no reasonable person could believe the claim had merit.[62] Objective baselessness is a threshold test; unless it is established, a court does not move to considering subjective intent. If a showing of objective baselessness is made, the antitrust plaintiff then must prove that the patentee *subjectively* intended to use the litigation *process*, itself, rather than the *outcome* of the litigation to interfere with the competitor's business relationships.[63] In *PRE*, the Court found that PRE's claim failed at the first step because Columbia's infringement claim was not objectively baseless; therefore, it did not consider the subjective intent issue.[64]

In the relatively straightforward case before it, the Court in *PRE* defined objective baselessness to mean that "no reasonable litigant could realistically expect success on the merits."[65] That is, if the litigant had "probable cause" to sue – a "reasonable belief that there [was] a chance that [its] claims may be held valid upon adjudication" – the suit was not objectively baseless.[66] The Court made clear that a patent assertion is not objectively baseless merely because it proves unsuccessful.[67] By contrast, if a patent assertion *is* successful, it is by definition not a sham.[68]

Courts, mindful of avoiding chilling of legitimate assertions of patent rights, have developed doctrines that make it challenging for sham litigation plaintiffs to prevail. For instance, the antitrust plaintiff bears the burden of proving objective baselessness by "clear and convincing evidence."[69] Moreover, a claim will not be deemed baseless if the party is advocating in good faith for a change to or extension of existing law.[70] Additionally, because sham is often "a legal question requiring the court to evaluate the reasonableness of [the antitrust defendant's] litigation positions

---

[61] *PRE*, 508 U.S. at 52.
[62] *Id.* at 60.
[63] *Id.* at 60–1.
[64] *Id.* at 62–4.
[65] *Id.* at 60.
[66] *Id.* at 62–3. Other courts have phrased the test even more stringently. See, e.g., *Cheminor Drugs, Ltd. v. Ethyl Corp.*, 993 F. Supp. 271, 281 (D.N.J. 1998) (case must be shown to have "absolutely no objective merit").
[67] 508 U.S. at 65.
[68] *Id.* at 58.
[69] *C.R. Bard, Inc. v. M3 Sys., Inc.*, 157 F.3d 1340, 1369 (Fed. Cir. 1998).
[70] *PRE*, 508 U.S. at 65.

vis-a-vis the patent laws," courts frequently resolve these claims on motions to dismiss or motions for summary judgment.[71]

Because plaintiffs often fail to prove objective baselessness by clear and convincing evidence, there is relatively little case-law addressing the subjective element of the *PRE* test. One recurring issue, however, is the interplay between the subjective prong – which calls for an investigation into the patentee's motives for bringing a suit – and the attorney-client privilege. A few rules have been established. On the one hand, most courts hold that the attorney–client privilege protects legal advice even if that advice is relevant to plaintiff's sham claim.[72] Thus, an antitrust plaintiff cannot make the patentee's attorney–client communications discoverable merely by asserting a claim for which attorney advice might be probative.[73] On the other hand, a patentee that affirmatively relies on the advice of its attorney to justify its suit waives the attorney–client privilege.[74] The difficulty lies in navigating situations that fall between these two poles.

In *Rhone-Poulenc*, the Third Circuit explained that waiver of attorney–client or work product privilege occurs when the privilege holder asserts a claim or affirmative defense that affirmatively puts the advice of counsel directly "at issue" by disclosing or describing it in support of the claim or defense.[75] However, the court cautioned:

> Advice is not in issue merely because it is relevant, and does not necessarily become in issue merely because the attorney's advice might affect the client's state of mind in a relevant manner ... Relevance is not the standard for determining whether or not evidence should be protected from disclosure as privileged, and that remains the case even if one might conclude that facts to be disclosed are vital, highly probative, directly relevant or even go to the heart of an issue.[76]

Under the *Rhone-Poulenc* framework, evidence that, for instance, a firm's businesspeople subjectively believed that a lawsuit was appropriate might not waive the attorney–client privilege. However, testimony about any actual legal advice would do so.

The sham litigation doctrine has attracted far less scholarly attention than the *Walker Process* doctrine. A key question is at what point in the litigation such claims are best evaluated. As Areeda and Hovenkamp have pointed out, a sham claim is very difficult to *win*, but "the conclusory allegation of a sham is easy, and such easy

---

[71] *Rochester Drug Co-op., Inc. v. Braintree Labs.*, 712 F. Supp. 2d 308, 321 (D. Del. 2008); see also *FilmTec Corp v. Hydranautics*, 67 F.3d 931, 938 (Fed. Cir. 1995); *Covad Commc'ns Co. v. Bell Atlantic Corp.*, 398 F.3d 666, 677 (D.C. Cir. 2005); *Schneck v. Saucon Valley Sch. Dist.*, 340 F. Supp. 2d 558, 574–77 (E.D. Pa. 2004).
[72] *Rhone-Poulenc Rorer Inc. v. Home Indem. Co.*, 32 F.3d 851, 863 (3d Cir. 1994).
[73] *In re Burlington Northern, Inc.*, 822 F.2d 518, 533 (5th Cir. 1987).
[74] *Berckeley Inv. Grp., Ltd. v. Colkitt*, 455 F.3d 195, 222 (3d Cir. 2006).
[75] *Rhone-Poulenc*, 32 F.3d at 862–4.
[76] *Id.*

allegations can themselves chill the exercise of the rights that *Noerr* would protect, for defending oneself against such claims can be burdensome indeed."[77] One solution is to bifurcate the underlying patent infringement action from the sham litigation claim, since any successful infringement action will necessarily preclude a sham claim. Even an unsuccessful, but closely contested, infringement action would strongly suggest that a claim was not objectively baseless, which would make it easier to resolve the sham claim through a dispositive motion rather than a second trial.[78] Although bifurcation can create some inefficiencies (e.g., when there is significant overlap between proof relevant to validity, enforceability, or infringement), it will nonetheless often be a sensible course given the enormous burden of discovery and litigation over sham claims, which will prove unnecessary if the patentee wins its infringement case (or even if it loses but the evidence shows that the assertion was clearly not a sham). The proposition that bifurcation will often be efficient is particularly compelling given how rarely sham litigation claims are ultimately successful.

In 2006, the staff of the Federal Trade Commission (FTC) issued a substantial report (FTC Staff Report (2006)) entitled "Enforcement Perspectives on the *Noerr-Pennington* doctrine," which remains the most significant agency commentary regarding the application of the doctrine. The FTC staff acknowledged that the values *Noerr-Pennington* immunity promotes "are significant and, when Constitutionally mandated, require deference," but expressed concern that "accommodating these values sometimes also imposes costs on consumers."[79] The FTC then offered recommendations for balancing consumer protection and First Amendment values in three areas where, in its view, "the risk to competition is great and antitrust enforcement need not impinge on the values underlying *Noerr*": (1) requests for ministerial acts, (2) misrepresentations to a government decision maker outside the political context, (3) serial requests for government action.[80] These remain important areas for further legal development.

(1). Requests for a ministerial government action raise issues that are distinct from requests for discretionary government actions. One of the first cases clearly to draw the distinction was *Litton Systems v. Am. Tel. & Tel. Co.* In that case, AT&T attempted to require customers to purchase AT&T's own specialized equipment in order to operate its competitors' products on AT&T's network using the mechanism of a mandatory (but normally unreviewed) tariff filing with the Federal Communications Commission (FCC).[81] The Second Circuit rejected AT&T's argument that AT&T's tariff filing was protected petitioning activity, explaining that "a mere

---

[77] Areeda and Hovenkamp (1997).
[78] See, e.g., *Repeat-O-Type Stencil Mfg. Corp. v. Hewlett-Packard Co.*, 141 F.3d 1178 (9th Cir. 1998).
[79] FTC Staff Report (2006).
[80] *Id.*
[81] 700 F.2d 785, 789–90 (2d Cir. 1983).

incident of regulation – the tariff filing requirement – [is not] tantamount to a request for governmental action akin to the conduct held protected in *Noerr* and *Pennington*."[82] As the court explained, the FCC had not actually made any substantive determination regarding AT&T's conduct; any anticompetitive effect was due to AT&T's decisions alone, not an exercise of the FCC's regulatory authority.

The FTC Staff Report observed that there is "little check on the truth or falsity of parties' representations" in the ministerial government action context, so it is particularly important to have additional checks on requests for ministerial action.[83] In recent years, this "ministerial action" issue has arisen most frequently in the context of branded/generic pharmaceutical litigation. Pharmaceutical companies are required to identify to the Food & Drug Administration (FDA) "any patent which claims [the relevant drug] or which claims a method of using such drug and with respect to which a claim of patent infringement could reasonably be asserted if a person not licensed by the owner engaged in the manufacture, use or sale of the drug."[84] The FDA in turn, without engaging in any independent assessment, lists those patents in a publication called the Orange Book. If a generic pharmaceutical company seeks FDA approval to market before expiration of an Orange Book-listed patent, that company must certify that the patent is either invalid or not infringed, or both. On notification of this certification, the patent-holding company may bring suit to enjoin the sale of the generic product, and that suit automatically triggers a thirty-month stay barring the FDA from granting final approval for that generic product. Thus, listing a drug in the Orange Book can have significant implications for competition. In *In re Buspirone Patent Litig./In re Buspirone Antitrust Litig.*, the plaintiffs brought various antitrust claims based on a company's Orange Book listing.[85] The court held that the act of submitting a patent for listing in the Orange Book was not petitioning activity within the meaning of the First Amendment.[86] Citing *Litton*, the court reasoned that *Noerr-Pennington* applies only when the "anticompetitive result" flows from the exercise of actual government discretion, and acts only to protect a party's right to influence the exercise of that discretion. Because the FDA's role with respect to the Orange Book is "merely ministerial" the act of submitting the patent for listing is not protected under *Noerr-Pennington* immunity.[87]

The FTC's Staff Report embraced both the *Litton* and *Buspirone* decisions, and drew from them the general principle that "[w]here the communication [to the government] furthers the exercise of governmental discretion or judgment, it likely warrants protection. Where it does not further – or even undermines – a valid and

---

[82] *Id.* at 807–8.
[83] FTC Staff Report (2006).
[84] 21 U.S.C. § 355(b)(1).
[85] 185 F. Supp. 2d 363 (S.D.N.Y. 2002).
[86] *Id.* at 372–3.
[87] *Id.*

independent government decision, it likely deserves no special treatment and should be subject to the antitrust laws."[88]

(2). Even blatant falsehoods in the political context are immunized from antitrust liability under *Noerr-Pennington*, which itself involved misrepresentations made as part of a lobbying campaign. It remains an open issue, however, whether misrepresentations made in the course of government petitioning but *outside* the political context are immune from antitrust scrutiny. Moreover, courts have not precisely delimited the types of activity that are considered "political." The FTC Staff Report takes the position that "[t]he core type of [political] activity that *Noerr* is meant to address is a request to a government decision maker to exercise its discretion to decide in a certain way."[89]

The FTC Staff Report, which relied extensively on the Commission's 2004 decision in *Union Oil Co. of Cal.* ("*Unocal*"),[90] argued that the potential for restricting First Amendment rights is strongly outweighed by the risks of anticompetitive harm if misrepresentations outside the core political context are immunized.[91] The Commission explained that the political arena differs from other contexts because any expectation that representations will be truthful are diminished there; there is greater governmental discretion and less need to rely on petitioners' factual assertions; and it is harder than in non-political contexts to determine the relationship between petitioning activity and the ultimate governmental action.[92]

Unocal involved allegations that the defendant had "made misrepresentations regarding its patent rights that induced the California Air Resources Board (CARB) to adopt an industry-wide standard reading on those patents."[93] In remanding to the administrative law judge for further factfinding, the Commission found that a misrepresentation or omission outside the core political context could lose *Noerr-Pennington* protection if it was "(1) deliberate (something more than mere error is necessary); (2) subject to factual verification; and (3) central to the legitimacy of the affected governmental proceeding."[94] In those limited circumstances, the antitrust enforcement would "not [be] challenging government action itself, but rather attacking a private misrepresentation that 'effectively supplanted government action.'"[95]

(3). The FTC report also addressed situations where a patentee allegedly engages in a pattern of repetitive, harassing petitioning meant to exclude a rival based on the

---

[88] FTC Staff Report (2006).
[89] *Id.*
[90] FTC Dkt. No. 9305, slip op. at 17–23 (2004) (opinion of the Commission). Available at www.ftc.gov/sites/default/files/documents/cases/2004/07/040706commissionopinion.pdf.
[91] FTC Staff Report (2006).
[92] *Id.*
[93] *Id.* (citing *Unocal*).
[94] *Unocal*, slip op. at 36.
[95] FTC staff Report (2006) (quoting *Unocal*, slip op. at 44).

process rather than the merits of the patent claim, but none of the petitions is – itself – objectively baseless. This issue has arisen in the context of serial lawsuits. Some cases, mostly notably the Ninth Circuit's *Contra Costa* decision, suggest that a plaintiff bringing a series of meritless patent litigations can lose *Noerr-Pennington* immunity even if the individual cases are not objectively baseless.[96] The law in this area remains unsettled and needs further development, including on questions such as how many lawsuits constitute a "series," or whether at least *some* of the lawsuits must be objectively baseless.[97]

The FTC Staff Report endorsed *Contra Costa*. The staff explained that lawsuits can be a way of inflicting considerable anticompetitive harm at low cost to the plaintiff. Although the FTC staff acknowledged that penalizing a party for bringing a lawsuit that is not objectively baseless raises First Amendment issues, it wrote that "[w]hen a pattern of petitioning is involved ... a court has more information and thus is likely to be in a better position to determine accurately whether a defendant's conduct is best characterized as a misuse of governmental processes that conceals an attempt directly to harm marketplace rivals and suppress competition."[98] Thus, given the lower "risk of error," the FTC staff opined that some relaxation of the strict "objective baseless" requirement from *PRE* is appropriate in the serial litigation context, and the courts should, in essence, have more leeway to examine the antitrust defendant's subjective motivations.[99]

14.3. ANTICOMPETITIVE SCHEMES ENCOMPASSING BUT GOING BEYOND PATENT ASSERTION

In some circumstances, a patent lawsuit does not give rise to a *Walker Process* or *Handgards* claim because the patent was not improperly obtained and the suit (by itself) is not objectively and subjectively baseless, yet the patent claim is alleged to be an important piece of a broader anticompetitive scheme. The courts are continuing to grapple with the question of whether exceptions to the general immunity for patent assertions should apply in these situations.

Justice Stevens warned in his concurrence in *PRE* that the objective baselessness test "may be hard to apply when there is evidence that the judicial process has been used as part of a larger program to control a market," and that "the distinction between 'sham' litigation and genuine litigation is not always, or only, the difference between lawful and unlawful conduct; objectively reasonable lawsuits may still

---

[96] See *USS-POSCO Indus. v. Contra Costa County Bldg. & Construction Trades Council*, 31 F.3d 800, 811 (9th Cir. 1994).
[97] See *In re Terazosin Hydrochloride Antitrust Litig.*, 335 F. Supp. 2d 1336, 1367 (S.D. Fla. 2004) (analyzing and expressing doubts about *Contra Costa*).
[98] FTC Staff Report (2006).
[99] *Id.*

break the law."[100] Neither the Supreme Court nor the Federal Circuit have addressed how *Noerr-Pennington* immunity would or would not apply in such a "larger program to control a market." Given the First Amendment concerns the Supreme Court articulated in *Noerr*, lower courts have been cautious in extending any sort of antitrust liability related to protected petitioning activity outside of the circumscribed realms of *Walker Process* and *Handgards* claims. Nonetheless, some have extended liability to certain new circumstances.

One recurring situation in recent years concerns standard setting organizations (SSOs) and declared standard essential patents. SSOs are industry groups that adopt technical specifications and standards for a given technology (e.g., cellular communications or Wi-Fi). SSOs typically require participants in standard setting activities to disclose if they have patents or applications covering technology that may be incorporated in the standard and to agree to license those patents on "fair, reasonable, and non-discriminatory" (FRAND) terms to all parties wishing to implement the standard. These patents are called "standard essential patents" or "SEPs."[101]

Disputes have frequently arisen when a standard setting participant is accused of (i) failing timely to disclose patents or applications when the standard is being developed, but later claiming that standard implementers are infringing its patents or (ii) refusing to meet its commitment to license its declared-essential patents on FRAND terms. Courts have found that a failure timely to disclose a standard essential patent or breach of a FRAND commitment can constitute exclusionary conduct for purposes of a Section 2 claim, as can a deceptive FRAND commitment.[102] Challenges to this conduct typically arise when the holder of the declared-essential patent sues the standard implementer for an injunction to remove allegedly infringing products from the market. Filing the lawsuit, then, is an integral part (indeed, the capstone) of the allegedly anticompetitive scheme. Yet the patent may well be valid, so *Walker Process* does not apply, and the allegations of infringement may be objectively reasonable, so *Handgards* does not apply.

In *Hynix Semiconductor Inc. v. Rambus, Inc.*, 527 F. Supp. 2d 1084, 1089 (N.D. Cal. 2007), a district court adopted the "causal connection" test as a way to reconcile this divergent case-law in the standard setting context. There, the alleged infringer argued that the patentee had engaged in a pattern of exclusionary conduct to obtain a monopoly: joining the relevant SSO to monitor development of the relevant standard; drafting claims in patent applications to cover the standard; failing to disclose those patent applications while the standard was under consideration; demanding exorbitant royalties after the standard was adopted (and product suppliers were locked into

---

[100] 508 U.S. at 73, 75.
[101] See generally *Hynix Semiconductor Inc. v. Rambus, Inc.*, 527 F. Supp. 2d 1084, 1089 (N.D. Cal. 2007) for a description of the process.
[102] See, e.g., *Broadcom Corp. v. Qualcomm, Inc.*, 501 F.3d 297, 314 (3d Cir. 2007); *Apple Inc. v. Samsung Electronics Co.*, No. 11-CV-01846, 2012 WL 1672493, at *2 (N.D. Cal. May 14, 2012).

practicing the standard); and then filing patent infringement suits in an effort to collect those royalties. The court found that the alleged deceptive conduct before the SSO could constitute exclusionary conduct for purposes of a Section 2 claim, but struggled with the question of whether the patentee's infringement lawsuit, which was not alleged to be objectively baseless, was protected activity under *Noerr*.[103]

After reviewing the case-law, the court found that the "causal connection" test most appropriately balanced competing imperatives of antitrust, First Amendment, and patent law principles.[104] The court in *Hynix* borrowed the "causal connection" concept from the Eighth Circuit's decision in *American Infra-Red Radiant Co. v. Lambert Industries, Inc.*, 360 F.2d 977, 996–97 (8th Cir. 1966), where the court held that a non-frivolous, good faith lawsuit could be the basis for an antitrust claim only if there was some causal link between the unprotected anticompetitive acts and the protected lawsuit, i.e., that the suit was inspired by or necessary for the anticompetitive scheme.[105] In *Hynix*, the court emphasized that "the other aspects of the scheme [must] independently produce anticompetitive harms" before a lawsuit could be deemed to violate the Sherman Act without meeting the elements of a sham litigation of *Walker Process*.[106] In other words, independently unlawful conduct must prompt the patent lawsuit before the lawsuit can be deemed part of the antitrust harm, and injuries resulting from the litigation (e.g., defense costs) can support a claim for antitrust damages. The court found that the allegations met that standard: deception during standard setting activities was independently anticompetitive conduct, and "the causal connection is that a patent 'ambush' or 'hold-up' is ineffective without the threat of litigation."[107]

The court in *Apple, Inc. v. Samsung Electronics Co.*, 2012 WL 2571719, at *26 (N.D. Cal. June 30, 2012), followed *Hynix* in another case involving allegations of anticompetitive conduct involving standard setting. The plaintiff alleged a pattern of misconduct by the patentee – culminating with an infringement lawsuit – had caused it antitrust injury in the form of litigation defense costs. The court held that "because Apple's litigation costs stem directly from Samsung's alleged anticompetitive behavior [i.e., deceptive behavior in the SSO], these litigation costs are a sufficient basis for a

---

[103] 527 F. Supp. 2d at 1098–9.
[104] *Id.*
[105] Historically, the courts of appeals had taken different views regarding whether a lawsuit or other petitioning activity might be deemed part of a broader anticompetitive scheme for purposes of imposing antitrust liability. In *Straus v. Victor Talking Machine Co.*, 297 F. 791, 799 (2d Cir. 1924), the Second Circuit announced a rule that foreshadowed *PRE*: a lawsuit with merit that was at least "debatable" could not be the basis for antitrust liability. In *Kobe, Inc. v. Dempsey Pump Co.*, 198 F.2d 416, 424 (10th Cir. 1952), the court held that even a good-faith patent lawsuit, combined with acts like advertising expired patents and notifying the alleged infringer's customers of the patents, could violate the Sherman Act. The court explained that "[t]he result of [the patentee's] infringement action, its verbal and written statements to the trade, was disastrous to the defendants . . . To hold that there was no liability for damages caused by this conduct, though lawful in itself, would permit a monopolizer to smother every potential competitor with litigation." The Eighth Circuit's causal connection test was an attempt to chart a middle ground between these two courses. *Am. Infra-Red Radiant Co. v. Lambert Indus., Inc.*, 360 F.2d 977, 996-97 (8th Cir. 1966).
[106] *Hynix*, 527 F. Supp. 2d at 1097.
[107] *Id.* at 1098.

potential award of antitrust damages," even though there was no allegation that the infringement lawsuit was a sham or that the patents were fraudulently obtained.[108]

Another court, however, has rejected the *Hynix* approach. In *Apple, Inc. v. Motorola Mobility, Inc.*, 2012 WL 3289835,*11-*14 (W.D. Wis. 2012), the court held that Apple could not make out an antitrust claim where the only antitrust injury alleged was the cost of defending the infringement lawsuit, because "Motorola's enforcement of its patents is privileged conduct protected by the First Amendment [and] the *Noerr-Pennington* doctrine applies." The Court observed, however, that *Noerr-Pennington* immunity would not protect Motorola from breach of contract claims, including claims that it would not enforce its FRAND-encumbered SEPs against willing licensees like Apple.[109]

This question of the scope of *Noerr-Pennington* protection where a pattern of anticompetitive conduct related to standard setting and FRAND commitments is alleged also arose in the context of the FTC's consent decree in *Motorola*. There, the Commission alleged that Motorola had violated Section 5 of the Federal Trade Commission Act by seeking injunctions against competing product suppliers based on patents that had been declared essential to an industry standard. One dissenting Commissioner wrote that "it is unclear how the seeking of injunctive relief, either in the courts or the ITC, on a patent – even a FRAND-encumbered SEP – would not be considered protected petitioning of the government under the *Noerr-Pennington* doctrine."[110] The majority disagreed, writing that "we are not persuaded by Commissioner Ohlhausen's argument that the conduct alleged in the Commission's complaint implicates the First Amendment and the *Noerr-Pennington* doctrine [because] we have reason to believe that [Motorola] willingly gave up its right to seek injunctive relief when it made the FRAND commitments at issue in this case."[111] In so holding, the majority cited case-law suggesting that most FRAND commitments carry an implicit promise to refrain from seeking injunctive relief, and found that the specific evidence before the Commission showed that Motorola had violated its particular FRAND commitments by seeking injunctive relief.[112]

---

[108] 2012 WL 2571719 at *27.
[109] 2012 WL 3289835, at *14.
[110] *In the Matter of Robert Bosch GmbH*, Dissenting Statement of Commissioner Maureen K. Ohlhausen, FTC File No. 121-0081 (Nov. 26, 2012), cited in *In the Matter of Motorola Mobility LLC and Google, Inc.*, Dissenting Statement of Commissioner Maureen K. Ohlhausen, FTC File No. 121-0120 (Jan. 3, 2013) ("The Commission announced this enforcement policy in *In re Robert Bosch GmbH*, stating that in 'appropriate circumstances' it will sue patent holders for seeking injunctive relief against 'willing licensees' of a SEP. I dissented then in large part because I question whether such conduct, standing alone, violates Section 5 and because the *Noerr-Pennington* doctrine precludes Section 5 liability for conduct grounded in the legitimate pursuit of an injunction or any threats incidental to it, outside of a handful of well-established exceptions not alleged there.... [T]oday's decision raise[s] many of the same concerns for me as did *Bosch*... I disagree with the majority's interpretation of the cases it relies on to preclude *Noerr*'s application here.").
[111] See *In the Matter of Motorola Mobility LLC and Google, Inc.*, Statement of the Federal Trade Commission at 4–5, FTC File No. 121-0120 (Jan. 3, 2013).
[112] *Id.* at 2 n. 7, 4.

Commentators Meyer and Thayamballi generally agreed with the majority's view, reasoning that imposing antitrust liability for suing based on declared standard essential patents punishes the patentee's unprotected deceptive conduct before the SSO, not its protected petitioning activity.[113] They further argue, however, that the entire problem may be overstated: "[a]t least in some sense price fixing and other agreements in restraint of trade entail both 'association' and 'speech' ... [but] [n]o court has suggested that the Constitution gives cartelists the privilege to utter 'I agree' without consequence."[114]

Particularly given the recent prevalence of antitrust disputes based on alleged abuses of declared standard essential patents, controversies regarding the scope of *Noerr-Pennington* protection for patent enforcement actions that are neither alleged to be based on fraudulently obtained patents nor objectively and subjectively baseless are likely to continue. This is another area that would benefit from additional normative commentary and development in the courts.

REFERENCES

Areeda, Phillip and Herbert Hovenkamp. 1997. *Antitrust Law*. New York: Wolters Kluwer.

Daniel, B. D. 2009. *Walker Process* Proof: The Proper Prescription. *Rutgers Law Journal*, 41, 105–61.

FTC staff Report. 2006. Ohlhausen, Maureen K. *et al.* 2006. Enforcement Perspectives on the *Noerr-Pennington Doctrine: An FTC Staff Report.*

Himes, Jay. 2009. When Caught with Your Hand in the Cookie Jar ... Argue Standing. *Rutgers Law Journal*, 41, 187–228.

Hovenkamp, Herbert, Mark Janis, Mark Lemley, and Christopher Leslie. 2014. *IP and Antitrust: An Analysis of Antitrust Principles Applied to Intellectual Property Law*. 2nd. edn. New York: Wolters Kluwer.

Leslie, Christopher. 2006. The Anticompetitive Effects of Unenforced Invalid Patents. *Minnesota Law Review*, 91, 101–83.

  2007. The Role of Consumers in Walker Process Litigation. *Southwestern Journal of Law & Trade in the Americas*, 13, 281–312.

  2008. Patents of Damocles. *Industrial Law Journal*, 83, 133–79.

Meyer, David and Fabien Thayamballi. 2014. Do First Amendment Principles Limit the Antitrust Agencies' Ability to Prohibit Enforcement of Standards-Essential Patents? *Competition: Journal of the Antitrust and Unfair Competition law Section of the State Bar of California*, 23, 142–55.

Steinman, David and Danielle Fitzpatrick. 2002. Antitrust Counterclaims in Patent Infringement Cases: A Guide to *Walker Process* and Sham-Litigation Claims. *Texas Intellectual Property Law Journal*, 10, 95–109.

Tokic, Stijepko. 2012. Enforcing the Duty of Disclosure After Therasense: Antitrust Implications. *American Intellectual Property Law Association Quarterly Journal*, 40, 221–65.

---

[113] Meyer and Thayamballi (2014).
[114] *Id.*

15

Does Antitrust Have a Role to Play in Regulating Big Data?

D. Daniel Sokol and Roisin Comerford

The collection of user data online has seen enormous growth in recent years. Consumers have benefited from the growth through an increase in free or heavily subsidized services, better quality offerings, and rapid innovation. At the same time, the debate about Big Data, and what it really means for consumers and competition, has grown louder. Many have focused on whether Big Data even presents an antitrust issue, and whether and how harms resulting from Big Data should be analyzed and remedied under the antitrust laws. The academic literature, however, has lagged behind the debate somewhat, and a closer inspection of existing scholarly works reveals a dearth of thorough study of the issue. Commentators generally split into two camps: one in favor of more proactive antitrust enforcement in the Big Data realm, and one opposing such intervention, considering antitrust inappropriate for regulation of Big Data. The academic case for the former has not, as yet, been fully developed, and is relatively light at present. Meanwhile, policy-focused work by academics and practitioners in this arena suggests that antitrust intervention in Big Data would be premature and misguided, especially considering the myriad pro-competitive benefits offered by Big Data.

In this chapter, we review the scholarly work on the implications of Big Data on competition, and consider the potential role of antitrust in the regulation of Big Data. Section 15.1 provides an overview of current, scarce, academic literature specifically addressing the role of antitrust in Big Data issues. Sections 15.2 and 15.3 delve into the policy issues surrounding Big Data and whether it poses a risk to competition that warrants antitrust intervention. Section 15.2 details the ways in which Big Data may prove procompetitive while Section 15.3 reviews and critiques the suggested potential harms to competition from Big Data. Section 15.4 discusses the suitability of antitrust as the institutional choice for Big Data issues, and Section 15.5 concludes that, at present, antitrust is ill-suited as the institutional choice. This conclusion is further borne out by the fact that thus far there have been no cases in the United States or Europe that have found Big Data itself to be a basis for a theory of harm on antitrust grounds for mergers or conduct cases. Further, the scholarly case for such harm has not yet been adequately established.

15.1. EXISTING ACADEMIC LITERATURE

A review of the academic literature addressing the intersection of Big Data and antitrust law reveals relatively few articles on the topic.[1] Scholars have yet to conduct an in-depth analysis of *why* Big Data issues are antitrust issues, and if so, *how* they may be best addressed by the antitrust laws as opposed to the consumer protection laws.[2] Work to date suggests instead that while antitrust and consumer protection laws are complementary, they still comprise distinct areas of law, and consumer protection remains the correct institutional choice to address potential Big Data harms.

Arguably the most comprehensive contribution to the academic debate on the topic of Big Data and competition is the work by Ohlhausen and Okuliar.[3] They present a three-part framework for analyzing Big Data concerns. First, they focus on the character of the harm – whether it is commercial, personal, or otherwise. They conclude that where there is harm to consumer welfare on the whole or to economic efficiency, antitrust should prevail over consumer protection law as a matter of institutional choice. Second, they examine the nature of the relationship between the user and the data collector, and determine that issues arising from the bargain between a firm and an individual consumer are more likely to fall within the realm of consumer protection law than antitrust. Third, they consider the nature of available remedies and their presumed efficiency in resolving particular violations. Ultimately, the authors advise that trying to fit consumer protection concerns within the antitrust framework is "unnecessary," "could lead to confusion and doctrinal issues in antitrust," and would not afford "true gains to consumer protection."[4] Ohlhausen and Okuliar also note four important features of Big Data that caution against an antitrust application over consumer protection law, which are explored in more detail in Section 15.4 below. First, Big Data creates efficiency gains. Second, an antitrust institutional choice would increase subjectivity into antitrust analysis. Third, using antitrust would create opportunities for strategic gaming by firms of the legal system. Finally, Ohlhausen and Okuliar warn that using an antitrust lens may threaten innovation for new products and services.

James Cooper echoes that antitrust law is an inappropriate tool to regulate Big Data. He writes:

---

[1] On online markets generally, see Goldfarb and Tucker (2011); Evans (2009). Much of the two sided online market work traces back to Rochet and Tirole (2002).
[2] For an overall analysis of how economics can better explain empirics in the age of Big Data, see Varian (2014).
[3] Ohlhausen and Okuliar (2015).
[4] *Id.*, p. 138.

> [E]ven if one were to accept the analogy between enhanced personal data collection and prices (or equivalently, lower quality) at face value, there is nothing in the antitrust laws to prevent a firm from unilaterally engaging in this conduct. Antitrust's longstanding aversion to price regulation means that a legal monopolist is free to charge whatever price the market will bear.[5]

Cooper also suggests that privacy in Big Data as an antitrust concern would raise certain First Amendment issues, as well as muddle the goal of enforcement, thereby introducing unnecessary subjectivity into the analysis, lending itself to Virginia School-styled rent seeking in antitrust.

Andres Lerner[6] argues that claims of Big Data presenting competitive concerns are unsupported by real-world evidence. In particular, Lerner argues that in practice the oft-cited "feedback loops" do not have the strong effects with which they are commonly credited. Lerner discusses the procompetitive rationales for collection and use of consumer data online, including the potential for improved services, and the ability of firms to monetize effectively on the paid side so as to provide better services at lower prices or for free. He dismisses the idea that firms may have the incentive or ability to use data to entrench their dominant position (e.g., user data is nonrivalrous and no one firm controls a significant share of data), citing similar attributes of data as Ohlhausen and Okuliar. Lerner maintains that there is a complete lack of evidence that online markets have "tipped" to dominant firms, due in most part to the differentiated nature of online offerings. He concludes that without strong real-world evidence of anticompetitive effects, aggressive antitrust enforcement would hamper competition and chill innovation, injuring consumer welfare in the process.

Although policymakers have dipped their toe into the antitrust in Big Data debate,[7] antitrust agencies and the courts have not found a Big Data competition problem. In fact, the Federal Trade Commission (FTC) and DG Competition have thoroughly considered Big Data as an antitrust problem and completely dismissed it. The agencies in the United States and Europe have moved cautiously so far, which is not only proper, but also serves as a reminder that the distinct issues addressed by antitrust and consumer protection law, and the solutions that may be applied by each set of laws to prohibited behavior, are distinct for good reason, and are complements rather than substitutes.[8]

---

[5] Cooper (2013).
[6] Lerner (2014).
[7] Preliminary Opinion of the European Data Protection Supervisor, Privacy and competitiveness in the age of big data: The interplay between data protection, competition law and consumer protection in the Digital Economy, March 2014, available at https://secure.edps.europa.eu/EDPSWEB/webdav/site/mySite/shared/Documents/Consultation/Opinions/2014/14-03-26_com petition_law_big_data_EN.pdf; Feinstein (2015); Ramirez (2015).
[8] Muris and Zepeda (2012); Averitt and Lande (1997).

15.2. CAN BIG DATA LEAD TO PROCOMPETITIVE BENEFITS?

Unprecedented consumer benefits have already been realized through the use of Big Data, chief among them free user services (as a number of the merger cases have noted), improved quality, and a rapid increase in innovation. Furthermore, fears surrounding Big Data and its use by large online firms are unwarranted, as the economic traits of Big Data ameliorate concerns that such data can be manipulated for anticompetitive gains.

15.2.1. *Monetization of Data Subsidizes Free Products for Consumers*

Perhaps the most obvious and pervasive benefit to be realized in the Big Data era has been the ability of firms to offer heavily subsidized, often free, services to consumers as consumers give those firms permission to monetize consumer data on the other side of their business.[9] In a competition law regime where lower prices for consumers are deemed highly desirable, this is undoubtedly a benefit to consumers.

The monetization of the data in the form of targeted advertising sales for antitrust purposes is not suspect or harmful, but rather "economically-rational, profit-maximizing behavior," that results in obvious consumer benefit.[10] Were online platforms to be prevented or restricted from collecting and monetizing consumer data, competition for users would be inhibited, and harm to consumers would result in the form of higher prices for services. Indeed switching costs are low regarding data and search.[11]

Some criticize the provision of free services, claiming that this makes it more difficult for rivals that cannot initially monetize as effectively to compete with established rivals,[12] but cases show that this argument misses the point of antitrust completely – the ability to offer high-quality services to consumers for free is a procompetitive effect of Big Data monetization, not an anticompetitive harm.[13] Also, the point is simply untrue – it is not more difficult for new entrants to compete with established rivals.[14]

15.2.2. *Improved Quality and Enhanced Innovation*

As an input, online firms use data to improve and refine products and services in a number of ways, and to develop brand new innovative product offerings. For

---

[9] Evans and Schmalensee (2014).
[10] Lerner (2014).
[11] Edlin and Harris (2013). See also Case COMP/M.7217, *Facebook*/WhatsApp, which noted data sets should not have an impact in a market for online advertising because there are so many different sources of user data available on the web.
[12] Newman (2014).
[13] Evans and Schmalensee (2014).
[14] Case COMP/M.7217, *Facebook*/WhatsApp.

example, search engines, both general and niche, can use data to deliver more relevant, high-quality search results. By learning from user search queries and clicks, search engines can identify what are the most relevant results for a particular query. "Click-and-query" data, as it is known, is a highly valuable input in delivering high-quality search results.[15] Outside of just relevant results, search engines can use data to provide additional "value-added" services to users. Travel search engines, for instance, can use data to forecast price trends on flights for specific routes. Amazon and multiple other e-commerce sites use past purchase information and browsing history to make personalized shopping recommendations for users.[16] Social networking platforms use data collected from users to suggest friends, celebrity or business pages, or articles that customers might be interested in. Online media outlets use browsing history and personal information to recommend other articles that a reader may be interested in.

15.2.3. Economic Characteristics of Big Data Protect Against Competitive Harm

In addition to the affirmative procompetitive benefits of Big Data expounded above, the economics of how Big Data works, as described below, damages claims that it should be feared, or reined in by antitrust. Additionally, the unique economic characteristics of data mean that its accumulation does not, by itself, create a barrier to entry, and does not automatically endow a firm with either the incentive or the ability to foreclose rivals, expand or sustain its own monopoly, or harm competition in other ways.[17] Lambrecht and Tucker explain that "[f]or there to be a sustainable competitive advantage, the firm's rivals must be unable to realistically duplicate the benefits of the strategy or input." As we suggest below, both theory and actual cases support a finding that the characteristics of data are such that rivals cannot be foreclosed from replicating the benefits of Big Data enjoyed by larger online firms, and Big Data in the hands of large firms does not necessarily pose a significant antitrust risk.

15.2.3.1. Low Barriers to Entry

Data-driven markets are typically characterized by low entry barriers, as evidenced by innovative challengers emerging rapidly and displacing established firms with much greater data resources than themselves.[18] While the existence or lack thereof of barriers to entry can, and will, differ from market to market, and a blanket determination cannot be made in the abstract, the history of the digital economy

---

[15] Salinger and Levinson (2015).
[16] Goldfarb (2014).
[17] Lambert and Tucker (2015).
[18] Tucker and Welford (2014).

offers many examples, like Slack, Facebook, Snapchat, and Tinder, where a simple insight into customer needs enabled entry and rapid success despite established network effects.

The data requirements of new competitors are far more modest and qualitatively different than those of more established firms. Little, if any, user data is required as a starting point for most online services. Instead, firms may enter with innovative new products that skillfully address customer needs, and quickly collect data from users, which can then be used toward further product improvement and success. As such, new entrants are unlikely to be at a significant competitive disadvantage relative to incumbents in terms of data collection or analysis.[19]

And, while a firm that has been operational for ten years may have a larger data store, lack of asset equivalence has never been a sufficient basis to define a barrier to entry in any cases as of yet. In brick-and-mortar retail, a new entrant may have a smaller showroom than an established competitor, but this does not render the need for a physical store location an insurmountable barrier to entry. Indeed, an established brick-and-mortar store could have much more data on local customer preferences, but that has never been viewed as prohibitive to entry.

15.2.3.2. Data is Ubiquitous, Inexpensive, and Easy to Collect

Data is ubiquitous, inexpensive, and easy to collect.[20] Users are constantly creating data – increased Internet and smartphone usage means customers are continuously leaving behind traces of their needs and preferences.[21] Data can be easily and quickly collected from consumers upon launch, and both data and the tools needed to store and analyze it are readily available from numerous third-party sources. Big Data has near-zero marginal costs of production and distribution.[22] There are many alternative sources of data available to firms, reflecting the extent to which customers leave multiple digital footprints on the Internet.[23] The fact that data can, therefore, be acquired from third-party sources, means that even on the first day of product launch, before any user has interacted with the platform, a provider can already have benefited from insights into consumer preferences, and designed a platform that can act quickly as data is collected and processed.

While some argue that the resources and effort expended by companies in pursuit of data is evidence enough that data collection and processing is both "costly" and "time-consuming,"[24] it is important to distinguish between the collection of raw data, and the analysis any given firm puts the data through, which is what makes the

---

[19] Id.
[20] Tucker (2013).
[21] Lambrecht and Tucker (2015).
[22] Shapiro and Varian (1999).
[23] Lambrecht and Tucker (2015).
[24] Stucke and Grunes (2015a).

data valuable. This is the firm's "secret sauce." It is also, incidentally, the part of a firm's Big Data usage that requires the most resources. There is also plenty of off-the-shelf and open source analytics software that could give small firms a head start.

15.2.3.3. Data is Nonexclusive and Nonrivalrous

Data is nonexclusive and nonrivalrous. No one firm can, or does, control all of the world's data. Collection of a piece of data by one firm does not occur at the expense of another firm. "Multi-homing" is the norm among Internet users – users can, and do, spread their data around the Internet, using multiple different providers for multiple different services, or sometimes the same service. While multi-homing, a user shares data with multiple providers.

Big Data has been likened to other inputs as it becomes an increasingly important asset. However, Big Data's nonrivalrous and nonexclusive nature sets it apart from other key inputs. If one provider has a piece of data, another provider is not prevented from collecting that very same piece of data. Similarly, while conceivably one provider could at least theoretically hold all of the world's oil resources, for example, no one provider can amass all available data. Furthermore, incumbent online providers do not have explicit or de facto exclusivity over user data. There are no exclusivity clauses in terms of service with users, and there are no structures (pricing or otherwise) that lock users into sharing their data with only one provider.

15.2.3.4. Data's Value is Short-Lived

Data has a limited lifespan – old data is not nearly as valuable as new data – and the value of data lessens considerably over time. Additionally, the returns on scale diminish over time. Therefore, any competitive advantage that data provides is fleeting, and entrants are unlikely to be significantly disadvantaged relative to incumbents in terms of data collection and analysis.[25] The need for fresh, differentiated data means that a firm with a large volume of stale or generalized data does not, necessarily, benefit the holder and disadvantage a potential challenger. Potential competitors do not need to create a data store equivalent to the size of the incumbent; rather they need to devise a strategy to accumulate highly relevant and timely data.[26]

15.2.3.5. Data Alone Is Not Enough

Data does not typically provide value on a stand-alone basis. Mere possession of data alone, therefore, even in large volume, does not secure competitive success – that

---

[25] Chiou and Tucker (2014).
[26] Shepp and Wambach (2015).

can only be achieved through engineering talent, quality of service, speed of innovation, and attention to consumer needs. As such, the firm with the most data does not necessarily win. Take the online dating application, Tinder, initially launched in September 2012, as an example. Data is of particular value in industries where personalized experience is important, such as online dating. When Tinder launched, it had no access to user data, but nevertheless it became the market leader within a couple of years. Lambrecht and Tucker explain that even in this highly data-driven industry, Tinder succeeded not through reliance on Big Data, but due to the strength of its underlying solution.[27] A simple user interface and a precise attention to consumer needs resulted in massive gains for the new entrant. Similarly, despite facing competition from long-established incumbents with access to huge volumes of data, amassed over years of customer service, WhatsApp was able to take on more established messaging and social networks because of its low cost and easy-to-use interface. Examination of these industries leads Lambrecht and Tucker to conclude that to build a sustainable competitive advantage from Big Data, a firm needs to focus on developing both the managerial toolkit and organizational competence that allows them to turn Big Data into value to consumers in previously impossible ways, rather than simply amassing tremendous amounts of data.

15.2.3.6. Highly Differentiated Platforms Need Highly Differentiated Data

Online platforms are highly differentiated, even in the provision of the same type of service, and as each entrant carves out a niche, the most useful data to them differs more and more from the data most useful to their rivals. Consumers are moving toward meeting more precise, niche consumer needs. A consumer looking to book a flight could use Kayak, Expedia, Orbitz, or a multitude of other travel-dedicated search engines. The same is true in Internet shopping, online dating, social networking, product and service reviews, and a host of other online markets. In today's online environment, successful firms must carve out their own niche, and increasingly, data that is useful (even crucial) to one firm may not be useful to its competitors.[28] An astute and innovative entrant will identify a niche where the incumbent does not have requisite data, and can very quickly "catch up" the incumbent in terms of valuable data amassed.

15.3. DOES BIG DATA POSE HARM TO COMPETITION?

Although data as a potential antitrust concern is not a particularly new issue, what has changed dramatically in recent years is the size and scope of the data that firms collect, store, and use.[29] With the growth in the amount of data, and the advent of

---

[27] Lambrecht and Tucker (2015).
[28] Schepp and Wambach (2015).
[29] Feinstein (2015).

Big Data, the importance of that data as an input in online platforms has also increased. The growing importance of Big Data as an input, and the consistent increase in the four Vs of data – volume, velocity, variety, and value[30] – has meant that companies are now more than ever undertaking data-driven strategies to gain operational efficiencies[31] and, some argue, to gain and sustain an unfair competitive advantage.[32]

This section describes a number of ways in which some have argued that Big Data can be used to perpetuate an unfair competitive advantage and consequently distort competition and harm consumers. These commentators argue that Big Data arms online providers with the incentive and ability to erect barriers to entry and maintain dominance by limiting their competitors' access to data, preventing others from sharing the data, and opposing data-portability policies that threaten data-related competitive advantages.[33] The resulting harm, according to such critics, is not necessarily higher prices (considering most of these services are provided for free), but rather a loss of quality, innovation, or privacy.

To properly assess the antitrust implications of Big Data, we must understand fully the ways in which online platforms use Big Data and the nature of competition among them. A crucial starting point in this endeavor is a solid understanding of two-sided platforms. A two-sided platform exists when one provider caters to two different customers groups on different sides of the same platform.[34] For example, social media platforms give users free access to social networking services on one side of the platform and rely on the provision of advertising services to businesses on the other side of the platform for revenue. A proper antitrust assessment of any two-sided platform must take into account competition on each side of the platform. It is important to recognize that certain actions may cause procompetitive effects for the platform as a whole, while initially appearing anticompetitive on one side of the platform. A comprehensive antitrust analysis cannot look at one side of the platform in a vacuum – it must weigh the benefits and harm to the platform as a whole.[35]

15.3.1. Loss of Quality and Innovation

While firms with access to troves of Big Data can use it to improve the quality of their products in several ways, a number of practitioners have argued that misuse of Big Data may result in a loss of quality. While the exact parameters of this proposition are open to debate, scale in data is, indisputably, important in improving the

---

[30] OECD (2013); Executive Office of the President (2014).
[31] Lambrecht and Tucker (2015).
[32] Stucke and Grunes (2015a).
[33] Stucke and Ezrachi (2015); Newman (2014).
[34] Evans and Schmalensee (2014).
[35] Salinger and Levinson (2015).

quality of online services. Smaller firms, the argument goes, often cannot adequately compete with larger firms because they lack access to the same volume of data as the larger firm. As the data gap, and consequently the quality gap, widens between the dominant firm and a smaller rival, the competitive constraint the rival poses to the dominant firm in terms of quality and innovation is diminished. The larger firm, in this scenario, is not driven to innovate or to maximize quality for the consumer.

Stucke and Ezrachi argue that inequality in access to data can lead to the potential degradation of quality for consumers in search engines in particular.[36] They claim that large search engines have the incentive and ability to prioritize paid advertising over more relevant, better-quality, organic search results. On a search engine, more advertisements, displayed more prominently, benefit both the advertiser and the search provider. More ads increase the opportunities for user clicks. This in turn means a greater likelihood of a pay-per-click conversion for the platform provider and a better chance of a product sale for the advertiser. Where this becomes an antitrust problem, Stucke and Ezrachi suggest, is where Big Data has widened the gap between large and small providers to the extent that a smaller provider cannot provide adequate quality competition to prevent its larger rival from sacrificing some degree of search quality in favor of expanding profits on the paid side. The fact that a large search engine has access to so much data, and therefore the ability to improve quality to such a high degree, means it can afford to sacrifice a higher level of search quality than a smaller search engine (which is already struggling on quality due to lower data levels) could. Additionally, the disparity in data volume means that users are generally unable to detect small degradations in quality - they "just know that Google is giving a better result than Bing," but not *how much* better.

In addition to the lack of real-world supporting evidence, this theory of harm also begs the question whether incremental degradation in quality by a search provider whose quality is still superior to rivals is an antitrust concern. Does a firm have an obligation to provide the absolute best quality product it can, even if not profit maximizing? No court or antitrust regulator has ever imposed such a requirement. And, in this example, how does a regulator measure the "best quality search results," since quality is relative and users are said not to be able to accurately assess quality?

These questions aside, this loss of quality theory also overlooks the importance of analyzing both sides of the two-sided platform. While no ads at all would certainly improve search quality, it would clearly be very detrimental to advertisers. Increasing ad space might be beneficial to advertisers, but could admittedly lead to search quality degradation for users. Antitrust analysis requires a balancing act and an understanding of the inherent tradeoffs between both sides of the platform. A holistic approach to the economic efficiency of the conduct is required, as opposed to delegating to antitrust the responsibility (instead of properly assigning

---

[36] Stucke and Ezrachi (2015).

to the market itself) to police whether search quality dips below "levels that consumers prefer," as Stucke and Ezrachi claim.

In addition to the alleged degradation in quality that can occur, Big Data can also, some allege, stifle innovation. Where a firm's value proposition is built on collecting and monetizing user data, if that firm collects so much user data that it becomes entrenched, it may gain both the ability and the incentive to use that data in a number of ways to eliminate potential challengers.[37] As this happens, smaller rivals are prevented from accessing necessary data, and the incentive for these firms to innovate and to compete with larger dominant firms is reduced. For example, a dominant firm with access to Big Data could conceivably look to trends in data to identify potential challengers and devise strategies to quickly stamp out any rising competition by limiting or preventing their access to necessary data, or by acquiring them. Where market leaders with deep pockets acquire potential or actual new entrants, a source of innovation is removed, and competition suffers. Of course, such a discernment of trends may also be beneficial to competition where it forces a market leader to further invest in innovation itself, as antitrust law fundamentals contemplate. It is also worth bearing in mind that acquiring a smaller rival is not, without proof that such acquisition is likely to substantially lessen competition, prohibited under the antitrust laws. Indeed, the potential for such acquisitions incentivizes entry.

15.3.2. Harm to Privacy

Proponents of antitrust involvement in Big Data suggest that consumers feel they do not have control over how their data is collected and used by online platform providers.[38] As users create more and more data, and firms continue to collect it, the safeguards protecting its collection and use may well become more important and more vulnerable to attack. The economics literature shows that in fact the collection of data may provide improved services,[39] product recommendations,[40] or free content.[41]

Privacy protections can be considered a form of nonprice competition, which is especially important in industries where the service itself is offered for free.[42] Firms may compete by offering tighter or more transparent privacy policies.[43] Yet Jones Harbour and Koslov argue that consumers can be harmed when a dominant firm has no incentive to invest in privacy protections.[44] Acquisti offers a literature review

---

[37] Stucke and Grunes (2015a).
[38] Stucke and Grunes (2015a); Jones Harbor and Koslov (2010).
[39] Acquisti and Varian (2005).
[40] Bennett and Lanning (2007).
[41] Goldfarb and Tucker (2010).
[42] Ohlhausen and Okuliar (2015).
[43] Evans (2009); Savage and Waldman (2015).
[44] Jones Harbour and Koslov (2010).

that provides a more nuanced view of the different ways of how consumers value privacy.[45] It is important to note, however, that harm to privacy does not, without more, equal harm to competition. And, as discussed in more detail below, antitrust is ill-equipped to solve consumer law problems.

15.3.3. *Data-Driven Mergers and Data-Driven Defenses*

The number of Big Data-related merger cases has increased over time. In this context, further potential harms could, it is argued, arise from data-driven mergers where the transaction rationale rests on the acquirer gaining access to the underlying data set of the target undertaking.[46] Some suggest that where privacy constitutes an important dimension of competition in a given market, or represents an important element of transaction rationale, antitrust agencies should closely examine transactions to determine whether the combination is likely to reduce incentives to compete in providing privacy protections to consumers. An early example of this argument can be found in then-Commissioner Harbour's dissenting statement in the investigation of Google's acquisition of DoubleClick, which suggested that privacy could be "'cognizable' under the antitrust laws," and should have been considered by the Commission "as part of its antitrust analysis of the transaction."[47] The former Commissioner's statement cited a possible theory that network effects could lead to fewer search engines, reducing "incentives of search firms to compete based on privacy protections or related non-price dimensions."[48]

These concerns have not been borne out yet by any actual cases. In the United States, the antitrust agencies have had occasion to consider the role of Big Data in a number of high-profile mergers, and merging parties have increasingly put forth data-driven efficiencies in defense of mergers, with varying degrees of success.[49] For example, with respect to ratings and reviews platform provider Bazaarvoice's 2012 acquisition of rival PowerReviews, both the Department of Justice and the trial court rejected the parties' efficiencies claims, citing a lack of evidence that the transaction had resulted in data gains, leading to an improved product, lower prices, or greater innovation.[50] On the other hand, during the Department of Justice's 2010 investigation into a search-related partnership between Microsoft and Yahoo!, the Justice Department did accept the parties' data-driven efficiency argument,

---

[45] Acquisti (2014).
[46] Graef (2015).
[47] Pamela Jones Harbour, Dissenting Statement, Google/DoubleClick, FTC File No. 071-0170, at 10 (2007), available at www.ftc.gov/sites/default/files/documents/public_statements/statement-matter-google/doubleclick/071220harbour_0.pdf.
[48] Id.
[49] Rather than being a concern for a court/regulator, Big Data combinations may be viewed as a potential justification for the mergers.
[50] *United States v. Bazaarvoice, Inc.*, Case No. 13-cv-00133-WHO, 2014 WL 203966, at *62–64 (N.D. Cal. Jan. 8, 2014).

suggesting that the transaction might be procompetitive where increased access to data enable more rapid improvements in Microsoft's search offering, thereby creating a more viable competitive alternative to Google.[51] In Europe, although the Commission did not ultimately opine on this particular issue, the parties to the *TomTom/TeleAtlas* merger argued that customer feedback data would allow the combined firm to produce better maps at a faster pace.[52] The *Telefonica/ Vodafone/Everything Everywhere* joint venture found no Big Data problem with regard to the data analytics services,[53] nor did *Publicis/Omnicom*, where the Commission noted competition from alternative providers of Big Data analytics.[54]

15.3.4. *The Perceived Strength of Scale, Network Effects, and Barriers to Entry*

Many, if not all, of the theories of harm attributed to Big Data rest on the perceived strength of the "feedback loop" and the consequential network effects enjoyed by large firms with access to tremendous amounts of data.[55] Big Data can give rise to network effects, and certainly, network effects can play a significant role in a sound antitrust analysis. However, agencies, policymakers, and scholars must resist any foregone conclusion that the presence of network effects in Big Data automatically results in anticompetitive harm.

Big Data can lead to economies of scale via the alleged "feedback loop." In search, some argue, "the availability of data on previous search queries is crucial" to competitive success.[56] There are two ways scale can be accomplished through the "feedback loop." The "user feedback loop" presumes that as a platform gains more users, it can collect more user data, leading to better insights into consumers and their needs, which can be used to improve quality, attracting even more users. The "monetization feedback loop" claims that as a platform gains more users and collects more user data, it is better able to target ads and therefore sell ads, and so is better able to monetize its platform, gaining revenues which can be invested in improving quality of service, thereby attracting more users.

Alongside these feedback loops, a number of distinct network effects come into play in online platforms that collect and use Big Data. Direct network effects occur when a product or service becomes more valuable to an individual user as more

---

[51] Press Release, U.S. Dep't of Justice, Statement of the Department of Justice Antitrust Division on its Decision to Close Its Investigation of the Internet Search and Paid Search Advertising Agreement Between Microsoft Corporation and Yahoo! Inc. (Feb. 18, 2010), available at www.justice.gov/opa/pr/statement-department-justice-antitrust-division-its-decision-close-its-investigation-internet.
[52] Case COMP/M.4854 – *TomTom/Tele Atlas*, Comm'n Decision, 2008 OJ C237, 53–54, ¶¶ 245–50.
[53] Case COMP/M.6314 – *Telefónica UK/Vodafone UK/Everything Everywhere/JV*, Comm'n Decision (Sept. 4, 2012).
[54] Case COMP/M.7023 – *Publicis/Omnicom* (Jan. 9, 2014).
[55] Graef (2015).
[56] Id.

people use that particular product or service. In a modern context, social networking platforms, photo sharing platforms, and chat applications may enjoy significant direct network effects. Indirect network effects occur when more users make the use of a product or service better or more attractive to consumers, though not because of direct interaction between users. Search engines benefit from indirect network effects as more users allow the search engine essentially to gain insight into what users want from user clicks, essentially learning by trial and error, and therefore improving the quality of search results.

Some argue that network effects are particularly strong in two-sided platforms. A firm operating a two-sided platform can, it is argued, benefit not only from traditional network effects, but also from cross-platform network effects, where more users on one side of the platform makes the platform more attractive to users on the other side of the market.[57] While entry barriers naturally vary from industry to industry, and indeed change over time, these practitioners suggest that the economies of scale and network effects that characterize data-driven markets lead to a "winner takes all" result, and present insurmountable barriers to entry.

In reality, the strength of the feedback loop may be grossly overstated. The feedback loop theory assumes smaller rivals and challengers will not be able to compete effectively as they lack comparable amounts of users, and therefore data, inhibiting their ability to improve quality and attract more users. As Lerner points out, however, these assumptions are unsupported by real-world evidence.[58] The economics characteristics of Big Data weaken the claimed strength of the feedback loop. Chief among these characteristics is the fact that online providers can gain scale in users in ways that do not involve user data, and that access to data alone is not enough to improve quality and gain scale in users. Additionally, firms can gather data from other sources than users (e.g. data brokers), and can gain scale in data in alternative ways, such as entering into strategic distribution arrangements.

As to network effects, even in classic cases of direct network effects such as social networking and communications applications, innovation can be strong enough to upend the market, and network effects have time and time again proven insufficient to prevent incumbents from disrupting established market leaders. In social networking, for example, Friendster, the original "market leader" was replaced quickly by MySpace, which has now been rendered almost completely obsolete by Facebook. An innovative product is enough to cause users to switch, notwithstanding any network effect enjoyed by the incumbent.

Among advertisers, network effects are diminished by the pricing structures employed by most online platforms, by advertiser multi-homing due to the low cost in advertising on multiple platforms, and by advertiser "congestion." The pay-per-click model means that while advertising on a "busier" platform may result in better

---

[57] *Id.*; Stucke and Grunes (2015a).
[58] Lerner (2014).

conversion rates for an advertiser, it also involves proportionally higher costs, and more clicks means the advertiser has to pay more. As such, it may actually not be as economically advantageous for an advertiser to choose a larger online platform over a smaller one (contrary to real-world platforms that are priced differently). Additionally, since fixed costs to advertise on any particular platform are low, advertisers may be incentivized to advertise on multiple different platforms as opposed to putting all their eggs in one basket. Finally, while more users on a platform might be good for advertisers, more advertisers on the platform can actually be detrimental. Limited available space for online ads and competition for users' attention means that advertisers may be better off on smaller platforms with less congestion.[59]

Perhaps most importantly, cross-platform network effects are also commonly overstated, and are actually one-sided. While advertisers certainly may flock to a search engine (or other online platform) with a strong user base with the hope of more impressions and hopefully more conversions, users, on the other hand, do not choose a search engine based on a greater number of advertisements. This weakening of the cross-platform network effects argument in turn weakens the potential for a strong "feedback loop" that locks users and advertisers into a dominant platform. If a smaller entrant offers a better product or service to users, users will switch, uninhibited by network effects, and advertisers will soon follow.[60]

The above discussion demonstrates how the feedback loop is not as effective as suggested in gaining scale, but the importance of scale is also misjudged by many. Big Data industries typically experience diminishing returns of scale. Statistically, as Lerner illustrates,[61] the value of user data in returning relevant results to user search queries is subject to quickly diminishing returns, as the advantages of scale weaken or disappear at a low level. While returns are greater for less frequent queries (known as "tail" queries), both large and small search providers are faced with queries they have never seen before on a daily basis, where both small and large platforms are at an equal disadvantage in delivering relevant results. Because of these rapidly diminishing returns, a larger provider may gain zero marginal value from incremental data after a certain point, and a smaller player may glean greater value from incremental data, incentivizing it to compete in attracting users at the margin by investing in quality and innovation.

Even if scale is crucial to competitive success, smaller rivals do maintain both the ability and the incentive to compete. As to *ability*, many online players are well-funded, or at least have access to additional funding from investors, with which they can improve quality and performance of their platform. Furthermore, all online players have access to stores of data from third parties, which is readily available and affordable, and can be deftly used to increase quality. As to *incentive*, economics tells

---

[59] Id.
[60] Id.
[61] Id.

us that investment incentive is based on marginal, not average effects. An investment in quality by a smaller firm will attract more incremental users than a similar investment by a larger firm. As such, the smaller firm's incentive to invest in quality may actually be greater than that of its larger rival.

15.4. IS ANTITRUST ENFORCEMENT THE RIGHT WAY TO REGULATE BIG DATA?

In order to consider whether antitrust is the most appropriate forum within which to explore, and potentially address, Big Data concerns, one should consider how antitrust case-law has treated Big Data issues to date, how Big Data might fit within the existing antitrust analysis framework or remedies, what legal or practical dangers might result from applying antitrust to Big Data, and whether an alternative framework is better suited to these issues.

15.4.1. *Case-Law Does Not Support the Contention that Big Data Is an Antitrust Problem*

A thorough search of case-law and agency actions does not reveal case-law, nor have the antitrust agency consents ever affirmatively concluded that consumer data constitutes a barrier to entry, and available precedent does not counsel in favor of using antitrust as a tool to right Big Data "wrongs."[62] While competition agencies and courts have concluded that data-related entry barriers *may* exist for the sale of data that cannot be sourced from consumers or Big Data marketplaces, they have yet to come to the same conclusion regarding data collected from consumers over the Internet. Over the last five to ten years, antitrust agencies, and to a lesser extent the courts, have considered a number of mergers and instances of conduct involving potential theories of harm built around Big Data. One of the earliest examples of this was Google's acquisition of DoubleClick in 2007. At the time, both parties were large players in the market for search advertising – Google was a large online advertising intermediary, and DoubleClick was a leading online ad server. Both parties had vast stores of data relating to user search and browsing history.

Similarly, both the FTC and the European Commission examined Facebook's 2014 acquisition of web-based messaging platform WhatsApp. Upon announcement of the transaction, several consumer groups complained to the FTC that the transaction would bolster Facebook's access to data which could be monetized through advertising, contradicting prior statements by WhatsApp.[63] The FTC cleared the transaction within two months, and sent a clear indication that the

---

[62] We can distinguish cases where data itself is the issue as a key input.
[63] See generally *Complaint, Request for Investigation, Injunction, and Other Relief*, Elec. Privacy Info. Ctr. & Ctr. for Digital Democracy, In re WhatsApp, Inc. (March 6, 2014). Available at www.centerfordigitaldemocracy.org/sites/default/files/WhatsApp%20Complaint.pdf.

issues raised rested squarely within consumer protection law. The FTC, upon clearing the transaction, sent a letter to the parties from the Director of the Bureau of Consumer Protection reminding them of their continuing obligations under privacy law.[64]

The European Commission also reviewed the Facebook/WhatsApp merger, in doing so provided an analytical framework for exclusionary behavior in Big Data industries, and ultimately cleared the transaction without conditions.[65] While the Commission acknowledged that network effects could sometimes pose a barrier to entry in communications markets, it concluded that this particular transaction was not likely to raise barriers to entry, noting "consumers can and do use multiple apps at the same time and can easily switch from one to another,"[66] and adding that "there are currently a significant number of market participants that collect user data alongside Facebook," including Google, Apple, Amazon, eBay, Microsoft, AOL, Yahoo, Twitter, IAC, LinkedIn, Adobe, and Yelp.[67] That investigation was significant as it recognized the factual inexistence of network effects as a barrier to entry in such a fast-moving online market. The basis for this conclusion was due to: (i) the Commission finding that messaging apps were a "fast-moving sector"[68] with low switching costs; therefore, "any leading market position even if assisted by network effects is unlikely to be incontestable."; (ii) the finding that usage of one particular messaging app did not "exclude the use of competing [messaging] apps by the same user;" in this context, multi-homing was common and facilitated by the "ease of downloading a consumer communications app;"[69] and (iii) acknowledgment that users of messaging apps "are not locked-in" to a given network.[70] The Commission found that even if Facebook were to begin collecting data from WhatsApp users, competitive harm would not result, as "there will continue to be a large amount of Internet user data that are valuable for advertising purposes and that are not within Facebook's exclusive control."[71] The Commission's decision also explicitly rejected the idea of considering a potential market for personal data in this case, citing the fact that the parties were not actually engaged in the sale of data to third parties.[72] In the US, there was a similar outcome with regard to *Nielson/Arbitron*, where the data

---

[64] Letter from Jessica L. Rich, Dir., Bureau of Consumer Protection, Fed. Trade Comm'n, to Erin Egan, Chief Privacy Officer, Facebook, Inc. & Anne Hoge, Gen. Counsel, WhatsApp Inc. (April 10, 2014). Available at www.ftc.gov/system/files/documents/public_statements/297701/140410facebookwhatappltr.pdf.
[65] See Case COMP/M.7217 – *Facebook/WhatsApp*, Comm'n Decision, 2014 OJ C 7239, 24–5, ¶ 134
[66] Press Release, Eur. Comm'n, Mergers: Commission Approves Acquisition of WhatsApp by Facebook (Oct. 3, 2014). Available at http://europa.eu/rapid/press-release_IP-14-1088_en.pdf.
[67] Tucker and Welford (2014) at 8; Case COMP/M.7217 – *Facebook/WhatsApp*, Comm'n Decision, ¶ 188 (March 10, 2014).
[68] *Id.* at ¶132
[69] *Id.* at ¶133
[70] *Id.* at ¶134
[71] Case COMP/M.7217 – *Facebook/WhatsApp*, Comm'n Decision, ¶ 189 (March 10, 2014).
[72] Case No COMP/M.7217 – *Facebook/WhatsApp*, October 3, 2014, ¶ 72.

was merely an input.[73] Such cases where data is merely an input are different from cases where data is a market that is sold to consumers.[74]

Outside the merger context, the FTC's 2011–2012 investigation of Google centered at least partially on the competitive significance of data. In a recent statement responding to the inadvertent release of portions of the FTC's Bureau of Competition Staff Report, Chairwoman Ramirez and Commissioners Brill and Ohlhausen noted that the Commission's "exhaustive" investigation into Google's Internet search practices, including agreements for syndicated search and advertising services, were not, "on balance, demonstrably anticompetitive."[75]

15.4.2. Big Data as Its Own Product Market

Market definition and market power still form the backbone of antitrust analysis under the current law. Some practitioners have suggested that data collection should form its own product market for the purpose of antitrust analysis.[76] The precise contours of such a market would be difficult, if not impossible, to define. In both the United States and Europe, substitution, via the hypothetical monopolist test, is an essential prerequisite to defining a market. The primary goal of defining a market is to measures a firm's ability to exercise market power and the relevant market determines goods or services that potentially compete, to the exclusion of those that do not. Data itself is not a relevant product in the sale of online advertising. Advertising services are the relevant product. Data is used (for the most part) by online providers as an *input* in their service, as opposed to actually being sold as a product to consumers. There is, therefore, no competition between providers for the actual sale of data, and no substitution. As such, under current antitrust law, no relevant market can be defined for the collection of consumer data. In reviewing the Facebook/Whatsapp acquisition, the European Commission overtly declined to define a market for Big Data since neither party was active in the provision of data to third parties.[77]

15.4.3. Consumer Protection Should Address Big Data Issues

The laws of consumer protection and antitrust serve different goals, protect consumers from different harms, and operate via different spheres of the same agency.[78] A review of the economics of privacy notes complexity as to how to regulate

---

[73] *In the Matter of Nielsen Holdings N.V. and Arbitron Inc.*, FTC File No. 131 0058 (September 20, 2013).

[74] Tucker and Welford (2014).

[75] Statement of Chairwoman Edith Ramirez, and Commissioners Julie Brill and Maureen K. Ohlhausen regarding the Google Investigation, March 25, 2015. Available at www.ftc.gov/news-events/press-releases/2015/03/statementchairwoman-edith-ramirez-commissioners-julie-brill.

[76] Jones Harbour and Koslov (2010).

[77] Case COMP/M.7217 – *Facebook/WhatsApp*, Comm'n Decision, ¶ 72 (March 10, 2014).

[78] Feinstein (2015).

privacy.[79] However, this recent review does not find a strand of academic literature in which on a theoretical basis or empirical basis antitrust should be used as a policy tool to address privacy concerns.

Such a finding is not surprising. Consider for a moment, though, other product elements such as product safety and efficacy that also constitute forms of nonprice competition. Those elements, though potentially affected by competition, are not primarily policed by the antitrust agencies but through consumer/data protection law.[80] The antitrust laws are not designed to address harm to privacy – an efficient market, bolstered by the consumer protection laws, provides adequate protection from those harms.

Suggested safeguards intended to prevent the misuse of Big Data by a dominant firm such as enabling the consumer to more easily select privacy preferences or to identify providers that match their privacy preferences, sit squarely within the remit of the consumer protection agencies.[81] Where an imbalance of power between users and online firms leads to diminished data portability, individual consumers or competitors might suffer but the mechanics of data collection is not for the antitrust laws to govern. Antitrust law is only a suitable choice where there is harm to competition. Antitrust's role is not to fill gaps in the privacy laws.

15.4.4. Are Antitrust Remedies Appropriate?

Some have suggested that antitrust remedies may be appropriate where a dominant firm has misused Big Data to gain or sustain an improper competitive advantage. The imposition of such remedies presents obvious problems. From an antitrust perspective, forced sharing of information with rivals infers the essential facilities doctrine, and such forced dealing with competitors in the Big Data environment is far beyond the limits of what a duty to deal would require. If Big Data were deemed an essential facility and a duty to deal imposed, the competitive dynamics of the market would be dramatically altered. Such an extreme and far-reaching remedy is out of line with current antitrust policy.[82]

Practically speaking, requiring affirmative user consent before data is collected may detract from the user experience and lessen quality; prohibiting or restricting data collection may stifle innovation and present users with lower quality services; and divesture or separation of distinct product lines may also stifle innovation and hinder a firm's ability to offer personalized services.[83]

Antitrust remedies haphazardly applied to the collection and use of consumer data may not only harm competition, but also may in fact raise separate, legitimate,

---

[79] Acquisti, Taylor, and Wagman (2016).
[80] Schepp and Wambach (2015); Ohlhausen and Okuliar (2015).
[81] Stucke and Grunes (2015b).
[82] Orbach and Avraham (2014).
[83] Tucker and Welford (2014).

privacy issues.[84] Antitrust remedies may also create privacy concerns as they would require data to be shared with rival firms even though consumers have not consented to their data being used in this way. Likewise, a forced sharing of data could violate a company's already existing consent decrees with the FTC.[85]

The FTC, in the closing statement from its investigation into the Google/DoubleClick merger, rejected the notion that antitrust remedies should be imposed to address privacy harms:

> [T]he sole purpose of federal antitrust review of mergers and acquisitions is to identify and remedy transactions that harm competition. Not only does the Commission lack legal authority to require conditions to this merger that do not relate to antitrust, regulating the privacy requirements of just one company could itself pose a serious detriment to competition in this vast and rapidly evolving industry.[86]

15.4.5. Practical and Legal Dangers of Antitrust Intervention

Using antitrust as a sword to address Big Data concerns risks reducing competition and innovation from new products.[87] Antitrust enforcement agencies are well advised to proceed cautiously in areas of rapid innovation, in order to avoid stifling competition, and the natural unfolding of the marketplace. While an industry is in its relative infancy, it can be difficult to distinguish between procompetitive innovation and changes that are designed to (or actually do) stifle competition. Even in established markets, antitrust should never be used as a replacement for sound business judgment. As the FTC's closing statement in the Google investigation explained: "Challenging Google's product design decisions in this case would require the Commission – or a court – to second-guess a firm's product design decisions where plausible procompetitive justifications have been offered, and where those justifications are supported by ample evidence."[88]

Consumer welfare is enhanced most dramatically by "leapfrog" competition, as opposed to incremental improvements. It is crucial that the antitrust laws cultivate and maintain an environment in which robust and rapid innovation is not only possible, but also incentivized. A paternalistic approach to Big Data will neither cultivate nor maintain such an environment, and may instead lead to stagnation and fear among platform providers.

---

[84] Goldfarb and Tucker (2012).
[85] Tucker and Welford (2014).
[86] Statement of Federal Trade Commission Concerning Google/DoubleClick, FTC File No. 071-0170, at 2 (Dec. 20, 2007).
[87] Ohlhausen and Okuliar (2015).
[88] Statement of the Federal Trade Commission Regarding Google's Search Practices, In the Matter of Google Inc. FTC File Number 111-0163 January 3, 2013. Available at www.ftc.gov/sites/default/files/documents/public_statements/statement-commission-regarding-googlessearch-practices/130103brillgooglesearchstmt.pdf.

15.5. CONCLUSION

This literature review suggests that antitrust law is ill-suited to police Big Data and its use by online firms. The empirical case regarding Big Data as an antitrust concern is still lacking. Further, from a theoretical perspective, not enough work has yet been done to thoughtfully study and analyze how antitrust could, or should, be applied to specific issues involving Big Data. In fact, the lack of empirical evidence, robust theories, or indeed legal precedent suggests that there is no cause for concern in this arena. All that is available at present are general theories of exclusion applied to this new area. Until theories of harm can be matched with specific factual circumstances and negative economic competitive harm can be shown, the antitrust case against Big Data is a weak one. The existing theories of harm conflict with the realities of Big Data (e.g., nonrivalrous, ubiquitous, low barriers to entry noted above) and consumer online behavior (e.g., multi-homing[89]). And while the case is weak, and the theories uncertain, antitrust authorities should proceed with caution. Antitrust intervention over market forces threatens consumer welfare, especially is fast-moving markets, and proposed remedies, such as limiting the collection and use of Big Data or forcing large firms to share with rivals, are likely to harm competition and innovation, and in fact may raise privacy concerns.

REFERENCES

Acquisti, Alessandro. 2014. From the Economics of Privacy to the Economics of Big Data. In Stefan Bender, Julia Lane, Helen Nissenbaum, and Victoria Stodden (eds.), *Privacy, Big Data, and the Public Good: Frameworks for Engagement*. Cambridge University Press, 76–95.

Acquisti, Alessandro and Hal R. Varian. 2005. Conditioning Prices on Purchase History. *Marketing Science*, 24(3), 367–81.

Acquisti, Alessandro, Curtis Taylor, and Liad Wagman. 2016. The Economics of Privacy. *Journal of Economic Literature*, 54(2), 442–92.

Averitt, Neil and Robert H. Lande. 1997. Consumer Sovereignty: A Unified Theory of Antitrust and Consumer Protection Law. *Antitrust Law Journal*, 65, 713–56.

Bennett, James and Stan Lanning. 2007. "The Netflix Prize," in Proceedings of KDD Cup and Workshop 2007 (San Jose, CA, August 12, 2007), 3–6.

Chiou, Lesley and Catherine Tucker. 2014. *Search Engines and Data Retention: Implications for Privacy and Antitrust*. Working paper.

Cooper, James C. 2013. Privacy and Antitrust: Underpants Gnomes, the First Amendment, and Subjectivity. *George Mason Law Review*, 20(4), 1129–46.

Edlin, Aaron S. and Robert G. Harris. 2013. The Role of Switching Costs in Antitrust Analysis: A Comparison of Microsoft and Google. *Yale Journal of Law and Technology*, 15, 169–213.

Evans, David S. 2009. The Online Advertising Industry: Economics, Evolution, and Privacy. *The Journal of Economic Perspectives*, 23(3), 37–60.

---

[89] Salinger and Levinson (2015).

Evans, David S. and Richard Schmalensee. 2014. The Antitrust Analysis of Multi-Sided Platform Businesses. In Roger Blair and Daniel Sokol (eds.), *Oxford Handbook on International Antitrust Economics*. Volume 1. Oxford University Press, 404–50.

Executive Office of the President President's Council of Advisors on Science and Technology. 2014. *Big Data and Privacy: A Technological Perspective.*

Feinstein, Debbie. 2015. The not-so-big news about Big Data, June 16, 2015, Competition Matters Blog. Available at www.ftc.gov/news-events/blogs/competition-matters/2015/06/not-so-big-news-about-big-data.

Goldfarb, Avi. 2014. What Is Different about Online Advertising?, *Review of Industrial Organization*, 44(2), 115–29.

Goldfarb, Avi and Catherine Tucker. 2010. Privacy Regulation and Online Advertising, *Management Science*, 57(1), 57–71.

 2011. Online Advertising. In Marvin V. Zelkowitz (ed.), *The Internet and Mobile Technology Advances in Computing* . Maryland Heights, MO: Academic Press.

 2012. Privacy and Innovation. In Josh Lerner and Scott Stern (eds.), *Innovation Policy and the Economy*. University of Chicago Press, 65–90.

Graef, Inge. 2015. Market Definition and Market Power in Data: The Case of Online Platforms. *World Competition*, 38(4), 473–505.

Jones Harbour, Pamela and Tara Isa Koslov. 2010. Section 2 in a Web 2.0 World: An Expanded Vision of Relevant Product Markets. *Antitrust Law Journal*, 76, 769–97.

Lambrecht, Anja and Catherine Tucker. 2015. *Can Big Data Protect a Firm from Competition?* Working paper.

Lerner, Andres V. 2014. *The Role of 'Big Data' in Online Platform Competition.* Working paper.

Muris, Timothy J. and Paloma Zepeda. 2012. The Benefits, and Potential Costs, of FTC-Style Regulation in Protecting Consumers. *Competition Law International*, 8, 11–17.

Newman, Nathan. 2014. Search, Antitrust and the Economics of the Control of User Data. *Yale Journal on Regulation*, 31: 401–52.

OECD. 2013. *Supporting Investment in Knowledge Capital, Growth and Innovation.*

Ohlhausen, Maureen K. and Alexander P. Okuliar. 2015. Competition, Consumer Protection, and the Right [Approach] to Privacy. *Antitrust Law Journal*, 80, 121–56.

Orbach, Barak and Raphael Avraham. 2014. Squeezing Claims: Refusals to Deal, Essentials Facilities, and Price Squeezes. In Roger Blair and Daniel Sokol (eds.), *Oxford Handbook on International Antitrust Economics*, Volume 2, Oxford University Press, 120–31.

Ramirez Edith. 2015. *Protecting Privacy in the Era of Big Data*, Remarks of FTC Chairwoman Edith Ramirez, International Conference on Big Data from a Privacy Perspective, Hong Kong, June 10.

Rochet, Jean Charles and Jean Tirole. 2002. Cooperation among Competitors: Some Economics of Payment Card Associations. *RAND Journal of Economics*, 33, 1–22.

Salinger, Michael A. and Robert J. Levinson. 2015. Economics and the FTC's Google Investigation. *Review of Industrial Organization*, 46: 25–57.

Savage, Scott J. and Donald M. Waldman. 2015. Privacy Tradeoffs in Smartphone Applications. *Economics Letters*, 137, 171–5.

Schepp, Nils-Peter and Achim Wambach. 2015. On Big Data and Its Relevance for Market Power Assessment., *Journal of European Competition Law & Practice*, 7(2) 120–4.

Shapiro, Carl and Hal R. Varian. 1999. *Information Rules: A Strategic Guide to the Network Economy*. Cambridge, MA, Harvard Business Press.

Stucke, Maurice E. and Ariel Ezrachi. 2015. When Competition Fails to Optimise Quality: A Look at Search Engines. *Yale Journal of Law & Technology*, 18, 70–110.

Stucke, Maurice and Allen Grunes. 2015a. Debunking the Myths over Big Data and Antitrust. *CPI Antitrust Chronicle*, May.
  2015b. No Mistake about It: The Important Role of Antitrust in the Era of Big Data. *Antitrust Source*.
Tucker, Catherine. 2013. The Implications of Improved Attribution and Measurability for Antitrust and Privacy in Online Advertising Markets. *George Mason Law Review*, 20, 1015–54.
Tucker, Darren and Hill Welford. 2014. Big Mistakes Regarding Big Data. *Antitrust Source*.
Varian, Hal R. 2014. Big Data: New Tricks for Econometrics. *Journal of Economic Perspectives*, 28(2), 3–28.

PART IV

Competitor Collaboration

16

Drug Patent Settlements

Michael A. Carrier

16.1. INTRODUCTION

One of the most pressing antitrust issues today involves settlements by which brand-name drug companies pay generic firms to settle patent litigation and delay entering the market. Whether this activity constitutes an antitrust violation is a subject that has occupied courts and enforcement agencies around the world.

This chapter focuses on the law in the United States, where the law is most developed. It also explores Europe, where the activity has started to receive attention, as well as Canada, India, and Korea.

16.2. UNITED STATES

16.2.1. Reverse Payment Settlements

Brands and generics often settle patent infringement cases. Most of these settlements do not present antitrust concern. Some do not delay entry at all. For example, the Federal Trade Commission (FTC) issued a report that found that nineteen out of 140 settlements between brands and generics in 2012 did not restrict generic entry.[1]

Other settlements do not involve payment. The FTC report found that eighty-one settlements contained a restriction on entry but did not provide compensation. Courts and the FTC have concluded that these "patent-term split agreements" do not violate the antitrust laws because they involve the parties dividing the patent term by selecting a date for generic entry based on the strength of the patent. The greater the likelihood that the patent is valid and infringed, the later in the period generic entry would be expected.

---

[1] FTC (2013).

It is the last category of payment and delayed entry (forty settlements in 2012) that presents concern. These payments have been called "reverse payments" because the consideration flows from patentee to alleged infringer (unlike typical settlements in which alleged infringers pay patentees). A brand is likely to gain additional exclusivity not warranted by the strength of the patent by supplementing the parties' entry date agreement with a payment to the generic. And the quid pro quo for the payment would appear to be the generic's agreement to stay out of the market beyond the date that otherwise reflects the parties' assessment of the patent's strength and likely outcome of litigation.

16.2.2. Pre-Actavis Case-Law

Drug patent settlements first received attention from the FTC, the US enforcement agency that takes the lead on the pharmaceutical industry, in 2000, when the agency entered into two consent orders. The first involved a payment of "millions of dollars" by brand Hoechst Marion Roussel (then Aventis) to generic Andrx to delay bringing its hypertension drug to the market.[2] The second involved an agreement by brand Abbott Laboratories to pay "substantial sums" to generic Geneva Pharmaceuticals to delay bringing its hypertension and prostate drug to the market.

In the first case in which a US appellate court examined the issue, the Sixth Circuit in *In re Cardizem CD Antitrust Litigation* explained that the settlement at issue guaranteed to the brand that "its only potential competitor" would "refrain from marketing its generic version of [the drug] even after it had obtained FDA approval."[3] The court was also concerned that the agreement prevented the generic from marketing products not covered by the patent. It concluded that the settlement was "a horizontal agreement to eliminate competition ... a classic example of a per se illegal restraint of trade."[4]

Courts, however, quickly retreated from such analysis, turning to a test that essentially immunized activity falling within the "scope of the patent." The Eleventh Circuit in *Schering-Plough Corp. v. FTC* upheld settlements that fell "within the protections of the ... patent" and stated that it "cannot be the sole basis for a violation of antitrust law" for a brand firm with a patent to pay a generic competitor.[5] The Second Circuit in *In re Tamoxifen Citrate Antitrust Litigation* concluded that as long as "the patent litigation is neither a sham nor otherwise baseless" or beyond the patent's scope, the patentee can enter into a settlement "to protect that to which it is presumably entitled: a lawful monopoly over the manufacture and distribution of the patented product."[6] The Federal Circuit in *In re Ciprofloxacin Hydrochloride Antitrust Litigation* found that settlements fell "well within" the patentee's rights,

---

[2] FTC (2000).
[3] 332 F.3d 896, 907 (6th Cir. 2003).
[4] Id. at 908.
[5] 402 F.3d 1056, 1076 (11th Cir. 2005).
[6] 466 F.3d 187, 208–09, 213 (2d Cir. 2006).

that patents bestowed "the right to exclude others," and that the crucial inquiry was "whether the agreements restrict competition beyond the exclusionary zone of the patent."[7] And the Eleventh Circuit in *FTC v. Watson Pharmaceuticals* stated that, in the absence of sham litigation or fraud in obtaining a patent, "a reverse payment settlement is immune from antitrust attack so long as its anticompetitive effects fall within the scope of the exclusionary potential of the patent."[8]

The judicial trend from 2005 until early 2012 thus favored deference to settlements.[9] In addition to relying on the scope of the patent, courts emphasized the importance of settlements, the link between settlements and innovation, the presumption of patent validity, and the "natural" status of reverse payments in upholding the agreements. In July 2012, however, in *In re K-Dur Antitrust Litigation*, the Third Circuit rejected the scope-of-the-patent test, explaining that it assumed the validity at issue in the case and was not relevant when the issue is infringement (on which the patentee bears the burden of proof).[10] The court also concluded that a reverse payment was "prima facie evidence of an unreasonable restraint of trade" that could be rebutted only if the settling parties could show that the payment was for a purpose other than delay or that it "offers some pro-competitive benefit."[11]

16.2.3. Actavis

To resolve the split among the courts, the US Supreme Court granted certiorari to review the Eleventh Circuit's *Watson* ruling, and in 2013, in the case of *FTC v. Actavis*,[12] rejected the scope-of-the-patent test. The Court found it "incongruous" to "determine antitrust legality by measuring the settlement's anticompetitive effects solely against patent law policy, rather than by measuring them against procompetitive antitrust policies as well."[13]

The Court found that the settlement at issue had the "potential for genuine adverse effects on competition" since "payment in return for staying out of the market ... keeps prices at patentee-set levels."[14] In addition, the Court highlighted the harms from a payment to a generic, which "in effect amounts to a purchase by the patentee of the exclusive right to sell its product, a right it already claims but would lose if the patent litigation were to continue and the patent were held invalid or not infringed by the generic product."[15]

---

[7] 544 F.3d 1323, 1332–33, 1336, 1337 (Fed. Cir. 2008).
[8] 677 F.3d 1298, 1312 (11th Cir. 2012).
[9] Carrier (2009).
[10] 686 F.3d 197 (3d Cir. 2012), judgment vacated sub nom. *Merck & Co. v. Louisiana Wholesale Drug Co.*, 133 S. Ct. 2849 (2013).
[11] *Id.* at 218.
[12] 133 S. Ct. 2223 (2013).
[13] *Id.* at 2231.
[14] *Id.* at 2235.
[15] *Id.* at 2234.

The Court revealed its strong preference for determining patent strength by examining the payment rather than the patent. The "size of the unexplained reverse payment can provide a workable surrogate for a patent's weakness, all without forcing a court to conduct a detailed exploration of the validity of the patent itself."[16] Even strong patents are not immune from the concern with payments, because an unexplained payment on a "particularly valuable patent ... likely seeks to prevent the risk of competition," with this consequence "constitut[ing] the relevant anticompetitive harm."[17] Finally, the Court found that the policy in favor of settlement did not immunize the agreements because of five arguments that centered on reverse payments' (1) anticompetitive effects, (2) lack of justification, and (3) market power, along with (4) the feasibility of judicial analysis and (5) parties' ability to settle without payment.

The Court concluded that "the FTC must prove its case as in other rule-of-reason cases."[18] And it instructed future courts to analyze payments' "size, ... scale in relation to the payor's anticipated future litigation costs, ... independence from other services for which it might represent payment, and ... lack of any other convincing justification."[19]

16.2.4. Post-Actavis Case-Law

Since the ruling in *Actavis*, courts have analyzed various issues, sometimes reaching divergent results. The most attention to date has been directed to the type of consideration that counts as payment. Other issues have included the patent merits, rule of reason, pleading requirements, causation, and state law.

16.2.4.1. Payment

The first issue is whether "payment" applies only to cash or whether it encompasses other forms of consideration. *Actavis* is consistent with an interpretation of payment that extends beyond cash. The *Actavis* opinion never uses the word "cash," but on five occasions, it uses the phrase "millions of dollars," and I have shown elsewhere that, at a minimum, four of the five instances anticipate an expansive interpretation of the term.[20] In addition, *Actavis* involved not a cash payment but alleged brand overpayments for generic services that "had little value."[21] Along these lines, the Court explained that the FTC "alleges that in substance, the plaintiff agreed to pay

---

[16] *Id.* at 2236–7.
[17] *Id.* at 2236.
[18] *Id.* at 2237.
[19] *Id.*
[20] Carrier (2015).
[21] *Actavis*, 133 S. Ct. at 2229.

the defendants many millions of dollars . . . ."[22] The Court's language indicates that it anticipated that its ruling would apply to transfers of consideration that *in substance* were equivalent to cash, in other words, non-cash transfers.

Every final decision that has considered the issue has found that "payment" is not limited to cash but extends to other forms of consideration.[23]

One of the most frequent settings in which the issue has arisen involves "authorized generics," which are approved by the US Food and Drug Administration (FDA) as brand drugs but marketed as generics.[24] Brands can introduce authorized generics during the 180-day exclusivity period reserved for the first generic to challenge a brand's patent, claiming that it is invalid or not infringed.[25] Given that first-filing generics often receive the "vast majority" of their profits during this 180-day period, and that, on average, first-filers lose 25 percent of their market share and suffer revenue reductions of roughly 40 to 52 percent when they compete with an authorized generic during the period, no-authorized-generic ("no-AG") promises can be immensely valuable to generics.[26]

For example, in an important federal appellate decision, the Third Circuit, in *King Drug Company of Florence v. Smithkline Beecham Corporation (Lamictal)*, held that no-AG promises could constitute payment.[27] The court explained that "[t]he anticompetitive consequences" of an agreement involving such promises "may be as harmful as those resulting from reverse payments of cash" since "[i]f the brand uses a no-AG agreement to induce the generic to abandon the patent fight, the chance of dissolving a questionable patent vanishes (and along with it, the prospects of a more competitive market)."[28] As a result, "a brand agreeing not to produce an authorized generic may thereby have 'avoid[ed] the risk of patent invalidation or a

---

[22] *Id.* at 2231.
[23] See *King Drug Co. v. Smithkline Beecham Corp.*, 791 F.3d 388 (3d Cir. 2015); *In re Loestrin 24 FE Antitrust Litig.*, 814 F.3d 538 (1st Cir. 2016); *In re Aggrenox Antitrust Litig.*, 2015 WL 1311352, at *12 (D. Conn. Mar. 23, 2015); *United Food & Commercial Workers Local 1776 v. Teikoku Pharma USA (Lidoderm)*, 2014 WL 6465235, at *11 (N.D. Cal. Nov. 17, 2014); *In re Effexor XR Antitrust Litig.*, 2014 WL 4988410, at *20 (D.N.J. Oct. 6, 2014); *Time Ins. Co. v. AstraZeneca*, 2014 WL 4933025, at *3 (E.D. Pa. Oct. 1, 2014); *In re Lipitor Antitrust Litig.*, 46 F. Supp. 3d 523, 542 (D.N.J. 2014); *In re Niaspan Antitrust Litig.*, 42 F. Supp. 3d 735, 751 (E.D. Pa. 2014); *In re Wellbutrin XL Antitrust Litig.*, No. 08-cv-02431, slip op. at 4 (E.D. Pa. Jan. 17, 2014); *In re Nexium (Esomeprazole) Antitrust Litig.*, 968 F. Supp. 2d 367, 392 (D. Mass. 2013).
[24] FTC (2009).
[25] *Mylan Pharms. v. FDA*, 2005 WL 2411674, at *7 (N.D.W. Va. Sept. 29, 2005); *Teva Pharm. Indus. v. Crawford*, 410 F.3d 51, 55 (D.C. Cir. 2005).
[26] FTC (2011).
[27] 791 F.3d 388 (3d Cir. 2015). See also *In re Loestrin*, 814 F.3d at 550 ("antitrust scrutiny attaches not only to pure cash reverse payments, but to other forms of reverse payment that induce the generic to abandon a patent challenge, which unreasonably eliminates competition at the expense of consumers"); *In re Cipro Cases I & II*, 2015 WL 2125291, at *18 n. 11 (Cal. May 7, 2015) (stating in dicta that courts "should not let creative variations in the form of consideration result in the purchase of freedom from competition escaping detection").
[28] 791 F.3d at 405.

finding of noninfringement.'"[29] A no-AG promise "presumably" results in the generic accepting an entry date "that is later than it would have otherwise accepted."[30] In short, the Third Circuit did "not believe the Court intended to draw such a formal line" or "limit its reasoning or holding to cash payments only."[31]

Not only are no-AG promises as concerning as cash payments, but they also bear the potential for even more competitive harm. The reason is that a cash payment, though it extends the brand's monopoly, does not affect generic competition when the generic enters the market. In contrast, a no-AG clause not only delays brand entry, but also reduces generic competition when it does take place. The promises to avoid entering markets can be viewed as a form of market division, with the brand and generic "allocat[ing] a part of the market to each of them, with their reciprocal agreements not to compete in each other's part of the market serving as a payment from one to the other."[32]

Another reason for concern with no-AG promises (and certain other non-cash conveyances) is that they provide more than the generic could have obtained from successful patent litigation. I have offered a test for determining exclusion payments that "asks if the brand conveys to the generic a type of consideration not available as a direct consequence of winning the lawsuit."[33] Under such a test, no-AG clauses, cash payments, and "acceleration clauses" (by which a settling generic can accelerate its entry when a non-settling, later-filing generic enters the market) satisfy the test for an exclusion payment that (assuming delayed entry) violates the antitrust laws. Such a test addresses Judge Richard Posner's concern that "any settlement agreement can be characterized as involving 'compensation' to the defendant, who would not settle unless he had something to show for the settlement."[34] If a generic obtains a type of consideration not available from winning patent litigation, exclusion results from the payment and not the patent.

16.2.4.2. Patent Merits

A second issue that courts have addressed after *Actavis* is whether settling parties are allowed to introduce evidence of the patent merits in arguing that a payment is justified by the strength of the patent. A hurdle to introducing such evidence is *Actavis* itself. In articulating the antitrust analysis of drug patent settlements, *Actavis* emphasized the payment rather than the patent. Large payments could themselves "provide strong evidence that the patentee seeks to induce the generic challenger to abandon its claim with a share of its monopoly profits that would otherwise be lost in

---

[29] *Id.* (quoting *Actavis*, 133 S. Ct. at 2236).
[30] *Id.*
[31] *Lamictal*, 791 F.3d at 405–6.
[32] *Id.*
[33] Carrier (2014).
[34] *Asahi Glass Co. v. Pentech Pharm., Inc.*, 289 F. Supp. 2d 986, 994 (N.D. Ill. 2003).

the competitive market."[35] The Court also connected an "unexplained large reverse payment" to "serious doubts about the patent's survival."[36] And "that fact, in turn, suggests that the payment's objective is to maintain supracompetitive prices to be shared among the patentee and the challenger rather than face what might have been a competitive market – the very anticompetitive consequence that underlies the claim of antitrust unlawfulness."[37]

The Supreme Court also rejected the argument that for a "particularly valuable" patent "even a small risk of invalidity justifies a large payment."[38] The concern was that "the payment (if otherwise unexplained) likely seeks to prevent the risk of competition" and "that consequence constitutes the relevant anticompetitive harm."[39] For this reason, the Court concluded that "it is normally not necessary to litigate patent validity to answer the antitrust question."[40] The size of the payment "provide[s] a workable surrogate for a patent's weakness, all without forcing a court to conduct a detailed exploration of the validity of the patent itself."[41]

Court decisions support analysis that does not rely on the patent merits. The court in *Time Insurance v. AstraZeneca* denied defendants' motion to dismiss because "without reaching the ultimate question of the validity of the patents, the risk that the generic manufacturers might enter the marketplace and demonstrate a reasonable likelihood of success in voiding the patents has an economic consequence which plaintiffs contend was blunted by unlawful agreements preventing that form of competition."[42] The court in *FTC v. Cephalon* ruled that a brand firm could not introduce evidence related to a patent's "potential validity, enforceability, or infringement" when the court had earlier found that the company had engaged in inequitable conduct.[43] In addition, the appropriate time for assessing competitive effects is the time of settlement, not afterwards, as the "fact that a patent's strength is a spectrum does not change simply because a judge later determines that the patent was, in fact, invalid all along."[44] Similarly, commentators have explained that allowing a defense based on patent validity and infringement "would defeat the Court's stated purpose of cutting to the chase in these cases" and would lead to the defense "invariably be[ing] raised," resulting in it, "directly contrary to the Court's repeated assertions ... normally be[ing] necessary to litigate patent validity and infringement."[45]

---

[35] *Actavis*, 133 S. Ct. at 2235.
[36] *Id.* at 2236.
[37] *Id.*
[38] *Id.* at 2236–7.
[39] *Id.*
[40] *Id.* at 2236.
[41] *Id.* at 2236–7.
[42] 2014 WL 4933025, at *6 (E.D. Pa. Oct. 1, 2014).
[43] 36 F. Supp. 3d 527, 537 (E.D. Pa. 2014).
[44] *Id.* at 533.
[45] Edlin *et al.* (2013) at 19.

16.2.4.3. Rule of Reason

A third issue that has arisen since *Actavis* is the type of antitrust analysis courts should apply. The Court anticipated that lower courts would play a crucial role in "structuring ... the ... rule-of-reason antitrust litigation."[46] Such a framework was to "consider traditional antitrust factors such as likely anticompetitive effects, redeeming virtues, market power, and potentially offsetting legal considerations."[47] The Court also instructed lower courts to "structure antitrust litigation so as to avoid, on the one hand, the use of antitrust theories too abbreviated to permit proper analysis, and, on the other, consideration of every possible fact or theory irrespective of the minimal light it may shed on the basic question – that of the presence of significant unjustified anticompetitive consequences."[48]

The Court's version, however, was more streamlined than the typical open-ended, exhaustive analysis. It offered shortcuts for plaintiffs in proving anticompetitive effects (payments have the "potential for genuine adverse effects on competition") and market power (the "size of the payment" can serve as "a strong indicator of [market] power").[49] Defendants then would have the burden to "show in the antitrust proceeding that legitimate justifications are present, thereby explaining the presence of the challenged term and showing the lawfulness of that term under the rule of reason."[50]

The Court limited the defendants' justifications, specifically mentioning only two: whether the payment "amount[s] to no more than a rough approximation of the litigation expenses saved through the settlement" and whether it "reflect[s] compensation for other services that the generic has promised to perform."[51] Each of those justifications has received attention.

16.2.4.3.1. LITIGATION COSTS. The first justification is based on litigation costs. The concept of litigation costs encompasses expenditures incurred in conducting litigation.[52] The most frequently cited survey of costs in intellectual property (IP) litigation in the United States is assembled every two years by the American Intellectual Property Law Association ("AIPLA"). The AIPLA defines litigation costs to include "outside legal and paralegal services, local counsel, associates, paralegals, travel and living expenses, fees and costs for court reporters, photocopies, courier services, exhibit preparation, analytical testing, expert witnesses, translators, surveys, jury advisors, and similar expenses."[53]

---

[46] *Actavis*, 133 S. Ct. at 2238.
[47] *Id.* at 2231.
[48] *Id.* at 2238.
[49] *Id.* at 2234, 2236.
[50] *Id.* at 2236.
[51] *Id.*
[52] Trubek *et al.* (1983).
[53] AIPLA (2013).

The figures from a recent AIPLA survey show that patent litigation in which there is more than $1 million at risk costs $2.6 to $5.5 million on average. In Hatch-Waxman litigation in particular, the figures range from $2.65 million to $6 million.[54] One study found that generics spent an average of $10 million for each challenge to a brand's patent.[55] In other words (and offering a conservative synthesis), a transfer of $5 million or $10 million seems to be a rough approximation of litigation costs. Transfers of less than this amount should be covered under this justification.

Kobayashi *et al.*[56] have criticized the focus on litigation costs, claiming that "rapid entry by multiple firms ... often follows the invalidation of a patent," which "implies [that] the payoff for the generic entrant that files the first [generic application entitled to 180-day exclusivity] and invalidates the patent is smaller than the monopoly-to-duopoly litigation payoffs generated in the single-entrant models." Not only does the first-filing generic allegedly receive a smaller payoff, but the patent holder faces higher losses from a collateral estoppel doctrine that prevents a brand firm with an invalidated patent from relitigating patent validity. As a result, there is "a significantly broader range of settlements" that "weakens the relationship between the strength of the patent and the size of the settlement," thereby lessening "calls to deem presumptively unlawful all payments greater than anticipated litigation costs."[57]

Edlin *et al.*[58] responded to this argument, stating that the authors' claims are "incorrect, inconsistent with [their] own model, or irrelevant to a faithful implementation of the Court's opinion in *Actavis*." As *Actavis* itself noted, first-filing generics often receive the "vast majority" of their profit during the period of generic exclusivity.[59] In fact, Edlin *et al.*[60] note that the "*Actavis* Inference" (that "a large and otherwise unexplained payment, combined with delayed entry, supports a reasonable inference of harm to consumers from lessened competition") is "even more important with multiple generic entrants than with just one generic entrant" since "delaying generic entry will boost profits even more, and harm competition even more, than with just one generic entrant."

16.2.4.3.2. PAYMENT FOR SERVICES. The second justification the *Actavis* Court acknowledged involves payment not for delayed entry but for unrelated services. For example, brands have paid generics for IP licenses, for supplying raw materials or finished products, and for helping to promote products.[61] As some scholars have contended, however, "[t]he parties to a payment for delay have ample reason to pack

---

[54] Id.
[55] Herman (2011).
[56] Kobayashi *et al.* (2015) at 89.
[57] Id. at 89–90.
[58] Edlin *et al.* (2015) at 613.
[59] *Actavis*, 133 S. Ct. at 2229.
[60] Edlin *et al.* (2015) at 606.
[61] FTC (2007).

complexities into the deal (such as relatively unimportant services) to conceal its genuine nature."[62]

Actavis itself is an example of a case in which the two sides disputed the reason for the payment. The settling parties alleged that the payment was for ancillary co-promotion services provided by the generic. The FTC, in contrast, alleged that the payment was for delayed entry. To pick one example, the brand "projected that it would pay [the generics] more than ... $300 per sales call," far more than a previous co-promotion deal that had "involv[ed] projected payments of around $30–$45 per sales call" and even more than the $150 per call that a senior Watson executive had called 'ridiculous.'"[63]

To similar effect, in entering into a $1.2 billion settlement with Cephalon and parent company Teva on the eve of trial, the FTC explained that side deals contemporaneous with settlement "were merely a mechanism for Cephalon to pay the generic challengers [more than $300 million] to forgo entry."[64] The FTC noted that Cephalon "entered into supply agreements with three of the generics or their partners to purchase the active pharmaceutical ingredient in [sleep-disorder drug] Provigil despite evidence that Cephalon already had adequate supply available at significantly lower prices."[65] In addition, Cephalon "agreed to pay up to $131 million to license intellectual property even though it had previously rejected any concerns about possible infringement risk" and the brand "entered into product development deals with Mylan, despite internal projections showing a negative net present value to Cephalon."[66]

A disguised payment could take the form not only of a brand's *overpayment* for generic services, but also a generic's *underpayment* for brand products. Along those lines, the FTC has claimed in one recent case that Abbott paid Teva to delay entering the market with a generic version of testosterone gel AndroGel by providing Teva with an authorized generic version of cholesterol drug TriCor at "a price that is well below what is customary in such situations."[67] In granting defendants' motion to dismiss, however, the court found that "the AbbVie Defendants are not making any payments to Teva," but that "[i]t is Teva which is paying Abbott for the supply of TriCor."[68] The court recognized that "the FTC correctly alleges that something of large value passed from Abbott to Teva," but was too formalistic in concluding that "it was not a reverse payment under *Actavis*."[69]

---

[62] Edlin *et al.* (2015) at 18.
[63] Second Amended Complaint for Injunctive and Other Equitable Relief at ¶ 82, *FTC v. Watson Pharm., Inc.*, No. 1:09-cv-00955-TWT (N.D. Ga. May 28, 2009).
[64] Statement of the Federal Trade Commission, *FTC v. Cephalon, Inc.*, at 2 n. 7 (May 28, 2015).
[65] *Id.*
[66] *Id.*
[67] *FTC v. AbbVie Inc.*, 2015 WL 2114380, at *6 (E.D. Pa. May 6, 2015).
[68] *Id.* at *7.
[69] *Id.*

16.2.4.3.3. "LARGE AND UNJUSTIFIED PAYMENT" THRESHOLD. Some lower courts and defendants have sought to impose a threshold requirement by which plaintiffs must show that a payment is "large and unjustified" before reaching the full antitrust analysis. The district court in *In re Loestrin 24 FE Antitrust Litigation*, for example, imposed a framework that required analysis of (1) whether "there [is] a reverse payment" and (2) whether "that reverse payment is large and unjustified" before addressing (3) the rule of reason. The *Loestrin* court borrowed this framework from the district court in *In re Lamictal Direct Purchaser Antitrust Litigation* And defendants have contended, for example, that "*Actavis* requires a plaintiff challenging a reverse-payment settlement ... to prove, as a threshold matter, that the ... payment was both large and unjustified"[70] and that "under *Actavis*, [plaintiffs] have to prove that [a] payment was 'large' (as well as unexplained)."[71]

A threshold for proving an "unjustified" payment (with the evidence of justifications in the defendants' possession) seems inconsistent with the shortcuts adopted in *Actavis*, would be duplicative, and in any event appears "nowhere in the *Actavis* opinion."[72]

Another question is how to define the terms "large" and "unjustified." "Unjustified" most naturally refers to a payment that cannot be explained by a procompetitive purpose but instead is for delayed entry. And the most natural definition of "large" is an amount higher than litigation costs. Some defendants have claimed that the amount should be measured relative to the patentee's expected profits. And the court in *In re Effexor XR Antitrust Litigation* asserted that payment must "appear [] to be large from the perspective of the brand company making the payment."[73] But while showings in relation to a brand's profits could be *sufficient* to show a "large" payment, they are not *necessary* because payments at lower levels can "induce the generic challenger to abandon its claim" and delay entering the market.[74] Any payments larger than litigation costs are made for a reason, and that reason typically is to delay generic entry.

16.2.5. Pleading Requirements

The fourth issue that has arisen since *Actavis* involves the pleading requirements plaintiffs must satisfy. Two courts have raised pleading requirements to very high levels. The *Effexor* court required plaintiffs to convert a non-monetary payment to "a concrete, tangible or defined amount which yields a reliable estimate of a monetary

---

[70] *King Drug Co. of Florence v. Cephalon, Inc.*, 2015 WL 356913, at *1 (E.D. Pa. Jan. 28, 2015).
[71] Teva's Memorandum in Support of Directed Verdict Motion on Threshold Issue of a Large and Unexplained Payment, *In re: Nexium (Esomeprazole Magnesium) Antitrust Litig.*, Case 1:12-md-02409-WGY, at 12–13 (D. Mass. Nov. 10, 2014).
[72] *Cephalon*, 2015 WL 356913, at *10.
[73] 2014 WL 4988410, at *23 (D.N.J. Oct. 6, 2014).
[74] *Actavis*, 133 S. Ct. at 2235.

payment."[75] Plaintiffs alleged that generic Teva, by not being required to compete with an authorized generic during its 180-day period, "would realize about double the volume of generic sales at significantly higher, supra-competitive prices," which – based on a comparison with a drug with similar sales – was valued "at over $500 million."[76] The court, however, found that "[s]imply alleging some sort of value of a no-authorized generic agreement, absent a reliable foundation supporting that value," is "vague and amorphous" and "does not establish the plausibility required by Rule 12(b)(6)."[77] Plaintiffs' references to the FTC's comprehensive reports on authorized generics were not sufficient, as "[t]he Complaint simply does not rely on any knowledge of business practitioners in the pharmaceutical industry."[78]

In *In re Lipitor Antitrust Litigation*, the same judge similarly required a non-monetary payment to be "converted to a concrete, tangible or defined amount which yields a reliable estimate of a monetary payment."[79] The court did not consider as a potential payment the brand's alleged forfeiture of hundreds of millions of dollars in damages in separate litigation because the plaintiff failed to allege facts as if it were "standing in the shoes of the parties at the time of settlement."[80] In particular, the plaintiff needed to introduce evidence of the type on which the brand would have relied in showing lost profits in the separate litigation: that the patent owner "would have made some or all of the infringers' sales" as well as "(1) demand for the product; (2) absence of noninfringing substitutes; (3) manufacturing and marketing capability; and (4) the amount of profit."[81]

On the other hand, other courts have applied more appropriate, flexible pleading standards. In *United Food & Commercial Workers Local 1776 v. Teikoku Pharma USA (Lidoderm)*, the court noted that a brand's provision of $96 million of Lidoderm pain patches is "not a complex, multifaceted payment" but "rather ... a simple transfer of a fungible product," for which "plaintiffs have plausibly alleged facts sufficient to support their calculations."[82] And in determining the value of a no-AG agreement, the court credited the plaintiff's calculation of "the difference between [the generic's] projected revenues with the agreement and ... projected revenues had it competed with [the] authorized generic from the start."[83] Similarly, the court in *In re Aggrenox Antitrust Litigation* explained that "very precise and particularized estimates of fair value and anticipated litigation costs may require evidence in the

---

[75] 2014 WL 4988410, at *19.
[76] *Id.*
[77] *Id.*
[78] *Id.*
[79] 46 F. Supp. 3d at 542.
[80] *Id.* at 544.
[81] *Id.* at 545.
[82] *Lidoderm*, 2014 WL 6465235, at *12.
[83] *Id.*

exclusive possession of the defendants, as well as expert analysis" and "these issues are sufficiently factual to require discovery."[84]

16.2.6. Causation

A fifth issue that has received attention involves causation. In the only trial since *Actavis*, a jury found that plaintiffs did not show that the settlement caused their injuries. Even though the jury found that brand AstraZeneca exercised market power, that the settlement included a "large and unjustified payment," and that it was "unreasonably anticompetitive," the jury found that "[h]ad it not been for" the settlement, AstraZeneca would not have "agreed with [generic] Ranbaxy that Ranbaxy might launch a generic version of Nexium before May 27, 2014."[85] The court had earlier raised concerns related to the plaintiffs' ability to show causation given the failure to offer "direct evidence that the FDA was likely to grant final approval to Ranbaxy's generic Nexium product within the proposed timeline" as well as evidence that Ranbaxy would "never" have launched generic Nexium "at risk" (before a court finding that the patent was invalid or not infringed).[86]

In contrast, causation was satisfied in *In re Niaspan Antitrust Litigation*, as the court found that plaintiffs successfully pleaded antitrust injury on the grounds that they showed that, but for the settlement, the generic would have prevailed in the patent litigation, based on (1) a large, unexplained payment showing "serious doubts about the patent's survival"; (2) the generic's expenditure of more than $2 million in research and due diligence concerning the potential invalidity and infringement of the brand's patents; and (3) the generic's willingness to launch at risk, which showed its "confiden[ce] that it would ultimately prevail ... in the infringement litigation."[87]

16.2.7. State Law

Antitrust analysis of reverse payment settlements has taken place in the context of not only federal law, but also state law. In the most important state law case to date, the California Supreme Court in *In re Cipro Cases I & II* applied a structured rule of reason similar to that applied by the US Supreme Court in *Actavis*. The *Cipro* court explained that California law did not support the scope-of-the-patent test. In noting the competitive harm of reverse payment settlements, the court made clear that "businesses may not engage in a horizontal allocation of markets, with would-be competitors dividing up territories or customers."[88]

---

[84] *Aggrenox*, 2015 WL 1311352, at *13.
[85] Jury Verdict in Favor of Defendants Against Plaintiffs Returned, *In re Nexium Antitrust Litig.*, No. 12-md-02409 (D. Mass. Dec. 8, 2014), ECF No. 1374.
[86] *In re Nexium (Esomeprazole) Antitrust Litig.*, 42 F. Supp. 3d 231, 272 (D. Mass. 2014).
[87] 42 F. Supp. 3d 735, 755–6 (E.D. Pa. 2014).
[88] *In re Cipro Cases I & II*, 2015 WL 2125291 (Cal. May 7, 2015) at *15.

The court then articulated the four elements of a plaintiff's prima facie case. First, the plaintiff must show "a limit on the settling generic challenger's entry into the market."[89] Second, it must establish "a reverse payment – financial consideration flowing from the brand to the generic challenger."[90] Third, "a plaintiff must establish [that] the consideration to the generic challenger exceeds the value of any other collateral products or services provided by the generic to the brand."[91] And fourth, the plaintiff must establish that "the amount of the payment, over and above the value of collateral products or services from the generic, also exceeds the brand's anticipated future litigation costs."[92] While the plaintiff bears the burden of proof on showing a "restraint on generic competition and a reverse payment to the generic in excess of both brand litigation costs and generic collateral products and services," the defendants "have the burden of coming forward with evidence of litigation costs and the value of collateral products and services."[93] Finally, the court rejected defendants' pre-emption arguments since the Court's rule was "in harmony with *Actavis*, which offered only broad outlines and explicitly left to other courts the task of developing a framework for analyzing the anticompetitive effects of reverse-payment patent settlements."[94]

16.3. EUROPE

The issue of settlements has taken a different path in Europe. There have been fewer enforcement actions and, until recently, less attention on the issue. One of the most important activities was the 2009 pharmaceutical Sector Inquiry.

The European Commission (EC or Commission) commenced this inquiry in January 2008, seeking to determine the reasons for delayed generic entry and for the decline in new medicines reaching the market.[95] The report found that 22 percent of settlements from 2000 to 2008 involved payments from the brand to the generic firm. The remainder involved agreements that did not limit the generic's ability to market the drug (52 percent) as well as settlements that limited generic entry but did not involve payment (26 percent).

The inquiry noted differences between settlements in the United States and the European Union. Most significant, the 180-day marketing period granted to the first-filing generic in the United States does not have a counterpart in the European Union. In addition, settlements in the European Union have not been marked by brand firms' promises not to launch authorized generics that would compete with

---

[89] *Id.* at *17.
[90] *Id.* at *18.
[91] *Id.*
[92] *Id.* at *19.
[93] *Id.*
[94] *Id.* at *24.
[95] European Commission (2009).

the generics. Finally, there have been fewer "side deals" in Europe in which the parties enter into agreements on products unrelated to the patent at the heart of the settlement.

As of the time this chapter went to press, the inquiry has been followed up by four monitoring exercises. The first, analyzing settlements between the middle of 2008 and end of 2009, found that the number of settlements that involved a restriction on generic entry and value transfer from the brand to the generic fell from 22 percent of settlements (between 2000 and 2007) to 10 percent.[96] In addition, the amount of money involved in the settlements fell from more than €200 million to less than €1 million. The report offered potential reasons for the decline, which included increased awareness of competition law scrutiny, continued monitoring of settlements, and the opening of proceedings in a case against France's Les Laboratoires Servier.

The second monitoring exercise, one year later, traced an even more significant decline in the most concerning agreements. Of the eighty-nine patent settlements in the period, only three limited the generic's ability to market its product and included a value transfer from brand to generic.[97] The report found that the companies continued to settle cases, but that settlements took different, less concerning, forms: 61 percent did not restrict generic entry and 36 percent limited entry but did not involve a value transfer to the generic. While the third monitoring exercise found an increase (to 11 percent) in settlements involving restrictions on generic entry and a value transfer, this was partially explained by a case (based on an "erroneously assumed expiry of exclusivity rights" caused by "an administrative error by public authorities") that did not "trigger potential competition law scrutiny."[98] The fourth monitoring exercise found a reduction to 7 percent of settlements involving payment for delayed entry.[99]

In addition to the monitoring exercises, the EC has also targeted individual companies. In June 2013, the Commission announced that it would fine Lundbeck (roughly) € 94 million and generic firms € 52 million for violating Article 101 of the Treaty on the Functioning of the European Union (TFEU) for agreeing "to delay the market entry of cheaper generic versions of Lundbeck's branded citalopram, a blockbuster antidepressant."[100] The Commission quoted from documents that referred to a "club" being formed and "a pile of $$$" that the participants would share.[101] In addition, "Lundbeck paid significant lump sums, purchased generics' stock for the sole purpose of destroying it, and offered guaranteed profits in a distribution agreement," with the agreements giving Lundbeck "certainty" that the

---

[96] European Commission (2010).
[97] European Commission (2011).
[98] European Commission (2012).
[99] European Commission (2013b).
[100] European Commission (2013a).
[101] Id.

generics "would stay out of the market for the duration of the agreements" without offering them "any guarantee of market entry thereafter."[102]

In January 2015, the EC published a 464-page non-confidential version of this decision. In this version, it made clear that the agreements constituted an "infringement by object" since they "were by their very nature injurious to the proper functioning of normal competition."[103] The Commission also found that the agreements prohibited entry and "contained a transfer of value"; that they "did not resolve any patent dispute" but "postponed the issue raised by potential generic market entry"; and that the agreements "obtained results for Lundbeck that [it] could not have achieved by enforcing its process patents before the national courts."[104]

In a second case, in July 2014, the EC fined Servier and generic rivals €427 million for settlements that delayed generic entry on perindopril, a blockbuster blood pressure medicine.[105] The Commission stated that "between 2005 and 2007, virtually each time a generic company came close to entering the market, Servier and the company in question settled the challenge."[106] On "at least five occasions," the generics "agreed to abstain from competing in exchange for a share of Servier's rent."[107] The Commission also quoted from documents in which one generic acknowledged being "bought out of perindopril" and another stated that "any settlement will have to be for significant sums," which it referred to as a "pile of cash."[108]

In July 2015, the EC released a non-confidential version of the decision. The Commission concluded that "Servier sought protection against generic entry by concluding five patent settlement agreements with the (most) advanced generic contenders" which "consisted of significant payments, or other inducements, to the generic companies, and the obligation for them not to challenge Servier's patents and not to enter the market (directly or indirectly) for a number of years."[109]

The Commission also revealed factors (similar to those articulated in its *Lundbeck* decision) for determining whether a settlement falls under the category of restriction by object: (1) whether the brand and generic "were at least potential competitors"; (2) whether the generic "committed itself in the agreement to limit, for the duration of the agreement, its independent efforts to enter one or more EU markets with a generic product"; and (3) whether there was a "transfer of value … as a significant inducement which substantially reduced the incentives of the generic undertaking to independently pursue its efforts to enter one or more EU markets with the generic

---

[102] *Id.*
[103] EC, Commission decision of 19 June 2013, Case AT.39226 – *Lundbeck* at ¶¶ 821, 1084.
[104] *Id.* ¶ 6.
[105] European Commission (2014).
[106] *Id.*
[107] *Id.*
[108] *Id.*
[109] EC, Commission decision of 9 July 2014, Case AT.39612 – *Perindopril (Servier)* ¶ 7. Available at http://ec.europa.eu/competition/antitrust/cases/dec_docs/39612/39612_11972_5.pdf.

product.""¹¹⁰ The Commission also considered as "other important factors" whether the restrictions "either lasted throughout the entire period of the patent term, or did not contain any commitment by Servier to refrain from infringement proceedings in case of independent entry with the relevant generic products after the expiry of the agreement"; whether the "value Servier transferred to generics took into consideration the turnover or the profit the generic undertaking expected if it had successfully entered the market"; and whether "the obligations on certain generic undertaking[s] in the respective agreements exceeded the scope of the underlying patent litigation/dispute, in particular as the restrictions went beyond what Servier could have legally obtained through successful enforcement of its patents in the underlying disputes/litigation.""¹¹¹

16.4. GLOBAL TREATMENT: CANADA, KOREA, INDIA

Most of the activity related to drug patent settlements has taken place in the United States and Europe. But other countries are starting to consider these issues. In September 2014, the Canadian Competition Bureau (Bureau) released a paper entitled "Patent Litigation Settlement Agreements: A Canadian Perspective."¹¹² In it, the Bureau explained the difference between the regulatory regimes in Canada and the US, such as

> (1) the lack of a notification system in Canada, (2) the absence in Canada of a 180-day period of exclusivity for the first generic to challenge a brand's patent, (3) particularities of [Canada's Patented Medicine Notice of Compliance Regulations (PM(NOC))] prohibition proceedings, and (4) the potential for generics to receive damages from brands in Canada.¹¹³

The Bureau concluded, however, that these differences do not "diminish the role of competition analysis in reviewing potentially anticompetitive settlements."¹¹⁴ And it stated that it would consider applying both civil and (for a more limited category of behavior) criminal liability to reverse payment settlements. As of the date this chapter went to press, the Bureau was in the process of revising its Intellectual Property Enforcement Guidelines, providing additional detail on how it planned to address this behavior.

Korea has also considered reverse payment settlements. In October 2011, it levied a $2.6 million fine against brand GlaxoSmithKline and $1.8 million fine against generic Dong-A for their agreement to delay generic entry on an anti-nausea drug. The two companies had settled patent litigation on terms requiring Dong-A to

---

[110] Id. ¶ 1154.
[111] Id. ¶ 1155.
[112] Canadian Competition Bureau (2014).
[113] Id. at 6.
[114] Id.

remove its generic drug from the market and to not develop, manufacture, or sell drugs that could compete with GSK's product. In return, GSK offered Dong-A the right to sell the drug in public/national hospitals and the exclusive right to sell a separate drug. The Korea Fair Trade Commission (KFTC) found that the agreement constituted an act of "delay" for which GSK paid Dong-A. As a result, it concluded that the agreement constituted an unreasonable restraint on competition.

Finally, at an earlier stage, the Competition Commission of India (CCI) reportedly is investigating two drug patent settlements: one between Hoffmann-La Roche and Cipla on lung cancer drug Tarceva, and one between Merck and Glenmark Pharmaceuticals on diabetes drug Januvia.

16.5. CONCLUSION

The introduction of generic drugs dramatically lowers price, and agreements by which brands pay generics to delay entering the market threaten significant harms to competition. In the aftermath of the *Actavis* decision, the United States is working through issues including payment, the patent merits, rule of reason, pleading requirements, causation, and state law. Europe is increasingly paying attention to this behavior with monitoring reports and cases against companies. And several other nations are beginning to focus on the conduct. As we go forward, the attention paid to this important issue shows no signs of abating.

REFERENCES

AIPLA. 2013. *Report of the Economic Survey*.
Canadian Competition Bureau. 2014. Patent Litigation Settlement Agreements: A Canadian Perspective.
Carrier, Michael A. 2009. Unsettling Drug Patent Settlements: A Framework for Presumptive Illegality. *Michigan Law Review*, 108,37, 60–6.
　2014. Payment After Actavis. *Iowa Law Review*, 100, 7
　2015. Eight Reasons Why "No-Authorized-Generic" Promises Constitute Payment. *Rutgers University Law Review*, 67, 697, 706–7.
Edlin, Aaron, Scott Hemphill, Herbert Hovenkamp, and Carl Shapiro. 2013. Activating Actavis. *Antitrust*, 28, 16, 19.
　2015. The Actavis Inference: Theory and Practice. *Rutgers University Law Review*, 67, 585, 613.
European Commission. 2009. *Pharmaceutical Sector Inquiry Final Report* (European Commission Staff Working Paper).
　2010. *First Report on the Monitoring of Patent Settlements (period: mid 2008–end 2009)*.
　2011. *Second Report on the Monitoring of Patent Settlements (period: January–December 2010)*.
　2012. *3rd Report on the Monitoring of Patent Settlements (period: January–December 2011)*, 25 July.
　2013a. *Antitrust: Commission fines Lundbeck and other pharma companies for delaying market entry of generic medicine*, 19 June.

European Commission. 2013b. *4th Report on the Monitoring of Patent Settlements (period: January–December 2012)*, 9 December.
  2014. *Antitrust: Commission fines Servier and five generic companies for curbing entry of cheaper versions of cardiovascular medicine*, 9 July.
EC, *Commission decision of 19 June 2013, Case AT.39226 – Lundbeck*.
  Commission decision of 9 July 2014, Case AT.39612 – Perindopril (Servier).
FTC. 2000. *FTC Charges Drug Manufacturers with Stifling Competition in Two Prescription Drug Markets*.
  2009. *The Evolving IP Marketplace*.
  2011. *The Evolving IP Marketplace: Aligning Patent Notice and Remedies with Competition*.
  2007. *Agreements Filed with the Federal Trade Commission Under the Medicare Prescription Drug, Improvement, and Modernization Act of 2003: Summary of Agreements Filed in FY 2006*.
  2013. *Agreements Filed with the Federal Trade Commission Under the Medicare Prescription Drug, Improvement, and Modernization Act of 2003: Overview of Agreements Filed in FY 2012*.
Herman, Michael R. 2011. The Stay Dilemma: Examining Brand and Generic Incentives for Delaying the Resolution of Pharmaceutical Patent Litigation. *Columbia Law Review*, 111, 1788
Kobayashi, Bruce H., Joshua D. Wright, Douglas H. Ginsburg, and Joanna Tsai. 2015. Actavis and Multiple ANDA Entrants: Beyond the Temporary Duopoly. *Antitrust*, 29, 89.
Trubek, David M. *et al*. 1983. The Costs of Ordinary Litigation. *UCLA L. Rev.* 31, 72, 91.

Cases

American Sales Co. v. Warner Chilcott Co., No. 14-2071 (1st Cir. filed Oct. 14, 2014).
Asahi Glass Co. v. Pentech Pharm., Inc., 289 F. Supp. 2d 986, 994 (N.D. Ill. 2003).
FTC v. AbbVie Inc., 2015 WL 2114380 (E.D. Pa. May 6, 2015).
FTC v. Actavis, 133 S. Ct. 2223 (2013).
FTC v. Cephalon, 36 F. Supp. 3d 527, 537 (E.D. Pa. 2014).
FTC v. Watson Pharmaceuticals, 677 F.3d 1298, 1312 (11th Cir. 2012).
In re Aggrenox Antitrust Litig., 2015 WL 1311352 (D. Conn. Mar. 23, 2015).
In re Cardizem CD Antitrust Litigation, 332 F.3d 896 (6th Cir. 2003).
In re Cipro Cases I & II, 2015 WL 2125291 (Cal. May 7, 2015).
In re Ciprofloxacin Hydrochloride Antitrust Litigation, 544 F.3d 1323 (Fed. Cir. 2008).
In re Effexor XR Antitrust Litig., 2014 WL 4988410 (D.N.J. Oct. 6, 2014).
In re K-Dur Antitrust Litigation, 686 F.3d 197 (3d Cir. 2012), judgment vacated sub nom. *Merck & Co. v. Louisiana Wholesale Drug Co.*, 133 S. Ct. 2849 (2013).
In re Lipitor Antitrust Litig., 46 F. Supp. 3d 523, 542 (D.N.J. 2014).
In re Nexium (Esomeprazole) Antitrust Litig., 968 F. Supp. 2d 367 (D. Mass. 2013).
In re Nexium (Esomeprazole) Antitrust Litig., 42 F. Supp. 3d 231, 272 (D. Mass. 2014).
In re Niaspan Antitrust Litig., 42 F. Supp. 3d 735 (E.D. Pa. 2014).
In re Tamoxifen Citrate Antitrust Litigation, 466 F.3d 187 (2d Cir. 2006).
In re Wellbutrin XL Antitrust Litig., No. 08-cv-02431 (E.D. Pa. Jan. 17, 2014).
Jury Verdict in Favor of Defendants Against Plaintiffs Returned, *In re Nexium Antitrust Litig.*, No. 12-md-02409 (D. Mass. Dec. 8, 2014), ECF No. 1374.
King Drug Company of Florence v. Smithkline Beecham Corporation (Lamictal), 791 F.3d 388 (3d Cir. 2015).

King Drug Co. of Florence v. Cephalon, Inc., 2015 WL 356913, at *1 (E.D. Pa. Jan. 28, 2015).
Loestrin, 45 F. Supp.3d 180, 189–91, 193 (D.R.I. 2014), appeal docketed, *City of Providence, RI v. Warner Chilcott Co.*, No. 15-1250 (1st Cir. filed Feb. 25, 2015).
Mylan Pharms. v. FDA, 2005 WL 2411674 (N.D.W. Va. Sept. 29, 2005).
Schering-Plough Corp. v. FTC, 402 F.3d 1056 (11th Cir. 2005).
Second Amended Complaint for Injunctive and Other Equitable Relief, *FTC v. Watson Pharm., Inc.*, No. 1:09-cv-00955-TWT (N.D. Ga. May 28, 2009).
Statement of the Federal Trade Commission, FTC v. Cephalon, Inc., at 2 n. 7 (May 28, 2015).
Teva Pharm. Indus. v. Crawford, 410 F.3d 51, 55 (D.C. Cir. 2005).
Teva's Memorandum in Support of Directed Verdict Motion on Threshold Issue of a Large and Unexplained Payment, *In re: Nexium (Esomeprazole Magnesium) Antitrust Litig.*, Case 1:12-md-02409-WGY, at 12–13 (D. Mass. Nov. 10, 2014).
Time Ins. Co. v. AstraZeneca, 2014 WL 4933025 (E.D. Pa. Oct. 1, 2014).
United Food & Commercial Workers Local 1776 v. Teikoku Pharma USA (Lidoderm), 2014 WL 6465235 (N.D. Cal. Nov. 17, 2014).

17

Copyright Licensing and the EU Digital Single Market Strategy

Pablo Ibáñez Colomo

17.1. INTRODUCTION

There is no such thing as an EU-wide copyright regime. Titles remain national in scope. In spite of the efforts toward the harmonization of Member States' legislation,[1] the fragmentation of intellectual property protection along national borders is at odds with the logic of the internal market that the European Union is expected to establish.[2] In the current state of copyright law in the European Union, consumers are not necessarily able to subscribe to the streaming service of their choice and may not be able to access content from a Member State other than the one where it was subscribed. The creation of a borderless market for digital services is one of the priorities of the current European Commission (hereinafter, "Commission"), which took office in November 2014.[3] The so-called Digital Single Market Strategy (hereinafter, the "DSMS"), launched in May 2015, encompasses several legislative and non-legislative initiatives, which span the regulation of the media, e-commerce, and data protection.[4] The overarching aim is to ensure that EU citizens fully benefit from the possibilities opened up by the Internet and, in general, digital technologies.

Ensuring access to – and the provision of – copyright-protected content across borders, in particular via the Internet, has emerged as one of the key areas of action in the context of the DSMS. Geo-blocking (that is, the use of technologies to limit the accessibility of a website from certain locations) is the most obvious

---

[1] A list of some of the most representative legislative instruments harmonizing national copyright regimes is included in the List of References.
[2] Article 3 Treaty on European Union (TEU).
[3] Juncker (2014).
[4] European Commission (2015a). The document in which the DSMS is sketched will be hereinafter referred to as the 'DSMS Communication'.

obstacle limiting the choice and availability of online services.[5] If there is tension between geo-blocking technologies and the objectives of the OSMS this is due, first and foremost, to the fact that the copyright titles have a limited territorial scope. Content providers may not hold the rights to offer content across the whole of the European Union, but only in one or several Member States. Against this background, it is not surprising that, shortly after the DSMS was announced, the Commission launched a consultation concerning the review of the principles applying, inter alia, to the online transmission of copyright-protected content. It seeks views, in particular, on whether it would be appropriate to allow the copyright holder in one Member State to lawfully offer content across the whole of the European Union.[6]

Right holders' strategies contribute to the fragmentation of markets along national borders. They tend to license their rights to a single provider in each territory. The terms and conditions found in these exclusive licensing agreements are being closely scrutinized by the Commission on grounds that they may be in breach of EU competition law. It has been clear since the 1960s that agreements giving absolute (that is, "airtight") territorial protection to distributors within the European Union are in principle restrictive of competition by object under Article 101(1) of the Treaty on the Functioning of the European Union (TFEU).[7] In its 2011 ruling in *Murphy*, the Court of Justice of the European Union (hereinafter, the "Court" or the "ECJ") held that this line of case-law is also "fully applicable" to the cross-border provision of television services.[8] Following the judgment, and in parallel with the launch of the DSMS, the Commission initiated a Sector Inquiry[9] into e-commerce.[10] The Commissioner for Competition openly acknowledged that the purpose of the initiative is to explore the potential contribution that EU competition law could make to the policy objectives of the DSMS.[11] She also identified access to digital content as one of the areas of focus.

---

[5] *Id*. The Commission defines geo-blocking as a set of 'practices used for commercial reasons by online sellers that result in the denial of access to websites based in other Member States'. In the DSMS Communication, the Commission announced that it would 'make legislative proposals in the first half of 2016 to end unjustified geo-blocking'.

[6] European Commission (2015b).

[7] Stothers (2007).

[8] Joined Cases C-403/08 and C-429/08, *Football Association Premier League Ltd and Others v. QC Leisure and Others* and *Karen Murphy v. Media Protection Services Ltd* ("*Murphy*"), EU:C:2011:631.

[9] Commission Decision of 6 May initiating an inquiry into the e-commerce sector pursuant to Article 17 of Council Regulation (EC) No 1/2003. C(2015) 3026 final.

[10] Kjølbye, Aresu, and Stephanou (2015).

[11] Vestager (2015a). The Commissioner explained in her speech that "the inquiry initiative is closely linked to the overall digital strategy of the European Commission" and added that "[s]everal Commission departments are working on the Digital Single Market at the moment. Understanding and facilitating cross-border online commerce is an important part of the contribution of the competition department."

Shortly after the launch of the Sector Inquiry, in July 2015, the Commission sent a Statement of Objections to the "Big Six" Hollywood major studios and to Sky UK.[12] The authority has come to the preliminary conclusion that the licensing agreements concluded between the right holders and the pay TV operator provide for territorial restraints that are contrary to Article 101(1) TFEU. The Statement of Objections relates to the transmission of content both via satellite and online. The preliminary view of the Commission is that geo-blocking provisions limiting access to Sky UK's content outside of the territory covered by the license amount to a restriction of competition, and this insofar as they prevent the pay TV provider from responding to unsolicited requests from consumers based in Member States other than the UK and Ireland, for which Sky UK holds a license.

This chapter provides an overview of the principles that apply to the territorial licensing of copyright under Article 101(1) TFEU. It does so against the background of the DSMS. The purpose of the piece is to show how EU competition law is enforced by the Commission to address concerns that are, first and foremost, the consequence of the nature and scope of national copyright regimes. The chapter is organized as follows. Section 17.2 explains that, as a matter of principle, it is not anticompetitive to grant an exclusive territorial license to a content provider, even when it amounts to absolute territorial protection. Section 17.3 shows how this conclusion might change depending on the underlying regulatory context and the nature of the restrictions at stake. This analysis is conducted in light of the *Murphy* ruling of 2011, without which ongoing administrative action by the Commission cannot be understood. Finally, Section 17.4 addresses the two-prong strategy to ensure that copyright-protected content is accessible across borders. The objectives of the DSMS cannot be achieved through the reform of national copyright regimes alone. The enforcement of EU competition law is also necessary.

17.2. COPYRIGHT LICENSING UNDER EU COMPETITION LAW: PRINCIPLES

17.2.1. *The Exhaustion Doctrine and the Right of Communication to the Public*

Even though copyright is national in scope, the exhaustion (or "first-sale") doctrine applies at the level of the European Union (and the European Economic Area (EEA) at large). Accordingly, it is not possible for an author to control subsequent sales of a copy of her work within the European Union once it has been put on the market in that territory by her or with her consent.[13] The case-law of the ECJ has

---

[12] European Commission (2015c).
[13] See in particular Case 78/70, *Deutsche Grammophon Gesellschaft mbH v. Metro-SB-Großmärkte GmbH & Co. KG*, EU:C:1971:59; Case 16/74, *Centrafarm BV and Adriaan de Peijper v. Winthrop BV*, EU:C:1974:115; and Case 15/74, *Centrafarm BV and Adriaan de Peijper v. Sterling Drug Inc*, EU:C:1974:114.

clarified that the exhaustion doctrine applies to the exploitation of the copyright in some forms, but not in others. It is clear, for instance, that it applies to the distribution of the tangible copies of a work.[14] This principle is enshrined in the InfoSoc Directive, which provides for a right of distribution, which is deemed exhausted with the first sale within the European Union.[15]

The exhaustion doctrine does not apply when the work is shown on television or in a cinema theater. This principle stems from the ECJ ruling in *Coditel I*.[16] The Court held in that case that the right to require fees for any showing of a film is part of the essential function of the copyright.[17] As a result, the exercise of the right of communication to the public by a licensee is not an infringement of the Treaty provisions dealing with the freedom to provide services, even when it leads to the partitioning of the internal market along national borders.[18] The logic of *Coditel I* is captured in Article 3 of the InfoSoc Directive. This provision requires Member States to provide for a right of communication to the public (which comprises, in turn, a right of "making available").[19] Article 3(3) clarifies that this right is not exhausted by "any act of communication to the public or making available to the public."[20]

17.2.2. Implications for EU Competition Law

It has been mentioned in the introduction that agreements partitioning the internal market along national borders are in principle contrary to Article 101(1) TFEU. Thus, agreements that give absolute ("airtight") territorial protection to a distributor

---

[14] The extent to which the doctrine applies to non-tangible copies of works remains an open question in EU law. The Court has held that this is the case for software, Case C-128/11, *UsedSoft GmbH v. Oracle International Corp*, EU:C:2012:407. The exhaustion of rights subject to the InfoSoc Directive has been addressed by the Court in Case C-419/13, *Art & Allposters International BV v. Stichting Pictoright*, EU:C:2015:27, which can be (and has been) interpreted as suggesting that the exhaustion doctrine only applies to tangible objects. For a discussion, see Rosati (2015).

[15] Directive 2001/29/EC of the European Parliament and of the Council of 22 May 2001 on the harmonisation of certain aspects of copyright and related rights in the information society OJ (2001) L 167/10 (hereinafter, the "InfoSoc Directive"). Pursuant to Article 4(2), "[t]he distribution right shall not be exhausted within the Community in respect of the original or copies of the work, except where the first sale or other transfer of ownership in the Community of that object is made by the rightholder or with his consent."

[16] Case 62/79, *SA Compagnie générale pour la diffusion de la télévision, Coditel, and others v. Ciné Vog Films and others* ("Coditel I"), EU:C:1980:84.

[17] *Id.*, para. 14.

[18] *Id.*, para. 16.

[19] Pursuant to Article 3(1) of the InfoSoc Directive, "Member States shall provide authors with the exclusive right to authorise or prohibit any communication to the public of their works, by wire or wireless means, including the making available to the public of their works in such a way that members of the public may access them from a place and at a time individually chosen by them."

[20] Bently and Sherman (2014). Article 3(3) of the InfoSoc Directive provides that the right of communication to the public "shall not be exhausted by any act of communication to the public or making available to the public as set out in this Article."

are prima facie prohibited as restrictive of competition by their very nature.[21] The same is true of agreements that provide for an export prohibition or that have an equivalent effect.[22] As a matter of principle, the fact that the product covered by the agreement incorporates an intellectual property right does not make a difference in this regard. This seems clear since *Consten-Grundig*, in which this principle was laid down. That case concerned the distribution of products bearing a trademark. In spite of this fact, the Court found the agreement gave absolute territorial protection and thus was in breach of Article 101(1) TFEU. That agreements aimed at limiting parallel trade within the European Union are in principle prohibited has been confirmed in a long line of case-law, which includes rulings like *Miller International*[23] (concerning the distribution of records), *Nungesser*[24] (concerning seed protected by plant breeders' rights) and *Glaxo Spain* (about restraints aimed at limiting the parallel trade of patented drugs).

The judgments mentioned in the preceding paragraph concerned the distribution of goods incorporating intellectual property rights. However, the restraints introduced in the agreement fell outside the scope of the protection afforded by the relevant intellectual property regime. As already explained, the right of distribution only covers the first sale of a product incorporating a copyright, a patent, or a trademark. This fact explains why the Court held them to amount to a restriction of competition by object within the meaning of Article 101(1) TFEU. It concluded in all four cases that the agreements hindered trade between Member States without their being necessary to preserve the essential function of the relevant intellectual property right.

Where the right of distribution is not exhausted, on the other hand, an agreement that provides for absolute territorial protection and/or an export prohibition will not be found to restrict competition under Article 101(1) TFEU. In such a case, the agreement remains within the scope of the relevant intellectual property right. This conclusion stems clearly from *Micro Leader*, where the Court ruled that measures aimed at limiting the export into France of Microsoft is not contrary to Article 101(1) TFEU where the distributor is based in Canada.[25] Because the exhaustion doctrine only applies at the EU-EEA level, the supplier in this case was in a position to achieve by means of an agreement what it would have been able to achieve through the exercise of its intellectual property right.

---

[21] Joined Cases 56/64 and 58/64 *Établissements Consten S.à.R.L. and Grundig-Verkaufs-GmbH v Commission* ("*Consten-Grundig*"), EU:C:1966:41.

[22] For export prohibitions, see Case 19/77, *Miller International Schallplatten v. Commission* [EU: C:1978:19 ("*Miller International*"). For practices having an equivalent effect, see Joined Cases C-501/06 P, C-513/06 P, C-515/06 P and C-519/06 P, *GlaxoSmithKline Services Unlimited v. Commission* ("*Glaxo Spain*"), EU:C:2009:610.

[23] *Miller International.*

[24] Case 258/78, *LC Nungesser KG and Kurt Eisele v. Commission*, EU:C:1982:211.

[25] Case T-198/98, *Micro Leader Business v Commission*, EU:T:1999:341..

The same conclusion applies in the case of rights that are not subject to the exhaustion doctrine. In *Coditel II*,[26] the Court held that an agreement providing for absolute territorial protection is not contrary to Article 101(1) TFEU by its very nature if it concerns the right of communication to the public. As a result, and as a matter of principle, a licensee is in a position to invoke its rights against a cable operator importing a foreign signal that offers the film covered by the license.[27] An agreement of this nature is only contrary to Article 101(1) TFEU if it is shown to have restrictive effects in the context in which it is implemented.[28] In *Murphy*, the Court confirmed that the mere fact that the right holder has granted to a sole licensee the exclusive right to broadcast protected subject-matter from a Member State, and consequently to prohibit its transmission by others, during a specified period is not sufficient to justify the finding that such an agreement has an anti-competitive object.[29]

17.3. THE CABLE AND SATELLITE DIRECTIVE AND THE *MURPHY* CASE

17.3.1. *Copyright Aspects of Transmission Via Satellite*

The copyright aspects of transmissions via cable and satellite are harmonized at the EU level. The fundamental objective of the Cable and Satellite Directive[30] is to ease the transmission across borders of copyright-protected content. The mechanism that is relevant for the purposes of this chapter relates to the provision of television content via satellite. The potential of satellite transmissions to reach viewers based in more than one territory was obvious at the time when the Directive was adopted. The legislator had to make a choice about the regime to which these transmissions would be subject. One option was to lay down the principle that the act of communication to the public is deemed to take place in every Member State reached by the satellite. An alternative approach was to make the transmission subject to the law of the Member State from which it originates. The Directive opted for the latter, which is known as the "country of uplink" principle (or "country of origin" principle, as it will be referred to hereinafter).[31]

---

[26] Case 262/81, *Coditel SA, Compagnie générale pour la diffusion de la télévision, and others v. Ciné-Vog Films SA and others* ("*Coditel II*"), EU:C:1982:334.
[27] Anderman and Schmidt (2011).
[28] *Coditel II*, paras 14 and 17–19.
[29] *Murphy*, para. 137.
[30] Council Directive 93/83/EEC of 27 September 1993 on the coordination of certain rules concerning copyright and rights related to copyright applicable to satellite broadcasting and cable retransmission OJ (1993) L 248/15 (hereinafter, the "Cable and Satellite Directive").
[31] Article 1(2)(b) of the Cable and Satellite Directive provides that "the act of communication to the public by satellite occurs solely in the Member State where, under the control and responsibility of the broadcasting organization, the programme-carrying signals are introduced into an uninterrupted chain of communication leading to the satellite and down towards the earth."

Insofar as the satellite transmission is only deemed to take place in the "country of origin," a satellite operator based in one Member State (Member State A) is in a position to reach viewers based in another Member State (Member State B) without infringing the exclusive rights of the licensee based in the latter. Thus, the principle enshrined in the Directive allows two different television operators to offer the same content to the same viewers without this leading to a dispute similar to the one underlying the *Coditel* saga. On the other hand, it is clear from the Directive that the "country of origin" principle does not affect the possibility of limiting the exploitation of the rights by licensees.[32] In other words, nothing in the text prevents right holders from granting territorial licenses to different television operators.

In a report reviewing the application of the Cable and Satellite Directive, issued in 2002, the Commission observed that the generalization of encryption technologies in satellite transmissions was frustrating the emergence of a single market for television services and, more generally, the provision of television services across borders.[33] The expectation of the Commission was that the adoption of the "country of origin" principle would allow television operators to offer copyright-protected content to viewers based in other Member States, and that right holders would negotiate compensation corresponding to the potential audience reached by the satellite footprint.[34] This is not the reality that emerged in the years that followed the adoption of the Directive. With the rise of encryption technologies, right holders licensed content to a local broadcaster in each Member State, thereby leading to the partitioning of the internal market along national borders.[35] This is the technological and regulatory background against which the facts and the ruling in *Murphy* must be understood.

17.3.2. *The* Murphy *Case*

17.3.2.1. Facts and Ruling

The ECJ examined two closely related questions in *Murphy*. First, it assessed the compatibility with Article 56 TFEU[36] of national legislation prohibiting the import

---

[32] Recital 16 of the Directive reads: "Whereas the principle of contractual freedom on which this Directive is based will make it possible to continue limiting the exploitation of these rights, especially as far as certain technical means of transmission or certain language versions are concerned..."
[33] European Commission (2002).
[34] *Id.*: "Complete application of the principle of the Directive, which involves moving beyond a purely national territorial approach, should therefore be encouraged in order to allow the internal market to be a genuine market without internal frontiers for rightholders, operators and viewers alike. The Commission will therefore conduct research and engage in consultations, in particular with the various sectors concerned, in order to determine how to reconcile the different interests involved with the principle of the free movement of television services."
[35] *Id.*: "A trend is thus emerging whereby producers sell their programmes to broadcasting organisations on condition that satellite transmissions are encrypted so as to ensure that they cannot be received beyond national borders. This encryption enables producers to negotiate the sale of the same programmes with broadcasting organisations in other Member States."
[36] Article 56 TFEU deals with the freedom to provide services within the internal market.

of foreign decoding devices giving access to satellite television services. Secondly, it ascertained whether a set of contractual provisions prohibiting a satellite television operator from selling decoding devices outside of the territory covered by the exclusive license was compatible with Article 101 TFEU. The Court ruled that national legislation such as that at stake in the case amounts to a restriction on the freedom to provide services. More importantly, it took the view that it would not be possible to justify this restriction on grounds that the relevant content is protected by intellectual property rights.[37] As far as Article 101 TFEU is concerned, the Court concluded that obligations relating to the export of decoding devices amount, by their very nature (that is, "by object"), to a restriction of competition, and that this restriction cannot be justified under Article 101(3) TFEU.[38]

The activities of the television operator on which the export prohibition was imposed were subject to the Cable and Satellite Directive. This aspect of the case seems to have proved decisive in its outcome. When examining the compatibility of national legislation with Article 56 TFEU, the Court took the view that the factual scenario at stake in the case could be distinguished from the one examined in *Coditel I*. In the latter, the cable provider had not acquired the rights to offer the film in question in the Member State in which it operated. In *Murphy*, by contrast, the satellite provider had been authorized by the right holder to perform an act of communication to the public in the "country of origin." The fact that its services reached viewers based in another Member State did not amount to an infringement of the copyright in the "country of destination". In addition, and in line with what has been pointed out above, nothing in the Directive prevents right holders and television operators from taking into account, when calculating the compensation, the potential audience reached by the operators in Member States other than the one where the communication originates.[39]

The analysis of the contractual restrictions under Article 101(1) TFEU is interesting insofar as the Court clarified that *Coditel II* remains good law.[40] Accordingly, prohibiting television operators from transmitting the licensed content in territories allocated to other licensees is not in itself contrary to Article 101(1) TFEU. At the same time, the Court made it clear that it was not ruling on the compatibility of an exclusive territorial license as such, but on whether providing for "additional obligations" (such as a prohibition on export of decoding devices outside the territory covered by the license) amounts to a restriction of competition by object.[41] In this regard, it noted that restrictions such as the one at stake in the case are in principle contrary to Article 101(1) TFEU by their very nature, and that the parties

---

[37] *Murphy*, paras 106–21.
[38] *Id.*, paras 141–5.
[39] *Id.*, paras 112–13.
[40] *Id.*, para. 137.
[41] *Id.*, para. 141.

had not put forward any evidence relating to the "economic and legal context" of the agreement that would lead to a different conclusion.[42]

17.3.2.2. Open Questions

Commentators have struggled to make sense of the scope of the *Murphy* ruling. It is not surprising that some of them have interpreted some of its passages as suggesting that something akin to the exhaustion doctrine now applies to the right of communication to the public. It is not unreasonable to infer from the analysis of the Court a general principle according to which an operator having received a license to offer copyright-protected content in one Member State is entitled to offer the same content anywhere in the European Union.[43] The fact that the same content has been licensed to other operators in other Member States would not make a difference in this regard. According to this interpretation of *Murphy*, the right of communication to the public has now ceased to exist as an exclusive right to authorize or prohibit any communication to the public. It would merely exist as a right to receive remuneration. In this regard, the Court noted, moreover, that the said right cannot be equated with the right to receive the "highest possible remuneration," but only "appropriate" remuneration.[44]

It is also possible to interpret the ruling as applying narrowly to the facts of the case.[45] From this perspective, the fundamental reason why the Court ruled the way it did has to do with the fact that, in accordance with the principles of the Cable and Satellite Directive, the act of communication to the public is only deemed to take place in the "country of origin." In such circumstances, it would not be possible to justify, on intellectual property grounds, any acts aimed at limiting the ability of the television operator to reach its potential audience. Any such restraints would fall outside the scope of the copyright. Subsequent developments suggest that this narrow interpretation of the ruling is accurate in the sense that EU institutions do not appear to have understood *Murphy* as creating something akin to the exhaustion doctrine in relation to the right of communication to the public. It would seem from the legislative proposals advanced in the context of the DSMS that the Commission is of the view that it is necessary to amend copyright legislation before extending the logic of *Murphy* to the online transmission of content.

The implications of the ruling for Article 101 TFEU are equally difficult to interpret.[46] The Court appears to suggest in some passages that the exclusivity given to a television operator cannot go as far as to lead to absolute territorial protection, and this irrespective of the underlying economic and legal context. Accordingly, it would be possible for a right holder to prohibit television operators from engaging in

---

[42] *Id.*, para. 143.
[43] Batchelor and Jenkins (2012).
[44] *Murphy*, para. 108. This was in fact the outcome suggested by the Commission in *Coditel I* (Keeling 2004).
[45] Graf (2011); Doukas (2012).
[46] Alexiadis and Wood (2012); Ibáñez Colomo (2014).

active selling in the territories allocated to other licensees. Passive sales (that is, to respond to unsolicited orders by end-users based in other territories), on the other hand, should always be allowed.[47] This interpretation would be problematic for several reasons. In particular, it would contradict the very principle laid down in *Coditel II*. In that case (which was deemed to remain good law by the Court), the licensee was entitled to prevent the import of television signals by cable operators, and this insofar as the transmission of content via cable would amount to a breach of its exclusive rights to authorize or prohibit any communication to the public of the film in question. Against the background of that precedent, it would seem that the statement made by the Court in *Murphy* needs to be qualified. It would seem that an agreement providing for absolute territorial protection is only restrictive of competition by object where the transmission of content across borders would not amount to a copyright violation in the "country of destination."

There are indeed good reasons to believe that the ruling in *Murphy* cannot as such be extended without qualification beyond the specific facts of the case. It was clear before the judgment that agreements providing for absolute territorial protection are restrictive of competition by object only "in principle."[48] Thus, it is possible for the parties to explain why, in the economic and legal context in which the agreement is implemented, the agreement does not violate Article 101(1) TFEU by its very nature.[49] As pointed out above, the parties had not put forward any evidence in this sense. In line with the above, this fact suggests that, in a different regulatory context (and, in particular, where the "country of origin" principle would not apply) the Court would not have ruled in the same way. This, however, remains an open question at the time of writing. The boundaries of *Murphy* are being tested by the Commission in the context of the ongoing proceedings against Sky UK and the major studios. As will be explained in greater detail below, these proceedings are based on the idea that the principles set out by the Court in the judgment can be extended to online transmissions and thus to a different economic and legal context.

17.4. TOWARD A DIGITAL SINGLE MARKET: A TWO-PRONGED STRATEGY

17.4.1. *Copyright Reform*

17.4.1.1. The Review of EU Copyright Rules

European Union institutions have long been concerned about the fragmentation of national markets resulting from the limited territorial scope of intellectual property

---

[47] These are the principles that apply to the sale of tangible goods. See in this sense Guidelines on vertical restraints, OJ (2010) C 130/1.
[48] See in particular *Glaxo Spain*, para. 59.
[49] *Murphy*, para. 140.

rights. Prior to the launch of the DSMS, the Commission discussed the approaches that could be taken to address obstacles hindering the accessibility of copyright-protected content across borders. In the Green Paper on the online distribution of audiovisual works,[50] it explored different legislative options, including the creation of an EU-wide copyright title. In particular, it discussed the possibility of extending to the online world the "country of origin" principle enshrined in the Cable and Satellite Directive. The Green Paper, which was issued some months before the *Murphy* ruling, does not dispute the possibility for the right holders and content providers to limit the territorial scope of their licenses.[51]

17.4.1.2. Toward the "Country of Origin" Principle for Online Transmissions

In the aftermath of *Murphy*, DG Markt commissioned an economic analysis of the territorial aspects of the exploitation of copyright online.[52] The different options contemplated by the authors are useful to understand the strategy that the Commission would follow at a subsequent stage. The study discusses several policy options to promote the cross-border provision of online works. One of the options is based on the idea that, following *Murphy*, absolute territorial protection for content providers is contrary to Article 101 TFEU and thus unlawful. According to this interpretation of the case-law, the right holder would only be entitled to provide contractual protection against active sales from rivals based in other Member States. Thus, it would be possible for content providers to respond to unsolicited orders from end-users based in other Member States. In other words, it would not be possible for right holders to limit, by means of an agreement, passive sales.

The other options discussed by the authors involve a regulatory shift toward the "country of origin" principle, in line with the position sketched in the Green Paper. A first possibility is one that combines the adoption of the said "country of origin" principle with freedom of contract. Pursuant to this approach, the communication to the public – and any other acts protected by copyright – would be deemed to take place in the Member State where the communication originates. On the other hand, it would be possible for the right holder to limit by means of an agreement, inter alia, the territorial scope of the license. This approach would make it easier, but not indispensable, for the right holder to grant licenses covering the whole of the European Union (the so-called pan-EU licenses). Instead of licensing its content on a territory-by-territory basis, the whole territory of the European Union could be covered by means of a single transaction. The other two options contemplated by the authors involve curtailing the parties' freedom of contract, at least to some

---

[50] European Commission (2011).
[51] In this regard, the Green Paper is in line with the principles of the Cable and Satellite Directive, which, in its Preamble, clarified that the Directive is based on the principle of contractual freedom.
[52] Langus, Neven, and Poukens (2014).

extent. Thus, not only would the "country of origin" principle apply but contractual restrictions limiting the cross-border provision of services would be prohibited, at least to some degree. Under these approaches, content would always be licensed on a pan-European basis.

The study concludes that the adoption of these policy options would not yield any obvious benefits for end-users. For instance, the adoption of the "country of origin" principle, coupled with the ability to impose contractual restrictions, is deemed to be a source of potential benefits. This policy option would make it easier for expatriates to access content provided by an operator based in their home Member State. On the other hand, some of the efficiency gains resulting from the ability of the supplier to control the conditions of distribution could be lost. Overall, the authors fail to see a clear gain in departing from the status quo. The study is more critical toward models resulting in the outright abolition of territorial licensing. Any potential gains, the authors argued in the study, would not be sufficiently robust to justify the adoption of an approach that is understood to be very risky.[53]

17.4.1.3. The DSMS and the Consultation on the Reform of EU Copyright

In the DSMS Communication the Commission identified "unjustified geo-blocking" and "better access to digital content" as two of the areas requiring action. Geo-blocking is seen with concern not only for reasons that relate to the exploitation of copyright-protected content. Practices such as redirecting an end-user toward another website or limiting the use of payment systems based on geographic criteria (in particular, the country in which a credit card is issued) are also deemed to require intervention. The Commission announced that the aspects of the DSMS that do not relate to cross-border access to content will be addressed both by means of legislative proposals and through the enforcement of competition rules.[54] This is one of the key objectives of the Sector Inquiry into e-commerce launched by the Commission.[55]

Regarding more specifically copyright-protected content, the Commission announced, inter alia, legislative proposals to ensure the portability of lawfully acquired content across borders and, similarly, the accessibility of services provided by operators based in other Member States.[56] Subsequent developments give a more clear idea of the form that these proposals could take in practice. In August 2015, the Commission launched a consultation on the review of the Cable and Satellite Directive. In line with the position at which it had already hinted in the Green Paper, it sought views on whether it is appropriate to extend to the online world the

---

[53] In this regard, the Green Paper is in line with the principles of the Cable and Satellite Directive, which, in its Preamble, clarified that the Directive is based on the principle of contractual freedom.
[54] DSMS Communication, p. 6.
[55] Vestager (2015a).
[56] DSMS Communication, p. 8.

"country of origin" principle that applies to satellite transmissions.[57] This consultation would lead to two legislative initiatives. In December 2015, the Commission issued a proposal to allow subscribers to access online content when temporarily present in another Member State.[58] In September 2016, it advanced the idea of extending the "country of origin" principle to "ancillary online services", including online simulcasts.[59]

In December 2015, the Commission presented a proposal for a Regulation to ensure the cross-border portability of online content in the internal market.[60] This piece of legislation would only apply to subscribers to online services that are temporarily present in a Member State other than the Member State of residence. It would provide for an obligation on content providers to enable the portability of subscriptions across borders. The copyright implications would be addressed by means of a limited expansion of the "country of origin" principle. Thus, the provision of copyright-protected services is deemed to take place only in the country where the subscriber is a permanent resident, and this includes acts subject to the InfoSoc Directive.

17.4.2. Competition Law Intervention

17.4.2.1. From the Pay TV Investigation to the Statement of Objections

Competition law intervention addressing contractual restrictions limiting the cross-border provision of copyright-protected content pre-dates the launch of the DSMS. In January 2014, Commissioner Almunia announced the launch of an investigation into pay TV services.[61] The investigation was originally directed at five of the major Hollywood studios and the leading satellite pay TV operators in the five largest EU Member States. The declared purpose of the investigation was to determine whether there were clauses in the agreements giving "absolute territorial protection" to pay TV operators. Commissioner Almunia's statement is notable in that it clarified that the purpose of the investigation was not to question the award of licenses on a territorial basis, or to promote pan-European licensing. Its more modest scope was to determine whether there are contractual arrangements between right holders and

---

[57] See European Commission (2015b). A set of questions included in the consultation document relates to the "extension of the principle of country of origin." See https://ec.europa.eu/eusurvey/runner/SatCabReview2015.
[58] European Commission (2015d).
[59] European Commission (2016). Article 1(a) of the Proposal defines an "ancillary online service" as "an online service consisting in the provision to the public, by or under the control and responsibility of a broadcasting organisation, of radio or television programmes simultaneously with or for a defined period of time after their broadcast by the broadcasting organisation as well as of any material produced by or for the broadcasting organisation which is ancillary to such broadcast".
[60] European Commission (2015d).
[61] Almunia (2014).

content providers preventing the latter from responding to unsolicited requests from end-users based in other Member States and, by the same token, whether end-users are prevented from accessing the services from Member States other than the one where they subscribed to them.

The Statement of Objections sent in July 2015, when the DSMS had already been announced, clarified several aspects of the scope of the case, which was still ongoing when this chapter was being completed. In a sense, the Statement of Objections is narrower than the original investigation, insofar as it is confined to a single pay TV operator, Sky UK. In another sense, it is broader, as it is addressed to all six major Hollywood studios. The Commission has now clarified that the scope of the investigation covers both online and satellite transmissions. In line with what has been pointed out above, the investigation seems to be based on the idea that *Murphy* is relevant beyond the scope of the Cable and Satellite Directive. In other words, the Commission seems to be of the view that "absolute territorial protection" is contrary to Article 101(1) TFEU irrespective of whether the "country of origin" principle applies to the underlying communication to the public. In a similar vein, it now seems clear that the case is based on the idea that "geo-blocking" provisions are restrictive of competition within the meaning of Article 101(1) TFEU. More precisely, such provisions are considered to be "additional obligations" within the meaning of *Murphy*.

17.4.2.2. The Interplay Between Competition Law and the Underlying Regulatory Context

The relatively broad understanding of the *Murphy* ruling raises important issues about the interplay between competition law and intellectual property and more precisely about how much the former interferes with the latter. In the press release issued when the Statement of Objections was sent to the firms, the Commission openly acknowledged that competition law intervention addressing geo-blocking provisions may not be enough to ensure that copyright-protected content can be accessed and provided across borders. The removal of the said restrictions does not guarantee that content providers would be able to lawfully offer their services to end-users based in other Member States. Doing so may amount to a breach of copyright in the "country of destination." This point was openly acknowledged by Commissioner Vestager in a speech given shortly after the Statement of Objections was issued.[62]

The Commission is in fact careful not to claim that geo-blocking provisions prevent Sky UK from offering its services across borders. It merely argues that such provisions are a source of concern insofar as they do not allow Sky UK to decide "on commercial grounds" whether to offer its services outside the UK and Ireland. Such

---

[62] Vestager (2015b).

a decision by the content provider would have to consider different factors, including the regulatory framework, and in particular national copyright laws.[63] This clarification suggests, in line with the interpretation of the ruling suggested above, that the principles of *Murphy* cannot be extended as such to the transmission of content online. The Commission acknowledges that, where the "country of origin" principle does not apply, the provision of content across borders would not be lawful. If it is accepted that the "country of destination" applies to the online transmission of copyright-protected content,[64] then Sky UK would not be able to offer its services across borders in other Member States without the authorization of the licensee. As a result, competition law does not seem sufficient to achieve the objectives of the DSMS. It would be necessary, in addition, to reform and harmonize copyright systems in line with the proposals mentioned above. This is something that the Commissioner pointed out in the speech mentioned above.[65]

---

[63] The issue was expressed in the following terms in the press release (European Commission 2015c): "Without these restrictions, Sky UK would be free to decide on commercial grounds whether to sell its pay-TV services to such consumers requesting access to its services, taking into account the regulatory framework including, as regards online pay-TV services, the relevant national copyright laws."

[64] Competition law enforcement before the Statement of Objections was based on the idea that the "country of destination" principle applies to the exploitation of content online. See in particular Commission Decision 2003/300/EC (Case No COMP/C2/38.014 – *IFPI "Simulcasting"*) OJ (2003) L 107/58. In para. 24 of the decision, the Commission acknowledged that the "country of destination" principle "appears to reflect the current legal situation in copyright law." Competition lawyers typically assume, in the same vein, that the "country of destination" principle applies to the online transmission of content (Graf 2011; Alexiadis and Wood 2012; Batchelor and Montani 2015). There is support for this interpretation. Since the InfoSoc Directive does not address the issue, and since copyright remains national in scope, it would seem that the issue remains subject to the legislation of each Member State (Triaille et al. 2013). In the same vein, it seems clear that, where the "country of origin" principle has been introduced in national legislation, the endorsement of the principle has been made explicit, as is true of the provisions transposing the Cable and Satellite Directive into national law (Shapiro 2011). It should be mentioned, in addition, that there are technical aspects that militate against the application of the "country of origin" principle to online transmissions. This is in part due to the fact that it is not always easy to identify the "country of origin" in this context (Aplin 2005). In addition, activities such as streaming involve acts that could qualify not only as a communication to the public within the meaning of Article 3 of the Directive, but also as reproductions within the meaning of Article 2 (Batchelor and Jenkins 2012). Similarly, some services may qualify as a rental within the meaning of the relevant Directive. On the other hand, there is case law on the online exploitation of protected works online that suggests that the relevant criterion is whether the website "targets" a particular country or group of customers. From this perspective, the communication to the public would only take place in the country of destination (Ginsburg 2014; Depreeuw and Hubin 2014a). This has led some authors to propose the targeting approach to identify where the "communication to the public" within the meaning of the Directive takes place (Depreeuw and Hubin 2014b).

[65] The Commissioner (Vestager 2015b) explained the following in her speech: "If such clauses did not exist, broadcasters would no longer be contractually prevented from responding to unsolicited requests coming from consumers from other countries. At the same time, broadcasters also have to take account of the applicable regulatory framework, such as, for online services, relevant national copyright laws. These aspects may need to be tackled by changes to copyright rules, for example as part of the copyright initiatives included in the Commission's Digital Single Market Strategy."

17.4.2.3. Challenges Raised from a Competition Law Standpoint

The expansion of *Murphy* to online transmissions is not uncontroversial from an EU competition law perspective. By accepting that the mere removal of "geo-blocking" provisions would not allow for the lawful provision of copyright-protected content across borders, the Commission is implicitly conceding that operators based in different Member States are not potential competitors, at least not so as far as online transmissions are concerned. The regulatory framework – and more precisely the territorial scope of copyright as enshrined in the InfoSoc Directive – would not allow for the emergence of intrabrand competition even in the absence of the contractual restraints challenged by the Commission. The question raised in this regard is therefore whether "geo-blocking" provisions can be said to amount to a restriction of competition in the technological and legal context of which they are part.

The position advanced by the Commission seems to be difficult to reconcile with some well-established principles in the case-law. It has been clear since *Société Technique Minière* that an agreement is only caught by Article 101(1) TFEU where it restricts (actual or potential) competition that would otherwise have existed. For instance, a distribution agreement cannot be said to be anticompetitive if it would not have been possible for the supplier to enter the relevant market in its absence. The Commission has expressed the view in the past that this principle applies even when an agreement gives absolute territorial protection to a distributor. In the Guidelines of vertical restraints, it takes the view that such agreements fall outside the scope of Article 101(1) TFEU altogether where they are necessary for the supplier to enter the market, and this, for a period of two years.[66] Against this background, it is not obvious to see how the Commission will be able to justify "geo-blocking" provisions restricting competition that would otherwise have existed or, alternatively, why the said provisions are in breach of Article 101(1) TFEU even though no lawful cross-border provision would have existed in their absence.

In a similar vein, the Commission would have to explain how administrative action in this case can be reconciled with the *Coditel* saga. In *Coditel I*, the Court held that the right of communication to the public is not subject to the exhaustion doctrine. As a result, a licensee based in one Member State would be entitled to prevent a cross-border communication to the public without this leading to a breach of Article 101(1) TFEU. In *Murphy*, the Court unambiguously held that an agreement whereby an operator is allowed to prohibit the transmission of the content subject to the licensee does not amount in itself to a restriction of competition by object.[67] It remains to be seen how "geo-blocking" provisions are distinguished from the mere prohibition of the transmission by the

---

[66] Guidelines on vertical restraints, paras 60–4.
[67] *Murphy*, para. 137.

Commission, and why it qualifies them as "additional obligations" within the meaning of *Murphy*.

17.5. CONCLUSIONS

It is not unusual to read that EU competition law is enforced to achieve regulatory objectives.[68] These claims are not particularly persuasive. It is difficult, if not impossible, to draw a clear line between the proper and improper interpretation and application of open-textured provisions like Article 101 TFEU, in particular considering the broad range of factual scenarios and industries that are potentially subject to scrutiny. Any boundaries defining the appropriate reach of EU competition law are likely to prove artificial or to be based on preconceptions about the nature and scope of the discipline. This chapter does not address these questions. Instead, it provides a concrete example of how EU competition law may be applied alongside legislation to contribute to a particular policy outcome. The Commission has been open about the fact that it is exploring how the enforcement of Article 101 can contribute to the objectives of the DSMS.

The policy aims of the DSMS seem clear. The Commission seeks, inter alia, to ensure that digital content can be accessed and provided across borders. This objective cannot be meaningfully achieved through the enforcement of EU competition law alone. As conceded by Commissioner Vestager, it appears that it is necessary to harmonize national copyright regimes to extend the "country of origin" principle to online transmissions. Only in such a way would it be possible to lawfully transmit content across borders. The converse is also true. The reform of copyright rules would be insufficient if right holders and operators continue to include "geo-blocking" provisions in their licensing agreements. Declaring that such provisions are unlawful under Article 101(1) TFEU therefore looks like an indispensable ingredient of the DSMS. This symbiotic relationship between copyright and EU competition law is interesting in itself.

The second issue discussed in this chapter relates to the impact that this relationship may have on the substance of Article 101 TFEU. It would seem that action by the Commission is grounded on a relatively expansive interpretation of the provision. It remains to be seen whether EU courts will endorse it.

REFERENCES

Literature

Alexiadis, Peter and David Wood. 2012. Free Market 1: Copyright 0 – UK Premier League Loses Away from Home. *Utilities Law Review*, 18, 243–9.

Almunia, Joaquin. 2014. Statement on opening of investigation into Pay TV services. Brussels, January 13.

---

[68] Larouche (2000); Dunne (2014) and (2015); Drexl and Di Porto (2015).

Anderman, Steven and Hedvig Schmidt. 2011. *EU Competition Law and Intellectual Property Rights: The Regulation of Innovation*. Oxford University Press.

Aplin, Tanya. 2005. *Copyright in the Digital Society*. Oxford: Hart.

Batchelor, Bill and Tom Jenkins. 2012. FA Premier League: The Broader Implications for Copyright Licensing. *European Competition Law Review*, 33, 157–64.

Batchelor, Bill and Luca Montani. 2015. Exhaustion, Essential Subject Matter and Other CJEU Judicial Tools to Update Copyright for an Online Economy. *Journal of Intellectual Property & Practice*, 10, 591–600.

Bently, Lionel and Brad Sherman. 2014. *Intellectual Property Law*. 4th edn. Oxford University Press.

Depreeuw, Sari and Jean-Benoit Hubin. 2014a. Of Availability, Targeting and Accessibility: Online Copyright Infringements and Jurisdiction in the EU. *Journal of Intellectual Property Law & Practice*, 9, 750–64.

2014b. Study on the Making Available Right and its Relationship with the Reproduction Right in Cross-Border Digital Transmissions. De Wolf & Partners (commissioned by DG Markt).

Doukas, Dimitrios. 2012. The Sky Is Not the (Only) Limit: Sports Broadcasting Without Frontiers and the Court of Justice: Comment on "Murphy". *European Law Review*, 37, 605–26.

Drexl, Josef and Fabiana Di Porto. 2015. *Competition Law as Regulation?* Cheltenham: Edward Elgar.

Dunne, Niamh. 2014. Commitment Decisions in EU Competition Law. *Journal of Competition Law & Economics*, 10, 399–444.

2015. *Competition Law and Economic Regulation: Making and Managing Markets*. Cambridge University Press.

European Commission. 2016. Proposal for a Directive of the European Parliament and of the Council on copyright in the Digital Single Market. COM(2016) 593 final. 2016/0280 (COD).

Ginsburg, Jane C. 2014. Where Does the Act of "Making Available" Occur? In Andrej Savin and Jan Trzaskowski (eds.), *Research Handbook on EU Internet Law*. Cheltenham: Edward Elgar, pp. 191–210.

Graf, Thomas. 2011. The Court of Justice Speaks on Licensing of Satellite Broadcasting. Kluwer Competition Law Blog. Available at http://kluwercompetitionlawblog.com.

Ibáñez Colomo, Pablo. 2014. The Commission Investigation into Pay TV Services: Open Questions. *Journal of European Competition Law & Practice*, 5, 531–41.

Juncker, Jean-Claude. 2014. A New Start for Europe: My Agenda for Jobs, Growth, Fairness and Democratic Change: Political Guidelines for the next European Commission. Opening Statement in the European Parliament Plenary Session. Available at http://ec.europa.eu/priorities/docs/pg_en.pdf#page=7.

Keeling, David T. 2004. *Intellectual Property Rights in EU Law. Volume 1 – Free Movement and Competition Law*. Oxford University Press.

Kjølbye, Lars, Alessio Aresu, and Sophia Stephanou. 2015. The Commission's E-Commerce Sector Inquiry: Analysis of Legal Issues and Suggested Practical Approach. *Journal of European Competition Law & Practice*, 6, 465–76.

Langus, Gregor, Damien Neven, and Sophie Poukens. 2014. Economic Analysis of the Territoriality of the Making Available Right in the EU. Charles River Associates (commissioned by DG Markt).

Larouche, Pierre. 2000. *Competition Law and Regulation in European Telecommunications*. Oxford: Hart.

Rosati, Eleonora. 2015. Online Copyright Exhaustion in a Post-Allposters World. *Journal of Intellectual Property Law & Practice*, 10, 673–81.

Shapiro, Ted. 2011. Directive 2001/29/EC on Copyright in the Information Society. In Brigitte Lindner and Ted Shapiro (eds.), *Copyright in the Information Society: A Guide to National Implementation of the European Directive*. Cheltenham: Edward Elgar, pp. 27–56.
Stothers, Christopher. 2007. *Parallel Trade in Europe: Intellectual Property, Competition and Regulatory Law*. Oxford: Hart.
Triaille, Jean Paul et al. 2013. Study on the Application of Directive 2001/29/EC on Copyright and Related Rights in the Information Society (The "InfoSoc Directive"). De Wolf & Partners and CRIDS (commissioned by DG Markt).
Vestager, Margrethe. 2015a. *Competition Policy for the Digital Single Market: Focus on e-commerce*. Berlin, March 26.
2015b. Intellectual Property and Competition. 19th IBA Competition Conference, Florence, September 11.

Legislation

Council Directive 93/83/EEC of 27 September 1993 on the coordination of certain rules concerning copyright and rights related to copyright applicable to satellite broadcasting and cable retransmission OJ (1993) L 248/15.
Directive 2001/29/EC of the European Parliament and of the Council of 22 May 2001 on the harmonisation of certain aspects of copyright and related rights in the information society OJ (2001) L 167/10.
Directive 2004/48/EC of the European Parliament and of the Council of 29 April 2004 on the enforcement of intellectual property rights OJ (2004) L 195/16.
Directive 2006/115/EC of the European Parliament and of the Council of 12 December 2006 on rental right and lending right and on certain rights related to copyright in the field of intellectual property OJ (2006) L 376/28.
Directive 2009/24/EC of the European Parliament and of the Council of 23 April 2009 on the legal protection of computer programs OJ (2009) L 111/16.
Directive 2014/26/EU of the European Parliament and of the Council of 26 February 2014 on collective management of copyright and related rights and multi-territorial licensing of rights in musical works for online use in the internal market OJ (2014) L 84/72.

Official documents

European Commission. 2002. Report from the European Commission on the application of Council Directive 93/83/EEC on the coordination of certain rules concerning copyright and rights related to copyright applicable to satellite broadcasting and cable retransmission. COM(2002) 430 final.
European Commission. 2011. Green Paper on the online distribution of audiovisual works in the European Union: opportunities and challenges towards a digital single market. COM (2011) 427 final.
European Commission. 2015a. A Digital Single Market Strategy for Europe. COM(2015) 192 final.
European Commission. 2015b. EU seeks views on the Satellite and Cable Directive. Available at https://ec.europa.eu/digital-agenda/.
European Commission. 2015c. Antitrust: Commission sends Statement of Objections on cross-border provision of pay-TV services.
European Commission. 2015d. Proposal for a Regulation of the European Parliament and the Council on ensuring the cross-border portability of online content services in the internal market. COM(2015) 627 final.

18

Patent Pools and Related Technology Sharing

Erik Hovenkamp and Herbert Hovenkamp

18.1. INTRODUCTION

A patent "pool" is an arrangement under which patent holders in a common technology or market commit their patents to a single holder, who then licenses them out to the original patentees and perhaps to outsiders. Structures vary, however, and may involve simple cross-licensing rather than a separate holder.[1] Licensors in a pool can include both practicing and nonpracticing entities. Licensees are of course practicing entities. The copyright blanket license, which has provoked some antitrust litigation,[2] is not really a pool because the artist members do little licensing to one another. Rather, artists and other copyright holders grant a nonexclusive license to a central clearinghouse, which then gives a blanket license to the entire portfolio to outside licensees such as radio stations. In addition there are several open source licensing arrangements, which largely involve agreements to share royalty-free access to copyrighted works.[3]

The payoffs to pooling include both revenue earned as a licensor, and technology acquired by the pool member as licensee. Public effects can also be significant. For example, technology sharing of complementary patents can improve product quality and variety. In some information technology markets, pools can prevent patents from becoming a costly obstacle to innovation by clearing channels of technology transfer. For "essential" patents, or patents incorporated into networks by a standard setting organization (SSO), a pool might enable competitive production by giving multiple firms access to the needed technology.[4] Pools can also be used to fix product prices or reduce output, however. Further, they can facilitate concerted

---

[1] Hovenkamp, *et al.* (2015); Mattioli (2014); Lerner, Strojwas, and Tirole (2007).
[2] E.g., *Broadcast Music v. Columbia Broadcasting Sys.*, 441 U.S. 1 (1979).
[3] One prominent example is the Android operating system. See https://source.android.com/source/licenses.html.
[4] Lerner and Tirole (2004).

refusals to deal, or boycotts directed against competing technologies outside the pool.[5] In sum, the private and social benefits to pooling can be substantial, but some threats to competition cannot be ignored.

To illustrate, suppose that one toaster maker has invented a device that detects accurately when bread is toasted just enough. Another maker has invented a device that senses how wide the bread is and moves the heating elements just the right distance away. These devices are complements in most scenarios. A toaster might have neither of them, either one, or both. Further, if customers value both, having both would make a toaster more valuable.

The two toaster makers might do one or more of the following: (1) they might pool or "cross-license," enabling both manufacturers to make toasters including both features. (2) They might additionally agree to issue licenses to other toaster makers, thus enabling third parties to make toasters with both features as well; in so doing, they might jointly agree on the price of a joint license; or they might agree to specify the price at which toasters produced under this joint license agreement are sold, or the number that can be made under the license. (3) They might make no agreement at all about licensing others, thus leaving each firm to decide whether to license out its particular feature. Or (4) they might agree with each other that neither will license its feature to third parties. While technology sharing between the two manufacturers is efficient, creating a more desirable product or lowering production costs, the various limitations on price or quantity or the limitations on third-party production might all pose competitive threats. Further, there might be less restrictive alternatives that can achieve similar goals but without the competitive injury. Most antitrust challenges to patent pools require analysis under the rule of reason, which requires proof of market power, or the absence of competitive alternatives, and anticompetitive effects.[6]

Historically, most pools involved situations where each patent was presumed to make a significant contribution to the final product. More recently, particularly in information technologies, individual patents have grown much smaller in relation to product value. Some products such as digital video or cell phones now arguably embody thousands of patents. In addition, much of the current debate over poor patent quality is focused on these areas.

18.2. TRADITIONAL CONSIDERATIONS – COMPLEMENTS VS. SUBSTITUTES

A traditional justification for patent pools is that they facilitate improved products by uniting complements, which are two things that are more valuable when used

---

[5] E.g., *United States v. Singer Mfg. Co.*, 374 U.S. 174 (1963).
[6] Areeda and Hovenkamp (2010).

together.[7] Sharing of complementary patents means that licensees can then employ all the patents in their product, rather than creating silos in which each manufacturer incorporates only its own patented features. In the toaster example, customers can buy toasters with both patented features. By contrast, if the patents are substitutes – i.e., if each one performs a version of the same thing – then manufacturers will ordinarily use one of the alternatives but not more. In that case cross-licensing of the patents is more suspicious and may facilitate price fixing.[8]

The substitutes/complements distinction is helpful for understanding the basic benefits and costs of cross-licensing. For example, it has been shown that patent pools increase welfare when included patents are perfect complements (i.e., they have no value unless they are used together), and decrease welfare when they are perfect substitutes (i.e., one is as good as the other but no one needs both).[9]

18.2.1. Avoiding Royalty Stacking and Holdup

"Stacking" occurs when multiple sellers of complementary or vertically related products each have some market power but are unable to coordinate their output. The result is double marginalization.[10] For example, if two patentees who each have market power should license individually, the sum of the two license fees will be greater than if they joined together and coordinated their price. Under traditional assumptions, double marginalization not only results in lower output and higher prices for consumers, it also injures sellers who are unable to coordinate their output. That is, the resulting price is higher than the monopoly profit-maximizing price. Pooling addresses this problem by permitting licensor participants to coordinate their prices and output.

The holdup problem is related, and occurs when licensing of essential technology occurs sequentially. For example, once a manufacturer has settled on technology A, its range of subsequent technology decisions narrows, and this may give the owners of a secondary technology power to charge more. The problem resembles that of the highway builder planning a route. Before any construction begins the land market is competitive. As soon as the builder makes a commitment to a certain route and starts building, however, the range of remaining options narrows. As a result the price can be expected to rise. A more competitive solution would be for the builder to "package" parcels of land along alternative routes in advance and then let the landowners bid against each other to sell the entire package. Patent pooling accomplishes the same thing by permitting packages of complementary technologies that make up a product to be sold simultaneously.

---

[7] See, e.g., *Standard Oil Co. v. United States*, 283 U.S. 163 (1931).
[8] Gilbert (2004); Santore, McKee, and Bjornstad (2010).
[9] Shapiro (2001).
[10] Heller and Eisenberg (1998); Lemley and Shapiro (2007); Hovenkamp and Hovenkamp (2015).

18.2.2. "Blocking" Patents

Two patents are said to be "blocking" when practice of one requires infringement (or a license) of the other. Blocking applies to individual patent claims, rather than the patent as a whole. Unlicensed practice of only one claim constitutes infringement. In a two-way block neither patent can be practiced without a license to the other. In the more common one-way block, patent A cannot be practiced without a license to B, but B can be practiced without a license to A.[11] So-called "improvement" patents are frequently one-way blocking patents.[12] That is, if B is an improvement patent on A, then the only way to make use of the B improvement would be to have a license to A as well. However, the owner of original A can still practice it without a license to B. The courts all agree that mandatory package licensing of blocking patents cannot be unlawful.[13] It follows that two separate owners of blocking patents should be able to pool them and insist on licensing them together. The *Line Material* case involved such a situation. The Supreme Court condemned the pool, but not because the members pooled. Rather, they additionally agreed to stipulate the price at which the licensed product could be sold.[14]

"Blocking" patents are necessarily complements with respect to blocking claims, but in other ways they may function as substitutes. A good, litigated example is the *Philips* decision, which involved a pool of technologies for rewritable compact discs.[15] One of the patents at issue employed an analog method for locating the stylus on the disc, while another patent employed a digital method. A manufacturer would use one of these technologies or the other, but not both. However, the analog method wrote on at least one claim in the digital patent, meaning that one could not practice the analog patent without licensing the digital patent. That is, the patents functioned as both substitutes and as one-way complements with respect to the blocking claim.

As the *Philips* case suggests, blocking patents can create a special case of patent complementarity. Two patented *inventions* may appear to be substitutes for most purposes, but a single conflicting claim can turn the patents into complements, at least for devices that practice that particular claim. Courts have observed that a large

---

[11] For a decision relying on the distinction, see *Carpet Seaming Tape Licensing corp. v. Best Seam, Inc.*, 616 F.2d 1133, 1138 (9th Cir. 1980).

[12] Lemley (1997).

[13] E.g., *Standard Oil Co. v. United States*, 283 U.S. 163 (1931); *International Mfg. Co. v. Landon, Inc.*, 336 F.2d 723 (9th Cir. 1964); *Int'l Mfg. Co. v. Landon, Inc.*, 336 F.2d 723, 729 (9th Cir. 1964) ("No commercially feasible device could be manufactured under one of the patents without infringing the other").

[14] *United States v. Line Material*, 333 U.S. 287 (1948).

[15] *U.S. Philips corp. v. ITC*, 424 F.3d 1179, 1193 (Fed. Cir. 2005). Bohannan and Hovenkamp (2011); Bohannan and Hovenkamp (2012).

information technologies pool might contain hundreds of potentially blocking relationships that are unlikely to be identified without costly and time-consuming litigation.[16]

18.2.3. Tying and "Unwanted" Tied Products

The patents in a pool are typically offered to outside licensees as a bundle. They are either not licensed separately or else separate licensing costs more per patent. The antitrust tying problem arises when a licensee claims that it can obtain the desired patents (the "tying" product) only by taking the remainder of the bundle (the "tied" product). Tying complaints are most likely when licenses from patentees into the pool are exclusive, meaning that outsiders cannot obtain a license except through the pool. If the licenses are nonexclusive then licensees might be able to avoid the pool altogether.

The antitrust law of tying requires a tie of "separate products," market power in the tying product, and some kind of harmful effect on competition.[17] An unlawful antitrust tie can also be found when a seller gives a lower price to someone who takes two products together than to those who make separate purchases.[18]

The general rule for establishing competitive harm from tying is that the tie must foreclose, or exclude, competition, typically in the tied product market. Most antitrust challenges to patent pool ties allege that the licensee really wants a smaller set of patents than the pool offers. He does not want the tied patents *at all*, from any seller.[19] But the purpose of the antitrust laws is to promote competition, not merely to regulate the size of packages, and thus these challenges typically fail. For example, it is not unlawful for a land owner to insist on selling her 100 acre farm as a single parcel rather than cutting out a single acre at a particular buyer's insistence.

The unwanted tied product rationale, however, also suggests that the patents in the pool are not all complements, at least not for this particular purchaser. If the patents were substitutes the licensee would want only one of them, and the others would clearly be "unwanted." If the patents were complements for that particular licensee, however, the licensee would want both.

An important qualifier is that although the licensee may want both of them he may not want both from the same seller. A bundle can sometimes violate the antitrust laws if the licensee wants complementary technology, but from a different seller. Suppose that a pool includes patents A and B, which are complements.

---

[16] *Nero AG v. MPEG LA, L.L.C.*, 2010 WL 4366448 (C.D. Cal. Sep. 14, 2010).
[17] Hovenkamp (2015b).
[18] *Collins Inkjet Corp. v. Eastman Kodak Co.*, 781 F.3d 264 (6th Cir. 2015), cert. petition filed (granting preliminary injunction).
[19] E.g., *Nero AG v. MPEG LA, L.L.C.*, 2010 WL 4366448 (C.D. Cal. Sep. 14, 2010).

However, an outside patentee has also created patented technology B', which performs the same function as B. In this case the pool can exclude B' by forcing the purchaser to take A and B together. To be sure, he might be able to license B' separately, but his incentive is lessened to the extent that the pool license already covers B. It effectively forces the licensor of B' to compete with a price of zero.[20] In this case even the pooling of complementary patents can cause competitive harm.

18.2.4. Limitations on the Complement/Substitute Distinction

The substitute/complement distinction has a robust history and is a helpful way of thinking about competitive problems in traditional patent pools. In many of the pools commonly seen today, however, the distinction is less important. Intermediate cases between perfect complements and perfect substitutes are far more numerous and also more complex. For example, patents may block one another in some uses but not others and even patents that are predominantly substitutes can have blocking claims, requiring that they be used together.[21]

The subject/complement relationship may depend not merely on the patents but also on the nature of the licensees' products. For example, the MPEG-LA pool licenses thousands of digital video patents to a wide variety of manufacturers of cell phones (both traditional and smart phones), cameras, televisions, computer displays, digital storage devices, video and other image editing and processing software, and so on. Some of these products both generate and display video content. Some do only one and a few do neither. Not all licensees from MPEG-LA use every patent or perhaps even more than a small number. Some may incorporate both of two particular patents; others may incorporate one but not the other. As a result, for one manufacturer two patents may operate as complements while for another they are either substitutes or unrelated.

Further, the products themselves may function as either complements or substitutes. For example, an operating system and a web browser are complementary in the sense that each performs some computer functions that the other cannot, and consumers often want both. On the other hand, consumers are increasingly able to perform tasks and run applications over the web, and this common ground makes them substitutable in some important respects. For example, a smartphone both takes and displays video, making it both a complement and a substitute for a large digital display.

---

[20] *Grid Sys. Corp. v. Texas Instruments, Inc.*, 771 F. Supp. 1033 (N.D. Cal. 1991) (sustaining antitrust complaint on this theory).
[21] Bohannan and Hovenkamp (2012).

18.3. REDUCING COSTS OF BOUNDARY DELINEATION AND DEFENSE

A more robust explanation for pooling in many markets comes out of the economics of transaction costs, which emphasizes the role of limited information and the costs of obtaining it, as well as uncertainty in bargaining and sharing.[22] Pooling is an efficient solution to problems of technology development and transfer when determining patents' validity or identifying their boundaries is costly.

While manufacturers typically know their own products well, they are often uncertain about how many patents their products write on and, if so, how many of those patents are valid. For example, very extensive litigation in the market for cellular phones and related hand-held electronic devices reveals a great deal of ambiguity about scope, narrow claim constructions, and high invalidity rates. By one estimate 90 percent of the claims in this market fail.[23] By another estimate the number of patents relevant to smartphone technology exceeds 250,000.[24]

In such cases the cost of identifying and defending boundaries can be higher than the cost of simply sharing access. Traditional resource pools often involve similar problems. These include fisheries, irrigation, grazing rights, and other low-technology alternatives that have been subject to private, shared production for centuries.[25] What these markets share in common is that the cost of sharing plus enforcing access and contribution rules is less than the cost of creating, identifying, and enforcing individual boundaries.

For example, common fisheries are created when many fishermen have the right to fish in certain waters, but the stock needs to be replenished and overfishing prevented. A traditional private property regime would address these problems by giving each fisherman an exclusive patch of water protected by a defensible boundary. If the fisherman fails to restock, his own yield suffers, as it will if he over-fishes. The advantage of private property rights is that they create proper incentives by internalizing both the costs and payoffs of exploiting a resource.

In this case, however, the costs of dividing the pool into many individual parcels would be catastrophically high. First the group would have to decide on boundaries, then they would have to build fences and maintain them against both human and ichthyan violators, perhaps with devastating consequences for the yield. A better approach is to give each rights-holding fisherman managed access to the entire pool, with defined catch limits and an obligation to contribute to maintenance.[26] This is simply a special case of Ronald Coase's *Nature of the Firm*, which showed that a

---

[22] Williamson (1975); Hovenkamp (2010); see also Orr (2013).
[23] Muller (2014).
[24] See www.techdirt.com/blog/innovation/articles/20121017/10480520734/there-are-250000-active-patents-that-impact-smartphones-representing-one-six-active-patents-today.shtml.
[25] Ostrom (1990).
[26] Bohannan and Hovenkamp (2012) at 325–64.

business firm maximizes profits by choosing as between more and less profitable ways of accomplishing something.[27] In this case the payoff to sharing is greater than the payoff to individual rights of exclusion. As Elinor Ostrom observed in her studies of common pool resources, the use of such commons is product (technology) specific. For example, in some regions farmers maintain individual parcels for their crops while using a commons for grazing.[28]

The patent pool gives members instant and indemnified access to all patents in the pool. The pool owners need not worry about which patents the licensee may be practicing, and the licensee need not worry about whether or not it is infringing any particular patent in the pool, because its pool license grants it access to all. Of course the solution is incomplete to the extent that a user might infringe one or more patents that have not been licensed to the pool.

Gilbert recommends that antitrust authorities concentrate their enforcement activities against pools on weak patents.[29] For many traditional pools that is sound advice. In information technologies pools with thousands of patents, however, it is probably not worth the enforcer's time to investigate every patent in a pool, except perhaps for identifying those that are clearly invalid on their face. Indeed, one might simply assume that several of these patents are quite weak and that many have indeterminate boundaries. The problem is that ex ante we do not know which ones, and the costs of claim construction and invalidity assessments can run to thousands of dollars per patent. One party in such litigation involving three patents spent $600,000 just through claim construction.[30]

18.3.1. Limited Importance of Delineating a Pool's Inner Boundaries

An individual patent's boundaries distinguish its protected embodiments from non-infringing technology. But when multiple patents are aggregated what matters are the outer boundaries that separate the portfolio as a whole from outside patents or the public domain. So long as the relevant rights are somewhere in the portfolio, the parties do not need to delineate the boundaries of individual patents in order to strike a deal.

For example, if a landowner wants to sell his neighbor a license to walk through his meadow, the contract need not specify the exact boundaries of the path; it is much simpler to include the entire meadow, even though this is more than the neighbor requires. This likely benefits the neighbor, as he need not worry about mistakenly traveling slightly outside a narrowly defined path, and it probably makes little difference to the landowner. Of course, if the parties were

---

[27] Coase (1937).
[28] Ostrom (1990) at 61–5.
[29] Gilbert (2004).
[30] *Eon-Net LP v. Flagstar Bancorp.*, 653 F.3d 1314 (Fed. Cir. 2011).

negotiating the *sale* of an exclusive strip of land for development, then they would want to specify precise boundaries.

Similarly, when a pool of patents is licensed, it is not necessary to trace out the exact set of pooled patents on which a licensee's product will read. For example, suppose a patent pool includes related patents X and Y, whose boundaries are difficult to distinguish (perhaps because they arguably overlap). A licensee wants to secure the rights to produce something that definitely reads on patent X, and may or may not also read on patent Y. But the value of the licensing application does not hinge on this question The parties might as well agree on a license for both X and Y, because there is no reason why the terms of trade would be any different if they hired a lawyer to discern the inner boundary between X and Y.

Of course, the portfolio would be worth less if a large number of its patents or its most desired ones were invalid or easily invented around. But these possibilities could just be priced into the license fee, and this would allow the parties to forgo the costs of boundary identification without affecting the *expected* terms of trade. Even if delineating a pool's inner boundaries were not prohibitively costly, it is not clear that the parties would derive a significant benefit from doing so. The value of the license to a product producer relates to the market value of its product in relation to costs: the product is not more valuable as it embodies more patents. The case-law has often neglected this insight. Value is determined by the technology's contribution to the value of the licensee's product or process as well as the cost of substitutes, regardless of how individual IP rights are divided up.

Nevertheless, a pool's inner boundaries are still important in determining how the pool's proceeds will be allocated among contributing members; a party who contributes more patents or whose patents are practiced more often may expect to get paid more. But this problem need not affect the fees charged by the pool. Indeed, a pool's objective is typically to choose the license fees that maximize aggregate licensing revenues, regardless of how those revenues are subsequently divided among members. This is best for each member, no matter its percentage-stake in the pool's revenues.

18.4. PRICING IN PATENT POOLS VS. CONVENTIONAL BUNDLING

A conventional bundle involves the sale of two or more physical components, such as a camera and film, both of which are rivalrous goods. By contrast, in a non-exclusive license to a patent pool the licensor retains the undiminished right to use or license the right. Indeed, even transaction costs need not be any higher when a patentee licenses a group of patents than when it licenses only one. These things are invariably true if the licensee does not practice these patents. Even if it does, however, practicing imposes costs on the licensor only if it adversely affects the licensor (as in the case of licensing a patent to a competitor) or if such use reduces other parties' willingness to pay for a license. In sum, if the licensee does not use all patents in the bundle, then the inclusion of the unused licenses has no impact on either party. Consequently, the price of the bundle will reflect the fact that many of the included licenses neither benefit the licensee nor impose costs on the licensor, and this suggests that a representative licensee is not necessarily being forced to pay extra for something it does not want.

The bundle of licenses in a patent pool resembles a restaurant buffet. Buffets are common even though ownership is clear, as are the "boundaries" between individual items. The entry fee gives the customer a nonexclusive right to select food options in any combination he wants, but the buffet serves other diners simultaneously. Each consumer takes a subset of the available options, and the price will reflect the average cost of such subsets – not the cost of providing every food option to every consumer. Of course, if the buffet adds a new food option, this will affect some consumers' choices; some will take the new option and a little less of something else, but this will not necessarily make consumers' choices more costly on average. A representative consumer is not being made to pay extra for anything he does not want, so a customer who chooses only beef does not have a "tying" claim against the restaurant for forcing it to pay for numerous unwanted products. Indeed, to the extent beef is more costly than other options, such a customer may actually impose higher costs on the restaurant. The more accurate characterization of buffets is that, on average, consumers pay only for what they like but also receive options for everything else. In the context of patent pooling, this suggests that prices do not reflect the value of all pooled patents, but rather the value of a representative application, most of which rely on only a subset of the patents.

18.5. POOLING AND COLLUSION THREATS

While most patent pools are socially beneficial overall, certain practices or structures can pose competitive problems. The biggest antitrust risk from pooling is collusion, and the threat of collusion depends on two things. First is the market structure and the power of the pool within its market. Second is the nature of pricing and exclusivity arrangements within the pool.[31]

On the first issue, cartels profit only if they have market power. Patent pools often cover an entire industry, but they can also be limited to a small number of firms in a larger market that have decided to share a unique technology. Whether that technology confers sufficient advantage to enable them to raise their prices above cost is an empirical question. Naked price fixing is generally said to be unlawful per se. The courts have been hesitant to apply the per se rule to agreements within a pool, however. Softer forms of price-affecting conduct, as well as conduct that is directly implicated in joint production, is generally treated under antitrust's rule of reason. This means that the pool's market power must be estimated.

---

[31] Merges (1996).

The antitrust issue is also complicated by the presence of the Patent Act, whose specific provisions generally control the much more general provisions of the antitrust laws. For example, the Patent Act authorizes patentees to grant licenses, including exclusive licenses.[32] It also states that a patentee's *unilateral* refusal to license cannot be unlawful, although it does not say the same thing about concerted refusals.[33] One can presume that the mere authorization to engage in a practice does not imply an authorization to do so *anticompetitively* – at least, that is how the Supreme Court has interpreted statutory authorizations in other contexts.[34] For example, while Section 271 authorizes a "patent owner" to refuse to license its patent, it says nothing about agreements among numerous patent owners to refuse to license outside the pool, nor does it extend that right to nonexclusive licensees.

Specific sharing agreements among pool members come in numerous varieties. They may simply agree to share technology without any royalty. Alternatively, one or any subset of pool members may charge other members a royalty – something that is likely to occur if the participants have patent portfolios with different values, or if some members have no patents and thus participate only as licensees. The royalty rate might be flat, annualized, or most commonly, an amount based on the number of units that a licensee produces or its revenue. In addition, members may agree to limit the number of units embodying pooled patents that any member may produce.

18.5.1. *Express Collusion*

Since market exchange cannot generally take place without an agreement on price, there is nothing inherently suspicious about a pool's agreements concerning how royalties on pooled patents are computed. Nothing is inherently suspicious about agreements in which members who have contributed smaller portfolios pay more than those with larger portfolios, or licensees with no portfolios at all pay more than contributing licensors. These are all essential characteristics of bargaining behavior. Further, Section 261 of the Patent Act places no limits on how royalties are computed.

What the Patent Act does not justify, however, is price fixing of the *products* that pool members sell. Nevertheless, the Supreme Court case-law in this area is troublesome. In its 1902 *Bement* decision the court permitted pool members to fix the price of agricultural harrows produced with shared technology.[35] The Court adhered to a version of that rule in the much criticized *General Electric* decision, upholding an agreement in which patentee GE licensed Westinghouse to produce

---

[32] 35 U.S.C. § 261.
[33] 35 U.S.C. § 271(d)(4).
[34] E.g., *FTC v. Phoebe-Putney*, 133 S. Ct. 1003 (2013) (state statute authorizing one hospital to acquire another did not authorize merger that violated antitrust laws).
[35] *Bement v. Nat'l Harrow Co.*, 186 U.S. 70 (1902).

incandescent bulbs under its patents but specified Westinghouse's selling price.[36] By contrast, in *Line Materials* the Court condemned product price fixing contained in a cross-licensing agreement involving complementary patents. More recently, dicta in the Supreme Court's majority opinion in *Actavis* expressly embraced *Line Materials* and expressly limited *GE* to a license price fix between a single licensee and single licensor.[37]

Express output limitations function much like price fixing agreements. Indeed, cartels often operate more successfully by limiting output and letting the price rise to the new market level, rather than by fixing the price directly.[38] Nevertheless, inherent in the concept of a license is specification of a certain amount, such as the right to make 1,000 units per year. In any event, it is important to remember that output limitations refer to the number of times a *patent* can be practiced, not to the number of times a good can be sold. For example, if a patentee licenses a manufacturer to make 1,000 toasters weekly with the patentee's toast-detection sensor, the manufacturer is still free to make as many toasters as it pleases that do not practice that patent. An attempt to limit those would be a per se unlawful output restriction.

The policy concerns respecting pool output agreements differ from those that govern more traditional common pool resources. The traditional resource placed in a common pool is rivalrous, or subtractive, meaning that each person's use diminishes what is left for others. For example, catch limits on a fishery commons are not inherently suspicious even though it resembles a cartel. The catch limit is necessary in order to prevent overfishing. Patents are nonrivalrous, however.[39] Absent a license restriction, if one of the two toaster makers in our previous illustration should increase its output by 50 percent that will not reduce the number of units that the competing manufacturer can make. Of course, it may produce a "congestion externality" to the extent that the remaining market for toasters is smaller. But that is nothing more than competition at work, and in fact social welfare is higher when patented inventions are more widely used. As a result, output limitations in patent pools should be regarded with greater suspicion than output limitations for traditional common pool resources.

18.5.2. *Pool Exclusivity*

Patent pool "exclusivity" can take several forms. *First*, it can refer to the contract that each licensor has with the pool, asking whether that licensor is free to license to

---

[36] *United States v. Gen. Elec. Co.*, 272 U.S. 476 (1926). For severe criticism of the rule, see Judge Posner's opinion in *Asahi Glass co. v. Pentech Pharm., Inc.*, 289 F.Supp.2d 986 (N.D. Ill. 2003).

[37] *FTC v. Actavis, Inc.*, 133 S. Ct. 2223, 2225, 2231–2 (2013), discussing *United States v. Line Material Co.*, 333 U.S. 287 (1948).

[38] Stigler (1962).

[39] Justice Story made this error in *Brooks v. Byam*, 2 Story 525, 4 F. Cas. 261 (C.C. Mass. 1843). See Bohannan and Hovenkamp (2012) at 328–9.

others outside of the pool. *Second* it can refer to the pool's willingness as licensee to accept an offered technology from an outsider for inclusion in the pool. *Third* it can refer to the pool's willingness as licensor to license to outsider manufacturers. *Fourth*, it can refer to field-of-use or other restrictions given to licensees from the pool. Nearly all of these arrangements are multilateral and thus are controlled by Section 1 of the Sherman Act. That does not mean, however, that they are unlawful.

No cartel can succeed unless it can reduce market-wide output. For this reason cartels are fastidious about policing the output of cartel members. For their part, the members have a strong incentive to cheat. At the cartel price, revenues are considerably higher than marginal cost, making cheating very profitable.[40] This is particularly true in IP markets, where marginal costs are low to begin with. Each cheating sale produces considerable additional profit.

For these reasons, nonexclusivity in pool licensing arrangements suggests that the pool is procompetitive,[41] provided that nonexclusivity really means what it says. That is, there might be small pools with public nonexclusivity but tacit exclusivity requirements. Aside from that, however, a clear and public nonexclusivity requirement, particularly when the pool has a large number of members, significantly reduces the collusion risk.

Exclusive agreements between individual licensors and the pool are more difficult to analyze. First, the pool may be willing to pay more for exclusive licenses to the extent that they commit technology suppliers to that particular pool's technology, giving the pool an advantage over alternative licensing outlets. This may also enable the pool to develop its own distinct brand, with technology that cannot be readily copied by outsiders.[42] In this sense they operate a little more like exclusive dealing or output contracts, which are competitive most of the time but can raise competition concerns when market power and foreclosure rates are sufficiently high.[43]

The second type of exclusivity concerns the outside developer of technology who wants to include its patents in the pool but is denied. One clearly procompetitive explanation for the pool's refusal to accept an offeror's technology is that the excluded technology is inferior to or no better than technology that the pool already has. There are good reasons for a pool not to accept an inferior substitute technology; namely, it will not make the pool products better and it might raise fears of collusion. If the prospective licensor's technology is simply unwanted by pool members, that also indicates lack of competitive harm. The firms in the pool have a right to make their own technology choices, even when they do so by agreement. Indeed, even a dominant network should not have an obligation to accept a set of

---

[40] Hovenkamp (2015b) at § 4.1a.
[41] Hovenkamp (2015b) at § 5.5c; US Dep't of Justice and FTC (2007); Greene (2010).
[42] E.g., *United States v. Sealy, Inc.*, 388 U.S. 350 (1967).
[43] Hovenkamp (2011) at ¶1821.

licenses to a technology that its members do not want, and no concern about competition arises.

Competitive harm can occur when some pool members already have technology in the pool that competes with the excluded firm's superior technology. In that case the pool can operate as an anticompetitive boycott.[44] Assessing the merits of such claims is difficult, because it may require the fact finder to assess the merits of the excluded technology. As a result it is essential to make sure that structural prerequisites are present. For example, an essential element in such a claim would be the existence of an influential pool member or members who own competing technology and have sufficient power to foist an inferior technology on others, perhaps by manipulating the process.[45]

The third type of exclusivity is the pool that licenses to others only selectively, perhaps only to member-licensors or perhaps to some others. Once again rule of reason treatment is generally appropriate. No general antitrust rule requires a joint venture to share its technology with rivals, particularly in small or specialized ventures whose formation required significant investment. The main reason is to limit free riding on the members' costly efforts. For example, three firms may agree with each other to cross-license their patents and develop a new product. A fourth stands aside, wishing not to assume the risk of failure. Then, after the venture succeeds, the fourth firms wants to join. An antitrust rule that permitted late joiners would operate as a disincentive to invest in the first place.

Larger, mostly open pools present a different issue. If the pool controls a network, access to the pool may be essential for operation on the network. Selective exclusion may be a way of disciplining price cutting by refusing pool access. As a result, a practice that selectively denies pool access to price cutters or other aggressive firms could be unlawful.[46] Relatedly, as with exclusivity agreements limiting licensor-membership, a concerted refusal to license to a firm may be unlawful if it amounts to a group boycott that serves to restrain competition in a downstream product market.

The fourth type of exclusivity really concerns market divisions in the pool, and several of these have been prosecuted under the antitrust laws. These are most likely to be a threat when individual pool members dominate different market segments. The pool management might then use field-of-use restrictions on member-licensees to divide the market, thus preventing each member from entering competition with a different member. For example, *Hartford-Empire* condemned such an arrangement among firms whose pool divided the market for different types of glass bottles.[47]

---

[44] E.g., *Golden Bridge Tech., Inc. v. Motorola, Inc.*, 547 F.3d 266, 270–1 (5th Cir. 2008) (rejecting such a claim); see also *TruePosition, Inc. v. LM Ericsson Tel. Co.*, No. 11-4574, 2012 WL 3584626, at *25 (E.D. Pa. Aug. 21, 2012) (sustaining a complaint); *Cryptography Research Inc. v. Visa Int'l Serv. Ass'n*, No. C 04–04143 JW, 2008 WL 5560873, at *6–7 (N.D. Cal. Aug. 13, 2008) (similar).

[45] Cf. *Allied Tube & Conduit Corp. v. Indian Head, Inc.*, 486 U.S. 492 (1988).

[46] Hovenkamp (2012) at ¶¶2220–4.

[47] *Hartford-Empire Co. v. United States*, 323 U.S. 386, 392–400 (1945).

Sometimes these various types of exclusivity can operate together and heighten antitrust concerns. For example, the pool's refusal to license a newcomer is potentially more harmful if the original licensor members of the pool also have exclusive license agreements. If these agreements are nonexclusive, then the manufacturer who is denied a pool license can still try to obtain essential patents from individual licensor members.[48]

18.5.3. A Brief Note on FRAND

Standard setting organizations often identify standards essential patents (SEPs) that cover an adopted and shared technology. In order to acquire SEP status participating patent owners must generally promise to license their patents to willing participants on fair, reasonable, and non-discriminatory (FRAND) terms.[49] As a result, these pools are generally nonexclusive in the sense that anyone can become a licensee simply by agreeing to pay the FRAND royalty. The principal litigation has involved royalty computation[50] and entitlement to an injunction.[51]

18.6. CONCLUDING NOTE: POOLING AND INNOVATION RATES

A large but quite inconclusive literature discusses the relationship between pooling and innovation. Conclusions are extremely sensitive to assumptions about patent strength and quality, about the relationship among the patents in a pool and the strength of alternatives outside the pool, about the impact on innovation of insiders vs. outsiders to the pool, and finally, about the strategic responses of participants. In addition there are difficulties in assessing innovation rates. Counting patents issued per time period is relatively easy, but is not a good measure of innovation if patents are of low quality or if firms respond to the pool by altering their patenting behavior rather than their innovation behavior.

The existence of a pool may distort innovation behavior. For example, a firm might decide to protect a particular innovation with a trade secret rather than a patent in order to avoid committing it to a pool. Alternatively, it may rely on simple first mover advantages in order to avoid patent disclosure requirements. Another distortion results from how royalties are divided up. For example, if the pool divides royalties by the number of patents, then pool members have an incentive to make patents thinner in order to increase their number.[52] Finally, one must distinguish

---

[48] E.g., *United States v. Associated Patents*, 134 F. Supp. 74 (E.D. Mich. 1955) (condemning agreement among manufacturers giving each exclusive rights to build particular types of machines and noting that pool members refused to license their patents outside the pool).
[49] Hovenkamp (2015a); Lemley (2002); Swanson and Baumol (2005).
[50] E.g., *Ericcson, Inc. v. D-Link Sys., Inc.*, 773 F.3d 1201 (Fed. Cir. 2014); *Microsoft Corp. v. Motorola, Inc.*, 2013 WL 2111217 W.D. (Wash. Apr. 25, 2013).
[51] *Apple, Inc. v Motorola, Inc.*, 757 F.3d 1286 (Fed. Cir. 2014).
[52] Baron and Delcamp (2015).

anticipatory innovation, which occurs when a patent pool is announced,[53] from equilibrium innovation, which occurs on an ongoing basis after the pool is in operation.[54]

A well-functioning patent pool should reduce the costs of sharing technology, which it can do by making it easier for manufacturers to assemble complements, eliminate double marginalization, or reduce the costs of defining or defending patent boundaries. Since all of these costs operate as a deadweight loss on innovation one would think that the impact of a well-designed pool is positive. On the other side, pooling may reduce the appropriation value of innovations, particularly if pool members must commit in advance to dedicate new patents to the pool. In that case, rather than obtaining the entire return, the pool member must contemplate sharing with others, and this may create common pool management problems. Each firm will have a tendency to shirk, profiting instead from the inventions of others.

The theoretical and empirical literature points in both directions, with most literature finding that most patent pools increase innovation rates,[55] although there is also literature concluding to the contrary.[56] Others find that pools of stronger patents tend to increase innovation while pools dominated by weaker patents more likely to be invalidated in litigation do not.[57]

As a general proposition a pool should increase the demand for innovation of complements to the pool. First of all, access to the existing technology by pool members should be guaranteed and cheaper. To the extent the pool reduces licensing costs and eliminates royalty stacking the cost of further improvements should decline. When innovation is cumulative the development of new technology may require the licensing of existing technology with multiple patent holders. Pooling can reduce these costs and thus facilitate cumulative innovation.[58]

REFERENCES

Areeda, Phillip E. and Herbert Hovenkamp. 2010. *Antitrust Law*, Vol. 7, New York: Wolters Kluwer.

Baron, Justus and Henry Delcamp. 2015. The Strategies of Patent Introduction into Patent Pools. *Economics of Innovation and New Tech*, 24, 776.

Baron, Justus and Tim Pohlmann. 2015. *The Effect of Patent Pools on Patenting and Innovation – Evidence from Contemporary Technology Standards* (Northwestern Univ. Working paper, 2015), available at www.law.northwestern.edu/research-faculty/searlecenter/innovationeconomics/documents/Baron_Pohlmann_effect_of_patents.pdf.

[53] Lerner and Tirole (2004); Dequiedt and Versaevel (2013).
[54] Baron and Pohlmann(2015).
[55] E.g., Baron and Pohlmann (2015); see also Shapiro (2001); Lerner and Tirole (2004); Mossoff (2011).
[56] E.g., Lampe and Moser (2010; 2013; 2016).
[57] Choi and Gerlach (2013).
[58] Scotchmer (2006).

Bohannan, Christina and Herbert Hovenkamp. 2011. Concerted Refusals to License Intellectual Property Rights. *Harv. Bus. L. Rev. Bull.*, 1, 21.
  2012. *Creation Without Restraint: Promoting Liberty and Rivalry in Innovation*. New York: Oxford University Press.
Choi, Jay Pil and Heiko Gerlach. 2013. *Patent Pools, Litigation and Innovation* (Cesifo Working Paper, Oct. 24). Available at http://papers.ssrn.com/sol3/papers.cfm?abstract_id=2344689.
Coase, Ronald H. 1937. The Nature of the Firm. *Economics*, 4 (n.s.).
Dequiedt, Vianney and Bruno Versaevel. 2013. Patent Pools and Dynamic R&D Incentives. *Int'l Rev. L. & Econ.*, 36, 59.
Gilbert, Richard J. 2004. Antitrust for Patent Pools: A Century of Policy Evolution. *Stan. Tech. L. Rev.*, 3.
Greene, Hillary. 2010. Patent Pooling Behind the Veil of Uncertainty: Antitrust, Competition Policy, and the Vaccine Industry. *B.U. L. Rev.*, 90, 1397.
Heller, Michael A. and Rebecca S. Eisenberg. 1998. Can Patents Deter Innovation? The Anticommons in Biomedical Research. *Sci.*, 280, 698.
Hovenkamp, Erik N. and Herbert Hovenkamp. 2015. Tying Arrangements. In *Oxford Handbook of International Antitrust Economics*. New York and London: Oxford University Press.
Hovenkamp, Herbert. 2010. Harvard, Chicago and Transaction Costs Economics in Antitrust Analysis. *Antitrust Bull.*, 55, 613.
  2011. *Antitrust Law*, Vol. 11. New York: Wolters Kluwer.
  2012. *Antitrust Law*, Vol. 13. New York: Wolters Kluwer.
  2015a. Antitrust and the Patent System: A Reexamination, *OSU L.J.*, 76, 467.
  2015b. *Federal Antitrust Policy: the Law of Competition and its Practice* (5th edn.). St. Paul, MN: West Academic.
Hovenkamp, Herbert, Mark D. Janis, Mark A. Lemley, Christopher R. Leslie, and Michael A. Carrier 2015. *IP and Antitrust: An Analysis of Antitrust Principles Applied to Intellectual Property Law*. New York: Wolters Kluwer.
Lampe, Ryan and Petra Moser. 2010. Do Patent Pools Encourage Innovation? Evidence from the 19th-Century Sewing Machine Industry. *J. Econ. Hist.*, 70, 898.
  2013. Patent Pools and Innovation in Substitutes – Evidence from the 19th-Century Sewing Industry. *RAND J. Econ.*, 44, 757.
  2016. Patent Pools, Competition, and Innovation – Evidence from 20 United States Industries under the New Deal. *J. L. Econ. & Organ.*, 32, 1.
Lemley, Mark A. 1997. The Economics of Improvement in Intellectual Property Law. *Tex. L. Rev.*, 75, 993.
  2002. Intellectual Property Rights and Standard-Setting Organizations. *Calif. L. Rev.*, 90, 1889.
Lemley, Mark A. and Carl Shapiro. 2007. Patent Holdup and Royalty Stacking. *Tex. L. Rev.*, 85, 2163.
Lerner, Josh and Jean Tirole. 2004. Efficient Patent Pools. *Am. Econ. Rev.*, 94, 691.
Lerner, Josh, Marcin Strojwas, and Jean Tirole. 2007. The Design of Patent Pools: The Determinants of Licensing Rules. *Rand J. Econ.*, 38, 610.
Mattioli, Michael. 2014. Power and Governance in Patent Pools. *Harv. J.L. & Tech.*, 27, 421.
Merges, Robert P. 1996. Contracting into Liability Rules: Intellectual Property Rights and Collective Rights Organizations. *Cal. L. Rev.*, 84, 1293.
Mossoff, Adam. 2011. The Rise and Fall of the First American Patent Thicket: The Sewing Machine War of the 1850s. *Ariz. L. Rev.*, 53, 1654.

Muller, Florian. 2014. Analysis of 222 Smartphone Patent Assertions. Available at www.fosspatents.com/2014/10/analysis-of-222-smartphone-patent.html.
Orr, Justin R. 2013. Patent Aggregation: Models, Harms, and the Limited Role of Antitrust. *Berkeley Tech. L. J.*, 28, 525.
Ostrom, Elinor. 1990. *Governing the Commons: The Evolution of Institutions for Collective Action*. New York: Cambridge University Press.
Santore, Rudy, Michael McKee, and David Bjornstad. 2010. Patent Pools as a Solution to Efficient Licensing of Complementary Patents? Some Experimental Evidence. *J.L. & Econ.*, 53, 167.
Scotchmer, Suzanne. 2006. *Innovation and Incentives*. Cambridge, MA: MIT Press.
Shapiro, Carl. 2001. Navigating the Patent Thicket: Cross Licenses, Patent Pools, and Standard-Setting. In A. Jaffe, J. Lerner, and S. Stern (eds.), *Innovation Policy and the Economy*, vol. 1. Cambridge, MA: MIT Press.
Stigler, George J. 1962. A Theory of Oligopoly. *J. Pol. Econ.*, 72, 44.
Swanson, Daniel G. and William J. Baumol. 2005. Reasonable and Nondiscriminatory (RAND) Royalties, Standards Selection, and Control of Market Power. *Antitrust L.J.*, 73, 1.
US Dep't of Justice and FTC. 2007. Antitrust Enforcement and Intellectual Property Rights: Promoting Innovation and Competition.
Williamson, Oliver E. 1975. *Markets and Hierarchies: Analysis and Antitrust Implications*. New York: The Free Press.

PART V

Vertical Relations

19

Bundling and High Tech Industries

Daniel J. Gifford and Robert T. Kudrle

From one generation to the next, innovation is undoubtedly a central determinant of the welfare of humankind. Economists studying individual projects, moreover, routinely find that the benefits of innovation to society as a whole greatly exceed the benefits to the firms that develop the innovation.[1]

Innovation stands at the heart of economic growth, and the high technology industries have been the source of the greatest economic innovations in recent decades. Nevertheless, antitrust has not always been good to industries involved in high technology. This inhospitality may be due to the persistence of antitrust rules and the slow pace of judicial change, on the one hand, and the rapidity of changing technology on the other. Bundling is peculiarly related to these potential incompatibilities because high technology products change quickly and are constantly in the process of being redesigned, and that redesign often takes the form of combining formerly separate products or at least of adding new dimensions or functionalities to an existing product. This combination or addition of features not infrequently runs up against antitrust rules directed against tying. We will use the term "bundling" and tying interchangeably as is sometimes done in case-law.[2] The design of high technology products is in a state of flux, but the antitrust rules governing tying were formulated during the industrial age and were based upon now outmoded understandings of economics. Because tying restrictions, like all legal rules, change slowly, they could be expected to pose continuing challenges to high technology industries.

---

[1] Baker (2007) quoted in Carlton and Heyer (2008). An innovating firm, even if it has a patent, generally captures only some of the value that it creates, the remainder appearing as consumer surplus. That surplus typically grows larger in the years following the patent's expiration.

[2] Motta (2004) at 460. Bundled sales are treated as a tie in *United States v. Loew's, Inc.*, (1962) 371 U.S. 38. Most economists use "bundling" to refer to "package tie-in sales" with fixed proportions of components and distinguish them from "requirements tie-in sales" that allow the proportions to vary. This is discussed in US Department of Justice and US Federal Trade Commission (2007).

19.1. PRODUCT DESIGN

Changes in high tech industries often involve swift alteration of product design. This was the case in the 1970s when IBM was leading the development of main-frame computers, and it was the case in the 1990s when Microsoft was developing its operating system and associated browser. In these examples a conflict (or potential conflict) arose when antitrust rules over tying arrangements were applied to the producer's product designs. In Europe the same conflict is currently generated by the European Commission's approach to tying in the media player element of the European Microsoft case and in its use of the essential facilities doctrine to deprive that company of the fruits of its integration activity in the server software part of the same case.

In these cases, the problem arose when the producer integrated formerly stand-alone items into its principal product. In IBM's case, the principal products were main-frame computers. IBM had been selling central processing units and separately selling peripheral devices such as disk drives, memory units, and printers.[3] IBM was the leading producer of central processing units, which were sophisticated and complex products. The market for peripheral devices, by contrast, was easier to enter and IBM was vulnerable to competition in this market. Because the peripheral devices were not necessarily used in a fixed ratio to the central processing units, there was room for IBM to vary prices of peripheral devices in accordance with the various segments of demand into which the purchasers of central processing units fell.

IBM introduced its new System 370 Model 145 in September 1970. At that time, IBM introduced the model 2319A disk drive with a control function that was integrated into the CPU. In February 1971, it integrated the disk control function into the CPU for its System 370 Models 158 and 168. In subsequent litigation, the Ninth Circuit upheld IBM's right to redesign its products against a contention by California Computer, an IBM rival in the peripheral market, that these design changes constituted monopolization.[4]

In the early 1990s, Microsoft began producing a graphical user interface for use with its then current MS-DOS operating system. Novel, which was then competing with Microsoft in the DOS (disc operating system) market, complained to the European Commission that Microsoft was tying its operating system to its Windows graphical user interface. After the Justice Department challenged Microsoft's licensing agreements with computer manufacturers, the tying issue raised by Novel was settled in negotiations involving the Justice Department, DG IV and Microsoft.[5]

---

[3] *Telex Corp. v. International Bus. Mach. Corp.* (1975) 510 F.2d 894, 899 (10th Cir. 1975) describing peripheral devices.

[4] *California Computer Products, Inc. v. International Bus. Mach. Corp.* (1979) 613 F.2d, (9th Cir. 1979). IBM's redesign (or integration) of its product is described in Fisher, McGowan, and Greenwood (1983).

[5] The negotiations are described in *United States v. Microsoft Corp.* (1998) 147 F.3d 935, 945–6 (D.C. Cir. 1998).

The resulting consent decree was later construed by the D.C. Circuit.[6] That court concluded that the parties intended to bar the tying of the separate graphical user interface to the operating system but further intended to allow genuine "integrations," of which the integration of the browser with the Windows operating system was one.[7]

To reach this result, the court carefully construed the concept of integration. In the court's understanding an integrated product is one that "combines functionalities (which may also be marketed separately and operated together) in a way that offers advantages unavailable if the functionalities are bought separately and combined by the purchaser".[8] The court explained that in Windows 95, the browser (Internet Explorer – "IE") is integrated with the Windows operating system by the manufacturer in a way that differs from the relation between the graphical user interface (Windows 3.1) and MS-DOS operating system. In the case of Windows 3.1 and MS-DOS, a customer could separately purchase each program, load them on his computer and would achieve the same result as if he purchased a Windows/OS package from the manufacturer. But Windows 95 was different. A customer could not bring about the integration of the IE browser and the Windows 95 operating system. The manufacturer achieved the integration of the two functionalities by using some of the same code to operate features of the IE and the operating system. This sharing of code between the two software devices meant that neither was independent from the other; the combination was therefore an "integrated" product.[9]

The D.C. Circuit's construction of the consent decree forms part of the background to the later antitrust suit which the Department of Justice brought against Microsoft in connection with Windows 98. The court issued two important rulings in the antitrust case that superficially seem to take opposite paths toward applying antitrust to product design. First, Microsoft was held to have violated the Sherman Act § 2 prohibition of monopolization by tying its browser to its operating system; and second, under Sherman Act § 1, the court created an exception to the per se rule governing tying arrangements applicable to platform software. We first discuss the monopolization ruling and then the new exception for platform software.

The basis for the court's monopolization ruling was the determination that Microsoft had tied its browser (IE) to the operating system in several ways that users could not reverse. The problem with this tie was that, because of Microsoft's practical monopoly in the operating system's market, Microsoft's primary browser rival (Netscape) was so disadvantaged that its ability to develop or maintain a critical

---

[6] The consent decree was approved in 1995 (*United States v. Microsoft Corp.* (1995) 1995-2 Trade Cases ¶ 71,096 (D.D.C. 1995)).
[7] *United States v. Microsoft Corp.* (1998) 147 F.3d 935, 945–6 (D.C. Cir. 1998).
[8] *Id.*
[9] *Id.*

user base was undermined. And preservation of Netscape's user base was essential to the preservation and development of a nascent technology that had the potential to challenge Microsoft's platform monopoly.[10]

The premise underlying the Department of Justice's suit and the D.C. Circuit's monopolization ruling was that the rival Netscape browser had the potential to develop into a "platform," to which software vendors could write their software programs. Microsoft's operating system was effectively the only platform usable by personal computers and by the software vendors seeking to offer programs to pc users. With further development, the Netscape browser had the potential of adding a sufficient number of application program interfaces (or APIs) to enable it to grow into a platform, rivaling the Windows platform and breaking its monopoly.

The reasoning behind the monopolization ruling is somewhat problematic. There was no likelihood that the Netscape Navigator was capable of developing into an alternative platform within the foreseeable future.[11] In the language of the court, it was a "nascent" threat.[12] Microsoft appears to have foreseen the threat, however, so that threat may have been real. Microsoft's principal product design actions relate to the ability of original equipment manufacturers (OEMs) and consumers to separate the constituent elements of the Windows 98 system, separating the browser from the operating system. The court identified three actions taken by Microsoft to bind the browser to the Windows operating system:

> excluding IE from the "Add/Remove Programs" utility; designing Windows so as in certain circumstances to override the user's choice of a default browser other than IE; and commingling code relating to browsing and other code in the same files, so that any attempt to delete the files containing IE would, at the same time, cripple the operating system.[13]

Microsoft justified overriding the user's browser choice in certain circumstances where Active X, an IE component that supported Windows 98 Help and Windows Update features, was involved because Navigator did not support Active X. Also Windows 98 did not invoke Navigator when a user accessed the Internet through the "My Computer" or Windows "Explorer" devices because one of the purposes of those features was to enable users to move seamlessly from local storage devices to

---

[10] According to the D.C. Circuit, there were two conditions that Netscape had to meet in order to develop into an alternative platform to Windows. First, Netscape had to add a sufficient number of application program interfaces (APIs) to provide a technical alternative platform. At the time of litigation, Netscape had 1,000 APIs and Windows had 10,000. Second, Netscape's user base had to be maintained in sufficient volume to be attractive to software developers.

[11] The district court found that it would take Netscape and other middleware products "several years" to change sufficiently to a form where they were alternative platforms competing with the Windows platform. And whether they would ever do so was uncertain (*United States v. Microsoft Corp.* (1999) 84 F. Supp. 2d 9, 17–18, at ¶¶ 28–9 (D.D.C. 1999)).

[12] *United States v. Microsoft Corp.* (1999) 84 F. Supp. 2d 9, 30 (D.D.C. 1999) (Finding 77).

[13] *United States v. Microsoft Corp.* (2001) 253 F.3d 34, 68 (D.C. Cir. 2001).

the Web in the same browsing window.[14] Because the government did not challenge Microsoft's justification, the court ruled in Microsoft's favor on this issue.

Microsoft lost on the other product design issues. The district court decided against it on the issue involving commingling of code, and the appeals court upheld that ruling as a plausible response to conflicting evidence; and Microsoft had no proffered justification for removing the browser from the Add/Remove program. Of course, if the court was correct that Microsoft could lawfully require the substitution of IE for the user's choice of another browser in certain circumstances, then it would seem necessary to retain IE and therefore to remove it from the Add/Remove program.

The basis of liability in the antitrust case involving Windows 98 is the integration of the IE browser with the operating system, which appears to meet most of the court's definition of integration as expounded in the consent decree involving Windows 95. Because the manufacturer intertwined the code for the browser and the operating system, they could not be separated. Only the manufacturer could have designed the products this way. This integration of the two functionalities here underlies the court's monopolization ruling, but there is a difference in the two situations. The later integration fails to meet the integration definition in the consent decree because it did not offer "advantages" otherwise unavailable.[15] The Ninth Circuit earlier had recognized that a legitimate integration was in some way superior to the mere combination of the components,[16] and the D.C. Circuit, in these cases, is following the approach of the Ninth Circuit. In the antitrust court's understanding, the "integration" of the browser into the operating system appears to be a sham: the functionalities cannot be separated, but there is no reason why they must be made inseparable. If Microsoft (instead of using separate code for the browser and separate code for the operating system located in the same files) had used portions of the same code to give instructions to both the browser and the operating system, then the program structure would have appeared less artificial and possibly more efficient. Had Microsoft established that it had so acted, it might have prevailed on this issue. Microsoft, however, had the opportunity to justify the commingling of code and it failed to do so.

The connection between Microsoft's actions, which the court ruled illegal, and the public harm they caused remains doubtful. The appeals court does not fault Microsoft for integrating the browser into the operating system in any way that is reversible by users or OEMs. So if the browser were removable by, say, subjecting it to the Add/Remove program and if the code essential for the browser had been stored in separate (removable) files, the product design would have passed muster. Yet if the product had been designed in the latter way, it seems likely that the user base of Netscape Navigator would nonetheless have shrunk over time as Windows 98 (with pre-installed, albeit removable, IE) gradually replaced older operating

---

[14] *Id.*
[15] *United States v. Microsoft Corp.* (1998) 147 F.3d 935, 945–6 (D.C. Cir. 1998).
[16] *California Computer Products, Inc. v. International Bus. Mach. Corp.* (1979) 613 F.2d (9th Cir. 1979)).

systems. Although, in this scenario, the IE browser would no longer be inseparable from the operating system, consumer inertia would likely produce the same result. In the absence of physical inseparability, the OEMs might have chosen to substitute another browser, perhaps Netscape Navigator, but it is far from clear.

In contrast to the context-specific Section 2 findings, the appeals court's decision on the Section 1 issue carries continuing important ramifications for high technology. At the time of the decision, the governing law treated ties imposed by firms with market power as per se illegal. And under the prevailing Supreme Court precedent, a separate product was anything for which there was a pre-existing demand. That precedent was *Jefferson Parish*,[17] which had held that anesthesiology was a separate product from the surgical services with which it was used because patients received separate bills from their anesthesiologists. Thus under this approach the browser would have been treated as a product separate from the operating system, and the joinder or "integration" of these two products would have been treated as a technological tie. Recognizing, however, that innovation in the software industry generally takes the form of adding a pre-existing functionality to an operating system, the court created an exception to the per se rule for platform software. This was a courageous act on the part of the D.C. Circuit, whose duty is normally to follow the Supreme Court's precedents. Here, however, the court was able to deviate from precedent on the theory that the Supreme Court would also have carved out the exception for the same reasons as the appeals court, had it been confronted with the situation faced by the appeals court.[18]

The D.C. Circuit's ruling on the Section 1 issue now stands as a precedent governing tying issues arising in the platform software context. That ruling has freed this sphere from the rigidities of the per se rule that had previously afflicted product development. That ruling may also have paved the way for future rulings further narrowing the per se rule. Indeed, the court's new exception to the per se rule can be seen as another step toward the ultimate demise of the per se rule governing tying. It follows the path marked earlier by Justice Sandra O'Connor's concurring opinion in *Jefferson Parish* in which three other Justices joined, where she called for the abolition of the per se rule.

19.2. EVANS AND SCHMALENSEE ON ANTITRUST AND HIGH TECHNOLOGY

Just before the D.C. Court's ruling in the *Microsoft* case, David Evans and Richard Schmalensee presented a paper expressing their concern that antitrust law was

---

[17] *Jefferson Parish Hospital District No. 2 v. Hyde* (1984) 466 U.S. 2, 32.
[18] This is the purport of the D.C. Circuit's reasoning. It concluded that "the sort of tying arrangement attacked here is unlike any the Supreme Court has considered (*United States v. Microsoft Corp.* (2001) 253 F.3d 34, 90 (D.C. Cir. 2001))." It also concluded that the application of the per se rule to software platform markets would run counter to the Supreme Court's criteria for the application of the per se rule.

ill-adapted to the demands of high technology.[19] Among their expressed concerns was the fear that the courts would apply traditional rules to Microsoft and thereby thwart the growth and development of technologically advanced industries, typified (in their view) by Microsoft. In particular, they feared that because Microsoft had been giving away its browser, the courts would construe this behavior as selling below cost and condemn it as predatory. They also feared that the appeals court would treat the integration of the browser into the operating system as a tie. As it turned out, these concerns were ill-founded. The courts did not treat giving away the browser as equivalent to a sale below cost.[20] And the D.C. Circuit appreciated the importance of innovation when it refused to let pre-existing rules governing tying interfere with the pattern of innovation that had emerged in the software industry. As just described, that court carved out an exception to the per se rule governing tying for software platforms in the interest of fostering innovation.

Broadly understood, Evans and Schmalensee were expressing their concerns over whether antitrust law was compatible with high technology as represented by Microsoft. At base, these authors believed that antitrust would conflict with the growth pattern of Microsoft, which they saw as typical of a knowledge-based industry with inadequate intellectual property protection. The company invested large sums up front in software development but incurred trivial marginal costs in producing its operating system. Companies in this situation would have to generate high returns over marginal costs to compensate for their up-front costs as well as providing a return for risk. The authors feared that government would take the large operating profits as signs of monopoly and use them as indicators of where antitrust actions should be brought. They were also concerned about Microsoft's vulnerability to attempts by rivals to appropriate their intellectual property.

Evans and Schmalensee's paper missed a critical element of Microsoft's market situation: the benefits obtained from network effects, which increase the value of a product as the network grows. The Windows operating system grew in value as its user base grew with an ever larger set of applications written to be compatible with it. The result was that rival operating systems posed a continually diminishing market threat to Microsoft as time passed, since Windows was so much more valuable to consumers than alternatives. Copyright, which protected the Windows operating system from identical or slavish copying but which has always contemplated differentiated-product competition,[21] was sufficient, in combination with network effects, to provide the needed protection. The combination of network effects and copyright produced a new and strong form of intellectual property protection in which each of these elements supplemented what was provided by the other: (1) copyright protected Windows from duplication by

---

[19] Evans and Schmalensee (2001, 2002).
[20] *United States v. Microsoft Corp.* (2001) 253 F.3d 34, 68 (D.C. Cir. 2001).
[21] Gifford and Kudrle (2011).

rivals; and (2) network effects ensured its success in differentiated-product competition.[22] Bundling (in its technical tying variant) informed its growth pattern. The US Microsoft case helped to reveal these interrelationships. Thus US antitrust policy related to bundling in high technology industries appears to be developing in ways that are more lenient and supportive of innovation than some early observers feared.[23]

In the European Union, the law is less friendly to innovation and bundling. There have been no analogs to the US software platform exception in the law governing tying arrangements. At their intersection, the EU competition law appears to trump copyright and other non-patent law protections. In the *IMS Health* case, for example, IMS Health was required to share a data-collection system with rivals, despite its copyright protection, and in the EU *Microsoft* case, Microsoft was required to share its server communications protocols with rivals, despite its claims that they were protected as intellectual property (presumably trade secrets). The ramifications of these rulings affect not only competition law, but also copyright, trade secret, and perhaps other intellectual property laws. One possible inference from these decisions is that the EU Courts believe that patent law should serve as the sole legal mechanism for generating incentives for technological improvements. In this view, the incentives provided by copyright have to do with copyright's traditional concern to foster artistic and literary works, and their role in protecting software should be limited to the literal protection of source and object code.[24] This view may have been stimulated by acknowledgments by American judges that copyright, although charged with software protection is ill-designed for the task.[25] Copyright does not work well as a mechanism for protecting software except from literal copying of source and object code, and neither copyright nor other non-patent types of intellectual property will be allowed to confer competitive advantages on utilitarian products or services.

In summary, antitrust claims involving the integration or bundling of products have included claims asserted against IBM in the 1970s, Novel's complaint against

---

[22] *Id.*

[23] Antitrust law is supportive of high technology industries in other ways as well. Research and development costs – which characterize those industries – are effectively sunk before the product reaches the market. See, e.g., Hovenkamp (2006) at 857 (observing that bundled discount cases are likely to arise in high tech markets because R&D constitutes fixed costs). Accordingly, these costs do not act as a constraint on pricing. This eases the path of the high tech firm to recoup its R&D costs. Moreover, because high tech firms often sell services (rather than products), they need not be concerned with the intricacies of the Robinson-Patman Act.

[24] This view can draw support from the acknowledgement by American judges that copyright, although charged with protecting software, is ill-designed for that task (*Computer Associates Int'l, Inc. v. Altai, Inc.*, 982 F.2d 693, 712 (2d Cir. 1992); *Sega Enterprises Ltd. v. Accolade, Inc.*, 977 F.2d 1510, 1524 (9th Cir. 1992, 1993)). Indeed, it does not work well as a mechanism for protecting software except from literal copying of source and object code.

[25] *Computer Associates Int'l, Inc. v. Altai, Inc.* 982 F.2d 693, 712 (2d Cir. 1992); *Sega Enterprises Ltd. v. Accolade, Inc.* 977 F.2d 1510, 1524 (9th Cir. 1992, 1993).

Microsoft's bundling of its Windows graphical user interface with its MS-DOS operating system in the 1990s, and the Justice Department's claims in the same decade based on the browser/operating system connection. Kodak's offering of its Instamatic camera with specially designed film in the 1970s should be included in this list. The only case in which a court ruled that the integration was unlawful was the D.C. Circuit's § 2 ruling in the Microsoft antitrust case. That ruling was based on the court's determination that Microsoft had affirmatively taken a range of actions to prevent Netscape, the rival browser, from developing into a rival software platform, thus "maintaining" its own platform monopoly in its operating system. The remedial decree prevented Microsoft from offering any "middleware" product in an integrated form that could not be undone by a computer manufacturer or by the user.

In the European Union, integration of high tech products has been repeatedly undone in the name of competition enforcement. Microsoft's communications protocols on its Windows 2000 server software were ordered to be shared with Sun and its other rivals in the server software market. The integration of Microsoft's media player into its operating system was treated as an illegal tie. There appears to be no accommodation of antitrust rules to high technology. The General Court has explicitly rejected any special consideration for knowledge-based (or dynamically competitive) industries. And the courts look at network effects as competitive-market aberrations that should be neutralized so far as possible.

Economists Dennis Carlton and Ken Heyer[26] have proposed that the proper goals for antitrust can be described as preventing the extension of market power by weakening or eliminating the competitive constraints on a firm provided by its rivals to the necessary detriment of consumer or total welfare, while tolerating the extraction of rents from surplus that the firm has created through its own investment, innovation, industry, or foresight. These goals seem consistent with the ultimate purpose of antitrust law, but operationalizing them remains a challenge for a set of US policies that are now de facto versions of the rule of reason in most dimensions. Moreover, the uncertainty facing high technology firms is greatly magnified because this line of commerce is quintessentially global, and other jurisdictions, most importantly the European Union, as the above discussion shows, take a very different approach in several central areas related to bundling. These areas include situations in which bundling appears to threaten competitors even where efficiencies seem clearly important, the balance between intellectual property rights and their restriction to boost the fortunes of competitors, the practice of price discrimination to increase the yield from innovation, the employment of bundled discounts, and the appropriate cost standard to apply when evaluating business practices challenged as inappropriate.

---

[26] Carlton and Heyer (2008).

19.3. DOUBLE MARGINALIZATION, OTHER EFFICIENCIES, AND PORTFOLIO EFFECTS

In merger evaluation, substitute products produced by the merging companies generate upward pricing pressure on both products because the newly merged firm will take into account that the loss on sales from a price increase of one of the constituent firm's products will simply drive sales to the other. That relation of close substitutability is thus a negative factor in the evaluation. For analogous reasons, a relation of complementarity between the products of the merging firms ought to be taken as a positive factor in merger evaluation.

Economists define complements as goods for which the increase (decrease) in the price of one leads to a decrease (increase) in demand for the other. That was the case in the GE/Honeywell merger of 2001,[27] which revealed divergent policy views between the United States and the European Union. A reduction in the price of jet engines could be expected to increase demand for avionics. In cases where two independent companies, both possessing some degree of market power, are marketing complements, each company will set its price to maximize its own profit. But when those companies merge, the merged company will set prices for a package of the complementary products at a level that maximizes profits on the two goods sold together. The package price will always be less than the total of the prices for the separate products that the constituent companies were charging prior to the merger. The output of the two companies will increase and their profits will also increase. In economic parlance, this is the avoidance of double marginalization.

When General Electric and Honeywell sought to merge, their merger was unopposed by the US antitrust authorities.[28] It was opposed, however, by the European Commission,[29] and the Commission's disapproval was upheld by the European Union's General Court. The European disapproval was partially grounded in what has become known as the doctrine of portfolio effects. In the context of merger evaluation, the term refers to the various anticompetitive effects that can emerge from a merger. (Logically, the term should be understood as referring also to the procompetitive effects that might result from a merger, but in practice the focus of those using this term seems to be on the anticompetitive effects.) In the proposed GE/Honeywell merger, the European authorities focused on GE's production of jet engines and Honeywell's production of avionics. These products were marketed to the same customers and were complements.

The European Commission understood the concept of avoiding double marginalization and that incentives to do so would arise from the merger of two companies

---

[27] Commission Decision of 03/07/2001 declaring the concentration to be incompatible with the common market and the EEA Agreement (Case No. COMP/M.2220 – *General Electric/Honeywell*); *Honeywell International Inc. v. Commission of the European Communities* (2005) (Case T-209/01).

[28] Majoras (2001).

[29] *Honeywell International Inc. v. Commission of the European Communities* (2005) (Case T-209/01).

producing complements. But it appeared to be pursuing a policy of preserving rivalry for its own sake rather than encouraging the kind of competition that fosters efficiency. Because the Commission believed that the merged company's European competitors would be unable to meet the merged company's bundle prices, it treated the merger as threatening its rivalry goal. In GE/Honeywell the issue was bundling that generated efficiencies and advanced consumer welfare. From a US perspective, bundling in such a case is procompetitive and is a reason to approve the merger. The Justice Department advanced this position in a post-litigation conference called by the Organization for Economic Co-operation and Development (OECD) on portfolio effects held in 2001.[30]

Efficiencies are currently generally recognized as procompetitive in both the United States and the European Union. In the case in which a merged company eliminated double marginalization, the costs to its customers would decline, just as a reduction in production costs would generate a reduction in prices, resulting in lower costs to customers. Although the role of efficiencies in European merger evaluation was less clear before the issuance of the revised Merger Regulation in 2004 (Council Regulation 2004), since that time efficiencies have been generally recognized in the European Union as positive factors in merger evaluation.

Despite the explicit concern for efficiency, the Commission's guidelines on nonhorizontal mergers[31] raises issues reminiscent of the Commission's position in GE/Honeywell. Paragraph 101 of these guidelines identifies as problematic a merged company's bundling of complementary goods in one of which it possesses market power. The guidelines suggest that in these circumstances, the merger could result in reduced sales of the competitive (and complementary) good by nonintegrated rivals (i.e., rivals producing only the "competitive" good). The Commission did not address the market effects of the bundling or tying, only the effects on rival companies. Neither did the Commission address the situation in which both merging companies possess market power over their respective products and the welfare-enhancing elimination of double marginalization. The provision may be innocuous if intended to be limited to the case involving market power over only one good, but it comes so close to reiterating the Commission's position in GE/Honeywell that clarification would have better served the public interest.

The guideline provision over the effect of bundling or tying on rivals reiterates the concern expressed by the Congress in enacting Clayton Act § 3 (Clayton Antitrust Act 1914) and that is addressed by the per se rule governing tying. So described, the provision could be seen as a movement in the direction of harmonizing EU law with its American counterpart. But the American law is moving toward an ultimate embrace of the rule of reason and away from per se rules. This movement is

---

[30] US Department of Justice, Antitrust Division Submission (2001).
[31] Guidelines on the Assessment of Nonhorizontal Mergers under Council Regulation on the Control of Concentrations between Undertakings (2008/C245/07).

evidenced, for example, in Justice O'Connor's call for the abolition of the per se rule in *Jefferson Parish* and in the D.C. Circuit's newly created exception to the per se rule for platform software. So an embrace of a now obsolete US position on tying (dating from 1914 to the mid-twentieth century) cannot be regarded as progressive harmonization.

The guideline provision under discussion also suggests that the presence of network effects might exacerbate the problem that it has identified. This may be an implicit reference to the Commission's 2004 ruling in the European *Microsoft* case that found Microsoft's integration of its media player into its operating system to be an abuse of a dominant position. The Commission's ruling on the media player is inconsistent with the D.C. Circuit's exception to the per se rule as applied to platform software. The future significance of paragraph 101 cannot be confidently forecast. One interpretation might view it as an attempt to reconcile pre- and post-2004 EU views of the role of efficiency in mergers. Alternatively, it could be seen as justification for future interference with mergers that improve both consumer and total welfare.[32]

19.4. ESSENTIAL FACILITIES

The essential facilities doctrine requires the owner of an "essential facility" to share its use with its rivals. The doctrine originated in the early years of the twentieth century in a case involving access to railroad terminal facilities at St. Louis, facilities that included the two railroad bridges across the Mississippi River at St. Louis and the only railroad ferry line.[33] Although twenty-four railroads terminated in St. Louis, the terminal facilities were in the control of a smaller group of railroads. It would have been impractical for any of the excluded railroads to build its own bridge and the existing bridges had the capacity to carry all of the railroad cars across the river. When the government had challenged the minority's control under Sections 1 and 2 of the Sherman Act, the Supreme Court ruled that the excluded railroads must be admitted into ownership of the facilities, or, if any railroad did not wish to become an owner, then it was entitled to use of the facilities at a reasonable charge. Although compulsory sharing was a novel remedy, the remedy was economically efficient because the facilities in question were too costly for any one railroad to build itself, and the facilities had the capacity to serve all of the railroads terminating at St. Louis. The facilities, in other words, had a sharply declining average cost up to capacity, and any other remedy would involve waste.

---

[32] This is not to deny that blocking mergers that improve efficiency through bundled selling could ever lower welfare. Our view, however, inclines to that of most economists who see future anticompetitive effects as in the *GE/Honeywell* case as speculative by comparison with substantial immediate gains. We are persuaded by the larger number of conditions for consumer or welfare losses to be likely that are outlined in US Department of Justice, Antitrust Division Submission (2001).

[33] *United States v. Terminal R.R. Association* (1912) 224 U.S. 383.

In Europe, the essential facilities doctrine has been expanded into intellectual property law. Intellectual property has sometimes been treated as an essential facility, and intellectual property owners have been required to share it with competitors. The leading cases applying the essential facilities doctrine to intellectual property are *Magill*,[34] *IMS*,[35] and *Microsoft*.[36]

In the mid-1990s, the Court of Justice ruled in *Magill* that copyrights over television schedules held by several Irish television stations would have to be shared with Magill, who wanted to publish a regional television guide. In the *IMS* case of 2004, the Court of Justice ruled that the owner of a copyrighted record-keeping system (referred to as a brick structure in the case) used to record uses of various pharmaceutical products could be ordered to share the system with rivals if certain conditions were met, and in *Microsoft*, also in 2004, the General Court ruled that communications protocols through which Microsoft's 2000 server software units communicated with other units of the same program would have to be shared with rivals, despite Microsoft's claim that these protocols were protected by trade secret law. In these cases the courts avoided using the phrase "essential facility," but their rulings followed the logic of that doctrine.

This use of the essential facility doctrine can sometimes be conceptualized as closely related to bundling, since the exercise of the doctrine may involve the deconstruction of a product or service that is composed of several bundled elements. The Microsoft server software, for example, was combined with the communications protocols. When the communicative abilities of the Microsoft server software were identified, the European Commission required the protocols to be unbundled from the rest of the server software and shared with rivals so that all companies shared in the advantages of the enhanced communication. Similarly, when the advantages of IMS's record-keeping system were identified, the Court of Justice was willing to require the system to be separated from the IMS package of services and to be shared with IMS's rivals if specified conditions were met. The most important of these conditions was that the firm seeking access to the intellectual property must use that intellectual property to offer new products or services not offered by the IP rights holder and for which there was a potential consumer demand. In *Microsoft* the General Court ruled that this condition could be satisfied by the rival offering a differentiated (but not identical) product in competition with the rights holder. This is a new product only in a technical sense and does not serve any demand not already served by the rights holder. Thus *Microsoft* made clear that the requisite conditions were easily met and that the essential facility inroads into intellectual property rights were substantial.

---

[34] *Radio Telefis Eireann (RTE) v. Commission of the European Communities* (1995) ECR I-743.
[35] *IMS Health GmbH & Co. OHG v. NDC Health GmbH & Co. KG* (2004) ECR I-5039.
[36] *Microsoft Corp. v. Commission of the European Communities* (2007) Case T-201/04.

19.5. BUNDLING AND INNOVATION

When Congress enacted Clayton Act § 3 in 1914[37] it was reacting to the practices of United Shoe Machinery Company requiring customers to take all or most of their machinery from that company. The terms of the Company's lease – defended by United's counsel and later Supreme Court Justice Louis Brandeis – used the customer's desire for some machines as a lever to force that customer to obtain its other machines from that same source, or so Congress thought. This conception – that United possessed market power over machinery and used that market power to coerce or otherwise force customers to obtain all of their machinery needs from United – underlay the provisions of Section 3 and the tying doctrine that Section 3 spawned. Indeed, the "policy" underlying Section 3 also caused the courts to create the "misuse" doctrine, independent from antitrust law, as part of patent and copyright law.[38]

Throughout much of the twentieth century, the courts treated tying as inherently anticompetitive.[39] This is the reason for the per se rule and why tying is treated as misuse under the patent law. Yet while the Supreme Court in mid-century was busy condemning tying as lacking any redeeming virtue, Congress, taking a somewhat different view, was simultaneously limiting the scope of the misuse doctrine. Eventually the courts began to understand that tying and bundling can generate positive effects, enhancing consumer welfare. This development took time and remains a work in progress. It appears, however, that the per se rule that has long been associated with tying will eventually be abandoned.

In practice bundling usually involves goods that are complements. A common instance of this relation involves a bundle or tie whose elements are used by customers in a fixed proportion with each other. This is the situation in which Chicago economists have argued that a bundle tying the two products together cannot increase the power of the seller. In a simple static model, if the seller has a monopoly over product A, for example, he cannot increase his power or his income by forcing buyers to take product B from him, rather than from some other seller, or by forcing buyers to pay a premium price for product B. These results occur because the two products are used in a fixed ratio so the buyers' goal is to obtain the whole bundle, and any price premium paid for product B reduces the amount buyers are willing to pay for product A.[40]

---

[37] Pub.L. 63-212, 38 Stat. 730, codified at 15 U.S.C. §§ 12–27, 29 U.S.C. §§ 52–53.
[38] *Motion Picture Patents Co. v. Universal Film Co.* (1917) 243 U.S. 502 (patent misuse); *Lasercomb Am., Inc. v. Reynolds* (1990) 911 F.2d 970, 976–7 (4th Cir. 1990) (copyright misuse).
[39] Tying agreements "are unreasonable in and of themselves whenever a party has sufficient economic power with respect to the tying product to appreciably restrain free competition in the market for the tied product and a 'not insubstantial' amount of interstate commerce is affected." The Court's carefully chosen language makes all tying by large sellers per se illegal (*Northern Pacific Ry. v. United States* (1958) 356 U.S. 1, 6).
[40] The simple argument presented here is based on two critical assumptions: the tying good must be used with the tied good and there are constant costs in the production of the tied good

The situation changes when the complements are not used in a fixed proportion. The old IBM case involving punched cards provides a good example.[41] In that case, IBM apparently offered its computers at low prices, probably not much in excess of its marginal cost. It then required users to buy computer cards from it at a significant markup over cost. Thus IBM made its profit from the sale of the cards. The beauty of this arrangement was that customers who placed a high value on the computers were likely to use them more and thus to require more cards. They paid more through their card purchases while those who valued the computers less used fewer cards and thus paid less. The cards were metering devices. The payments were effectively user charges.

The computer and punch card metering system is an example of the simplest form of second-degree price discrimination: all users paid an identical lump sum (nominally for the computer) and users then paid a fixed price for the cards, but could choose the number of cards to be taken. Were both the price of the computer and each unit of card sales adjusted to draw the maximum willingness to pay for each purchaser (without any possibility of resale by purchasers) one would have an instance of perfect price discrimination (or first-degree price discrimination). It involves no resource misallocation and thus no deadweight loss in contrast with other forms of monopoly pricing. Perfect price discrimination furthers aggregate social welfare, but because it also eliminates consumer surplus, transforming consumer surplus into seller profits (producer surplus), it does not further consumer welfare. Accordingly, perfect price discrimination either is compatible with antitrust policies or is not, depending upon which of two contending antitrust goals (aggregate social welfare or consumer welfare) is ultimately determined to underlie antitrust law.[42]

While perfect price discrimination does not increase consumer welfare, second-degree price discrimination may well do so. Professor Hovenkamp has stressed that some buyers may be willing to buy ink cartridges at a premium price but be unwilling to pay the market price of the printer unencumbered by a tied-ink obligation.[43] In this case, the tie makes possible the sale of additional printer units, demonstrating (in this case) its procompetitive character. This is a real possibility in many cases but not a general result about consumer welfare because it depends on the relative magnitudes of the production costs of the tying and tied goods as well as the configuration of demand.

(Whinston 1990, 837–59). In addition, when sales are made over time, upgrades and switching costs can make tying profitable even when both of those conditions hold (Carlton and Waldman 2011, 675–706). All of this means that real-world cases can only infrequently be sorted decisively on the basis of simple theoretical ideas.

[41] *International Business Machines Corp. v. United States* (1936) 298 U.S. 131.
[42] Blair and Sokol (2012).
[43] Hovenkamp and Hovenkamp (2010).

We conclude that ties are sometimes useful devices with no necessary anticompetitive effect on the market. For that reason, a per se rule is inappropriate for ties. The existence of such a rule is inconsistent with the Supreme Court's stated criteria that limits per se rules to restraints "that would always or almost always tend to restrict competition and decrease output."[44] Because tying does not always or almost always decrease output, the per se rule governing tying will probably be abolished within the foreseeable future. For the present, however, business firms must plan their activities around this dysfunctional rule.

19.6. BUNDLED DISCOUNTS AND REBATES

Bundled discounting involves a reduced price for the purchase of a package of goods from a seller's offerings; it is related to "mixed bundling" because all of the goods are also available from the seller at higher prices outside of the discounted package.[45] It differs from the usual textbook example of "mixed bundling," however, because in addition to the availability of all of the goods without the discount, the composition of the bundle is often highly flexible.[46] Antitrust problems may arise when the seller of the bundle is competing with a rival that does not sell all of the products in the bundle and, in order to stay competitive, must reduce the price of his products, perhaps below production cost, to offset the discount on the bundle.

Bundled discounts and rebates are widely used but may be especially important in high technology sales. The reported cases in this area involve significant numbers from the health industry where the development, sale, and use of high-level technology is commonplace. Indeed, two of the early bundled discount cases involved the health industry: *SmithKline Corporation v. Eli Lilly & Co.*[47] and *Ortho Diagnostic Systems, Inc. v. Abbott Laboratories, Inc.*[48] *SmithKline*, decided by the Third Circuit in 1978, involved a package discount offered by Eli Lilly on purchases of its cephalosporin drugs. Lilly competed with SmithKline on a generic cephalosporin and also marketed two other cephalosporins on which Lilly held patents in a price-discounted package with the generic product. The court ruled that Lilly was monopolizing by linking the patented products with the generic one, forcing SmithKline to pay rebates on its competitive generic product equal to the rebates

---

[44] *Leegin Creative Products, Inc. v. PSKS, Inc.* (2007) 551 U.S. 877, 886.
[45] Carlton and Perloff (2005) at 324–8.
[46] Bundled discounting may be quite similar to a general volume discount offered on a particular product or it can be tailored to the purchases of particular buyers. (Whether applied to single or multiple products, such a discount when offered to repeat purchasers is often called "loyalty discounting.") Tailored discounts are regarded with suspicion in the EU (EU Commission 2005; EU Commission 2009).
[47] 575 F.2d 1056 (3d Cir. 1978).
[48] 920 F. Supp. 455 (S.D.N.Y. 1996).

paid by Lilly on volume sales of three products, and jeopardizing SmithKline's continued presence in the market.[49]

In 1996, a federal district court in New York also confronted bundled discounts in the market for blood screening tests. That court required a plaintiff asserting a Sherman Act § 2 claim for bundled pricing against a monopolist to prove either that the monopolist had priced below its average variable cost or that the plaintiff was at least as efficient a producer of the competitive product as the defendant, but that the defendant's pricing makes it unprofitable for the plaintiff to continue to produce.[50] Because the plaintiff could not prove either prong of the court's test, the case was dismissed.

It was in low tech sales, however, that the Third Circuit, a decade ago, captured widespread attention in a decision involving bundled discounts.[51] In *LePage's Inc. v. 3M*, the court determined that the 3M Company's bundled discounts constituted monopolization and left the business community with ill-defined criteria for distinguishing lawful from unlawful bundled discounts.

The 3M decision involved across-the-board rebates to large purchasers of 3M products, with broad freedom among the purchasers to choose the products in the bundle. LePage's, which competed with 3M in the sale of transparent tape, asserted that it was impossible for it, as the producer of a single product, to match the aggregate discount on large packages. The court, sitting en banc, agreed and ruled in favor of LePage's on its monopolization claim. Although the Third Circuit asserted that the large market share represented by sales of 3M's Scotch brand tape gave it an effective monopoly,[52] the discount that adversely affected LePage's was on the whole package, a package composed mostly of competitive (i.e., non-monopolized) products. The bundled pricing, by seriously injuring LePage's, 3M's primary rival in transparent tape, constituted monopoly maintenance. The 3M decision was widely criticized because the court did not develop criteria to define the antitrust offense that it created.

The Third Circuit was correct that LePage's (the plaintiff) was unable to drop its price on its one product sufficiently to offset a small percentage discount on a large dollar-volume of goods. But the parameters of 3M's offense are unclear. In *Smith-Kline*, for example, the bundle was composed of patented products that were leveraged to foster sales of the competitive product. In *LePage's*, it was the aggregate sales of the competitive products that (in the court's view) were leveraged to foster sales of the monopolized product. Does it make a difference whether 3M's pricing was profit maximizing in the short run (i.e. without excluding an equally efficient

---

[49] 575 F.2d at 1065.
[50] 920 F. Supp. at 469.
[51] *LePage's Inc. v. 3M*, 324 F.3d 141 (3d Cir. 2003).
[52] This has produced confused commentary because LePage's was not competing directly against "Scotch" tape; rather it supplied unbranded tape for resale as the customer's house brand, and it was competing against 3M's "fighting brand," "Highland" as well as 3M's own private label tape.

competitor) or not? What is the competitive significance of a bundle's size? What is the significance of the fact that most of the products in the bundle were competing with similar products produced by others?[53] Is it relevant that those rival producers would be pressed to lower their own prices to compete with the lower prices that 3M was offering? Could LePage's have organized competing bundles? Could the other suppliers have offered them?

A competing bundle composed of LePage's tape and independently produced competitive products could have neutralized the attractiveness of the 3M bundles. The 3M discounts, moreover, were not offered to everyone; they were offered to only a handful of large buyers. So the remaining market was unaffected. The court attempted to deal with this issue by arguing that large buyers were critical to achieving scale economies and that the loss of sales drove up LePage's costs. But the court was short on supporting data. The court asserted that entry barriers protected 3M's Scotch tape monopoly. Yet the technology involved is straightforward. LePage's had entered the private label tape market and expanded rapidly to gain an 88 percent share of that market. And Walgreens and Dollar General were dealing with foreign producers.[54] So the court's assertion of entry barriers is another area that needed to be backed up with data.

The Antitrust Modernization Commission (AMC) criticized the *LePage's* ruling and developed its own set of rules for determining the lawfulness of bundled discounts.[55] The AMC gave us a set of three rules for evaluating the lawfulness of bundled discounts: (1) The first was a so-called discount attribution rule. The court is to apply the whole discount on the defendant's entire bundle to the competitive product (i.e., the product in which the plaintiff and the defendant compete). After so applying the discount, the court is to determine whether the competitive product is being sold below cost. If it is not, the case can be dismissed.[56] That rule seems to create a safe harbor, rather than proving legality. (2) Then the court determines whether the defendant's "loss," so computed, could be recouped. (3) Finally, the court determines the impact on the market. The last determination is (or should be) the most critical.

The Ninth Circuit in its *Peacehealth* decision[57] seems to have converted the discount attribution rule into a test of legality. Then it rejected the AMC's Rule 2 as

---

[53] Daniel Crane (2005, 31–2) argues that a discounted bundle composed of a monopoly product and numbers of "competitive" (i.e., nonmonopolized products) would likely (but not certainly) trigger corresponding discounts from suppliers in the competitive product markets.

[54] 324 F.3d at 172–3.

[55] Antitrust Modernization Commission (2007).

[56] Versions of the discount attribution rule can also be found in *Ortho Diagnostic Systems v. Abbott Laboratories*, 920 F. Supp. 455 (S.D.N.Y. 1996) at 469 and *Virgin Atlantic Airways Ltd. v. British Airways PLC*, 69 F. Supp. 2d 571, 580 n. 8 (S.D.N.Y. 1999), aff'd, 257 F.3d 256 (2d Cir. 2001).

[57] *Cascade Health Solutions v. Peacehealth*, 515 F.3d 883 (9 Cir. 2008). In *Masimo Corp. v. Tyco Health Care Group*, 350 Fed. Appx. 95, 2009 WL 3451725, 2009-2 Trade Cases ¶ 76,780 (9th Cir. 2008), the 9th Circuit affirmed the dismissal of a bundled discount claim in which the product price was above cost, even as adjusted by the discount attribution rule.

inappropriate because in bundled discounting there is often no loss to recoup; and it rejected the third rule on the ground that the court would already have considered the market effects when it determined the plaintiff's standing. Thus the Ninth Circuit transformed the AMC's first rule from a safe harbor into a test of legality. In our view, the discount attribution rule should be treated as a safe harbor and not as a test of legality, and if the practice is not protected by the safe harbor, its legality should turn on market effects: If the practice alters the market structure by significantly increasing concentration or erecting entry barriers, then it should be treated as prima facie anticompetitive.

We believe that the above approach will ultimately prevail in the courts. Right now, however, we have disarray in the official approach to bundled discounting: each of the three official approaches (of the two circuit courts and the AMC) is wrong and is inconsistent with the other two. The analytical problem faced by the courts has given rise to a large literature. Among the most interesting academic contributions are papers by Patrick Greenlee, David Reitman and David Sibley,[58] Einer Elhauge,[59] Barry Nalebuff,[60] and Daniel Crane.[61] The Greenlee et al. article shows how bundles and ties can potentially enhance welfare of both sellers and buyers as well as decrease welfare. The Elhauge article, among other insights, presents an analysis of tying and bundling keyed to the present state of the law. The Nalebuff article develops a useful (and theoretically precise) analysis of exclusionary bundling using the price/cost test. The 2005 Crane article, the earliest of his several contributions to bundling and loyalty analysis, offers, inter alia, a consideration of bundling as a long-term strategy, a perspective often missing from the bundling literature.

The European Commission treats bundled discounts in its Guidance, apparently adopting an "as efficient competitor" test as the criterion of lawfulness. According to the Guidance:

> If the incremental price that customers pay for each of the dominant undertaking's products in the bundle remains above the LRAIC [long run average incremental cost] of the dominant undertaking from including that product in the bundle, the Commission will normally not intervene since an equally efficient competitor with only one product should in principle be able to compete profitably against the bundle.

There is no reference to a discount attribution rule and no discussion of how an equally efficient competitor that produces only one product will be able to compete profitably against a discount calculated on a multi-product bundle, the issue that has troubled the American courts and the AMC.

---

[58] Greenlee, Reitman, and Sibley (2008).
[59] Elhauge (2009).
[60] Nalebuff (2005).
[61] Crane (2005; 2006; 2013).

The Commission has developed an analysis that could have been applied to deal with this issue in its approach to loyalty discounts. There the Commission uses an "effective price" concept. As explained in the next section, the effective price to the buyer (being continually reduced by the growing accumulated but contingent discounts) falls throughout the contract period (until the buyer's purchases reach the contractually specified target). A similar approach could have been taken to bundled discounts, with the effective price for a competitive product falling as more items are added to the bundle. That effective price then could be compared with the seller's costs. In a case in which the effective price fell below the seller's costs, an equally efficient competitor would be unable to compete. This result would, of course, resemble the result produced by the discount attribution rule. We have observed above, however, that the discount attribution rule has generated significant confusion among the US authorities. So it may have been wise to avoid importing it into Europe.

19.7. LOYALTY DISCOUNTS

Antitrust evaluations of bundled discounts are closely related to antitrust evaluations of loyalty discounts: A loyalty discount is granted for multiple purchases from a seller, and multiple purchases could sometimes be described as a bundle. Ordinarily we think of a bundled discount as a discount applicable to the simultaneous purchase of several products from a seller, and think of a loyalty discount as a discount applicable to periodic purchases of the same product from the same seller. Yet these terms are sufficiently flexible as to embrace instances of overlap.[62] Indeed, the EU Guidance applies the same analysis to both kinds of discounts.

American courts have had little trouble with loyalty discounts, generally analyzing them similarly to the way that they would analyze exclusive-supply contracts. Unless the supplier possesses significant market power and ties up a "substantial share" of the market (at least 40 percent), exclusive-supply contracts are treated as innocuous; and the same approach generally applies to contracts conferring loyalty discounts. *Allied Orthopedic Appliances Inc. v. Tyco Health Care Group LP*,[63] a case from the Ninth Circuit, is instructive. In that case, purchasers claimed that they were overcharged on pulse oximetry sensors by Tyco under two types of agreements (market share discount agreements and sole-source agreements). Although the court repeatedly refers to them as exclusive-supply contracts, they appear to bear the characteristics widely

---

[62] *Masimo Corp. v. Tyco Health Care Group*, 2006 WL 1236666 C.D. Cal. 2006, aff'd, 350 Fed. Appx. 95 (9th Cir. 2009). This case also illustrates the overlap. The plaintiff Masimo charged Tyco with violating the Sherman Act through bundled discount agreements, sole source agreements and market share agreements. Tyco's sole source and market share agreements were similar to typical loyalty discount agreements, in that the discount apparently was earned when the customers actually purchased a high percentage of their requirements from Tyco, and because the agreements involved multiple products they were properly describable as bundled discount agreements.

[63] 592 F.3d 991 (9th Cir. 2010).

associated with loyalty discount agreements. According to the court, the plaintiff hospitals were not obliged to buy any sensors from Tyco; rather they lost their discount if they purchased from other suppliers. The court did not see this loss of discount as particularly important, and it saw the freedom of the buyers to purchase from others as negating the existence of an exclusive contract. If the buyers, however, were to lose discounts on previously purchased products when they purchased from others, this would subject them to an increasing incentive to continue with their original supplier. The arrangement then would be equivalent to exclusive-supply contracts, except for the growing incentive to continue with the original seller. This incentive is described as a "suction effect" in Europe and underlies the European Commission's Guidance provisions dealing with loyalty and bundled discounts.[64]

Since the Supreme Court's decision in *Brooke Group*, some defendants have invoked the "as efficient competitor" or price/cost test as a defense to alleged antitrust infractions involving, inter alia, exclusive-supply agreements and bundled and loyalty discounts. Under this approach, the key question is whether the defendant excluded the plaintiff by pricing below cost.[65] 3M, asserting that all of its prices were above cost, unsuccessfully raised this defense in *LePage's*. By contrast in *Nicsand*,[66] where exclusive-supply contracts were in issue, 3M raised that same defense successfully. There it contended that it won exclusive agreements with large retailers because it offered more attractive (above-cost) prices that could have been met by equally efficient competitors. In *ZF Meritor, LLC v. Eaton Corp*,[67] the defendant Eaton raised the same defense. Eaton, a manufacturer of heavy-duty truck transmissions, defended its exclusive-supply agreements with the four North American heavy-duty truck manufacturers on the ground that its prices were above cost and therefore could have been met by an equally efficient competitor. The Third Circuit rejected that defense on the ground that price was not the critical factor used by Eaton to exclude its rival, Meritor.

Because loyalty rebate contracts are normally equivalent (apart from the penalties for nonperformance) to exclusive-supply contracts, the decision in *Eaton* denying the applicability of the price/cost test needs explanation. On the one hand, if the defense had been accepted, it would have raised the question of whether all

---

[64] EU Guidance ¶40. Because the conditional rebates grow with buyer purchases, the amount rebated to the buyer who purchases the target amount grows as well. And at any given time, the amount of sales necessary to reach the target amount is declining. Thus the amount to be rebated is growing and the amount of additional purchases needed to reach the target is declining. The combined effect of these two factors is to reduce the effective per unit sales price of additional units.

[65] The reader will recognize the discount attribution rule employed by the AMC and the Ninth Circuit to evaluate bundled discounts as a version of the price/cost test discussed earlier in this chapter. In its *Eaton* decision, the Third Circuit indicated that where price is the critical factor in exclusion in a loyalty-discount context, the price/cost test is applicable. See 696 F.3d at 274 n. 11. This approach carries the potential for incorporating the EU "suction effect" analysis.

[66] *Nicsand, Inc. v. 3M Co.*, 507 F.3d 442 (6th Cir. 2007).

[67] 696 F.3d 254 (3d Cir. 2012).

exclusive-supply contracts could be defended on the price/cost test. And the rejection of that test would have raised the question of whether the price/cost was inapplicable to all exclusive-supply contracts. Neither result, we believe, holds. The price/cost test is widely, but not universally, applicable. The Third Circuit was correct to reject it in the *Eaton* case, because, although Eaton was not shown to have priced below its costs, the percentage share of the market tied up was close to 100 percent and the public interest would be better served by ensuring that the truck manufacturers had alternatives open to them. These conclusions do not depend on the level of the supplier's prices.

In approaching loyalty discounts, the European Commission distinguishes between "noncontestable" and contestable portions of customer demands. The former reflect the demand of a buyer that can only be satisfied by the dominant supplier's brand (because, for example, the buyer's own customers insist on the particular brand) and the latter reflect that part of a buyer's demand that is not brand specific. Ordinarily, a buyer has freedom to choose among substitutes in satisfying the contestable portion of its demand. But loyalty rebates can enable a dominant seller to leverage the power inherent in the noncontestable portion of a buyer's demand to force the buyer to satisfy the contestable portion with purchases from the dominant supplier. This could result when the dominant supplier's effective price[68] to a buyer of additional purchases falls to levels that are beneath the supplier's cost and thus beyond the ability of an equally efficient supplier to meet.[69]

The European case-law on loyalty discounts reflects a difference between the European Commission and the European Union courts, i.e., the Court of Justice and the General Court. These courts have employed a so-called form-based analysis in applying the EU Treaty provisions on competition policy. This form-based analysis is akin to the practice of American courts in applying a per se rule. The court determines whether the behavior before it falls within a category that it has previously identified as unlawful. If so, the behavior is treated as presumptively unlawful. By contrast, the Commission has been developing an effects-based analysis, in which the Commission examines the facts of the particular cases to reach a conclusion about the lawfulness of the behavior in question. The Commission's approach thus appears to be more concrete and fact specific while the approach of the courts is more abstract and category dependent.

The European Commission, in carrying out its effects-based analyses, has adopted criteria for evaluating loyalty discounts (including the "equally efficient competitor" standard) and incorporated those criteria in its Guidance. The Commission's rulings, however, are reviewable first by the General Court and subsequently by the Court of Justice. These courts appear to take a form-based analysis to loyalty

---

[68] The effective price that the dominant supplier's rival will have to match is the normal (list) price less the rebate the customer loses by switching. See Guidance ¶41 and 324 F.3d at 172–3.
[69] Guidance ¶39.

discounts in which retroactive rebates are treated as presumptively illegal.[70] But the Commission chooses the enforcement cases that it will bring. Thus the courts and the Commission take somewhat different stances to determining the legality of loyalty rebates and they possess different and somewhat offsetting powers. It is not clear how these interrelated but very different approaches will develop.

19.8. CONCLUSION

Several lessons can be drawn from the consideration of bundling in high technology industries. First, the stakes are high: virtually all observers agree that a determination of legality based on traditional restrictions against tying could be highly detrimental to innovation and overall economic progress. Second, the most defensible approach to applying antitrust to high technology industries cannot substitute one simple approach for another: shifting to an "anything goes" paradigm is neither warranted on logical or empirical grounds nor is it politically feasible. Despite its complexities, a rule of reason approach is the only feasible path to applying antitrust to high technology industries. The ordinary version of the rule, which places the burden of proof on the plaintiff, seems keyed to respecting the necessary autonomy of these industries, avoiding the application of rigid rules to inherently dynamic behaviors, and minimizing misunderstanding by the courts. Third, a rule of reason approach means that the responsibility for ensuring the workability of the system will fall on policymakers and the courts. Finally, the policy differences between the United States and the European Union in antitrust loom larger in high technology than perhaps in any other, and efforts to achieve a common view may be most consequential.

REFERENCES

Allied Orthopedic Appliances Inc. v. Tyco Health Care Group LP 592 F.3d 991 (9th Cir. 2010).
Antitrust Modernization Commission. 2007. *Report and Recommendations*. Washington, DC: The Antitrust Modernization Commission.
Baker, Jonathan B. 2007. Beyond Schumpeter vs. Arrow: How Antitrust Fosters Innovation. *Antitrust Law Journal*, 73, 575–602.
Blair, Roger D. and D. Daniel Sokol. 2012. The Rule of Reason and the Goals of Antitrust: An Economic Approach. *Antitrust Law Journal*, 78, 471–504.
California Computer Products, Inc. v. International Business Machines Corp. (1979) 613 F.2d (9th Cir. 1979).
Carlton, Dennis and Ken Heyer. 2008. Extraction vs. Extension: the Basis for Formulating Antitrust Policy towards Single-firm Conduct. *Competition Policy International*, 285–305.

---

[70] *Tomra Systems ASA v. Commission* (Case 549/10); *Manufacture française des pneumatiques Michelin v. Commission* (Case T-203/01); *NV Nederlandsche Banden Industrie Michelin v. Commission* [1983] ECR 3461 (Case 322/81); *Hoffmann-La Roche v. Commission* [1879] ECR 461, ¶91 (Case 85/76).

Carlton, Dennis and Harvey Perloff. 2005. *Modern Industrial Organization*, 4th edn. Boston, MA: Pearson Addison Wesley.

Carlton, Dennis and Michael Waldman. 2011. Upgrades, Switching Costs and the Leverage Theory of Tying. *The Economic Journal*, 122, 675–706.

Cascade Health Solutions v. Peacehealth 515 F.3d 833 (9th Cir. 2008).

Clayton Antitrust Act 1914 § 3, Pub. L. 63-212, 38 Stat. 731.

Commission Decision of 03/07/2001 declaring a concentration to be incompatible with the common market and the EEA Agreement (2001) (Case No COMP/M.2220 – *General Electric/Honeywell*).

Computer Associates Int'l, Inc. v. Altai, Inc., 982 F.2d 693, 712 (2d Cir. 1992).

Crane, Daniel A. 2005. Multiproduct Discounting: A Myth of Nonprice Predation, *University of Chicago Law Review*, 72, 27–48.

2006. Mixed Bundling, Profit Sacrifice, and Consumer Welfare. *Emory Law Journal*, 55, 423–96.

2013. Bargaining Over Loyalty. *Texas Law Review*, 92, 253–300.

Elhauge, Einer, 2009. Tying, Bundled Discounts, and the Death of the Single Monopoly Profit Theory. *Harvard Law Review*, 123, 397–450.

EU Commission. 2005. Discussion Paper on the Application of Article 82 of the Treaty to Exclusionary Abuses.

2009. Guidance on the Commission's Enforcement Priorities in Applying Article 82 of the EC Treaty to Abusive Exclusionary Conduct by Dominant Undertakings. 2009. 42 (C 45/02).

EU Council Regulation No. 139/2004 of 20 January 2004 on the control of concentrations between undertakings (the EC Merger Regulation). Official Journal L 24, no. 29.01.

Evans, David S. and Richard Schmalensee. 2001. 2002. Some Economic Aspects of Antitrust Analysis in Dynamically Competitive Industries. NBER Working Paper w8268. www.nber.org/papers/w8268, subsequently published in Adam B. Jaffe, Josh Lerner, and Scott Stern (eds.). 2002. *Innovation Policy and the Economy National Bureau of Economic Research Book* 2. Cambridge, MA: MIT Press: 1–50.

Fisher, Franklin M., John J. McGowan, and Joen E. Greenwood. 1983. *Folded, Spindled, and Mutilated: Economic Analysis and USV IBM*. Cambridge, MA: MIT Press.

Gifford, Daniel J. and Robert T. Kudrle. 2011. Antitrust Approaches to Dynamically Competitive Industries in the United States and the European Union. *Journal of Competition Law and Economics*, 7, 695–731.

Greenlee, Patrick, David Reitman, and David S. Sibley. 2008. An Antitrust Analysis of Bundled Loyalty Discounts. *Journal of Industrial Organization*, 26, 1132–52.

Guidelines on the Assessment of Nonhorizontal Mergers under Council Regulation on the Control of Concentrations between Undertakings (2008/C245/07) (2008).

Hoffmann-La Roche v. Commission [1879] ECR 461, ¶91 (Case 85/76).

Honeywell International Inc. v. Commission of the European Communities (2005) (Case T-209/01).

Hovenkamp, Erik N. and Herbert J. Hovenkamp. 2010. Tying Arrangements and Antitrust Harm. *Arizona Law Review*, 52, 925–76.

Hovenkamp, Herbert. 2006. Discounts and Exclusions. *Utah Law Review*, January, 841–62.

IMS Health GmbH & Co. OHG v. NDC Health GmbH & Co. KG (2004) ECR I-5039.

International Business Machines Corp. v. United States (1936) 298 U.S. 131.

Jefferson Parish Hospital District No. 2 v. Hyde (1984) 466 U.S. 2, 32.

Lasercomb Am., Inc. v. Reynolds (1990) 911 F.2d 970, 976–77 (4th Cir. 1990).

Leegin Creative Products, Inc. v. PSKS, Inc. (2007) 551 U.S. 877, 886.

LePage's Inc. v. 3M (2003) 324 F.3d 141 (3d Cir.).
Majoras, Deborah Platt. 2001. GE-Honeywell: The US Decision. Remarks of Deputy Assistant Attorney General Antitrust Division US Department of Justice on November 29.
Manufacture française des pneumatiques Michelin v. Commission (Case T-203/01).
Masimo Corp. v. Tyco Health Care Group, 2006 WL 1236666 C.D. Cal. 2006), aff'd, 350 Fed. Appx. 95 (9th Cir. 2009).
Microsoft Corp. v. Commission of the European Communities (2007) Case T-201/04.
Motion Picture Patents Co. v. Universal Film Co. (1917) 243 U.S. 502.
Motta, Massimo. 2004. *Competition Policy: Theory and Practice*. Cambridge University Press.
Nalebuff, Barry. 2005. Exclusionary Bundling. *Antitrust Bull.*, 50, 321–70.
Nicsand, Inc. v. 3M Co., 507 F.3d 442 (6th Cir. 2007).
Northern Pacific Ry. v. United States (1958) 356 U.S. 1, 6.
NV Nederlandsche Banden Industrie Michelin v. Commission [1983] ECR 3461 (Case 322/81).
Ortho Diagnostic Systems v. Abbott Laboratories, 920 F. Supp. 455 (S.D.N.Y. 1996).
Radio Telefis Eireann (RTE) v. Commission of the European Communities (1995) ECR I-743.
Sega Enterprises Ltd. v. Accolade, Inc., 977 F.2d 1510, 1524 (9th Cir. 12, 1992, 1993).
SmithKline Corporation v. El Lilly, 575 F.2d 1056 (3d Cir. 1978).
Telex Corp. v. International Business Machines Corp. (1975) 510 F.2d 894, 899 (10th Cir. 1975).
Tomra Systems ASA v. Commission (Case 549/10) ECLI:EU:C:2012:221.
United States v. Loew's, Inc. (1962) 371 U.S. 38.
United States v. Microsoft Corp. (1995) 1995–2 Trade Cases ¶ 71,096 (D.D.C. 1995).
United States v. Microsoft Corp. (1998) 147 F.3d 935, 945–946 (D.C. Cir. 1998).
United States v. Microsoft Corp. (1999) 84 F. Supp. 2d 9, (D.D.C. 1999).
United States v. Microsoft Corp. (2001) 253 F.3d 34, 68 (D.C. Cir. 2001).
United States v. Terminal R.R. Association (1912) 224 U.S. 383.
US Department of Justice Antitrust Division Submission. 2001. OECD Roundtable on Portfolio Effects in Conglomerate Mergers. *Range Effects: The United States Perspective*.
US Department of Justice and United States and Federal Trade Commission. 2007. *Antitrust Enforcement and Intellectual Property Rights: Promoting Innovation and Competition*.
Virgin Atlantic Airways Ltd. v. British Airways PLC, 69 F. Supp. 2d 571, 580 n. 8 (S.D.N.Y. 1999), aff'd, 257 F.3d 256 (2d Cir. 2001).
Whinston, Michael D. 1990. Tying, Foreclosure, and Exclusion. *American Economic Review*, 80, 837–59.
ZF Meritor, LLC v. Eaton Corp 696 F.3d 254 (3d Cir. 2012).

20

Tying Arrangements and Intellectual Property

Christopher R. Leslie

The application of antitrust law to intellectual property rights is sometimes viewed as an emerging niche in the law, as IP owners develop new methods of exercising the exclusionary power associated with their IP rights. IP owners, however, have been using tying arrangements to exploit their rights for well over a century. A tying arrangement exists when a seller will not sell one product (the "tying product") unless the buyer agrees to purchase another separate product (the "tied product"). Given IP owners' long use of tying arrangements, Congress and the courts have wrestled with how antitrust law should treat tying arrangements involving IP for a similarly long time. During that century, economic theories about the effects of tying have changed and the law has correspondingly evolved.

This chapter examines how antitrust law limits the ability of IP owners to impose tying requirements on their customers, licensees, and franchisees. It begins by discussing why antitrust law cares about tying arrangements, including the competing theories about the possible anticompetitive effects of tying agreements. After noting the relationship between tying law and patent misuse, this chapter presents the current legal test for determining the legality of challenged tying arrangements, including possible defenses.

20.1. THE POTENTIAL EFFECTS OF TYING ARRANGEMENTS: WHY ANTITRUST LAW CONDEMNS TYING ARRANGEMENTS

Courts condemn tying arrangements based on the assumption that firms are leveraging their market power in one market (the "tying product market") in order to monopolize a second market (the "tied product market"). According to this leverage theory, a tying seller employs a tie-in to suppress competition in the market for the tied product. The Supreme Court has explained that "the vice of tying arrangements lies in the use of economic power in one market to restrict competition on the merits in another."[1] Consumers are injured, according to the Court, because tying arrangements "deny competitors free access to the market for the tied product, not because the party imposing the tying requirements has a better product or a lower price but because of his power or leverage in another market."[2] Concerned about these potential anticompetitive effects, in 1949 the Supreme Court famously asserted that "tying arrangements serve hardly any purpose beyond the suppression of competition."[3] This philosophy led the Supreme Court to condemn tying arrangements as per se illegal.[4] Leverage Theory has been criticized by both judges and scholars as too sweeping in its claims and too quick to condemn tying arrangements.[5]

Modern economists have explained how firms can use tying arrangements to deter entry.[6] The foreclosure associated with tying arrangements can cause rivals to exit the market or to reduce their output.[7] If a rival is forced to reduce output, it may be unable to achieve the necessary economies of scale to be an efficient competitor that can price discipline the tying seller.[8] If the tying seller prevents competitors in the tied product market from reaching minimum efficient scale, then the tying seller can charge a monopoly price for the tied product and can earn additional profits from buyers of the tied product who do not purchase the tying product.[9]

In addition to directly making entry into the tied product more difficult, tying arrangements can increase the costs of entry by forcing a rival in the tied product market to enter the tying product market as well.[10] The Supreme Court has recognized this conundrum, noting that "one of the evils proscribed by the antitrust laws is the creation of entry barriers to potential competitors by requiring them to enter two markets simultaneously."[11] Because "entry into both markets is significantly more expensive than simple entry into the tied market," the dual-entry problem shows another way in which tying arrangements operate as a barrier to entry.[12]

Tying can also diminish innovation by reducing competitors' incentives and abilities to innovate in the tied product market.[13] When a dominant firm imposes a tying requirement on its customers, the competitors in the tied product market have less incentive to innovate because even if their product is better than the dominant firm's tied product, consumers are contractually prevented from buying

---

[1] *Northern Pacific* 1958.
[2] *Fortner* 1969.
[3] *Standard Oil* 1949.
[4] *International Salt* 1947.
[5] Leslie (2004).
[6] Whinston (1990); Nalebuff (2004).
[7] Hovenkamp *et al.* (2016).
[8] *Id.*
[9] Leslie (2011).
[10] Hovenkamp *et al.* (2016).
[11] *Eastman Kodak* 1992.
[12] *Jefferson Parish* 1984.
[13] Leslie (2011).

the improved product.[14] The anti-innovation effects of tying arrangements can be illustrated by the American and European government prosecutions of Microsoft. Microsoft possessed monopoly power in the market for Intel-compatible PC operating systems, but it was concerned that Internet browser technology could evolve so as to undermine its monopoly. If computer applications could run on browsers, as well as on operating systems, consumers would no longer be locked in to using Microsoft's operating system in order to maximize the number of applications that they can use. Consumers could use any operating system and access the applications through their browser. In order to prevent the evolution of browser technology, Microsoft imposed a tying arrangement that forced purchasers of Microsoft's operating system to use its browser, Internet Explorer, as well. The tie-in was initially contractual in nature but Microsoft later made it a technological tie-in by commingling the code for its operating system and browser. Microsoft succeeded in stifling innovation in browser technology and the federal government brought an antitrust suit that found Microsoft liable. Similarly, the European Commission challenged Microsoft's tying of Windows to its media player because "the Commission feared that the tying of Windows Media Player (WMP) and Windows would foreclose competition and stifle innovation in the related media software encoding and management markets, because WMP would become the preferred choice for complementary content and application providers."[15] In both instances, competition law authorities condemned a tying arrangement for reducing innovation in the tied product market. Finally, to the extent that competitive markets are more likely to spur innovation, those tying arrangements that are anticompetitive can reduce long-term innovation as well.[16]

Not all tying arrangements injure competition. Whether a particular tying arrangement is likely to be anticompetitive depends on several factors, including: the form of the tie-in (e.g., one-to-one bundling or a requirements contract), the seller's market share in the tying product market, the seller's market share in the tied product market, the number of customers subject to the tying arrangement, and the competitive conditions in the tied product market.[17]

Some scholars, particularly those associated with the Chicago School of Law and Economics, have argued that sellers impose tying requirements in order to price discriminate. Chicago School scholars, such as Robert Bork, argued that it was "impossible" to use tying arrangements to leverage monopoly power in one market to monopolize a second market.[18] These scholars posit price discrimination as an alternative rationale for tying arrangements.[19] For example, if a manufacturer sells

[14] Id.
[15] Schmidt (2009).
[16] Leslie (2011).
[17] Leslie (2004).
[18] Bork (1978).
[19] Id. Bowman (1973); Posner (2001).

copiers to consumers with two different profiles – large paper-intensive businesses (such as large law firms) and small businesses that have fewer copying needs (such as small real estate offices) – the manufacturer might like to charge the law firm $50,000 and the real estate firm $5,000. Direct price discrimination may not be possible because of arbitrage whereby the real estate firm could, for example, buy the copier for $5,000 and then sell it to the law firm for $30,000. The manufacturer can use a tying arrangement to eliminate this risk of arbitrage by charging $4,000 across the board for its copiers and requiring all of its customers to buy all of their paper from the manufacturer at a supracompetitive price. The low-volume user, who does not use much paper, will end up spending $5,000 over the life of the copier. Conversely, the high-volume user, who uses substantial quantities of copy paper, will end up paying closer to $50,000 for the copier and paper supplies over the life of the copier. Both low- and high-volume users will pay a price at or below their reservation price and the manufacturer will extract maximum consumer surplus from different consumer groups.[20] The price discrimination explanation assumes that high-intensity users are more willing to pay a higher price for the tying product and supplies for it (i.e., the tied product).[21]

Without using economic terminology, the Supreme Court has essentially rejected the price discrimination defense of tying arrangements. For example, the Court in *Motion Picture Patents* condemned when a patent "owner intends to and does derive its profit, not from the invention on which the law gives it a monopoly, but from the unpatented supplies with which it is used."[22] Furthermore, the price discrimination explanation cannot explain tying arrangements that involve one-to-one bundling because these do not measure intensity of use.[23]

20.2. A BRIEF HISTORY OF INTELLECTUAL PROPERTY AND TYING ARRANGEMENTS

The Supreme Court initially gave patent owners great leeway in imposing restrictions on licensees, including tying requirements. In the earliest tying case to reach the Supreme Court, the seller of a patented mimeograph machine required its licensees to purchase all of their paper, ink, and other supplies from the licensor. In *Henry v. A.B. Dick Co.*, the Supreme Court held that use of a competitor's ink constituted infringement of the patented machine, despite the fact that the ink was unpatented. The decision essentially made patent holders immune from antitrust liability for engaging in tying. Two years later, Congress repudiated *A.B. Dick* with the enactment of Section 3 of the Clayton Act, which provides:

---

[20] Leslie (1999).
[21] Leslie (2004).
[22] *Motion Picture Patents* 1917.
[23] Leslie (2004).

> It shall be unlawful for any person ... to lease or make a sale or contract for sale of goods, wares, merchandise, machinery, supplies, or other commodities, *whether patented or unpatented* ... or fix a price charged therefor, or discount from, or rebate upon, such price, on the condition, agreement, or understanding that the lessee or purchaser thereof shall not use or deal in the goods, wares, merchandise, machinery, supplies, or other commodities of a competitor or competitors of the lessor or seller, where the effect of such lease, sale, or contract for sale or such condition, agreement, or understanding may be to substantially lessen competition or tend to create a monopoly in any line of commerce.

Prior to the Clayton Act, patent holders could impose tying arrangements on their customers in order to leverage their market power associated with IP rights into markets for unpatented goods, all without fear of antitrust liability. Tying was just another way for patentees to monetize the value of their patents. The Clayton Act forbids patent holders from engaging in tying that "substantially lessen[s] competition." The statute specifically excludes the mere possession of a patent as a defense to an otherwise illegal tying arrangement. Although not stated in the statute, the mere possession of copyrights or trademarks also does not provide a defense to otherwise illegal tying arrangements.[24]

Soon after the passage of the Clayton Act, the Supreme Court in *Motion Picture Patents Co. v. Universal Film Mfg. Co.* noted that, because of the congressional enactment, A.B. Dick "must be regarded as overruled." In *Motion Picture Patents*, the patentee licensed its patented film projectors subject to the licensee agreeing to use the licensed projector only to show the patentee's films. The defendant was employing tying arrangements in an attempt "to monopolize the entire movie industry by using its technologically superior projector as a bottleneck through which it could control the films that were exhibited."[25] In evaluating the legality of this tying condition, the Court held that patent laws did not confer upon patentees the right to "prescribe by notice attached to a patented machine the conditions of its use and the supplies which must be used in the operation of it, under pain of infringement of the patent." *Motion Picture Patents* essentially treated tying as a species of patent misuse.[26] The Court ultimately held that tying arrangements could also violate the Sherman Act.[27] Although the legal tests under the Sherman Act and the Clayton Act are identical, the reach of the two statutes is different because, by its text, the Clayton Act is limited to goods, while the Sherman Act is not. This means that some tying arrangements involving intellectual property licenses may violate the Sherman Act, but not the Clayton Act.[28]

---

[24] Hovenkamp et al. (2016).
[25] Hovenkamp (2005).
[26] Congress has since narrowed the reach of the opinion by amending the federal patent statute to provide that tying can be misuse only if the patentee has market power in the tying product market. 35 U.S.C. § 271(d).
[27] Leslie (2011).
[28] *Tele Atlas v. NAVTEQ Corp.* 2005.

Over time, tying doctrine became intertwined with the patent misuse defense.[29] Patent misuse is an equitable doctrine that provides that a patent infringement defendant is not liable for infringement damages so long as the patent holder is misusing its patent. Early courts treated the imposition of a tying requirement as a type of patent misuse. Consequently, tying arrangements became a cause of action in antitrust jurisprudence as well as a defense in patent litigation. The origin of the modern patent misuse doctrine can be found in *Morton Salt Co. v. G. S. Suppiger Co.*[30] In that case, the patentee sold a patented machine for injecting salt tablets into canned goods. It required its customers to also purchase all of their salt tablets from the patentee as well. When the patentee sued a rival salt-injecting machine manufacturer for infringement, the Court held that the patentee could not enforce its patent because it was misusing the patent by imposing the tying requirement. Similarly, in *Mercoid*, the Supreme Court treated tying arrangements as patent misuse because the patentee was "secur[ing] a limited monopoly of an unpatented material."[31]

Morton and its patent progeny were not antitrust cases, yet the Supreme Court relied upon them when constructing antitrust law's treatment of tying arrangements. In *International Salt* – another case involving a manufacturer of patented salt machines requiring its customers to purchase their salt from the patentee – the Supreme Court cited the *Morton Salt* and *Mercoid* opinions for the proposition that from patents, a patent holder

> derives a right to restrain others from making, vending or using the patented machines. But the patents confer no right to restrain use of, or trade in, unpatented salt. By contracting to close this market for salt against competition, International [Salt] has engaged in a restraint of trade for which its patents afford no immunity from the anti-trust laws.[32]

Built on a foundation of patent misuse cases, *International Salt* played an important role in the development of modern antitrust jurisprudence regarding tying arrangements.

20.3. ELEMENTS OF ILLEGAL TYING AND THEIR APPLICATION TO INTELLECTUAL PROPERTY

Despite early Supreme Court rhetoric to the contrary,[33] tying arrangements do not inherently injure competitive markets. Some do; some don't. Although the Supreme Court has condemned tying arrangements as per se illegal, the per se rule against

---

[29] Bohannan and Hovenkamp (2012).
[30] *Morton Salt* 1942.
[31] *Mercoid Corp.* 1944.
[32] *International Salt* 1947.
[33] *Standard Oil* 1949.

tie-ins does not follow the format of the traditional per se rule in antitrust law. The traditional per se rule condemns certain trade restraints as a matter of law; the plaintiff need not show market power or anticompetitive effects, and the defendant cannot argue that it had a legitimate business justification for its conduct. The tying per se rule requires the plaintiff to prove elements that relate to the defendant's market power and the tie-in's effects; courts also generally entertain defenses for tying arrangements, which would not be allowed in a true per se scenario. The Supreme Court has not articulated a formal legal test for tying arrangements. Lower courts, however, have converted the Court's writings on tying arrangements into a usable element-driven legal test. One common version of the legal test for a per se tying claim requires the plaintiff to prove:

> (1) the tying and the tied products are actually two distinct products; (2) there is an agreement or condition, express or implied, that establishes a tie; (3) the entity accused of tying has sufficient economic power in the market for the tying product to distort consumers' choices with respect to the tied product; and (4) the tie forecloses a substantial amount of commerce in the market for the tied product.[34]

Some courts have required the plaintiff to prove a fifth element: the tying arrangement created anticompetitive effects. The above test assumes that one seller is unilaterally imposing a tying requirement. If two or more sellers agree with each other to impose tying arrangements on their respective customers, that agreement among the sellers could itself constitute an illegal conspiracy.[35]

As noted in Section 20.1, not all tying arrangements have the same effects on competition. The per se rule applied to tying arrangements implicitly acknowledges this because it condemns only those tie-ins where the seller has market power over the tying product and a not insubstantial volume of commerce in the tied product market is affected. Each element of the test is intended to distinguish between conduct that is benign (or procompetitive) and conduct that is potentially anticompetitive. This is most plainly seen in the anticompetitive effects element employed by some courts.

The following sections discuss each element of a tying claim individually and note when the presence of IP rights affects the antitrust analysis.

20.3.1. Two Separate Products

Plaintiffs alleging illegal tying must first prove that the defendant did, in fact, link two separate products. Because tying arrangements are condemned for harming competition by leveraging market power across markets, this element ensures that market power in one market (the tying product market) is being used to distort competition in another separate market (the tied product market). If two separate

---

[34] *Borschow Hosp. & Med. Supplies* 1996.
[35] Leslie (2007).

products are not being linked, then the defendant is not leveraging its economic power in one market to injure competition in another market.[36] The element is applied in order to determine whether the tying seller has "foreclosed competition on the merits in a product market distinct from the market for the tying item."[37]

The legal test for determining whether the defendant is selling one product or tying together two separate products is determined by consumer demand, and not by the functional relationship between the alleged tying and tied products. Rejecting the functional approach, the Supreme Court in *Jefferson Parish* noted that the Justices "often found arrangements involving functionally linked products at least one of which is useless without the other to be prohibited tying devices." Instead, the Court held that existence of two separate products depends upon "the character of the demand for the two items."[38] As interpreted by lower courts, the plaintiff "must identify the products at issue in each tie and demonstrate that there is sufficient demand for the purchase of the tied product separate from the tying product to identify a distinct product market in which it is efficient to offer the tied product separately from the tying product."[39]

Before discussing how this element applies to tying arrangements involving intellectual property, a note on language is necessary. Tying law has a linguistic quirk in that the law refers to tying and tied *products* even when the tying and/or tied "product" is a service or a license to use intellectual property. Even though services and licenses are not technically products, most courts still refer to them as such in order to parallel the language of tying law. For example, the Supreme Court in *Eastman Kodak Co. v. Image Technical Services., Inc.* explicitly referred to service as a product.[40]

The "two separate products" inquiry has varying degrees of complexity depending on the type of intellectual property at issue. Tie-ins involving patents present the fewest difficulties with this element. Patent tying generally entails the seller of a patented tying product requiring its customers to buy their (usually unpatented) tied product, which is often used in conjunction with the tying product. For example, in *Henry v. A.B. Dick Co.*, the tying seller had a patent on the tying product, a mimeograph machine, and required its customers to purchase several different tied products, including paper and ink, which were unpatented.[41]

The two-products element can be more complicated in the context of tying together copyrighted works. The Supreme Court has treated block booking as a form of tying arrangement. Studios that own copyrighted works engage in block booking when they condition the licensing of one movie (or television program) on

---

[36] Leslie (2015).
[37] *Jefferson Parish* 1984.
[38] *Id.*
[39] BookLocker.com 2009.
[40] *Eastman Kodak* 1992.
[41] *Henry* 1912.

the exhibitor's agreement to pay for licenses on other copyrighted works as well. When considering film studios' imposition of block booking on movie theaters, the Court held that "a refusal to license one or more copyrights unless another copyright is accepted" violates antitrust law.[42] The Court later extended this holding to movie studio efforts to engage in block booking against television stations.[43] The Court in these cases implicitly treated each copyrighted movie as a separate product.

Owners of copyrighted television programs sometimes engage in block booking by requiring television stations to take multiple television programs. Courts have generally been amenable to the argument that each copyrighted program can be considered a product market unto itself to the extent that consumers demand one product but not another.[44] Individual episodes, however, should not generally be treated as separate products.[45]

Courts have found the greatest difficulty in determining whether a trademark is a separate "product" from the physical product – or products – covered by the trademark. In *Chicken Delight*, the Ninth Circuit considered the policy of a fast-food franchise that charged no franchise fee or royalty for use of the Chicken Delight trademark but required franchisees to purchase from the franchisor cooking equipment, dry-mix food items, and packaging bearing the Chicken Delight trade-marked logo. The Ninth Circuit held that the Chicken Delight trademark constituted a tying product that was a separate product from the various tied products – the equipment, mixes, and packaging – that franchisees were required to buy from Chicken Delight. The court reasoned that the franchise license was separate from the common articles that "the public does not and has no reason to connect with the trade-mark."[46] In determining that the trademark and the attendant products were separate, the court used a functional-relationship test, which *Jefferson Parish* later rejected. Nevertheless, *Chicken Delight* proved influential in the short term. For example, the Eighth Circuit cited *Chicken Delight*, among other cases, for the proposition that "[a] franchise license constitutes a separate and distinct marketable item. The weight of judicial authority supports the proposition that if prospective franchisees are compelled to purchase equipment or other tied products in order to obtain the franchise and trademark, an illegal tying arrangement exists."[47]

Other courts took issue with the *Chicken Delight* holding and, consequently, rejected or distinguished the case. Most notably, eleven years after *Chicken Delight*, a different panel of Ninth Circuit judges distinguished that precedent. In *Krehl v. Baskin-Robbins Ice Cream Co.*, franchisees challenged Baskin-Robbins' policy of conditioning the grant of a franchise upon the purchase of ice cream exclusively

---

[42] United States v. Paramount Pictures 1948.
[43] Loew's 1962.
[44] MCA Television 1999.
[45] Hovenkamp et al. (2016).
[46] Siegel v. Chicken Delight 1971.
[47] Northern v. McGraw-Edison Co. 1976.

from Baskin-Robbins as an illegal tying arrangement. The court reasoned that the *Chicken Delight* opinion

> distinguished between two kinds of franchising systems: 1) the business format system; and 2) the distribution system. A business format franchise system is usually created merely to conduct business under a common trade name. The franchise outlet itself is generally responsible for the production and preparation of the system's end product. The franchisor merely provides the trademark and, in some cases, supplies used in operating the franchised outlet and producing the system's products. Under such a system, there is generally only a remote connection between the trademark and the products the franchisees are compelled to purchase ... Under the distribution type system, the franchised outlets serve merely as conduits through which the trademarked goods of the franchisor flow to the ultimate consumer. These goods are generally manufactured by the franchisor or, as in the present case, by its licensees according to detailed specifications. In this context, the trademark serves a different function. Instead of identifying a business format, the trademark in a distribution franchise system serves merely as a representation of the end product marketed by the system ... The desirability of the trademark and the quality of the product it represents are so inextricably interrelated in the mind of the consumer as to preclude any finding that the trademark is a separate item for tie-in purposes.[48]

The court concluded that Baskin-Robbins represented a distribution system and, thus, its trademark did not constitute a separate tying product. Baskin-Robbins did not, therefore, impose a tying arrangement. This approach has proven persuasive.[49]

20.3.2. Condition or Coercion

Courts have articulated the second element in various ways – conditioning, coercion, and forcing, for example. However styled, the second element requires the defendant to have improperly compelled purchasers of the tying product to also purchase the tied product. This element ensures that the tying seller is manipulating the market and not merely responding to consumer demand for particular product bundles. Early on, the Supreme Court explained that "[b]y conditioning his sale of one commodity on the purchase of another, a seller coerces the abdication of buyers' independent judgment as to the 'tied' product's merits and insulates it from the competitive stresses of the open market."[50]

Three different forms of tying arrangement can satisfy this element. First, and most commonly, sellers sometimes contractually require purchasers of the tying product to promise to buy the tied product from that seller as well. Such clauses can be put in licensing agreements or sales contracts involving goods over which the

---

[48] *Krehl v. Baskin-Robbins Ice Cream Co.*
[49] Hovenkamp et al. (2016).
[50] *Northern Pacific Railway* 1958.

seller has intellectual property rights. In addition to directly requiring one's customers to buy the tied product, a tying seller can effect a tying requirement by including a clause in the sales contract that voids the customer's warranty on the tying product if the customer uses a rival's version of the tied product in conjunction with the tying product.[51]

Second, a seller can effect a tying arrangement through product design changes that make a rival's version of the tied product incompatible with the seller's tying product. This is often referred to as a technological tie-in. A "technological tie" exists when the tying and tied product are bundled together physically, or produced in such a way that they are compatible only with each other. The classic example is Kodak's simultaneous introduction of its new cartridge-loaded film and its new camera for using the film.[52] Because consumers could not use the desirable new film unless they had a camera the size of the new film, many consumers felt compelled to buy the compatible camera from Kodak.

Courts may give more deference to a technological tie-in than a contractual tying requirement because courts do not want to discourage beneficial product design changes.[53] For example, the Ninth Circuit has opined that:

> As a general rule, ... the development and introduction of a system of technologically interrelated products is not sufficient alone to establish a *per se* unlawful tying arrangement even if the new products are incompatible with the products then offered by the competition and effective use of any one of the new products necessitates purchase of some or all of the others. Any other conclusion would unjustifiably deter the development and introduction of those new technologies so essential to the continued progress of our economy.[54]

Nevertheless, the D.C. Circuit in its *Microsoft* opinion cautioned sellers that "[j]udicial deference to product innovation, however, does not mean that a monopolist's product design decisions are per se lawful."[55] Legality generally turns on whether the integration "has been designed for the purpose of tying the products, rather than to achieve some technologically beneficial result."[56] Firms imposing technological tie-ins may also be able to argue an efficiency defense, discussed below, that would be unavailable to justify a contractual tie-in. Given the seemingly different standards applied to contractual and technological tying arrangements, sellers may have a strong incentive to use technological integration to achieve a tying requirement in order to evade the more stringent judicial approach applied to contractual tie-ins.[57]

---

[51] Hovenkamp et al. (2016).
[52] Id.
[53] Meurer (2003).
[54] *Foremost Pro Color* 1983.
[55] *United States v. Microsoft* 2001.
[56] *Response of Carolina* 1976.
[57] Hylton (2003).

Third, in addition to imposing contractual requirements and manipulating technical capability, a seller can also create a de facto tie-in through its pricing policies. This form of tying occurs when a seller prices the bundled tying and tied products below the combined price of each individual product being bought separately, so that consumers are economically compelled to buy the bundle. The Second Circuit has explained that

> where there is no quality or distinguishing desideratum between a product offered singly or in a package, the seller cannot charge substantially higher for the individual product if the price differential has the effect of conditioning the sale of the single product to the sale of the entire package and if the difference in price cannot be legitimately justified by cost considerations.[58]

As the Sixth Circuit held, "[w]hen a defendant adopts a policy that makes it unreasonably difficult or costly to buy the tying product (over which the defendant has market power) without buying the tied product from the defendant, it 'forces' buyers to buy the tied product from the defendant and not from competitors."[59] Thus, a tying arrangement exists "where the buyer's only economically viable option is to purchase both the tying and the tied product."[60] A seller may not escape the charge of illegal tying by claiming that one of the products in the bundle is being given away because, as the Tenth Circuit has held, "[w]here the price of a bundled product reflects any of the cost of the tied product, customers are purchasing the tied product, even if it is touted as being free."[61]

20.3.3. Sufficient Economic Power

The tying plaintiff must prove that the defendant possessed "sufficient economic power with respect to the tying product to appreciably restrain free competition in the market for the tied product ..."[62] If the defendant does not possess economic power, it is unlikely to be injuring competition.[63]

The application of this element to tying arrangements involving intellectual property has proven controversial and shifting. The earliest tying cases concerning intellectual property rights primarily involved patent owners. In *International Salt*, the Supreme Court implicitly presumed that the tying seller had sufficient economic power when the tying product is patented. The Court made this presumption explicit in *Loew's*, a case involving block booking of copyrighted movies. The *Loew's*

---

[58] American Manufacturers 1967.
[59] Collins Inkjet 2015.
[60] Shamrock Marketing 2011.
[61] Multistate Legal Studies 1995.
[62] Northern Pacific Railway 1958.
[63] Leslie (2015).

Court first articulated the presumption that patents confer sufficient economic power to impose an anticompetitive tie-in:

> Since one of the objectives of the patent laws is to reward uniqueness, the principle of these cases was carried over into antitrust law on the theory that the existence of a valid patent on the tying product, without more, establishes a distinctiveness sufficient to conclude that any tying arrangement involving the patented product would have anticompetitive consequences.[64]

The Court then expanded this presumption to copyrighted tying products, holding that the "requisite economic power is presumed when the tying product is patented or copyrighted."[65] Lower courts extended the *Loew's* presumption beyond copyrighted works of entertainment to software.[66]

The Supreme Court never considered whether this presumption of market power extended to trademarks. Some courts have applied *Loew's* to trademarked tying products. For example, the Ninth Circuit in *Chicken Delight* opinioned that:

> Just as the patent or copyright forecloses competitors from offering the distinctive product on the market, so the registered trade-mark presents a legal barrier against competition. ... [W]e see no reason why the presumption that exists in the case of the patent and copyright does not equally apply to the trade-mark.[67]

Other courts, however, declined to extend *Loew's* to trademarks, instead distinguishing trademarks from patents and copyrights. For example, the Fifth Circuit explained:

> Unlike a patent or copyright which is designed to protect the uniqueness of the product or process itself, a trademark protects only the name or symbol of the product. This is a basic conceptual difference. Thus, ... we ... hold that a legally unique name alone cannot demand a presumption of economic power.[68]

These presumptions proved problematic because most IP rights do not confer market power. Some lower courts essentially rejected the Supreme Court's holdings creating a presumption of market power for tying products protected by intellectual property. The Sixth Circuit found "the pronouncement in *Loew's* to be overbroad" and it "reject[ed] any absolute presumption of market power for copyright or patented product..."[69]

In *Illinois Tool Works Inc. v. Independent Ink, Inc.*, the Supreme Court reversed the presumption of market power created by *International Salt* and *Loew's*, and recognized in many other Supreme Court opinions. The *Illinois Tool Works* Court

---

[64] *Loew's* 1962.
[65] *Id.*
[66] *Digidyne* 1984.
[67] *Chicken Delight* 1971.
[68] *Carpa* 1976.
[69] *A.I. Root* 1986.

began by noting that "the presumption that a patent confers market power arose outside the antitrust context as part of the patent misuse doctrine." After tracing the presumption's development, the Court observed that Congress had amended the patent law so that, in order to establish misuse through tying, market power had to be proved, not presumed. The Court concluded that the congressional elimination of the presumption in the context of patent misuse necessitated rejecting the presumption in the context of antitrust law. The Court also cited antitrust scholars and the US Department of Justice and FTC Antitrust Guidelines for the Licensing of Intellectual Property, which argued against maintaining the presumption. The Justices ultimately held "that, in all cases involving a tying arrangement, the plaintiff must prove that the defendant has market power in the tying product."[70] Although the case dealt explicitly with patented tying products, the holding is not so limited and courts have applied *Illinois Tool Works'* reversal of the presumption of market power to cases involving copyrighted tying products.[71]

20.3.4. *Effect on Commerce*

The fourth element requires the plaintiff to prove that the defendant's tying arrangement had an effect on a substantial – or not insubstantial – dollar volume of commerce. The required dollar volume is exceedingly small. The Supreme Court has held that "normally the controlling consideration is simply whether a total amount of business, substantial enough in terms of dollar-volume so as not to be merely de minimis, is foreclosed to competitors by the tie ..."[72] Although most courts seem to employ the commerce element as a proxy for anticompetitive effects, given the low dollar-volume requirement, the commerce element is not an appropriate proxy for anticompetitive effects.[73] This could explain why some courts seem to treat the element as primarily serving a jurisdictional function.[74] However the commerce is articulated – and whatever its intended function – the presence of intellectual property does not affect the analysis of this element.

20.3.5. *Anticompetitive Effects*

Several courts claim to require tying plaintiffs to prove "anticompetitive effects in the tied market."[75] However, courts that list anticompetitive effects as a separate element do not actually interpret and apply the element with any vigor.[76] Professors

---

[70] *Illinois Tool Works* 2006.
[71] *Paramount Pictures* 2006.
[72] *Fortner* 1969.
[73] Leslie (2015).
[74] *Id.*
[75] *Hack* 2000.
[76] Areeda and Hovenkamp (2004).

Areeda and Hovenkamp have explained: "foreclosure shares were not measured and assessed by the lower courts that purported to require 'anticompetitive effects' ... Instead, the courts either ignored that additional requirement or employed it very circumspectly to eliminate a specified class of relatively harmless tie-ins from per se condemnation."[77] In those circuits that have articulated this element, courts have not required any special treatment for tying arrangements involving intellectual property.

20.4. POTENTIAL DEFENSES

Tying arrangements are unlike other per se violations of the Sherman Act in that courts generally allow a defendant to argue that it imposed the tying arrangement for a legitimate business reason. This section will briefly review some of the common legitimate business justifications that tying defendants have argued, some successfully and some not. As an initial matter, it is worth highlighting that the mere possession of IP rights is not a defense to a tying claim. Section 3 of the Clayton Act explicitly prohibits patentees from employing tying arrangements when the effect is to substantially lessen competition. The patent-inclusive language in Section 3 is not an afterthought; indeed, the original Senate version of the bill reached *only* those tie-ins involving a patented tying product.[78] The Congressional focus on patent tying represented the legislators' repudiation of the Supreme Court's earlier decision in *A.B. Dick* to allow patentees to impose tying requirements. Arguments that IP owners use tie-ins to facilitate efficient price discrimination or to generate profits that reward past innovation and fund future research do not provide an appropriate defense of otherwise illegal tying arrangements.[79] However, IP owners can attempt to defend their tying arrangements with other potential legitimate business justifications.

20.4.1. *Quality Control and Good Will*

Tying defendants often argue that tying is necessary to protect the proper functioning of the tying product. For example, in *IBM v. United States*, the government challenged IBM's policy of requiring users of its tabulating machines to purchase from IBM the cards to be used with their machines. IBM argued that the tying clause was lawful "because its purpose and effect are only to preserve to [IBM] the good will of its patrons by preventing the use of unsuitable cards which would interfere with the successful performance of its machines."[80] IBM asserted that use

---

[77] Id.
[78] Bilicki (1984).
[79] Leslie (2011).
[80] IBM 1936.

of cards of an improper size and thickness or contaminated by carbon spots or other defects could cause the machines to generate inaccurate results. Problems with quality would, according to IBM, diminish the good will associated with the company's name.

Over the decades, the Justices have given conflicting views on the availability of a good will defense for tying arrangements. In her *Jefferson Parish* concurrence, Justice O'Connor noted that tying arrangements "may protect the reputation of the tying product if failure to use the tied product in conjunction with it may cause it to misfunction."[81] But the majority noted that the Court had "uniformly rejected similar 'goodwill' defenses for tying arrangements, finding that the use of contractual quality specifications are generally sufficient to protect quality without the use of a tying arrangement."[82] Several lower courts, however, have entertained these defenses in tying cases. In the context of trademarks, for example, courts have held that "[t]he protection of goodwill is a reasonable basis to justify an otherwise unlawful tying agreement."[83] Similarly, the Eleventh Circuit has held that "[e]ven if plaintiffs can show the existence of an illegal tie, defendants can still defeat the claim by showing that the tie constitutes a necessary device for controlling the quality of the end product sold to the consuming public by a franchisee."[84]

Even when entertaining the defense, courts have often been suspicious of the quality control justification for tying. This suspicion is well grounded to the extent that tying sellers can easily assert quality control whether true or not. For example, in *International Salt*, the Supreme Court noted the tying seller's lack of evidence to support its claim that use of inferior salt would interfere with the functionality of its salt-injecting machine.[85]

Plaintiffs can rebut a quality control defense by showing that quality can be protected through a less restrictive alternative than tying.[86] For example, the Supreme Court rejected IBM's good will defense because "others are capable of manufacturing cards suitable for use in [IBM's] machine, and ... paper required for that purpose may be obtained from the manufacturers who supply [IBM]." Furthermore, the Supreme Court suggested that IBM had alternatives to imposing a tie-in, including "proclaiming the virtues of its own cards or warning against the danger of using, in its machines, cards which do not conform to the necessary specifications, or even ... making its leases conditional upon the use of cards which conform to them."[87]

---

[81] *Jefferson Parish* 1984.
[82] *Id.*
[83] *Heatransfer Corp.* 1977.
[84] *Midwestern Waffles* 1984.
[85] *International Salt* 1947.
[86] The Eleventh Circuit put the burden of proof on the defendant (*Midwestern Waffles* 1984).
[87] *IBM* 1936.

While characterizing quality control as "[t]he justification most often advanced" to defend tying arrangements, the Supreme Court has observed that "the protection of the good will of the manufacturer of the tying device [] fails in the usual situation because specification of the type and quality of the product to be used in connection with the tying device is protection enough."[88] Such specifications may not constitute a less restrictive alternative in at least two situations. First, "the protection of good will may necessitate the use of tying clauses ... where specifications for a substitute would be so detailed that they could not practicably be supplied."[89] Second, quality specifications do not work when they would require the seller to share its trade secrets with all of its franchisees.[90]

20.4.2. *Infant Industries and New Products*

Courts have also created a new product defense to tying claims. First recognized by lower court judges during the beginning of the era of talking pictures,[91] the Supreme Court implicitly accepted the defense when it affirmed per curiam the lower court opinion in *United States v. Jerrold Electronics Corp.*, which reasoned that tying may be permissible when the seller is creating a new product. The Supreme Court later more explicitly recognized this infant industry defense when it noted that "*Jerrold* also indicates that tying may be permissible when necessary to enable a new business to break into the market."[92] This defense may be useful to IP owners, who are often innovators that may be creating new products and new markets. However, once the market is no longer "new," the tying arrangement may become illegal and courts have provided little guidance on how to determine when this threshold is crossed.

20.4.3. *Efficiency*

Some technological tie-ins can enhance efficiency by lowering costs, which can benefit consumers by decreasing price. If a technological tie-in represents an improvement over prior products, that can be a defense to otherwise illegal tying.[93] One district court's order articulated the defense as follows:

> An affirmative defense to liability will be recognized if the responding party demonstrates that the tying arrangement can reasonably have been expected, ex ante, to achieve efficiency gains that benefit customers, such as increased quality,

---

[88] *Standard Oil* 1949.
[89] *Id.*
[90] Hovenkamp *et al.* (2016).
[91] Wood (1942).
[92] *Jefferson Parish* 1984.
[93] *Caldera* 1999.

significantly lower costs, or the introduction of new or better products, and that these efficiencies substantially outweigh any competitive harms caused by the tying arrangement.[94]

Although an efficiency defense to tying has not been uniformly recognized, it is sufficiently established that tying sellers should be prepared to make efficiency arguments to justify their tying arrangements.

20.4.4. Price-Protection Clause

A patent holder may include a price-protection clause in its tying arrangement. A price-protection clause provides that, as a condition of buying the tying product, the consumer must purchase the tied product from the tying seller as long as the consumer cannot find the tied product from another supplier at a lower price or on better terms.[95] If the consumer can find the tied product at a lower price, then she is not bound by the tying provision. International Salt, for example, employed a price-protection clause in the leases for its machines.[96] Thus, if one of International Salt's competitors charged a lower price for salt, customers could purchase their salt from that seller instead of from International Salt. This provision ensured lessees a competitive price for salt. If the lessee ever found a better price for salt, International Salt would have to meet that price or the lessee was free to purchase the salt from another vendor. Consumers could accept the tying provision in the contract with confidence that they would not be locked into paying above-market prices for salt in the future. Price-protection clauses can protect consumers against incorrectly analyzing or predicting the downstream costs of the tied product.[97] The Supreme Court in both *International Salt* and *Northern Pacific Railway* rejected the argument that the price-protection clause saved the tie-in from condemnation and such a defense is unlikely to work for IP owners accused of illegal tying.

20.5. RULE OF REASON

If the plaintiff cannot prove that a tying arrangement is per se illegal, the plaintiff can still argue that the alleged tie-in violates the rule of reason by showing that it represents "an unreasonable restraint on competition in the relevant market."[98] Courts have neither clearly defined the contours of the rule of reason as applied to tying arrangements nor adequately explained how it differs from the so-called per

---

[94] *Valassis* 2011.
[95] Leslie (2004).
[96] *International Salt* 1947.
[97] Craswell (1982); Leslie (2004).
[98] *Jefferson Parish* 1984.

se rule against tie-ins. It is unlikely that a plaintiff unable to satisfy the per se test for tying arrangements could somehow prove that same tie-in violated the rule of reason.

20.6. CONCLUSION

IP owners risk antitrust liability if they impose tying requirements on their customers or licensees. Tying arrangements are nominally per se illegal. The legal test for tying, however, is more properly described as a structured rule of reason because the plaintiff must prove that the defendant has market power over the tying product, that the tie-in affected commerce, and, in some jurisdictions, that the tie-in had anti-competitive effects. Furthermore, courts allow tying defendants to argue that they have a legitimate business justification, which may provide a defense for IP owners that engage in otherwise illegal tying arrangements.

REFERENCES

Statutes

Clayton Act, Section 3, 15 U.S.C. § 14
Sherman Act, Section 1, 15 U.S.C. § 1
Sherman Act, Section 2, 15 U.S.C. § 2

Cases

A.I. Root Co. v. Computer/Dynamics, Inc., 806 F.2d 673 (6th Cir. 1986)
American Manufacturers Mutual Insurance Co. v. American Broadcasting-Paramount Theatres, Inc., 388 F.2d 272 (2d Cir. 1967)
BookLocker.com, Inc. v. Amazon.com, Inc., 650 F. Supp. 2d 89 (D. Me. 2009)
Borschow Hosp. & Med. Supplies, Inc. v. Cesar Castillo Inc., 96 F.3d 10 (1st Cir. 1996)
Caldera, Inc. v. Microsoft Corp., 72 F. Supp. 2d 1295 (D. Utah 1999)
Carpa, Inc. v. Ward Foods, Inc., 536 F.2d 39 (5th Cir. 1976)
Collins Inkjet Corp. v. Eastman Kodak Co., 781 F.3d 264 (6th Cir. 2015)
Digidyne Corp. v. Data Gen. Corp., 734 F.2d 1336 (9th Cir.1984)
Eastman Kodak Co. v. Image Technical Services, Inc., 504 U.S. 451 (1992)
Foremost Pro Color, Inc. v. Eastman Kodak Co., 703 F.2d 534 (9th Cir. 1983)
Fortner Enterprises, Inc., v. U.S. Steel Corp., 394 U.S. 495 (1969)
Hack v. President & Fellows of Yale College, 237 F.3d 81 (2d Cir. 2000)
Heatransfer Corp. v. Volkswagenwerk, 553 F.2d 964 (5th Cir. 1977)
Henry v. A. B. Dick Co., 224 U.S.1 (1912)
IBM v. United States, 298 U.S. 131 (1936)
Illinois Tool Works Inc. v. Indep. Ink, Inc., 547 U.S. 28 (2006)
International Salt Co. v. United States, 332 U.S. 392 (1947)
Jefferson Parish Hospital District No. 2 v. Hyde, 466 U.S. 2 (1984)
Krehl v. Baskin-Robbins Ice Cream Co., 664 F.2d 1348 (9th Cir. 1982)
MCA Television Ltd. v. Public Interest Corp., 171 F.3d 1265 (11th Cir. 1999)

Mercoid Corp. v. Mid-Continent Investment Co., 320 U.S. 661 (1944)
Midwestern Waffles, Inc. v. Waffle House, Inc., 734 F.2d 705 (11th Cir. 1984)
Morton Salt Co. v. G. S. Suppiger Co., 314 U.S. 488 (1942)
Motion Picture Patents Co. v. Universal Film Mfg. Co., 243 U.S. 502 (1917)
Multistate Legal Studies, Inc. v. Harcourt Brace Jovanovich Legal & Professional Publications, Inc., 63 F.3d 1540 (10th Cir. 1995)
Northern v. McGraw-Edison Co., 542 F.2d 1336 (8th Cir. 1976)
Northern Pacific Railway Co. v. United States, 356 U.S. 1 (1958)
Paramount Pictures Corp. v. Johnson Broad. Inc., 432 F. Supp. 2d 707 (S.D. Tex. 2006)
Response of Carolina, Inc. v. Leasco Response, Inc., 537 F.2d 1307 (5th Cir. 1976)
Shamrock Marketing, Inc. v. Bridgestone Bandag, Inc., 775 F. Supp. 2d 972 (W.D.Ky. 2011)
Siegel v. Chicken Delight, Inc., 448 F.2d 43 (9th Cir. 1971)
Standard Oil Co. of California v. United States, 337 U.S. 293 (1949)
Tele Atlas v. NAVTEQ Corp., 397 F. Supp. 2d 1184 (N.D. Cal. 2005)
United States v. Jerrold Electronics Corp., 187 F.Supp. 545 (E.D.Pa.1960), aff'd per curiam, 365 U.S. 567 (1961)
United States v. Loew's, Inc., 371 U.S. 38 (1962)
United States v. Microsoft Corp., 253 F.3d 34 (D.C. Cir. 2001)
United States v. Paramount Pictures, 334 U.S. 131 (1948)
Valassis Comm'ns, Inc. v. News Am. Inc., 2011 WL 2420048 (E.D. Mich. Jan. 24, 2011) report and recommendation adopted, 2011 WL 2413471 (E.D. Mich. June 15, 2011)

Secondary Sources

Areeda, Phillip E. and Herbert Hovenkamp. 2004. *Antitrust Law: An Analysis of Antitrust Principles and Their Application* Vol. 9 (2nd edn). New York: Aspen.
Bilicki, Byron A. 1984. Standard Antitrust Analysis and the Doctrine of Patent Misuse: A Unification Under the Rule of Reason. *University of Pittsburgh Law Review*, 46, 209–39.
Bohannan, Christina and Herbert Hovenkamp. 2012. *Creation without Restraint: Promoting Liberty and Rivalry in Innovation.* New York: Oxford University Press.
Bork, Robert. 1978. *The Antitrust Paradox.* New York: Basic Books.
Bowman, Ward S. 1973. *Patent and Antitrust.* University of Chicago Press.
Craswell, Richard. 1982. Tying Requirements in Competitive Markets: The Consumer Protection Issues. *Boston University Law Review*, 62, 661–700.
Hovenkamp, Herbert. 2005. *The Antitrust Enterprise: Principle and Execution.* Cambridge, MA: Harvard University Press.
Hovenkamp, Herbert, Mark Janis, Mark Lemley, Christopher Leslie, and Michael Carrier. 2016. *Antitrust and IP: An Analysis of Antitrust Principles Applied to Intellectual Property Law.* 3rd edn. New York: Wolters Kluwer.
Hylton, Keith N. 2003. *Antitrust Law: Economic Theory and Common Law Evolution.* Cambridge University Press.
Leslie, Christopher R. 1999. Unilaterally Imposed Tying Arrangements and Antitrust's Concerted Action Requirement. *Ohio State Law Journal*, 60, 1773–876.
  2004. Cutting Through Tying Theory with Occam's Razor: A Simple Explanation of Tying Arrangements. *Tulane Law Review*, 78, 727–825.
  2007. Tying Conspiracies. *William & Mary Law Review*, 48, 2247–312.
  2011. Patent Tying, Price Discrimination, and Innovation. *Antitrust Law Journal*, 77, 811–54.
  2015. The Commerce Requirement in Tying Law. *Iowa Law Review*, 100, 2135–60.

Meurer, Michael J. 2003. Vertical Restraints and Intellectual Property Law: Beyond Antitrust. *Minnesota Law Review*, 87, 1871–912.

Nalebuff, Barry. 2004. Bundling as an Entry Barrier. *The Quarterly Journal of Economics*, 119, 159–87.

Posner, Richard A. 2001. *Antitrust Law*. University of Chicago Press.

Schmidt, Hedvig. 2009. *Competition Law, Innovation and Antitrust*. Cheltenham: Edward Elgar.

Whinston, Michael D. 1990. Tying, Foreclosure, and Exclusion. *American Economic Review*, 80, 837–59.

Wood, Laurence I. 1942. *Patents and Antitrust Law*. New York: Commerce Clearing House.

21

Online RPM

John B. Kirkwood

21.1. INTRODUCTION

In *Leegin* (2007), the Supreme Court of the United States reversed almost a century of precedent and ruled that resale price maintenance (RPM) is not invariably anticompetitive. While RPM can be used to suppress price competition and injure consumers, it can also be used to induce the provision of valuable services, enhance rivalry, and benefit customers. To allow for the consideration of both possibilities, the Court held that vertical price fixing, unlike hard-core horizontal price fixing, should not be illegal per se. It must be evaluated under the rule of reason.[1]

Unlike per se illegality, the rule of reason allows a court to examine a practice's procompetitive effects as well as its anticompetitive effects and reach a judgment as to whether, on balance, it enhances or reduces competition. In *Leegin*, the Supreme Court identified ways in which RPM could lead to either outcome. It might, for example, be used by a group of manufacturers to reduce the incentive to cheat on a cartel they had established.[2] Alternatively, it might be used by a manufacturer to prevent discount retailers from free riding on the promotional services of full-price retailers, undercutting their motivation to provide those services.[3] But the Court did

---

I want to thank Danny Sokol for inviting me to contribute to this Handbook and Kelly Kunsch for his quick and adept research assistance.

[1] In the European Union, the legal standard is different. This chapter focuses exclusively on the treatment of RPM under US law. The relevant economic principles, however, are the same in both jurisdictions, whether RPM is applied to online retailers or brick-and-mortar retailers. While online retailers may object to RPM, major online platforms like Amazon frequently insist on most-favored-customer clauses. For an analysis of these clauses under European competition law, see Akman (2015).

[2] RPM blunts the incentive to cheat because it makes it more difficult for a cheating manufacturer to gain sales through a price cut. If the manufacturer did not use RPM, its dealers could pass on the price cut by lowering their resale prices, increasing sales of the manufacturer's product. With RPM in place, this avenue is blocked.

[3] RPM can strengthen the incentive of retailers to provide valuable services by preventing retailers who do not provide those services from taking business from those who do by offering

not lay down a structure or framework for the application of the rule of reason to RPM. To the contrary, it invited the lower courts to develop one over time, as they gain experience with RPM cases. The Court noted that such a structure might include presumptions, since, if justified, they would make the rule of reason more efficient in identifying anticompetitive instances of RPM and protecting procompetitive instances. *Leegin* itself, however, did not endorse any presumptions.

In this chapter, I first address whether any presumptions would be particularly warranted when RPM is used to eliminate price cutting by *online* retailers. Either a presumption of illegality or a presumption of legality might be especially apt. Because online retailers typically offer much lower prices than brick-and-mortar retailers, RPM would reduce their principal competitive advantage, inhibit the growth of a major channel of distribution, and force consumers to pay significantly higher prices. As a result, online RPM may be more anticompetitive than traditional RPM, and a presumption of illegality may be especially appropriate. On the other hand, online retailers may be prone to engage in free riding, since they not only offer very low prices but cannot supply many of the point-of-sale services that brick-and-mortar retailers provide. Even if no free riding is involved, the low prices of online retailers tend to depress the margins of brick-and-mortar retailers, reducing their incentive to provide desirable services. For both reasons, online retailers may be especially likely to disrupt the efficiency of a manufacturer's distribution system. If so, a presumption of legality may be particularly apt. In Section 21.2, I examine this issue.

In Section 21.3, I return to the larger question I have considered before: If all RPM, whether used online or elsewhere, was subject to the same legal standard, what should it be? As commentators have noted, there are various possibilities: Courts might employ the full rule of reason; they might impose elaborate proof requirements on plaintiffs; they might utilize presumptions, either generally or in specific circumstances; or they might establish safe harbors. In an earlier article,[4] I urged courts to adopt a presumption of illegality coupled with two safe harbors. I argued that this approach was more cost-effective than any of the principal alternatives. It would make it easier for the courts to dispose of the easy cases, in which RPM is either plainly anticompetitive (because the manufacturer cannot offer a plausible justification) or obviously procompetitive (because it was used by a new entrant or by a firm without market power[5]). And it would be at least as good as the full rule of reason, or a truncated alternative based on the factors mentioned in *Leegin*, at resolving the more difficult, mixed cases, in which RPM has both procompetitive and anticompetitive effects.

---

lower prices. If the manufacturer enforces RPM, consumers cannot obtain the services at a full-price retailer but purchase the product at a lower price from a retailer that does not offer the services.

[4] Kirkwood (2010).
[5] And whose retailers did not have market power.

In Section 21.2, I conclude that neither theory nor evidence allows us to determine, at present, whether either a presumption of illegality or a presumption of legality would be especially appropriate in the online context. As a result, absent more information, it appears that online RPM should be subject to the same legal standard that applies to traditional RPM. In Section 21.3, I reiterate my judgment that this standard ought to be a rebuttable presumption of illegality with two safe harbors.

21.2. PRESUMPTIONS FOR ONLINE RPM

Some legal scholars have argued that online RPM is particularly dangerous and that a presumption of illegality is warranted. As Marina Lao notes, one of the "competitive strengths" of online retailers is their ability to charge lower prices than brick-and-mortar retailers.[6] RPM would prevent them from taking advantage of this strength, which would not only reduce their profits and discourage new entry but raise the prices that consumers must pay. Moreover, in Lao's view, the most commonly offered justification for RPM – free riding – appears to be less valid online. Thus, "the case for RPM as a means of controlling free riding problems is probably ... weakened, not strengthened, by the advent of Internet retailers."[7]

Lao acknowledges that online retailing makes free riding easier in the case of sensory-experience products like cosmetics. For example, if a consumer wants to buy a facial foundation, she needs to visit only one brick-and-mortar store, try various brands and shades, make a selection, and then purchase the product online. But in the case of many other products, Lao believes, the Internet reduces the need for consumers to shop at a store before buying online. She explains: "This is because of the wealth of information, such as detailed product features and specifications, professional product reviews and user opinions that can now be found online on practically any product or service, both from online retailers' Web sites and from numerous independent sources."[8] Indeed, given this "wealth of information," several marketing studies have found that more consumers report looking online and then purchasing at a brick-and-mortar store than the reverse.[9] In short, on many products, free riding on online information is more likely than free riding on in-store assistance.

Even if free riding on brick-and-mortar retailers were a problem, Lao argues that it can be solved without RPM. If some incentive is necessary to induce brick-and-mortar retailers to provide in-store assistance or other promotional services, the manufacturer can use promotional allowances. Promotional allowances are less

---

[6] Lao (2011) at 14.
[7] Id. at p. 15.
[8] Id. at pp. 16–17.
[9] For a review of the marketing literature, see Gundlach, Cannon, and Manning (2010); see also Lao (2011), n. 10 (citing specific studies).

restrictive than RPM because they do not prevent a retailer from lowering its price. They are likely to be as effective as RPM because they can be set high enough to cover the typical retailer's cost of providing the desired services. To be sure, the manufacturer will have to monitor its retailers to reduce cheating. But a manufacturer also faces the possibility of cheating when it uses RPM: "a retailer receiving an RPM margin [may] ignore the manufacturer's service expectations and simply keep the margin."[10] Moreover, Lao contends, it is "difficult to see how the costs of policing the retailers' service obligations in the two scenarios could be any different."[11]

Lao recommends, therefore, that the law adopt a "rebuttable presumption of illegality."[12] Because online RPM would deprive consumers of the low prices offered by Internet retailers, and because a manufacturer can obtain many, if not all, the services it desires through a less restrictive alternative, Lao believes that antitrust law should presume that online RPM is anticompetitive. A defendant could rebut that presumption by showing that there was in fact no cost-effective alternative to RPM – that RPM creates stronger incentives or requires less monitoring than any other approach. But without such a showing, a plaintiff should not have to establish either market power or anticompetitive effects. If the plaintiff can prove there is a less restrictive alternative, power and effects can be presumed from the price-raising impact of RPM.

Economist William Comanor maintains that "there are two economics of vertical restraints."[13] The first arises when the restraint is imposed by the manufacturer acting on its own. The second occurs when the manufacturer is coerced to employ the restraint by a large distributor. In the second case, Comanor believes, a strong, if not irrebuttable, presumption of illegality is warranted. Comanor illustrates the second case by reviewing the facts and economics of a lawsuit in which he testified for the plaintiffs. In this action, consumers sued Babies "R" Us ("BRU"), a major retailer of baby supplies, for allegedly forcing many of its suppliers to impose RPM on Internet retailers. In the only published decision in the case, Judge Brody, relying heavily on Comanor's testimony, allowed it to go forward as a class action.[14]

In his article, Comanor describes why a large brick-and-mortar retailer like BRU would have the motivation and the ability to force suppliers to reduce the competitive threat created by online retailing:

> Established retailers faced a set of new rivals with lower distribution costs that could offer services that in some ways were superior to those they provided. At the same time, the leading established retailers retained a substantial market advantage in their dealings with suppliers. If this advantage could be leveraged to disadvantage

[10] Lao (2011) at 19.
[11] Id. at p. 19.
[12] Id. at p. 21.
[13] Comanor (2013) at 127.
[14] McDonough 2009.

the new rivals, the new distribution forms could be impeded, if not prevented altogether.[15]

Thus, when online retailers moved into the baby supply space in the 1990s, BRU, concerned about the impact on its prices and profits, allegedly demanded that existing and future suppliers impose RPM on all their customers, including online retailers. If a supplier would not acquiesce, BRU would not deal with them. In one instance, according to Comanor, to

> comply with BRU's demands, Medela [a leading breast pump producer] terminated seventeen Internet retailers that had priced below BRU's prices. With BRU accounting for a very large proportion of its sale volume, Medela believed it had no other choice and even acknowledged that in doing so, it had "accepted considerable legal risk."[16]

Like Lao, Comanor argues that suppliers could not justify the RPM that BRU demanded on the ground that online retailers were free riding on BRU's promotional services. Surveys reported that consumers were more likely to gather product information online and then shop at brick-and-mortar stores than the reverse. In the case of baby products, greater than 60 percent of consumers behaved this way.[17] Thus, on balance, brick-and-mortar retailers were the beneficiaries, not the victims, of free riding. In the second place, baby product suppliers used promotional allowances to obtain the majority of the services they wanted. "For the most part, [Judge Brody found,] manufacturers paid BRU directly for advertisements and promotions of their products."[18] Finally, Comanor's regression analysis showed that when suppliers temporarily abandoned their restraints, their sales almost always fell, a telling sign that vertical price fixing was not improving the efficiency of their distribution systems.[19]

In short, when RPM was imposed on Internet retailers at the behest of a powerful brick-and-mortar retailer, competition was reduced, not enhanced. Prices rose and few, if any, additional services were provided.

Economist Benjamin Klein agrees that RPM would be anticompetitive if it were forced on a manufacturer by a powerful retailer. In that case, RPM would serve the interests of the retailer, not the interests of the manufacturer or consumers. Klein asserts, though, that this anticompetitive scenario is unlikely to occur unless the retailer has a "dominant" market share.[20] He does not explain why a "dominant" market share is required. The issue is whether the retailer has sufficient buyer power to coerce the manufacturer into adopting a vertical restraint that it would not

---

[15] Comanor (2013) at 111.
[16] Id. at 115 (citing *McDonough*).
[17] Id. at 117.
[18] Id. at 123.
[19] Id. at 119–20.
[20] Klein (2015) at 294.

otherwise adopt. In *McDonough*, the evidence indicated that BRU was able to accomplish this with a substantial but not overwhelming share. In the case of Maclaren brand strollers, BRU's share of dollar purchases was just 32 percent. In the case of Britax brand car seats, it was 45 percent.[21] While BRU's share in each instance was significantly larger than the share of any other retail purchaser (and was "dominant" in that sense), it was not dominant in the sense of being significantly greater than 50 percent. BRU did not need a market share of 60 or 70 percent to force suppliers to adopt an anticompetitive restraint.

For Klein, this is a side issue. The main purpose of his 2015 book chapter is not to explore the anticompetitive explanations for RPM but to explain why he believes that online retailing is likely to *increase* the procompetitive justifications for RPM. As he puts it, the "growth of online retailing has magnified the competitive economic forces that are likely to lead manufacturers to adopt resale price maintenance."[22] He supports this thesis by analyzing whether the two principal justifications for RPM – (1) free riding and (2) service compensation in the absence of free riding – are particularly applicable to online retailing.

Klein asserts that online retailing substantially increases "the free riding problem by permitting consumers to more easily buy products at a lower price online after first obtaining desired services at a full-service brick-and-mortar retailer."[23] Before the advent of online retailing, a free riding consumer had to visit a full-service brick-and-mortar retailer to receive the desired services and then make a second trip – to a discount brick-and-mortar retailer – to buy the product. Now, the consumer can avoid the second trip by purchasing online, lowering the cost of free riding. Klein notes that one study found that "26 percent of online purchasers visited a brick-and-mortar store before their online purchase."[24]

Klein also acknowledges, however, that Gundlach, Cannon, and Manning concluded, after surveying the marketing literature, that it was *more common* for consumers to gather information online and then make a purchase at a brick-and-mortar retailer.[25] This reversal of the traditional free riding flow suggests that imposing RPM on online retailers is unnecessary, since brick-and-mortar retailers benefit more than they lose from online retailing. Klein rejects this reasoning:

> These two forms of free-riding ... are separate "distortions" that should not merely be netted out against one another. Manufacturers wish to mitigate the malincentive effects from both types of free-riding, preventing both (1) an insufficient amount of face-to-face retailing services provided by brick-and-mortar retailers, and (2) an insufficient amount of information provided for their products on the Internet.[26]

---

[21] Comanor (2013) at 113.
[22] Klein (2015) at 277.
[23] *Id.* at p. 277.
[24] *Id.* at p. 279, citing Van Baal and Dach (2005).
[25] Gundlach, Cannon, and Manning (2010).
[26] Klein (2015) at 279.

Klein's remedy is to impose RPM on online retailers and possibly on brick-and-mortar retailers, sharply raising the prices of the former and potentially raising the prices of the latter.[27]

This prescription is understandable. A brick-and-mortar retailer's incentive to furnish promotional services would be reduced by consumers who obtain the services at its store but then purchase online. And that "distortion" would not be alleviated by consumers who look online first but actually buy from the retailer, since they would purchase from the retailer whether or not it provided the promotional services. In short, there is a procompetitive rationale for online RPM despite the substantial amount of "reverse free riding" that Gundlach, Cannon, and Manning highlight. But there is also a drawback. If reverse free riding more than compensates a brick-and-mortar retailer for the sales losses it incurs from traditional free riding, it would earn a competitive return on its operation without RPM. If the manufacturer imposes RPM on online retailers, the brick-and-mortar retailer would earn even more money, since RPM would enable it to raise prices, increase sales, or both. As a result, its profits would be *above* the competitive level, at least for a time. Such a supracompetitive margin would not serve the interests of the manufacturer or consumers, if it could be avoided. And it might be avoided through a more targeted device, such as promotional payments, discussed below.

In short, it is not clear that the traditional free riding justification is stronger in the case of online retailing. To the contrary, given the presence of substantial *reverse* free riding, RPM may raise the margins of full-service brick-and-mortar retailers above the competitive level, at least temporarily. As a result, if there is a less restrictive alternative to RPM, it would be particularly desirable to use it rather than force online retailers to increase their prices.

The second principal procompetitive justification for RPM is to ensure that retailers are adequately compensated for providing the promotional services manufacturers desire, even if free riding is not present. Here, too, Klein believes that the rationale for RPM is stronger where brick-and-mortar retailers compete with online retailers:

> Resale price maintenance often is used by manufacturers even when there is not a consumer free-riding problem as a way to preserve and promote the manufacturer's products. This non-free-riding procompetitive motivation for controlling retailer price discounting is particularly significant when there is online retailing.... . Uncontrolled online retailer price discounting has the potential to significantly disrupt a manufacturer's efficient retail distribution network by reducing retailer compensation below the level required ... to effectively distribute the manufacturer's products.[28]

---

[27] He seems to think, plausibly, that online retailers need less protection than brick-and-mortar retailers since online retailers are likely to provide product information in any event (i.e., whether or not brick-and-mortar retailers engage in discounting).

[28] Klein (2015) at 278.

Klein asserts that online discounting is especially likely to take sales from brick-and-mortar retailers – and deprive them of the revenue they need to supply desired services – for two reasons. First, online retailers are likely to be able to charge lower prices than discount brick-and-mortar retailers.[29] Second, it is easier for a consumer to buy online than to make a purchase at a discount brick-and-mortar retailer.[30]

RPM could be a solution to this problem, but it is not the only possible solution, since there are other ways to compensate a retailer for performing desired services. It is certainly reasonable to assume that a brick-and-mortar retailer would not provide a service unless it expects to earn higher profits from that service. It is also reasonable to assume that the manufacturer would need to provide at least some compensation to the retailer, since many brand-specific services, such as promoting a specific product line, would tend to shift sales among the products carried by the retailer, not expand its total sales. In other words, if a retailer promotes a manufacturer's brand, it is likely to cannibalize sales from its other brands, reducing the profits it earns.[31] As Klein acknowledges, however, a manufacturer can supply the necessary compensation through means other than RPM.[32] It could pay the retailer directly for furnishing brand-specific services such as newspaper advertising or product demonstrations. It could also compensate the retailer for stocking its brands.

The issue is whether alternatives such as these (promotional allowances and slotting fees) are preferable to RPM. That depends on how efficient they are, relative to RPM, in inducing retailers to provide valuable services and whether they are less likely than RPM to reduce competition. Klein argues that RPM is more efficient in inducing the desired services because it increases the retailer's margin on each sale that it makes, giving it an incentive to employ the services to expand sales of the manufacturer's brands. In contrast, a flat payment for stocking a brand or performing a specific promotional service would not vary with sales and thus would not provide the same motivation to *increase* sales. But that incentive problem could be rectified by basing promotional payments or slotting fees on the retailer's sales of the manufacturer's brands – the more it sold, the larger the payment. Such payments would align the manufacturer's and the retailer's incentives.

Klein also suggests that RPM reduces the manufacturer's monitoring costs. If a manufacturer were to use flat (e.g., annual or quarterly) payments for stocking or promoting its brands, the retailer would receive the same payment whether it furnished a large quantity or a small quantity of these services during the relevant

---

[29] *Id.* at p. 290 ("the online retailer price discount is likely to be substantially larger [than the brick-and-mortar discount] because online retailers can take advantage of their relatively low retailing costs").

[30] *Id.* ("the likely demand response from an individual online retailer price discount is likely to be substantially greater than an individual brick-and-mortar retailer price discount because of the increased ability of consumers to shop alternative retailers online").

[31] See Klein and Murphy (1988); Kirkwood (2010); see also Blair and Haynes (2010) at 253 ("RPM is a means of correcting an incentive alignment problem ... The distributor's focus solely on his own profit results in too few promotional services being provided to consumers").

[32] Klein (2015) at 285.

time period. To ensure that the retailer provided enough, the manufacturer would have to monitor the retailer. With RPM, the need for monitoring would be reduced, since, as just noted, RPM provides an incentive to furnish services that increase sales. But that would also be true with payments based on sales volume. As Lao argues, therefore, it is hard to see why the monitoring required with RPM would be any less than the monitoring required with stocking or promotional payments geared to the retailer's sales. In both cases, there is some risk that the retailer would not provide the desired services. Moreover, the retailer's ability to skimp on services would be greatest where competition among retailers is limited, and that is precisely the case where Klein asserts that RPM is most needed.[33]

The other critical issue is whether RPM poses a greater danger to competition than alternatives like promotional payments or slotting fees. At first glance it does, since RPM, unlike these payments, restricts an online retailer's ability to discount. That adverse effect on price competition may be insignificant, however, if the manufacturer has no anticompetitive motivation to use RPM. If the structure and dynamics of the relevant market do not provide the manufacturer – or any of its retailers – with the opportunity to make supracompetitive profits through RPM, then the adverse effect of RPM on prices is likely to be outweighed by its beneficial effect on services. Although alternatives like promotional payments may seem less restrictive, they are unlikely to be if there is no anticompetitive motivation to choose RPM over those alternatives. Instead, the choice is likely to be based on the manufacturer's judgment as to which is more efficient. And of course, antitrust law should not ordinarily second-guess that choice without a credible theory of how RPM would harm consumers.

Thus, the question of whether RPM is more likely to be desirable or harmful when applied to online retailing may ultimately depend on the structure and dynamics of the markets in which online retailing is significant. Are those markets more conducive than other markets to the anticompetitive use of RPM?

There are five principal ways in which RPM can have anticompetitive effects. The practice may (1) facilitate collusion at the manufacturer level, (2) facilitate collusion at the dealer level, (3) suppress more efficient or more innovative dealers, (4) lead to excessive resale services, or (5) result in misleading promotion.[34] There are no systematic studies that attempt to determine whether these scenarios are more likely or less likely in markets in which brick-and-mortar retailers compete with online retailers. On the surface, the second theory (facilitating dealer collusion) seems less likely where online retailing has a significant share of the market. Such a market is likely to be less concentrated than it would be if there were no online retailers. In addition, the market would be composed of two different types of retailers – brick-and-mortar retailers and online retailers – with sharply different

---

[33] *Id.* at p. 281.
[34] Kirkwood (2010).

competitive strategies, and that heterogeneity would further reduce the likelihood of effective collusion.

On the other hand, the third theory (suppressing more efficient or innovative dealers) seems more probable in markets where online retailers have a significant presence. Given their propensity to cut prices, these new and (in some respects) more efficient rivals would pose a more severe threat to the leading traditional retailers than their smaller, brick-and-mortar rivals ever did. As a result, these retailers would be more likely to pressure manufacturers to use RPM to suppress them, as BRU assertedly did in the case described above.

Without systematic empirical data, however, it is hard to say whether RPM is more likely to be procompetitive or anticompetitive in markets with significant online retailing. As noted, there has been no systematic study of online RPM and there are virtually no reported judicial decisions.[35] Absent more information, it is difficult to conclude that online RPM is more likely to be desirable or harmful than traditional RPM. At this point, then, it appears that the proper legal standard for online RPM ought to be the same as the proper legal standard for traditional RPM – an appropriate version of the rule of reason. In Section 21.3, I describe such a standard: a presumption of illegality with two safe harbors.

21.3. APPROPRIATE RULE OF REASON FOR ALL RPM

As I have explained,[36] I believe the best way to structure the rule of reason in an RPM case is to establish a presumption of illegality and two safe harbors. If administered fairly, and there is good reason to believe it would be, this approach would be superior to all the major alternatives – per se illegality, per se legality, the full rule of reason, and a truncated rule of reason triggered by the factors identified in *Leegin*. Unlike any of these alternatives, this approach would resolve both sets of easy cases (those in which RPM is plainly harmful and those in which it is clearly beneficial) correctly and at low cost. Moreover, the approach would be superior to the two per se rules and at least as good as the two types of rule of reason in resolving the harder, mixed cases, where RPM has both anticompetitive and procompetitive effects.[37]

---

[35] There has been one new empirical study of RPM (McKay and Smith 2014), but it does not examine online retailing or compare markets with a significant online presence to other markets. Instead, it uses scanner data from brick-and-mortar retailers. It does, however, find statistically significant evidence that prices are higher – and output is lower – in states that permit RPM as compared to states that do not permit it, findings that suggest that RPM is likely to harm, rather than benefit, consumers. To be sure, the study was unable to link these general findings of anticompetitive effect with any specific theory of competitive harm. But despite this weakness, the study's results suggest that a presumption of illegality, rather than a presumption of legality, would be appropriate for RPM.

[36] Kirkwood (2010).

[37] For a detailed comparison of these alternatives, see Kirkwood (2010) at 545–70.

21.3.1. Burdens of Proof and Production

Under this proposal, the plaintiff would establish a prima facie case by proving that the defendant entered into a vertical agreement to maintain resale prices. The plaintiff would not have to show, at this stage of the litigation, that the defendant had market power, that the restraint was likely to have anticompetitive effects, or (in lieu of market power and likely effects) that it had actual anticompetitive effects. If the plaintiff meets this burden, then the RPM agreement would be presumed illegal and the burden would shift to the defendant to rebut the presumption.

The defendant could do so by producing substantial evidence that its conduct is entitled to a safe harbor or that it has a procompetitive business justification. Under either route, the plaintiff would be allowed to dispute the defendant's evidence, but if the defendant proves the existence of a safe harbor, then the presumption would be conclusively rebutted. If the defendant advances a justification, the plaintiff could introduce evidence that the requirements for the justification are not satisfied or that its efficiency benefits could be achieved through a less restrictive alternative. If the plaintiff puts forward an alternative, the defendant could respond with evidence that the alternative would have been significantly more costly or less effective than RPM. If so, the ultimate burden of establishing the existence of an equally cost-effective but less restrictive alternative would fall on the plaintiff.

If the plaintiff fails to undermine the defendant's justification – by showing that it is invalid, insignificant, or unnecessarily restrictive – the burden would shift back to the plaintiff to show that the restraint's anticompetitive effects exceeded its procompetitive benefits and therefore harmed consumers. To make this showing, the plaintiff would have to set forth a theory of anticompetitive harm – for example, that the RPM facilitated collusion at the dealer level – and show that the preconditions for this theory existed. In nearly every instance, that would entail proof that a manufacturer or dealer, or a group of manufacturers or dealers, had market power.[38] In addition, the plaintiff would have to demonstrate that the anticompetitive effects that flowed from its theory of harm were likely to outweigh the procompetitive effects of the defendant's justification.

21.3.2. Safe Harbors

My approach would create a limited safe harbor for established manufacturers and a broad safe harbor for new entrants.

---

[38] The exception would occur if the plaintiff's theory is that RPM caused one or more dealers subject to it to engage in deceptive promotion. That practice can cause consumer harm, at least for a time, even though the dealers engaging in it, and the manufacturer whose RPM stimulated it, do not have significant market power.

21.3.2.1. Established Manufacturers

An established manufacturer that uses RPM would be entitled to a safe harbor if it proves that its does not have significant market power, that none of its dealers has significant market power, and that most other manufacturers in the market (measured by sales volume) do not maintain resale prices. Proof of these three conditions would entirely or almost entirely remove the possibility of significant harm from any of the anticompetitive effects normally attributed to RPM. It is highly unlikely, for example, that RPM facilitated collusion at the manufacturing level if the great majority of manufacturers did not engage in RPM or otherwise maintain resale prices.[39] Likewise, excessive or misleading dealer promotion is unlikely to have occurred when there was intense competition at both the manufacturing and dealer levels, and a manufacturer that induced excessive or misleading promotion by its dealers was likely to face both price and nonprice competition in response.

To be sure, it will not always be easy for a manufacturer to show that neither it nor its dealers had significant market power and that maintained prices, whether accomplished by agreement or otherwise, were not pervasive in the market. Litigation of the issues of market power and pervasiveness can often be a challenging and expensive process, one that may discourage some manufacturers from asserting this safe harbor, even when they would be entitled to it. But there will be clear cases, and the existence of this safe harbor will make it easier to dispose of them. Where the manufacturer is, by widespread agreement, a small factor in the market, where its dealers are numerous and small, and where dealers in rival brands discount them regularly, the safe harbor could be established without difficulty. In *Nine West*, for example, the FTC resolved several of these issues without apparent hesitation, concluding that the company's "use of resale price maintenance is not likely to harm consumers."[40] In clear cases, in short, this approach would readily eliminate non-meritorious suits.

The safe harbor is also subject to dynamic conditions beyond a manufacturer's control. If other producers adopt RPM, the first manufacturer in the market to use RPM might no longer be eligible for the safe harbor. Similarly, if one of its dealers becomes dominant in a geographic market, the manufacturer could no longer rely on the safe harbor. This would not mean, of course, that the manufacturer's RPM had become anticompetitive; it would only mean that the manufacturer would have to present evidence of a procompetitive justification in order to rebut the

---

[39] A manufacturer can maintain resale prices without RPM by refusing to sell to off-price dealers or by cutting off full-price dealers when they engage in discounting. While these alternatives may not be as effective as RPM, they can have a significant impact on resale prices.

[40] See *Nine West* 2008, p. 6 ("On the record before us, it appears that Nine West has only a modest market share in any putative relevant products market in which it competes. This suggests *prima facie* that it lacks market power, and there is no reason to believe that there is collective market power in any putative market. There is also no evidence of a dominant, inefficient retailer in this market").

presumption of illegality. Over time, as courts gain experience with this safe harbor, it could be made more precise; e.g., by picking specific market shares and concentration levels below which market power would be conclusively presumed to be absent.[41] The safe harbor could also be enlarged if warranted. What is important at this stage is establishing the principle that cases of plainly innocuous RPM should escape the presumption of illegality.

21.3.2.2. New Entrants

For the same reason, a manufacturer that employs RPM to help it enter a new market would be entitled to a safe harbor. In this setting, RPM may be procompetitive because it is the most efficient way to induce dealers to promote the new product, but even if it is not – even if a less restrictive alternative exists – RPM is highly unlikely to be anticompetitive.

New entrants often face a free rider problem. When they enter a new market, they frequently need dealers to promote their new brand in order to help it become established, but if those promotions are successful, other dealers will also want to carry the product and those dealers did not incur the promotional costs incurred by the original dealers. As a result, the latter can undercut the former and prevent them from recouping their outlays. RPM can solve this free rider problem by, as Steiner put it, inhibiting "Johnny-come-lately" stores from siphoning off the rewards that pioneering dealers need for their "missionary work."[42]

Like all free rider arguments, however, this one is a valid justification for RPM only if the new entrant could not achieve its promotional goals in other, less restrictive ways. Many commentators have observed that free riding on pioneering

---

[41] The task is somewhat complicated because a manufacturer is most likely to use RPM for procompetitive reasons if its products are differentiated from those of its closest rivals. As a result, the manufacturer is likely to have some market power because of this differentiation. Nevertheless, a court might conclusively presume that the manufacturer lacked significant market power if its share of a properly defined relevant market was under 15% and market concentration was below 1,500. The same figures might be applied to dealer markets, though higher ceilings would be appropriate (say 25% and 2,500) if the manufacturer proved that its dealers sold only its brands and were easily replaced.

[42] Steiner (1997) at 430; see also Mathewson and Winter (1998) at 60 ("In markets where extensive distribution systems are necessary, RPM is often used in the early part of a product's life cycle to aid in the establishment of the distribution system. In this situation, which holds for markets as diverse as those for stereo components and jeans, RPM lowers the barriers [to] entry into upstream markets"). Even when dealers do not need to engage in expensive promotional activity on behalf of a new product, RPM may facilitate new entry by reducing the risks that dealers face in carrying an untried product. If the product is unsuccessful, the dealers may have to mark it down sharply in order to sell off their inventory. If the product is successful, the dealers may be unable to earn high profits on it because other dealers decide they want to carry it too. RPM can reduce both risks by making it more likely that dealers earn significant profits on a new product, whether it is successful or not. See Lambert (2010). Of course, a manufacturer may be able to alleviate those risks in other ways: it can help dealers unload inventory by accepting returns of unsold units at full price, and it can protect dealer margins on successful products by limiting the number of authorized dealers.

dealers' promotional efforts can be reduced, if not eliminated, through nonprice restraints such as exclusive distribution,[43] or through up-front payments or other incentives to the pioneering dealers.[44] In addition, rather than relying on dealers to create demand, manufacturers may create demand for a new product themselves, through advertising or other consumer-directed marketing. If any of these alternatives is as effective and efficient as RPM, a new entrant would not need to employ RPM in order to gain a foothold in a new market.

Despite these objections, a safe harbor for new entrants appears appropriate for two reasons. First, the likelihood of significant, persistent anticompetitive effects is very low. After all, new entry is, by definition, a time-limited process: it ends as soon as the firm becomes established in the market.[45] Moreover, new entrants are easy to identify, they are unlikely to gain market power in the near term, and there will be only one or a few of them at any given time, reducing the danger of collusion. For these reasons, even prominent opponents of RPM, who generally favor a per se ban, have been willing to create an exception for new entrants.[46] Second, in this setting, where the prospects of anticompetitive harm are so low, a firm is more likely to choose the distribution arrangement that is most efficient. If the entrant cannot reasonably expect to gain significant market power during the entry process, then its margin for error is small. It is unlikely to enjoy any cushion – any supracompetitive margin – that would protect it if its distribution strategy is not the most cost-effective. Although entrants often make mistakes, and most of them perish, there is reason to presume in this setting that RPM is the least restrictive alternative.

21.3.3. Justifications

Under this approach, a defendant could rebut the presumption of illegality by establishing a safe harbor, demonstrating a significant procompetitive justification, or both. In order to demonstrate a significant procompetitive justification, a defendant would have to show that the justification was theoretically valid, that it was significant in magnitude, and that if the plaintiff asserts a less restrictive alternative, there are genuine problems with it. Areeda and Hovenkamp describe these requirements in a particularly thoughtful way:

---

[43] See Areeda and Hovenkamp (2004) at ¶ 1617a3 (while the new entry rationale makes sense as a justification for exclusive territories, it "seems presumptively inapplicable to resale price maintenance").

[44] See, e.g., Peeperkorn (2008) at 212 ("it seems more efficient, both for the manufacturer and for the consumers, if the manufacturer rewards the investments made by the first distributor through a lump sum payment").

[45] Id. at p. 211 ("this is not a justification to allow RPM for a long period, or ... for established brands").

[46] See, e.g., Pitofsky (1983) at 1495; see also Leegin 2007, pp. 917–18 (Breyer J., dissenting) ("And if forced to decide now, at most I might agree that the per se rule should be slightly modified to allow an exception for the more easily identifiable and temporary condition of 'new entry'").

To define a justification in terms that can realistically be proved, we can reasonably expect at least substantial evidence that the manufacturer has a legitimate business problem, that resolution of that problem would confer a nontrivial benefit, that the restraint can be reasonably effective for the claimed purpose, and that less restrictive alternatives would be significantly more costly or significantly less effective.[47]

This formulation, while sensitive to the practical limits of proof, is nevertheless demanding, and it is likely that many defendants would not be able to satisfy it. But that is appropriate, for RPM is likely to be anticompetitive in a substantial number of cases. At the same time, though, both economic theory and empirical evidence suggest that RPM is likely to be procompetitive in a substantial number of cases. It is important, therefore, that the approach be administered in a way that gives manufacturers a reasonable opportunity to overcome the presumption of illegality. This presumption should not become, as it may have become in Europe, a conclusive presumption of illegality, a per se ban in disguise.

Fortunately, there is considerable reason to expect that American courts would administer a presumption of illegality in a balanced way. In recent decades American courts have become more cautious about finding antitrust liability and more sympathetic to economic justifications of business practices. Approximately half of federal appellate court judges were appointed by Republican presidents,[48] and judges at all levels recognize that economics plays a crucial role in antitrust analysis.[49] Both perspectives are likely to cause courts, in their findings of fact and instructions to juries, to take an open-minded – rather than dismissive – approach to economic arguments on behalf of RPM. In addition, the Supreme Court has now declared, on three separate occasions, that vertical intrabrand restraints are frequently procompetitive and thus should be evaluated under the rule of reason, not condemned per se.[50] This consistent message, reiterated in *Leegin* itself, is likely to make courts receptive to assertions that RPM is procompetitive and skeptical of claims that it is harmful. Finally, in most cases the defendant is likely to be able to advance a plausible justification for RPM, since there are several well-known, theoretically valid, procompetitive explanations for the practice. To be sure, the plaintiff is bound to assert that the manufacturer's procompetitive goals could be achieved in a less restrictive way, but on that issue the plaintiff has the initial burden of production and the ultimate burden of proof. Moreover, if the plaintiff cannot show that there was a practical, cost-effective alternative to RPM, the plaintiff would be forced to prove that the anticompetitive effects of the restraint outweighed its beneficial effects, a task that would ordinarily require the plaintiff to undertake the challenging project of showing market power. All these reasons make it unlikely that the approach would simply become a rule of per se illegality.

---

[47] Areeda and Hovenkamp (2004) at ¶ 1633.
[48] At the beginning of the Obama Administration, the figure was 56 percent (Kingsbury 2009).
[49] Kirkwood (2004).
[50] See *Continental T.V.* 1977; *State Oil* 1997; *Leegin* 2007.

21.4. CONCLUSION

Scholars differ over whether online RPM is more likely to be anticompetitive or procompetitive than traditional RPM. They agree that online retailers have two significant competitive advantages over traditional brick-and-mortar retailers. Online retailers typically charge lower prices than traditional retailers, and online purchases are typically quicker – for consumers with access to the Internet – than purchases that require visiting a physical store. Scholars disagree, however, over whether these characteristics (lower prices and easier access) mean that vertical price fixing RPM is more likely, or less likely, to be justified when it is applied to online retailers. It might be more justified because online retailers are particularly effective in taking sales from traditional retailers, reducing their ability to provide services that benefit manufacturers and consumers. It might be less justified because online retailers would no longer be able to offer the very low prices that so many consumers desire.

As this chapter explains, the resolution of this debate turns on two key issues: whether there are likely to be less restrictive alternatives to online RPM and whether, in markets where online retailing is significant, the structure of the market is conducive to anticompetitive effects.[51] At this point, however, it is difficult to decide either issue. There are no systematic studies of online RPM and virtually no reported opinions. Without more evidence, we cannot say with confidence that online RPM is generally unnecessary (because there are less restrictive alternatives) or generally procompetitive (because the markets where it is used would not permit the manufacturer or its retailers to exercise market power). Absent more research, in short, it is not possible to determine whether online RPM is more likely to be anticompetitive or procompetitive than traditional RPM.

The prudent course at this time, then, is to apply the same legal standard to online RPM as to traditional RPM. There ought to be one legal standard for all RPM, and as I have urged before, that standard ought to be a rebuttable presumption of illegality combined with two safe harbors. This approach would resolve the easy cases cheaply and correctly, and would be at least as good as the full rule of reason in resolving the harder, mixed cases.

REFERENCES

Akman, Pinar. 2015. A Competition Law Assessment of Platform Most-Favoured-Customer Clauses. Working Paper, http://ssrn.com/abstract=2669395.

Areeda, Phillip E. and Herbert Hovenkamp. 2004. *Antitrust Law*. Vol. 8 (2nd edn.). Boston, MA: Little, Brown & Co.

Blair, Roger D. and Jessica S. Haynes. 2010. The Plight of Online Retailers in the Aftermath of Leegin: An Economic Analysis. *Antitrust Bulletin*, 55, 245–69.

---

[51] In other words, are market shares, barriers, and other market characteristics – at the manufacturing level, the retail level, or both – likely to permit the exercise of market power?

Comanor, William S. 2013. Leegin and Its Progeny: Implications for Internet Commerce. *Antitrust Bulletin*, 58, 107–27.
Continental T.V., Inc. v. GTE Sylvania Inc., 433 U.S. 36 (1977).
Gundlach, Gregory T., Joseph P. Cannon, and Kenneth C. Manning. 2010. Free Riding and Resale Price Maintenance: Insights from Marketing Research and Practice. *Antitrust Bulletin*, 55, 381–422.
Kingsbury, Alex. 2009. The Long Road to Remaking the Courts, U.S. News & World Report, June 1, at 54.
Kirkwood, John B. 2004. Consumers, Economics, and Antitrust. *Research in Law and Economics*, 21, 1–62.
  2010. Rethinking Antitrust Policy toward RPM. *Antitrust Bulletin*, 55, 423–72.
Klein, Benjamin. 2015. Resale Price Maintenance of Online Retailing. In Roger D. Blair and D. Daniel Sokol (eds.), *The Oxford Handbook of International Antitrust Economics* 2. Oxford University Press, 277–303.
Klein, Benjamin, and Kevin M. Murphy. 1988. Vertical Restraints as Contract Enforcement Mechanisms. *Journal of Law and Economics*, 31, 265–97.
Lambert, Thomas A. 2010. A Decision-Theoretic Rule of Reason for Minimum Resale Price Maintenance. *Antitrust Bulletin*, 55, 167–224.
Lao, Marina. 2011. Internet Retailing and "Free-Riding:" A Post-Leegin Antitrust Analysis. *Journal of Internet Law*, 14(1), 15–24.
Leegin Creative Leather Products, Inc. v. PSKS, Inc., 551 U.S. 877 (2007).
MacKay, Alexander and David Aron Smith. 2014. The Empirical Effects of Minimum Resale Price Maintenance. Working Paper, http://ssrn.com/abstract=2513533.
Mathewson, Frank and Ralph Winter. 1998. The Law and Economics of Resale Price Maintenance. *Review of Industrial Organization*, 13, 57–84.
McDonough v. Toys "R" Us, Inc., 638 F. Supp. 2d 461 (E.D. Pa. 2009).
Nine West Group Inc., No. C-3937, 2008 WL 2061410 (FTC May 6, 2008).
Peeperkorn, Luc. 2008. Resale Price Maintenance and Its Alleged Efficiencies. *European Competition Journal*, 4, 201–12.
Pitofsky, Robert. 1983. In Defense of Discounters: The No-Frills Case for a Per Se Rule Against Vertical Price Fixing. *Georgetown Law Journal*, 71, 1487–95.
State Oil Co. v. Kahn, 522 U.S. 3 (1997).
Steiner, Robert L. 1997. How Manufacturers Deal With the Price-Cutting Retailer: When Are Vertical Restraints Efficient? *Antitrust Law Journal*, 65, 407–48.
Van Baal, Sebastian and Christian Dach. 2005. Free Riding and Customer Retention Across Retailers' Channels. *Journal of Interactive Marketing*, 19, 75–85.

PART VI

Mergers in High Technology

22

US Merger Enforcement in the Information Technology Sector

Jeffrey A. Eisenach

22.1. INTRODUCTION

United States merger policy seeks to prevent transactions that would result in a substantial lessening of competition in a relevant product market. Competition is harmed when a transaction results in a significant increase in market power, defined as the ability of a firm, or group of firms, to set and maintain prices above (or reduce quality below) the competitive level, thereby harming consumer welfare; or, in an increase in the incentive and ability of a dominant firm to engage in anticompetitive activities, such as raising rivals' costs. Under prevailing jurisprudence, transactions which significantly increase concentration in a relevant market above certain levels are presumed to increase market power and substantially lessen competition. However, merging parties seeking to win approval of transactions that significantly increase concentration may rebut the presumption of illegality by demonstrating through other evidence that the transaction will not actually harm competition, or that any harms are offset by increases in transaction-specific efficiencies.

The application of these concepts to transactions in information technology (IT) markets[1] is complicated by the fact that such markets have distinct economic characteristics. Specifically, IT markets typically exhibit *dynamism, modularity,* and *demand- and supply-side economies of scale and scope.*[2] Each of these characteristics introduces challenges and complexities into the assessment of the competitive effects of mergers.

The views expressed in this chapter are mine and should not be attributed to any of the firms, clients or other institutions with which I am associated. I am grateful to Janusz Mrozek for extensive research assistance. Any remaining errors are my own.

[1] For purposes of this chapter, I define IT markets as markets for digital content, software, information services, and communications services. Such markets are sometimes referred to as "Internet ecosystem" markets (Eisenach 2012). Many of the issues discussed herein are also encountered in other "high tech" markets, such as markets for pharmaceutical products.

[2] Eisenach (2012); Eisenach and Gotts (2015a).

For example, dynamism (i.e., competition through innovation) combines with economies of scale and scope to create high levels of concentration in many markets, often accompanied by the presence of difficult-to-replicate assets (sometimes embodied in intellectual property rights). Rapidly changing technologies and markets, and extensive product differentiation, raise challenges to market definition and to the assessment of market power. Modularity (the presence of strong complementarities among the various inputs that make up a final product) further complicates market definition and forces consideration of the impact of transactions on vertical relationships. The role of efficiencies – always difficult to quantify, and thus traditionally an afterthought in merger analysis – cannot easily be discounted in high tech markets characterized by rapid innovation and strong economies of scale and scope.

These and similar issues are the subject of active debate and discussion among antitrust practitioners, and they are increasingly central in US antitrust enforcement activity and litigation involving IT transactions. The goal of this chapter is to describe the state of play in both doctrine and practice.

The remainder of this chapter is organized as follows. Section 22.2 presents a brief summary of contemporary US antitrust policy as it relates to mergers in general and IT-sector mergers in particular. Section 22.3 discusses the competitive dynamics of high tech markets, organized around the three characteristics described above, dynamism, modularity, and economies of scale and scope, and describes in broad terms how these characteristics affect merger enforcement. Section 22.4 provides some examples of how these issues have manifested themselves in US agency reviews of recent transactions. Section 22.5 concludes.

22.2. MERGER POLICY AND JURISPRUDENCE IN THE UNITED STATES

The primary US statute governing mergers and acquisitions is the Clayton Act. Specifically, Section 7 of the Clayton Act (15 USC § 18) prohibits a person "engaged in commerce or in any activity affecting commerce" from acquiring "the whole or any part" of a business's stock or assets if the effect of the acquisition "may be substantially to lessen competition, or to tend to create a monopoly." The Clayton Act is enforced by both the Antitrust Division of the US Department of Justice (DOJ or "the Division") and by the Federal Trade Commission (FTC; together, "the agencies"). To successfully block a merger in court, the agencies must demonstrate by a "preponderance of the evidence" that the transaction is "reasonably likely to result in a substantial lessening of competition."[3] Such effects may result from either horizontal or vertical mergers.

---

[3] See e.g., Oracle-PeopleSoft Order. *United States, et al v. Oracle Corporation*. 2014. No. C 04-0807. United States District Court for the Northern District of California. As discussed further below, mergers involving communications companies are also assessed under the "public interest" standard of the Communications Act.

This section first addresses current doctrine as it relates to horizontal and vertical transactions. Next it describes the role of efficiencies in assessing the effects of mergers, the use of remedies to address concerns about anticompetitive effects, and the merger review process itself.

22.2.1. Horizontal Transactions

In assessing the competitive effects of horizontal mergers, the agencies have for many years relied on principles embodied in the jointly issued horizontal Merger Guidelines. First issued in 1968, the Guidelines were most recently updated in 2010.[4] The 2010 revisions have important implications for mergers in the high tech sector.

Traditionally, both the agencies and the courts have begun their assessment of mergers by examining the impact of the transaction on the level of market concentration, as measured by the market share of the leading firm in a relevant product market, the combined shares of the top few firms, or by the Herfindahl-Hirschman Index (HHI). A merger resulting in the merging firms possessing a post-merger share of 30 percent or higher in a relevant market creates a presumption of illegality.[5]

Measurement of market share depends on the definition of a relevant market, defined as a product or group of products sold in a particular geographic area.[6] Market definition, in turn, focuses on an analysis of demand-side substitution or "interchangeability"[7] as conceptualized in the "hypothetical monopolist test," which "requires that a product market contain enough substitute products so that it could be subject to post-merger exercise of market power significantly exceeding that existing absent the merger."[8] In practice this concept is embodied in the so-called SSNIP test, in which the analysis considers, for the market whose definition is being tested, whether a single firm (the "hypothetical monopolist") "likely would impose at least a small but significant and non-transitory increase in price."[9] A relevant market encompasses the widest group of products for which a SSNIP would be profitable.

---

[4] Importantly, the Guidelines are neither laws nor regulations, and do not bind either the agencies or the courts: "Under general principles of administrative law, the Merger Guidelines are a statement of agency enforcement policy that is not binding on the courts." However, "[i]n practice, courts have relied heavily on prior versions of the Merger Guidelines, quickly adopting new analytic tools promoted therein" (Brannon and Bradish 2010).

[5] *Philadelphia National Bank*, 374 US at 364; *United States v Bazaarvoice* No. 13-cv-00133 slip opinion at ¶ 148.

[6] Merger Guidelines (2010) at § 4.

[7] A product market is "determined by the reasonable interchangeability of use or the cross-elasticity of demand between the product itself and substitutes for it." *Brown Shoe*, 370 U.S. 325.

[8] Merger Guidelines (2010) at § 4.

[9] *Id.* A SSNIP is "most often" defined as 5 percent of the price paid by customers, but may vary depending on the nature of the market (*id.* at 4.1.2).

Once the market is defined, traditional analysis typically moves to assessing the effects of the transaction on concentration. The Merger Guidelines define three levels of market concentration based on the pre-merger level of the HHI: "unconcentrated" (HHI < 1500); "moderately concentrated" (1500 < HHI < 2500); and "highly concentrated" (HHI > 2500). Mergers in unconcentrated markets "are unlikely to have adverse competitive effects and ordinarily require no further analysis." However, mergers in moderately concentrated markets that raise the HHI by 100 points or more, and mergers in concentrated markets that raise the HHI by between 100 and 200 points, "potentially raise significant concerns and warrant further scrutiny." Mergers in highly concentrated markets that raise the HHI by more than 200 points "will be presumed likely to enhance market power."[10]

Courts also continue to give weight to market concentration. In general, for mergers that result in significant increases in concentration, "[t]he burden shifts to the defendant to rebut the presumption [of anticompetitive effects] by showing why the merger is unlikely to substantially lessen competition, or by discrediting the data underlying the presumption in the government's favor,"[11] or by demonstrating efficiencies sufficient to outweigh any anticompetitive effects.

In recent years, both the agencies and the courts have begun to place less emphasis on market structure and more on directly assessing the competitive effects of mergers,[12] focusing on two analytical categories: "unilateral effects," in which anticompetitive outcomes arise because the merging parties in markets no longer compete against each other, and "coordinated effects," in which anticompetitive outcomes arise because increased market concentration enables the merged entity to more readily coordinate its decisions with those of other participants in the market.

While unilateral effects theories have been recognized in the Merger Guidelines since 1992, their role has expanded significantly in subsequent updates. Recent developments in unilateral effects analysis, leading up to and reflected in the 2010 Merger Guidelines, de-emphasize market definition in favor of "a more integrated approach" that "more directly answers the 'ultimate inquiry in merger analysis,' i.e., 'whether the merger is likely to create or enhance market power or facilitate its exercise.'"[13] Unilateral effects theories are especially important in

---

[10] *Id.* at § 5.3.
[11] *United States v. Bazaarvoice* at ¶ 122. See also *H.J. Heinz Co.*, 246 F.3d at 715 ("If the defendant successfully rebuts the presumption of illegality, the burden of producing additional evidence of anticompetitive effect shifts to the government, and merges with the ultimate burden of persuasion, which remains with the government at all times").
[12] See Baker and Shapiro (2008) at 29 ("The structural presumption remains in force, but it is dramatically weaker. Courts and enforcers today place less weight on market structure, pay closer attention to possible expansion by smaller suppliers and entry by new ones, and exhibit less hostility to merger efficiencies.")
[13] Merger Guidelines (2010) at § 1.

mergers in the high tech markets with differentiated products and the potential for exclusionary conduct.[14]

Models of unilateral effects range from simple models of upward pricing pressure (UPP), depending on only a few pieces of information,[15] to full-fledged complex merger simulation models that capture price effects across many differentiated products in a market. Analyses rely on models demonstrating how the incentives of the firms change after combining due to the internalization of cross-firm externalities. Unilateral effects may include higher prices, reductions in product quality, or the enhanced ability or incentive to engage in exclusionary conduct.[16]

Economic models of unilateral effects – while still controversial – are well developed in the literature. By contrast, coordinated effects are more difficult to analyze, as they "involve conduct by multiple firms," and "include a range of conduct." For example, the coordination may be explicit or may involve "common understanding" or "parallel accommodating conduct." Thus, coordinated effects can include "conduct not otherwise condemned by the antitrust laws."[17] While coordinated effects theories sometimes play a role in high tech mergers, they are seldom the central issue, as the characteristics of such markets (including product differentiation and rapid technological change) are not generally conducive to collusion.[18]

Other factors enter into the agency's consideration of the market power effects of horizontal mergers, including the possibility of entry as a countervailing force to any possible anticompetitive effects. Initially, entry is considered in assessing the existing market: "Firms that would rapidly and easily enter the market in response to a SSNIP are market participants and may be assigned market shares."[19] However, the agencies also evaluate "entry or adjustments to pre-existing entry plans that are induced by the merger."[20] Entry must be "timely, likely, and sufficient in its magnitude, character, and scope to deter or counteract the competitive effects of concern."[21] "Failing firm" defenses for mergers, by which a merger is considered not to be anticompetitive if one of the parties would otherwise exit the market, are sometimes invoked, but tend to be held to a high standard.

---

[14] See Shapiro (2010) at 706 ("The introduction of unilateral effects in the 1992 Guidelines reflected and anticipated a shift in merger enforcement away from relatively homogeneous industrial commodities and towards more differentiated products. While the Guidelines necessarily apply to all industries, the 1992 Guidelines were a major step in the evolution of antitrust enforcement from the industrial age to the information age.")
[15] Farrell and Shapiro (2010).
[16] See Shapiro (2010).
[17] Merger Guidelines (2010) at § 7.
[18] See generally Harrington, Jr. (2013).
[19] Merger Guidelines (2010) at § 9.
[20] Id.
[21] Id.

Importantly for consideration of high tech mergers, the new Guidelines place strong emphasis on innovation, with an entire section[22] now devoted to "whether a merger is likely to diminish innovation competition by encouraging the merged firm to curtail its innovative efforts."[23] The agencies will consider both current "efforts to introduce new products" and "capabilities that are likely to lead it to develop new products in the future."[24] In general,

> [t]he Agencies evaluate the extent to which successful innovation by one merging firm is likely to take sales from the other, and the extent to which post-merger incentives for future innovation will be lower than those that would prevail in the absence of the merger. The Agencies also consider whether the merger is likely to enable innovation that would not otherwise take place, by bringing together complementary capabilities that cannot be otherwise combined or for some other merger-specific reason.[25]

22.2.2. Vertical Transactions

Vertical transactions involve entities which produce complementary inputs. While the term "vertical" evokes the notion of a linear ("upstream" and "downstream") production process (e.g., coal and iron ore being "upstream" and steel production being "downstream"), vertical theories of anticompetitive effects also apply to "platform" markets in which multiple firms may be consumer facing (e.g., providers of smart phones and of smart phone apps or content). Because such relationships are both commonplace and economically significant in high tech markets, and because they play an important role in the competitive dynamics of such markets, mergers which implicate vertical issues are receiving increasing scrutiny by the agencies.[26] However, the last time the agencies issued guidelines for assessing vertical transactions was in 1984, causing some commentators to suggest that an update is in order.[27]

22.2.3. The Role of Efficiencies

One way parties to a transaction can rebut a presumption of illegality is to demonstrate that the efficiency benefits of a merger exceed any anticompetitive effects. In the Merger Guidelines, the agencies consider transaction-specific efficiencies arising from a transaction, and merging parties commonly make efficiency arguments. However, as noted above, efficiencies seldom play a determinative role either in the agencies' enforcement decisions or in the courts.

In part, the lack of weight given to efficiencies can be attributed to the difficulty in measuring or accurately predicting the dynamic effects of transactions, e.g.,

---

[22] Merger Guidelines (2010) at § 6.4.
[23] Id. at § 23
[24] Id.
[25] Id.
[26] See generally Salop and Culley (2016).
[27] Id.

increased innovation arising from synergies between the transacting firms' research and development activities, or the ability to more efficiently combine inputs to bring new products more rapidly to market. These difficulties arguably are of particular importance for high tech mergers given the importance of innovation and the presence of strong economies of scale and scope.[28]

22.2.4. Remedies

Remedies, embodied in consent decrees between agencies and the transacting parties (or, in the case of communications mergers before the Federal Communications Commission (FCC or Commission), in the Commission orders that grant formal approval to proceed) are often applied to ameliorate potential anticompetitive effects of mergers. In general, remedies fall into two categories. Structural remedies involve restructuring the transaction so as reduce its impact on levels of market concentration, for example by "spinning off" some operating component into a new firm, or by divesting assets (or customers) to an existing competitor to enable it to more effectively compete in the post-transaction market. Conduct (or "behavioral") remedies involve ongoing restrictions (or affirmative obligations) on the parties. In IT industries, conduct remedies often take the form of "open access" or "sharing" obligations for non-replicable assets, such as patent portfolios or spectrum licenses.

In 2011, the Antitrust Division issued revised guidelines outlining its approach to merger remedies, updating the version previously issued in 2004. One notable difference from the prior version is a more favorable view toward conduct remedies,[29] effectively codifying the increasing use of conduct remedies in recent merger cases, especially in information technology mergers.[30]

---

[28] See Creighton (2010) at 2, citing Antitrust Modernization Comm'n, Report and Recommendations 40 (2007) ("To improve the application of antitrust in new economy industries, antitrust enforcers should give further consideration to efficiencies that lead to more rapid or enhanced innovation. The potential benefits to consumer welfare from such efficiencies are great, thus warranting careful assessments of the potential for certain business conduct to create more rapid or enhanced innovation"). Available at http://govinfo.library.unt.edu/amc/report_recommendation/chapter1.pdf). See also Merger Guidelines (2010) at § 10 ("Research and development cost savings may be substantial and yet not be cognizable efficiencies because they are difficult to verify or result from anticompetitive reductions in innovative activities").

[29] See US Department of Justice. June 2011. *Antitrust Division Policy Guide to Merger Remedies*, at 6 ("Conduct remedies are a valuable tool for the Division. They can preserve a merger's potential efficiencies, and, at the same time, remedy the competitive harm that otherwise would result from the merger. Conduct relief can be a particularly effective option when a structural remedy would eliminate the merger's potential efficiencies, but, absent a remedy, the merger would harm competition").

[30] Kwoka and Moss (2012). ("The earlier Remedies Guide, issued in 2004, emphasized structural remedies such as divestitures as the preferred approach to resolving competitive problems with mergers. In contrast, the 2011 revision is considerably more favorably disposed toward the use of behavioral remedies that proscribe specified anticompetitive behavior by the merged companies ... This policy revision is reflected in three recent mergers that in quick succession have all been permitted subject to consent orders with substantial behavioral remedies: Ticketmaster-Live Nation, Google-ITA, and Comcast-NBCU. The 2011 Remedies Guide essentially codifies much of the approach adopted in these merger cases.")

TABLE 22.1: *Hard-Scott-Rodino filings, IT-sector transactions, 2010–2014*

NAICS	Industry	Filed	Reviewed	Second Request
334	Computer and Electronic Products	257	65	7
516	Internet Publishing and Broadcasting	18	1	1
517	Telecommunications	138	27	13
518	ISPs, Web Search Portals, and Data Processing	165	21	5
519	Other Information Services	58	6	3
Total		636	120	29

Source: FTC and DOJ. 2011–2014. Hart-Scott-Rodino Annual Reports, Fiscal Years 2010–2014.

22.2.5. The Merger Review Process

As a practical matter, the merger review process generally is applied to mergers meeting guidelines established under the Hart-Scott-Rodino Antitrust Improvements Act of 1976.[31] Specifically, parties engaged in transactions valued in excess of $305.1 million must pre-file information with the government, which then has thirty days to determine whether to issue a request for additional information ("second request"); and, if a second request is issued, an additional thirty days in which to determine whether to challenge the transaction in court.[32]

As shown in Table 22.1, in Fiscal Years 2010–2014 there were approximately 636 transactions in NAICS industry categories associated with the IT sector, of which 120 were reviewed and twenty-nine were deemed worthy of deeper investigation.[33]

As Table 22.1 shows, transactions involving telecommunications firms were the most likely to receive a second request from the reviewing agency. As noted above, in addition to being reviewed by either the DOJ or the FTC, communications transactions require affirmative approval by the FCC. Specifically, Sections 214 and 310 of the Communications Act (47 USC §§ 214(a), 310(d)) provide that acquisitions

---

[31] 15 U.S. Code § 18a.
[32] Formally, notification is required if three tests are satisfied: the commerce test, the size of transaction test, and the size of person test. See FTC Premerger Notification Office. 2008. *To File or Not to File: When You Must File a Premerger Notification Report Form*. 2.
[33] Data tabulated from FTC and DOJ, Hart-Scott-Rodino Annual Reports, Fiscal Years 2010–2014. The agencies report transactions in an "Information Technology" groups for purposes of constructing a pie chart (see DOJ and FTC. 2014. Hart-Scott-Rodino Annual Report: Fiscal Year 2014). This category is not defined but for 2014 appears to consist of the following 3-digit NAICS industries: 334, 514, 516, 517, 518, and 519. For the tabulation in this paragraph and the accompanying table, the NAICS of the acquired entity is reported. Note that NAICS industries 514 and 516 do not exist in the 2012 NAICS definitions but the agencies continue to report transactions using those codes. As the high tech elements in the definition of NAICS 514, Information Services and Data Processing Services, have been transferred to other codes, and as only three acquisitions were reported for 2010–2014, that category is excluded from the statistics and table reported here. Transactions for NAICS 2016 are reported, however.

TABLE 22.2: *Major mergers and acquisitions reviewed by FCC, 2010–2015*

Year	Transaction	Value ($ millions)	Result	Conditions
2010	Frontier/Verizon	8,600	Allowed	Yes
2011	AT&T/Qualcomm	1,925	Allowed	Yes
2011	AT&T/T-Mobile	39,000	Denied	N/A
2011	CenturyLink/Qwest	22,400	Allowed	Yes
2011	Comcast/NBC Universal	30,000	Allowed	Yes
2011	Cumulus/Citadel	2,500	Allowed	Yes
2011	EchoStar/Hughes	2,000	Allowed	No
2011	Level 3/Global Crossing	3,000	Allowed	Yes
2012	Verizon/SpectrumCo/Cox	3,600	Allowed	Yes
2013	Gannett/Belo	2,200	Allowed	No
2013	Softbank/Sprint/Clearwire	21,600	Allowed	Yes
2013	T-Mobile/MetroPCS	1,500	Allowed	Yes
2013	Tribune/Local TV	2,730	Allowed	Yes
2014	AT&T/Leap Wireless	1,200	Allowed	Yes
2014	Frontier/AT&T	2,000	Allowed	No
2015	AT&T/DirecTV	67,000	Allowed	Yes
2015	Comcast/TWC/Charter	45,200	Denied	N/A

Source: FCC's *Major Transaction Decisions. 2010–2015.* Federal Communications Commission. Available at www.fcc.gov/encyclopedia/major-transaction-decisions. Data limited to mergers and acquisitions from 2010 to 2015 with transaction value greater than $1 billion.

involving transfers of transmission lines and/or wireless licenses must receive advance approval by the Commission. In order to approve a transaction, the Commission must find "that the public interest, convenience, and necessity will be served thereby" (Sec. 310(d)) or find "that the present or future public convenience and necessity require" it (Sec. 214(a)). Thus, unlike the Clayton Act, the Communications Act places the burden of proof on the merging parties to demonstrate that the transaction is "in the public interest."

As shown in Table 22.2, the FCC reviewed seventeen major transactions between 2010 and 2015, with a total transaction value of more than $250 billion. Of these, fifteen were approved and two, AT&T's proposed acquisition of T-Mobile and Comcast's proposed acquisition of Time Warner Cable, were denied.

22.3. COMPETITIVE DYNAMICS OF INFORMATION TECHNOLOGY MARKETS

IT markets exhibit at least three meaningful distinguishing characteristics: *dynamism*, *modularity*, and *demand-side effects*.[34]

---

[34] For a more extensive discussion of these phenomena and their implications for competition analysis, see Eisenach (2012). See also Eisenach and Gotts (2015a).

Dynamism refers to the significance of innovation as a measure of market performance. In dynamic markets, the ability of a firm to offer new and improved products plays at least as significant a role in its success (i.e., its profitability) as the ability to produce and sell existing products at lower prices.[35] In such markets, firms incur significant sunk cost investments to create new products, causing average costs to exceed marginal costs over the relevant range of output, but resulting in product differentiation (innovation being simply product differentiation over time) that allows sellers to recoup their investments by earning high margins (relative to marginal cost).

Modularity results from strong complementarities in production or consumption: Operating systems are strong complements with personal computers; online music stores are strong complements with smart phones; smart phones are strong complements with communications networks, etc. Complementarity, in turn, creates demand for compatibility or "interconnection." Competition in such markets takes place both within and among clusters of mutually compatible complementary products, referred to as "platforms." Intra-platform competition sometimes takes the form of efforts by firms to increase their bargaining power within the platform and thus their share of the economic rents it generates. It is not uncommon for merger transactions to result in shifts in bargaining power among platform participants. Whether, or under what circumstances, such shifts raise antitrust concerns is a topic of debate.

Lastly, IT markets are characterized by significant demand-side effects, including economies of both scale and scope. Demand-side economies of scale, also known as network effects, occur when a product is more valuable to consumers as the *number* of users increases.[36] Demand-side economies of scope exist when a product's value depends on having different *types* of users (e.g., radio listeners and advertisers).[37]

These characteristics of IT markets have important implications for antitrust enforcement generally and merger enforcement in particular.

One obvious implication of dynamism is the need to consider how competition will develop in the future. Rapid technological change may make monopoly power more ephemeral,[38] but it also suggests the need to give weight to the potential that a transaction might limit future competition by, for example, reducing the number of innovators.[39] Further, the essence of competitive success in dynamic industries lies

---

[35] Baumol (2002) at 4 ("Innovation has replaced price as the name of the game in a number of important industries. The computer industry is only the most obvious example, whose new and improved models appear constantly, each manufacturer battling to stay ahead of its rivals"). See also Schumpeter (1942).
[36] See Katz and Shapiro (1994).
[37] See e.g., Wright (2004).
[38] See, e.g., Ginsburg and Wright (2012). See also Sokol (2014) ("As high tech markets change rapidly, market power may be transient").
[39] Striking this balance may require predicting future events, a difficult task in dynamic industries. See generally Gotts and Rapp (2004). See also Hovenkamp (2012) ("Innovation can produce sudden and dramatic shifts in prices or output and almost instantly expand the range of consumer choices. As a result, predicting and managing competitive processes in highly innovative industries is much more difficult than in markets where technology is very largely constant and most movements affect only the output and price of a set of unchanging products").

in successful product differentiation, i.e., the ability to create unique product characteristics that are both valued by consumers and not easily replicable by competitors, perhaps as a result of ownership of difficult-to-replicate physical assets (such as communications infrastructures) or intellectual property (patents, copyrights), or due to network effects arising from a large user base or the possession of large amounts of "Big Data."[40] Mergers which combine non-replicable assets often require regulators to balance potential synergies against the increased likelihood or effectiveness of exclusionary conduct.

Lastly, the presence of strong economies of scale and scope, including network effects, raises a multitude of issues. Most obviously, markets characterized by strong network effects tend to be highly concentrated, yet concentration in such markets is not a reliable indicator of market power, nor are price-cost margins or the "Lerner Index," which is central to UPP models.[41] Multisided markets (demand-side economies of scope) pose their own special concerns, forcing regulators to consider the effects of mergers on both downstream "consumers" and upstream "suppliers."[42] Economists have only recently begun to develop the tools necessary to assess such effects.

22.4. KEY ISSUES IN RECENT IT-SECTOR MERGER REVIEWS

Recent US agency reviews of IT-sector mergers highlight the salience of the issues raised above. In several important cases, the agencies have either conditioned or blocked IT-sector transactions based on concerns about the incentive and ability of the combined firm to engage in exclusionary conduct. Regulators have also been faced with the need to assess the impact of transactions on market power in rapidly changing markets with differentiated products; to make difficult judgments about competition in markets for future goods (i.e., markets for products that are still under development); and to evaluate the welfare effects of shifts in bargaining power. This section provides some examples of how these issues have been addressed in recent transactions.[43]

22.4.1. *The Incentive and Ability to Foreclose Competition*

One important and repeated theme in recent US IT-sector merger reviews is concern about the impact of transactions on the incentive and ability of the merging firms to engage in exclusionary conduct; and, concomitantly, the growing use of

---

[40] Eisenach and Gotts (2015b).
[41] Elzinga and Mills (2011).
[42] See, e.g., Evans and Schmalensee (2013) (warning against "basing judgments about market power on analysis of only a single side of a multi-sided platform").
[43] For more complete discussions of recent cases, see Gotts (2015); Eisenach and Gotts (2015a; 2015b).

conduct remedies to address such concerns while allowing transactions to go forward.

Concerns about exclusionary conduct have been especially significant in recent mergers involving communications carriers, including Comcast/NBCU (approved 2011), Comcast-TWC-Charter (disapproved 2015), AT&T-DirecTV (approved 2015), and Charter-TWC-Brighthouse (approved 2016).[44] Specifically:

- Comcast/NBCU was primarily a vertical transaction involving the acquisition by the nation's largest cable television operator of one of its largest providers of video content. The agencies expressed concern that the combined firm "would have both greater incentive and greater ability to raise prices for its popular video programming to disadvantage Comcast's rival multichannel distributor rivals [as well as] the incentive and ability to hinder the development of rival online video offerings and inhibit potential competition from emerging online video distributors that could challenge Comcast's cable television business."[45] To address these concerns, the agencies imposed significant conduct remedies, including requiring the merged firm to make content available to competing cable companies and online video providers (OVDs) on non-discriminatory terms, and preventing Comcast from discriminating against OVD's traffic in its provision of broadband services.[46]
- Comcast-TWC-Charter was primarily a horizontal transaction involving the acquisition by the largest US cable operator (Comcast) of the fourth largest (Time Warner Cable). Despite the fact that the companies did not compete in the downstream market for video distribution (because their service territories did not overlap), both reviewing agencies expressed opposition, and the parties ultimately withdrew their application. DOJ officials later explained that the merger "would have placed Comcast in a stronger position to frustrate the rise of online video competitors, who provide new competitive alternatives to traditional cable service that could become substitutes for Comcast's video business."[47]
- AT&T-DirecTV involved the purchase by AT&T of DirecTV, a satellite-based, nationwide distributor of video programming. The merger was both horizontal (as the firms competed directly in the portions of the

---

[44] All four matters were reviewed by both the DOJ and the FCC. The author was engaged by Charter Communications in the Comcast-TWC-Charter and Charter-TWC transactions.
[45] FCC Comcast/NBCU Order. *Applications of Comcast Corporation, General Electric Company and NBC Universal, Inc.* Before the Federal Communications Commission. MB. Docket No. 10–56. 2011, at ¶ 3.
[46] Sher and Kemp (2014); Salop and Culley (2016).
[47] See Hesse (2016).

United States served by AT&T's wireline video distribution business) and vertical (in that it combined AT&T's broadband business with DirecTV's video business). Again, the agencies expressed concerns that the transaction would enhance the firms' incentives to "hamper competition from online video content or online video distribution services," and imposed restrictions preventing "discriminatory usage-based allowances" and requiring mandatory review of its interconnection agreements with OVDs.[48]

- Charter-TWC involved the acquisition by Charter Communications – the nation's sixth largest cable operator – of the fourth largest, Time Warner.[49] Unlike Comcast-TWC, where Comcast's ownership of NBCU played a significant role in the analysis, neither Charter nor TWC owned significant content assets. However, the agencies nevertheless expressed concern that the combined firm would have an increased incentive and ability to discriminate against actual and potential competitors. As the FCC explained:

    > Because of New Charter's increased MVPD and broadband footprint, and its increased number of homes passed, it will capture a greater share of the benefits that would accrue to MVPDs should New Charter take actions that reduce the competitive viability of OVDs. For the reasons stated above, we find that New Charter is likely to have a greater incentive to take such actions following the transaction.[50]

- Accordingly, the agencies imposed on New Charter stringent conditions preventing it from engaging in a variety of practices they believed to be potentially anticompetitive, including imposing data caps on broadband services, charging interconnection fees for OVDs and other online content providers, and entering into contracts with content providers limiting OVD's access to content.[51]

One common characteristic of all of these transactions is the relatively low market shares involved. For example, even after its proposed acquisition of TWC, Comcast (the largest cable and broadband operator) would have served fewer than 30 percent of video subscribers, and while its share of broadband subscribers arguably would have been higher (depending on how broadband was defined), even there it faced

---

[48] Memorandum Opinion and Order. *In Applications of AT&T Inc. and DIRECTV*. Federal Communications Commission. July 24, 2015. Available at www.fcc.gov/document/fcc-releases-order-approving-att-directv-transaction.
[49] Charter also acquired a much smaller regional firm, Brighthouse Communications.
[50] Charter-TWC Order. *Charter Communications, Inc., Time Warner Cable Inc., and Advance/ Newhouse Partnership*. Before the Federal Communications Commission. Docket N. 15-149. 2016. ¶ 47.
[51] *Id* at ¶¶ 9–11.

competition from telephone companies and from a variety of new entrants (such as Google fiber) and technologies (i.e., fiber to the home and wireless).

High market shares, however, do not appear to be a prerequisite for agency concerns about the incentive and ability to engage in foreclosure in information technology markets. Indeed, in its 2011 review of Google's (ultimately successful) acquisition of ITA Software, the DOJ conceded that both the upstream and downstream markets were unconcentrated.[52] ITA owned the leading airfare pricing and shopping system (QPX), which collected information on fares and availability and licensed it to online travel agents like Kayak and Orbitz. Google argued that the acquisition would allow it to offer an improved flight search and price comparison service in competition with such firms.[53] The DOJ concluded, however, that Google – despite the fact that it did not yet offer such services – could have a post-merger incentive to foreclose competition by denying QPX's offerings to the incumbent providers. Accordingly, it imposed an array of conditions on the transaction, including an obligation to continue licensing QPX (including any upgrades) to competitors on fair, reasonable, and non-discriminatory terms, forgoing contracts with airlines limiting competitors' access to information, and continuing R&D of new functions competitors arguably needed to be competitive with Google's new service.

As one analysis of the case concluded, the Google/ITA transaction broke new ground in at least two respects:

> First, the agency may challenge a transaction when one of the parties is a likely entrant into a market in which the other merger party buys or sells. Second, the agency may conclude that a vertical or potential vertical transaction presents a risk of input foreclosure if the merging party's upstream product is superior to that of its rivals.[54]

Further, the same analysis concludes, "the Google/ITA case seems to indicate that a concentrated market is no longer a prerequisite to input foreclosure concerns in vertical transactions reviewed by the DOJ."[55] As the subsequent transactions discussed above demonstrate, Google/ITA did indeed signal an increased willingness by the DOJ to intervene in both horizontal and vertical transactions based on concerns about the merged firms' incentive and ability to engage in exclusionary conduct.

22.4.2. Assessing Competitive Effects in Horizontal Mergers

Another recurrent theme in recent transactions is the diverse set of analytical approaches and frameworks being relied upon by the agencies to assess the

---

[52] Rosch and Tucker (2011) at 5.
[53] Eisenach and Gotts (2015a).
[54] Rosch and Tucker (2011) at 6.
[55] Id. at p. 7.

competitive effects of horizontal mergers. While the 2011 guidelines appear to de-emphasize structural analysis in favor of direct assessments of competitive effects (i.e., UPP-based models), structural factors continue to be cited in the agencies' analyses of IT-sector transactions. And other factors, such as the importance of "disruptive" competitors (or "mavericks") and the need to consider the effects of transactions on innovation and future competition, are also playing prominent roles.

The complexities of horizontal merger analysis in dynamic industries were highlighted by the agencies' review of the 2011 AT&T-T-Mobile transaction, which would have combined two of the four leading wireless carriers in the United States.[56] The Department of Justice eventually sued to block the transaction, alleging a loss of competition in retail wireless services and arguing that T-Mobile acted as a maverick competitor. For its part, the FCC issued a lengthy staff report indicating its objections to the transaction, which included concerns regarding unilateral and coordinated effects in the retail market for mobile voice and data services as well as lessened competition in related markets for roaming, wholesale services, resale services, backhaul, and handsets/devices.[57] Ultimately the transaction was abandoned on December 19, 2011.

While the AT&T-T-Mobile transaction is notable for many reasons, one interesting characteristic for the current purpose is the multifaceted nature of the agencies' competitive analyses. The FCC Staff Report, for example, begins by assessing the effect of the transaction on market concentration (measured by subscribership) at both the national and local level, concluding that the transaction would result in the "top two firms by share (AT&T and Verizon Wireless) together serving three-fourths of all US mobile wireless subscribers."[58] It supplements this subscriber-based structural analysis with an assessment based on the firms' spectrum holdings. Next, it conducts a full-fledged unilateral effects analysis based on estimated diversion ratios. Finally, it puts forward a qualitative competitive effects analysis, which argues that the transaction would eliminate T-Mobile as a "disruptive firm" and that, partly as a result, "the retail mobile wireless services market would be more vulnerable to coordination post transaction."[59]

It is also notable that both the FCC's Staff Report and the DOJ's complaint focus on the need to protect innovation. The Staff Report expends significant effort rebutting AT&T and T-Mobile's arguments that T-Mobile was struggling in the marketplace, finding instead that "T-Mobile has played an important role in the development of a more competitive mobile services marketplace by engaging in both

---

[56] *AT&T to Acquire T-Mobile USA from Deutsche Telekom.* March 20, 2011. AT&T Press Release, Available at www.att.com/gen/press-room?pid=19358&cdvn=news&newsarticleid=31703.

[57] *AT&T Inc. Staff Analysis and Findings.* November 29, 2011. Federal Communications Commission. Docket No. 11-65. See also DOJ Complaint. 2011. *United States v. AT&T, T-Mobile, Deutsche Telekom.* Second Amended Complaint. No. 11-01560.

[58] *Id.* at ¶ 42.

[59] *Id.* at ¶ 76.

pricing and technical innovation."[60] The DOJ's complaint also emphasizes T-Mobile's role as a disruptor: "By eliminating T-Mobile as an independent competitor, the proposed transaction likely will reduce innovation and product variety."[61] As one senior DOJ official explained: "[W]e are concerned by evidence that shows that a firm being acquired has been a particularly innovative or disruptive competitor. Such concerns have contributed to the division's decisions to challenge mergers involving technology companies, such as ... AT&T's proposed acquisition of T-Mobile."[62]

Other recent horizontal transactions also highlight the complexities of assessing horizontal effects in rapidly changing IT markets in which the agencies are focused on protecting innovation. In *United States v. Bazaarvoice*,[63] for example, the DOJ sued to unwind a consummated merger between two providers of "rating and review platforms" (Bazaarvoice, Inc. and PowerReviews, Inc.) used by online retailers to collect and display product rating information. The DOJ's case focused heavily on internal documents in which Bazaarvoice executives commented that the merger would eliminate its main competitor, which had "suppressed prices," and that the two firms had "pushed each other to innovate."[64] Bazaarvoice, in response, argued that market conditions had changed since the merger was consummated, with new firms entering the market and others repositioning their offerings. The Court rejected Bazaarvoice's dynamism arguments, finding that "while Bazaarvoice indisputably operates in a dynamic and evolving field, it did not present evidence that the evolving nature of the market itself precludes the merger's likely anticompetitive effects."[65]

Another transaction in which innovation played a central role was a combination of two media rating services, Nielsen (which provides ratings for television programming) and Arbitron (which rates radio programming). While the two firms did not offer competing services at the time of the acquisition, the FTC concluded that they were each in the process of developing "national syndicated cross-platform audience measurement services, which allow audiences to be measured accurately across multiple platforms, such as TV and online" which, once complete, would compete directly. The Commission found that "the elimination of future competition between Nielsen and Arbitron would likely cause advertisers, ad agencies, and programmers to pay more for national syndicated cross-platform audience measurement services."[66] To remedy this concern, the Commission required Nielsen to

---

[60] *Id.* at ¶¶ 22ff.
[61] DOJ complaint at ¶ 38.
[62] See Hesse (2014).
[63] *United States v. Bazaarvoice.* No. C-13-0133JSC. Jan. 1, 2014. Opinion.
[64] See e.g., *United States v. Bazaarvoice.* No.13-cv-00133. Jan. 8, 2014. Plaintiff United States of America's Post-Trial Proposed Findings of Fact (Filed Oct. 31, 2013). Available at www.justice.gov/atr/cases/f301400/301437.pdf. ¶¶ 198–216.
[65] *United States v. Bazaarvoice, Inc.*, No. C-13-0133JSC. Jan. 1, 2014. Opinion.
[66] *FTC Puts Conditions on Nielsen's Proposed $1.26 billion Acquisition of Arbitron.* September 20, 2013. FTC Press Release. Available at www.ftc.gov/news-events/press-releases/2013/09/ftc-puts-conditions-nielsens-proposed-126-billion-acquisition.

license key elements of the technology to an FTC-approved third party for up to eight years, thereby effectively creating a new competitor. The FTC's Chairwoman, Edith Ramirez, explained:

> [T]his acquisition would eliminate future competition in an emerging market for national syndicated, cross-platform audience measurement services. To remedy the likely loss of future competition, we required Nielsen to divest or license certain technological assets and data, including relevant intellectual property, to a Commission-approved buyer, so that the buyer could offer a competing service.[67]

In an indication of the unsettled nature of the law in this area, Commissioner Josh Wright dissented from the decision, arguing that a higher standard of evidence should be required before sanctioning a transaction based on competitive effects in a market that does not yet exist.[68]

The issue of future competition arose in a somewhat different way in the Commission's 2010 review of Google's acquisition of AdMob.[69] AdMob was at the time the leading provider of advertising services for mobile phones, including those running Google's Android operating system and Apple's iPhone platform, while Google was also developing a mobile platform that worked on both Android and iPhone. The Commission initially considered challenging the merger on the basis of this overlap, but decided against it at the last minute when Apple announced that it had acquired a nascent competitor and was entering the market for online mobile advertising on the iPhone with a new service called iAd. In its closing statement, the Commission explains that "[a]s a result of Apple's entry (into the market), AdMob's success to date on the iPhone platform is unlikely to be an accurate predictor of AdMob's competitive significance going forward, whether AdMob is owned by Google or not."[70]

While competition in the market for mobile online advertising has proven robust, it did not develop in the way the Commission envisioned: Today, the primary competitor to Google is not Apple but Facebook; the iAd network never captured more than 3 percent of the market,[71] and was shuttered in June 2016.[72] Thus, the AdMob case arguably demonstrates the hazards inherent in staking enforcement decisions on predictions about the future of competition in rapidly changing markets. As one FTC official later commented, "[h]ad Apple announced its own

---

[67] See Ramirez (2015).
[68] Eisenach and Gotts (2015a).
[69] *FTC Closes its Investigation of Google AdMob Deal.* May 21, 2010. FTC Press Release. Available at www.ftc.gov/news-events/press-releases/2010/05/ftc-closes-its-investigation-google-admob-deal.
[70] Id.
[71] *U.S. Digital Advertising Landscape And Key Players (Part 2).* July 9, 2015 Forbes. Available at www.forbes.com/sites/greatspeculations/2015/07/09/u-s-digital-advertising-landscape-and-key-players-part-2/#10b8f3ca73ed.
[72] *iAd App Network will be Discontinued.* January 15, 2016. Apple Press Release. Available at https://developer.apple.com/news/?id=01152016a.

mobile advertising plans just months later, the course of the merger investigation might have been very different."[73]

22.4.3. *The Role of Bargaining Power*

A third theme running through recent IT-sector cases involves the agencies' efforts to assess the impact of changes in bargaining power on consumer welfare. As discussed above, in platform markets involving complementary inputs, firms collaborate to create value by exploiting synergies between their products and services, but compete over the division of the resulting profits.[74] In markets with economies of scale and scope (on either the demand or supply side of the market), larger firms are generally supposed to have the ability to extract superior terms, i.e., to appropriate a larger share of rents. Hence, even a purely horizontal merger can shift rents among platform participants.[75] The question is when, if ever, a shifting of profits in this way creates a cognizable effect under the antitrust laws; and if so how to take that effect into consideration in assessing the effects of mergers.

Bargaining power played an important role in all four of the major communications sector transactions mentioned above: Comcast-NBCU, Comcast-TWC and Charter-TWC, and AT&T-DirecTV. In the two transactions involving Comcast, the alleged shift in bargaining power was regarded as harmful; in Charter-TWC and AT&T-DirecTV, the agencies identified benefits as well as costs.

In assessing the Comcast-NBCU transaction the agencies conducted an extensive empirical analysis of bargaining power issues, employing a Nash bargaining framework.[76] The analysis examined both horizontal and vertical theories associated with the potential that the combined firm would gain and exploit enhanced bargaining power as a seller of video content.

With respect to horizontal effects, the analysis that the combined firm's bargaining power would be enhanced by combining the programming assets of NBCU with those of Comcast, which included both broadcasting stations and regional sports networks (RSNs). As explained in the FCC's Order:

> If failing to reach an agreement with the seller will result in a worse outcome for the buyer – if its alternatives are less attractive than they were before the transaction – then the buyer's bargaining position is weakened and it can expect to pay more for the products. In this case, for example, prior to the transaction, if an MVPD did not reach an agreement with Comcast to carry the RSN, the NBC network programming would still be available; and if the MVPD did not reach an agreement to carry

---

[73] See Shelansky (2013).
[74] Bresnahan (1999); Eisenach (2012).
[75] Nevo (2014); Rose (2015).
[76] FCC Comcast-NBCU Order. *United States v. Comcast, General Electric, and NBC Universal.* 2013. United States District Court for the District of Columbia. Case 1:11-cv-00106. Appendix B.

NBC, it could still carry the RSN. Post-transaction, if the MVPD does not reach an agreement with Comcast-NBCU, it will not be able to carry either.[77]

In addition to this horizontal effect, the Commission noted that the vertical aspect of the merger would also contribute to the combined firm's bargaining power, because "failure to reach an agreement [with rival MVPDs] means that some of the rival's subscribers will shift to Comcast." As a result, the combined firm would be able to "extract higher prices from rival MVPDs than pre-transaction NBCU ... These higher programming prices to rivals would ultimately result in higher consumer prices for MVPD service."[78] The Commission concluded that its concerns about these alleged harms, which are conceptually separate from and in addition to the potentially exclusionary effect of raising rivals' costs, were not sufficient to justify blocking the merger "because the program-access restrictions we impose will prevent Comcast-NBCU from using any increased bargaining power it might obtain to raise rates above market levels."[79]

A different form of bargaining power was of concern in the Commission's reviews of the Comcast-TWC and Charter-TWC transactions. In reviewing those transactions, the Commission (and the DOJ) collected and analyzed data on the fees charged by broadband firms for interconnection with over-the-top providers. As the DOJ's chief economist explained (referring to the Comcast-TWC transaction):

> What we found is that there was very strong evidence of substantial bargaining leverage in this industry by larger cable firms. Not only did larger firms get better deals in both of these distribution channels, but the implication with putting together these two is that they would have bargaining leverage and be able to raise prices significantly as a result of that. So that was something we were quite concerned about.[80]

The Commission reached a similar conclusion in its review of the Charter-TWC transaction, finding that "[w]hile New Charter's subscriber base would be 12 percent smaller than Comcast's, it will have a sufficiently large subscriber base and control over interconnection to have the ability to impose higher interconnection prices." Moreover, it concluded that "Because New Charter would not face substantial competition for BIAS subscribers, it would not be incented to pass through to its subscribers a significant proportion of these additional fees."[81] As noted above, the Commission addressed these concerns by conditioning the transaction on Charter's agreement not to impose such fees at all.

---

[77] *Id.* at ¶ 136.
[78] *Id.* at Appendix B, ¶ 37.
[79] *Id.* at ¶ 138.
[80] Rose (2015).
[81] Charter-TWC Order. *Charter Communications, Time Warner Cable and Advance/Newhouse Partnership*. 2016. Before the Federal Communications Commission. Docket No. 15-149. ¶ 121.

Yet another aspect of the bargaining power issue was evident in both the Charter-TWC and AT&T-DirecTV transactions: the potential for consumer benefits resulting from the ability of larger firms to bargain for lower input costs. In both cases, the FCC considered and at least partly accepted the parties' arguments that their larger size would allow them to obtain lower prices on video programming, and that (unlike the higher interconnection fees it feared would be extracted in the Comcast-TWC and Charter-TWC cases), at least a portion of these savings would be passed through to consumers. In each case, the finding was based on the Commission's assessment of merger simulation analyses submitted by the parties, which estimated the pass-through rate from reductions in unit costs.[82]

In each case, the Commission also considered the possibility that the combined firm's increased bargaining power would harm the market for programming, and concluded it would not:

> Commenters have not provided adequate empirical evidence to show that the reduction in programming rates that the combined entity might achieve would curtail investment in content production. Thus, we find that the record here does not allow us to conclude that a decrease in programming rates would have the net effect of lowering the quality or quantity of programming.[83]

Accordingly, the Commission concluded, "[w]e do not find that any increased bargaining power of the combined entity is likely to harm consumer welfare."[84]

One take-away from these cases is that the agencies appear to be prepared to take into account the impact of bargaining power shifts when they can be shown to affect directly downstream consumer prices (for better or worse), but not are not prepared to consider – at least not without compelling empirical evidence – the potential for dynamic effects associated with the transfer of rents among the platform participants (e.g., the potential for reduced content quality resulting from lower payments to video programmers).

22.5. CONCLUSIONS

Competition analysis of IT markets presents particular challenges for academics and antitrust enforcers alike. Their dynamic nature, the presence of strong economies of scale and scope, and the significance of complementarities and "platform competition" complicate efforts to understand the competitive effects of mergers, and call into question some of the basic precepts of traditional antitrust doctrine and jurisprudence. As the cases discussed here demonstrate, US authorities are working

---

[82] AT&T-DirecTV Order. *AT&T Inc and DIRECTV*. 2015. Before the Federal Communications Commission. Docket No. 14-90. Appendix C, Section V; compare Charter-TWC Order at ¶¶ 321–46.
[83] AT&T-DirecTV Order at ¶ 235; compare Charter-TWC Order at ¶ 277.
[84] AT&T DirecTV Order at ¶ 220; compare Charter-TWC Order at ¶ 278.

hard to develop the tools and analytical frameworks needed to accurately assess the effects of both horizontal and vertical transactions in the IT space, but the task is far from complete.

REFERENCES

Baker, Jonathan B. and Carl Shapiro. 2008. Detecting and Reversing the Decline in Horizontal Merger Enforcement. *Antitrust* 22, 29–36.

Baumol, William. 2002. *The Free Market Innovation Machine: Analyzing the Growth Miracle of Capitalism*. Princeton University Press.

Brannon, Leah and Kathleen Bradish. 2010. The Revised Horizontal Merger Guidelines: Can the Courts Be Persuaded? *The Antitrust Source*, 10(1) 7–10.

Bresnahan, Timothy. 1999. New Modes of Competition: Implications for the Future Structure of the Computer Industry. In J. Eisenach and T. Lenard (eds.), *Competition, Innovation and the Microsoft Monopoly: Antitrust in the Digital Marketplace*. Norwell, MA: Kluwer Academic Press, pp. 155–208.

Creighton, Susan. 2010. 2010 Horizontal Merger Guidelines: The View from the Technology Industry. *The Antitrust Source*, 10(1) 15–18.

Eisenach, Jeffrey. 2012. *Broadband Competition in the Internet Ecosystem*. American Enterprise Institute.

Eisenach, Jeffrey and Ilene Gotts. 2015a. In Search of a Competition Doctrine for Information Technology Markets: Recent Antitrust Developments in the Online Sector. In D. Cugia di Sant'Orsola, F. Noormohamed, and R. Guimarães (eds.), *Communications and Competition Law*. Alphen aan den Rijn: Wolters Kluwer Legal 69–90.

2015b. Looking Ahead: The FTC's Role in Information Technology Markets. *George Washington Law Review*, 83, 1876–901.

Elzinga, Kenneth and David Mills. 2011. The Lerner Index of Monopoly Power: Origins and Uses. *American Economic Review: Papers & Proceedings*, 101(3), 558–64.

Evans, David and Richard Schmalensee. 2013. The Antitrust Analysis of Multi-Sided Platform Businesses. In R. Blair and D. Sokol (eds.), *Oxford Handbook on International Antitrust Economics*. Oxford University Press, p. 423.

Farrell, Joseph and Carl Shapiro. 2010. *Antitrust Evaluation of Horizontal Mergers: An Economic Alternative to Market Definition*. Berkeley Electronic Press.

Ginsburg, Douglas and Joshua Wright. 2012. Dynamic Analysis and the Limits of Antitrust Institutions. *Antitrust Law Journal*, 78(1), 1–22.

Gotts, Ilene. 2015. Summary of Recent U.S. Enforcement Decisions in Communication/Entertainment Industry Transactions. In D. Cugia di Sant'Orsola, F. Noormohamed, and R. Guimarães (eds.), *Communications and Competition Law*. Alphen aan den Rijn: Wolters Kluwer Legal, pp. 13–26.

Gotts, Ilene and Richard Rapp. 2004. Antitrust Treatment of Mergers Involving Future Goods. *Antitrust*, Fall, 100–5.

Harrington, Jr., Joseph. 2013. Evaluating Mergers for Coordinated Effects and the Role of "Parallel Accommodating Conduct." *Antitrust Law Journal*, 78, 651–68.

Hesse, Renata. 2014. At the Intersection of Antitrust & High-Tech: Opportunities for Constructive Engagement. Remarks as Prepared for the Conference on Competition and IP Policy in High-Technology Industries.

2016. Remarks at Global Competition Review. Available at www.justice.gov/opa/speech/principal-deputy-assistant-attorney-general-renata-b-hesse-delivers-remarks-global.

Hovenkamp, Herbert. 2012. Antitrust and the Movement of Technology. *George Mason Law Review*, 19(5), 1119–45.
Katz, Michael and Carl Shapiro. 1994. Systems Competition and Network Effects. *The Journal of Economic Perspectives*, 8(2), 93–115.
Kwoka, John and Diana Moss. 2012. Behavioral Merger Remedies: Evaluation and Implications for Antitrust Enforcement. *Antitrust Bulletin*, 57(4), 979–1011.
Merger Guidelines (Horizontal). 2010. US Department of Justice and the Federal Trade Commission.
Nevo, Ari. 2014. *Mergers that Increase Bargaining Leverage*, Remarks as Prepared for the Stanford Institute for Economic Policy Research and Cornerstone Research Conference on Antitrust in Highly Innovative Industries, Stanford Institute for Economic Policy Research.
Ramirez, Edith. 2015. Keynote Remarks. The Hal White Antitrust Conference.
Rosch, J. Thomas and Darren Tucker. 2011. Emerging Theories of Competitive Harm in Merger Enforcement. *The Antitrust Source*, 11(1) 1–7.
Rose, Nancy. 2015. Deputy AAG for Economic Analysis, Speaking on Bargaining Leverage and Competitive Effects. Audio Tape. Available at www.americanbar.org/tools/digitalassetabstract.html/content/dam/aba/multimedia/antitrust_law/20150625_at150625_mo.mp3 (membership required).
Salop, Steven and Daniel Culley. 2016. Revising the US Vertical Merger Guidelines: Policy Issues and an Interim Guide for Practitioners. *The Journal of Antitrust Enforcement*, 4(1), 1–41.
Schumpeter, Joseph. 1942. *Capitalism, Socialism and Democracy*. New York: Harper and Brothers.
Shapiro, Carl. 2010. The 2010 Horizontal Merger Guidelines: From Hedgehog to Fox in Forty Years. *Antitrust Law Journal*, 77, 712–37.
Shelanski, Howard. 2013. Information, Innovation, and Competition Policy for the Internet. *University of Pennsylvania Law Review*, 161, 1673–705.
Sher, Scott and Kellie Kemp. 2014. A Comparative Analysis of the Use of Merger Remedies in Technology Industries. *CPI Antitrust Chronicle*, December.
Sokol, Daniel. 2014. The Broader Implications of Merger Remedies in High Technology Markets. *CPI Antitrust Chronicle*, 12(1) 1–7.
Wright, Julian. 2004. One-Sided Logic in Two-Sided Markets. *Review of Network Economics*, 3(1), 44–64.

23

Competition Assessment of IPRs in China's Merger Control

Liyang Hou

23.1. INTRODUCTION

It is commonly accepted that antitrust law[1] and intellectual property rights (IPRs) law share a common objective, i.e. protecting innovation and competition.[2] Although the two are complementary in nature, they nonetheless do come into conflict. This disharmony stems from the fact that the two legal mechanisms protect the same end, albeit by different means. Antitrust law sets its focus on the loss of total social welfare, and concerns itself little with individual gains or sufferings. By comparison, IPRs law grants individuals exclusive rights to use a property-equivalent right, such as copyright, patent, trade secret or trademark, and is less concerned with the social interest as a whole. It hence transpires that certain legitimate usages of IPRs might not be good for the whole of society, and should hence be prohibited. However, the tricky point is two-fold. First, since IPRs remain exclusive their harm is temporary. The question is then whether and how such a temporary anticompetitive effect should be regulated by competition law. Secondly, antitrust intervention always limits the actual or potential economic value of IPRs. This contravenes a long-standing assumption in modern society that only the owner knows best how to maximize the value of property. An additional question is subsequently raised as to whether competition law offers any added value to regulate IPRs. Consequently, many argue that antitrust law that traditionally more or less targeted only tangible property, though justified in being applied to IPRs, must strike a different balance.[3]

It goes without saying that the balance between antitrust law and IPRs law is achieved in divergent ways across jurisdictions. This chapter aims to shed some light on the relevant practice within the domain of merger control in the People's

---

[1] Depending on the terminology used across jurisdictions, antitrust law may variously be called competition law or anti-monopoly law. These are considered synonymous in this chapter.
[2] Coco (2008).
[3] Hou (2012).

Republic of China (China). Section 23.2 gives an overview of the competition rules in China and their enforcement in past years. The following two sections focus on the practice of merger review, with Section 23.3 touching upon cases of standard essential patents (SEPs) while Section 23.4 reviews cases of non-SEPs. Section 23.5 explores related practice within the domain of abusing dominance that may serve as guidance for Mofcom in the future. The final section concludes.

23.2. AN OVERVIEW

The People's Republic of China adopted its Anti-Monopoly Law (AML) on August 30, 2007 and it entered into effect on August 1, 2008.[4] It is comprised of eight chapters and fifty-six articles. The prohibitions of the AML are no different from those in other parts of the world. It also features anticompetitive agreements (Chapter 2), abusing dominance (Chapter 3) and merger control (Chapter 4). One minor difference lies in the fact that while competition law in other jurisdictions mainly focuses on anticompetitive conduct of private enterprises, the AML includes in its scope of application the conduct of governments that may limit or restrict competition, which in China is often referred to as administrative monopoly.

Moreover, as a matter of fact, the AML itself does not directly appoint any competition authorities. This may be due to the fact that China possibly foresaw the difficulties in doing so, and thus left it to be dealt with by the State Council at a later stage to facilitate speedy adoption. Later developments seem to verify this impression. The process of institutional arrangements was indeed not easy. One of the main reasons was the fight among three ministries,[5] i.e. the National Development and Reform Commission (NDRC), the State Administration of Industry and Commerce (SAIC), and the Ministry of Commerce (Mofcom). Although the AML was adopted in 2007, prototype competition rules did exist and were dispersed in different legislation. The three authorities were roughly the former "competition authorities" active in different areas. During the pre-AML era, the NDRC, as a central planning authority, took charge of price-related regulation; the SAIC dealt with non-price-related anticompetitive market conduct or activities, such as tying; and Mofcom was responsible for conducting merger control review involving international companies and cross-border transactions.[6] The fight for power began even when the AML was being drafted, and continued after adoption. Only on March 21, 2008, almost seven months after the adoption of the AML, did the State Council release the Notice on Institutional Establishment[7] based on the decision of

---

[4] Anti-Monopoly Law of the People's Republic of China, adopted at the 29th meeting of the Standing Committee of the 10th National People's Congress of the People's Republic of China, August 30, 2007.
[5] Wang (2007).
[6] Zhang (2005).
[7] State Council of China, Notice on Institutional Establishment, Guofa [2008] Nr. 11, March 21, 2008. Available at www.gov.cn/zwgk/2008-04/24/content_953471.htm.

the National People's Congress of China. This notice in principle confirmed the traditional power division between the three authorities.[8] Consequently, three competition authorities were designated, with the NDRC dealing with price-related monopolistic behavior,[9] the SAIC handling cases of non-price-related anticompetitive agreements, abuse of dominant positions and abuse of administrative power to constrain competition,[10] and Mofcom reviewing concentrations.[11] The expectation of those ministries as well as of scholars of establishing a unified competition authority failed to materialize.[12]

The AML, similar to the US Sherman Act and competition rules in the EU Treaties, contains only crude provisions and cannot be immediately enforced. Consequently, the first problem confronted by those authorities was to mitigate legal uncertainties by publishing enforcement guidelines. So far about nineteen guidelines have been published with regard to almost every aspect of the AML. Among those, merger control has received particular attention. This domain has been supported by a total of nine guidelines, including some highly important ones such as Pre-notification Procedures for Concentrations,[13] Review Procedures for Concentrations,[14] Interim Measures of Divesting Assets or Business for Concentrations,[15] Interim Measures of Competition Assessment for Concentrations,[16] and Interim Measures of Applicable Standards for Speedy Procedure of Concentrations.[17]

The AML sets a pre-notification mechanism for merger review. Under the requirement of Article 30 of the AML, Mofcom has published all the decisions of conditional approval and disapproval on its website.[18] As of the end of 2015, Mofcom had reviewed close to 1,400 merger notifications, with twenty-seven conditional approvals and two blocks[19] (Figure 23.1). Of all the twenty-nine decisions, twenty-six concerned horizontal mergers and accounted for about 80 percent. In addition, Mofcom has committed itself to generating transparency by publishing quarterly the information on unconditional approval cases.[20] Nevertheless, the information

---

[8] Hao (2013).
[9] NDRC, "Rules on the Main Responsibility, Internal Institutions and Staff in NDRC." Available at www.sdpc.gov.cn/gzdt/t20080821_231802.htm.
[10] SAIC, "Rules on the Main Responsibility, Internal Institutions and Staff in SAIC." Available at www.gov.cn/gzdt/2008-07/26/content_1056531.htm.
[11] Mofcom, "Rules on the Main Responsibility, Internal Institutions and Staff in SAIC." Available at www.mofcom.gov.cn/aarticle/ae/ai/200808/20080805739577.html.
[12] Wang (2000).
[13] Mofcom [2009] Nr. 11, November 21, 2009.
[14] Mofcom [2009] Nr. 12, November 24, 2009.
[15] Mofcom [2010] Nr. 41, July 5, 2010.
[16] Mofcom [2011] Nr. 55, August 29, 2011.
[17] Mofcom [2014] Nr. 12, February 11, 2014.
[18] All the decisions are available at: http://fldj.mofcom.gov.cn/article/ztxx.
[19] The two blocked mergers were Coke-Cola/Huiyuan (Mofcom [2009] Nr. 22) and Maersk/MSC/CMA (Mofcom [2014] Nr. 46).
[20] See the unconditionally approved cases at http://fldj.mofcom.gov.cn/article/zcfb.

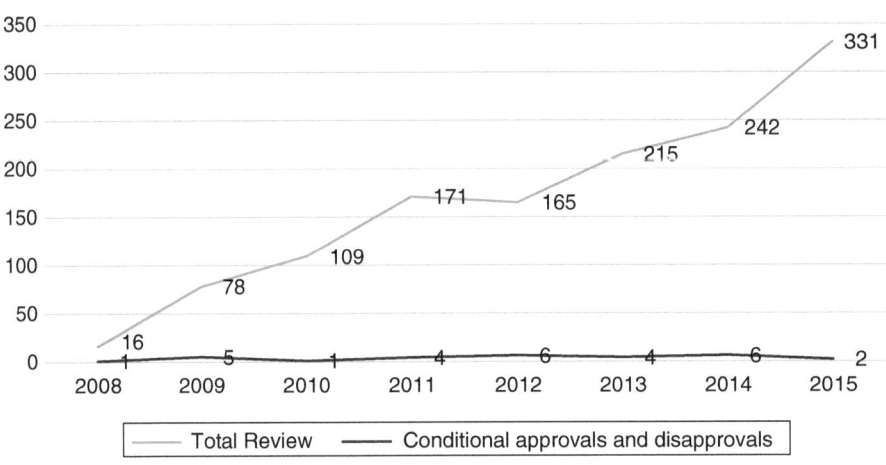

FIGURE 23.1: Annual number of cases reviewed by Mofcom before 2016

released by Mofcom only contains names of the merged entities and thus an in-depth analysis of unconditional approval cases is not possible.

More than half (literally sixteen) of the twenty-nine decisions adopted by Mofcom were IPR related. This offers us a more or less solid basis for probing into Mofcom's views on IP-related mergers, more specifically how IPRs may affect competition in a positive or, and more importantly, negative way. Those sixteen decisions can be categorized into three groups, depending on the different major themes involved, which are respectively (1) decisions concerning SEPs, (2) decisions imposing obligations on de facto standards, and (3) decisions related to other patents.

23.3. SEPS-RELATED MERGERS

23.3.1. *Cases*

In three cases Mofcom specifically analyzed how SEPs may affect competition. These were *Google/Motorola*, *Microsoft/Nokia*, and *Nokia/Alcatel Lucent*. All of them concerned SEPs with regard to mobile communications technologies.

The earliest case was the *Google/Motorola* case in 2012.[21] The relevant product markets comprised mobile operating systems, mobile smart phones, and SEPs for mobile communications. At that time, Google's Android OS had a 73.99 percent share of the market for mobile operating systems, far greater than those of its competitors, such as Symbian's 12.53 percent and iOS's 10.67 percent. While Motorola was no longer a strong competitor in the market for mobile smart phones, it did own many SEPs for mobile communications. Mofcom observed that the entry

---

[21] Case *Google/Motorola*, Mofcom [2012] Nr. 25, May 19, 2012.

barriers for mobile operating systems and mobile communications standards were rather prohibitive. Consequently, post-merger competition concerns were raised from two perspectives. First, Google might reserve the latest version of its Android OS only to Motorola, thus handicapping other producers of mobile smart phones. Secondly, after the merger, Google, when licensing its patents related to mobile communications standards, might attach unreasonable conditions, thereby harming competition and impairing consumer interests.

The Microsoft acquisition of Nokia was the second.[22] This case was similar to that of *Google/Motorola*. The products involved included mobile operating systems, mobile smart phones, and SEPs for mobile communications. The major problem was still SEPs of mobile communications on the one hand and Google's Android OS for which Microsoft owned SEPs on the other. Similar to the previous case, Mofcom was concerned that Microsoft might increase the royalty fees or might refuse to license its SEPs in relation to mobile communications and the Android OS.

The last case was a horizontal merger between Nokia and Alcatel Lucent.[23] This case was different from the previous two. Nokia and Alcatel Lucent were direct competitors in many aspects, such as the radio access network device, the core network system device, the network infrastructure device, and SEP licensing. Regarding the first three products, the merged entity was confronted with fierce competition. Their combined market share was squeezed to a low level, and would not limit or restrain competition. However, the major concern of Mofcom was still about SEP licensing, just as in the previous two cases.

23.3.2. Relevant Markets for SEPs

Mofcom has not been able to define SEPs in any legal document. This term was actually first defined by the SAIC as patents that are indispensable for practicing a standard in the Regulation of Abusing Intellectual Property Rights to Eliminate or Restrict Competition.[24] There are two problems with regard to this definition. First, it does not suggest what could be considered as standards. Secondly, this regulation as a matter of fact is not binding on Mofcom. Although Mofcom has not adopted any guideline on SEPs mergers, it did give a definition in two decisions.[25] Mofcom's definition of SEPs is exactly the same as that of the SAIC. But Mofcom gives a more specific definition of standards, namely binding documents that are adopted by standard setting organization (SSO) and are commonly repeatedly applicable.

---

[22] Case *Microsoft/Nokia*, Mofcom [2014] Nr. 24, April 8, 2014.
[23] Case *Nokia/Alcatel Lucent* [2015] Nr. 44, October 19, 2015.
[24] Regulation of Abusing Intellectual Property Rights to Eliminate or Restrict Competition, SAIC Order [2015] Nr. 74, April 7, 2015 (hereinafter referred to as "the Regulation"), Article 13.
[25] Case *Microsoft/Nokia*, Mofcom [2014] Nr. 24, April 8, 2014; and Case *Nokia/Alcatel Lucent*, Mofcom [2015] Nr. 44, October 19, 2015.

Unlike the definition of the SAIC, which might be sufficiently broad to cover de facto standards, Mofcom's definition includes only standards set by SSOs.

Until now, the only industrial sector reviewed by Mofcom with concerns over SEPs has been the standards for mobile communications, as presented by the above three cases. As far as market definition for SEPs is concerned, there is a discrepancy between Mofcom and other authorities. As will be explored in detail in Section 23.5, other authorities maintained that relevant markets should be defined as containing every SEP of mobile communications standards. In comparison, Mofcom seems not to endorse such an approach. While Mofcom in Case *Google/Motorola* did not touch upon the relevant markets and market power of Motorola with regard to SEPs, it did made some endeavors to further its analyses in the other two cases. It first indicated that all the SEPs for mobile communications standards might be defined into sub-markets from demand-side substitution based on different standards, such as 2G, 3G, and 4G. In the end, Mofcom defined a separate market containing all SEPs. The exact boundaries of the relevant markets were not defined since a narrower market definition would affect the final results.

23.3.3. *Assessing Market Power*

As will also be introduced in Section 23.5, when proceeding to the evaluation of market power subsequently, other authorities rapidly came to the conclusion that every SEP holder has dominance for the reason of no actual or potential competition. By comparison, Mofcom took a different analysis, in particular in Case *Nokia/Alcatel Lucent*. Within this horizontal merger case, Mofcom had to assess whether post merger the merged entity would acquire or strengthen the ability, incentive, or possibility to limit market competition.[26] Factors examined included concentration ratio, countervailing buying power, and indispensability of these SEPs to others. There is little doubt that SEPs for mobile communications standards can easily fulfill the second and the third requirements. With regard to the first, Mofcom looked especially at the market shares of Nokia and Alcatel Lucent. Their combined market shares would be increased from 25–35 percent to 35–45 percent for 2G and 3G, and jump from second place to first place for 4G. The conclusion was drawn afterwards that this merger indeed limits competition.

Such an evaluation first implies that Mofcom probably considers that relevant markets for SEPs may be comprised of three markets, i.e. SEPs for 2G, 3G, and 4G. Secondly, it also suggests that Mofcom takes market share as an important factor for evaluating market power. However, since post merger the merged entity

---

[26] Interim Measures of Competition Assessment for Concentrations, Mofcom Order [2011] Nr. 55, August 29, 2011, Article 4.

would become the biggest company on the relevant market, the question remains whether SEP holders whose market shares are not significant can be considered as possessing market power.

23.3.4. Remedies

The merits of the three cases might be different as *Google/Motorola* and *Microsoft/Nokia* were vertical mergers while *Nokia/Alcatel Lucent* was a horizontal merger. However, the competition concern was exactly the same. In all three cases, Mofcom raised the same issue that the merged entity would attach unreasonable conditions when licensing SEPs. This might partly answer the question raised in the previous section as to whether SEP holders, even if their market shares are not the highest on the relevant market, might enjoy market power. Moreover, the remedies imposed by Mofcom in all three cases illustrate a clear trend of moving from the general to the specific.

First of all, in all the cases the principal remedy was that the merged entity must continue to respect the FRAND (Fair, Reasonable and Non-discrimination) obligations to which they gave a commitment to the relevant SSOs. This element of the remedy may hardly be considered innovative because all those companies must respect their commitments to the SSOs. In Case *Microsoft/Nokia* and Case *Nokia/Alcatel Lucent*, the FRAND requirements went further. Mofcom required the merged entities to accept that a license under FRAND should not be conditioned on licensees accepting that the merged entities were not bound by the FRAND obligations promised to the relevant SSOs. Moreover, the merged entities must ensure that third parties who acquire those SEPs shall also respect the same set of obligations.

Secondly, Mofcom in Case *Microsoft/Nokia* and Case *Nokia/Alcatel Lucent* set some limits on the charging of royalty fees. When setting royalty fees, the merged entities should take into account the following factors (but not limited to): SEPs or SEP combinations included, licensing period, licensing products, business models related to retail and distribution, standards covered, acceptance of the market to the standardized function, structure of licensing agreements, value of grantback and other non-monetary compensation, payment methods, applicable industries, etc. Post merger the merged entities should not deviate from the commonly applicable royalty charge at the time, unless the aforementioned factors are different.

Last but not least, Mofcom put limitations on seeking injunctions in addition to the FRAND obligation in Case *Microsoft/Nokia* and Case *Nokia/Alcatel Lucent*, echoing the European Commission's practice in Case *Samsung*.[27] Most importantly, Mofcom specified the conditions in which the merged entity shall not seek injunctions. Injunction can only be sought where SEP holders have made offers

---

[27] European Commission (2014).

compatible with the FRAND obligations and the potential licensees nevertheless do not accept and respect the offers in good faith. Moreover, injunctions may not be appropriate where potential licensees, without undue delay, agree to file disputes with merged entities regarding failure to respect FRAND obligations with an independent authority, and promise to be bound by the decision of that authority, to sign licensing agreements based on the decision, and to pay royalty fees and damages based on the decision.

23.4. NON-SEP-RELATED MERGERS

The above three cases illustrate Mofcom's approach to SEP-related mergers. It has been found that the involvement of SEPs was the major reason for Mofcom to impose remedies in those cases, such as FRAND obligations and limits in seeking injunctions. In this section Mofcom's handling of IPR issues in non-SEP-related mergers will also be discussed. Depending on the importance of IPRs to the competition assessment, two sub-categories of cases can be observed, mergers involving de facto standards, and those with normal patents.

23.4.1. De Facto Standards

De facto standards are standards not set by SSOs but developed by companies by themselves. Unlike SEPs, de facto standards are not legally binding. However, due to the strong market power of the companies concerned, they may produce similar anticompetitive effects as SEPs. The Mofcom has delivered its opinion on de facto standards in the following three cases.

In the first case, *General Electric (GE)/Shenhua Group*, China's largest coal producer proposed to set up a 50-50 gasification joint venture (JV) that would specialize in converting coal into a synthesis gas.[28] This was a vertical merger. GE owned the technology for such a conversion while Shenhua Group, a state-owned enterprise (SOE), produced and supplied coal, electricity, and heating. Mofcom discovered that the upstream market for the technology licensing was highly concentrated with only three competitors, as well as high entry barriers. GE had the highest market share. On the downstream market, Shenhua Group was the biggest supplier of coal for synthesis gas in China. Mofcom was concerned that after the merger the JV would not be restrained by competitors, and would thus possibly limit competition on the upstream market. In the end, the merged entity promised not to leverage its market power to the licensing market. Moreover, it was specifically prohibited from conditioning the supply of coal on the use of GE's technology by other companies, or increasing the costs of using competing technologies.

---

[28] *General Electric/Shenhua Group*, Mofcom [2011] Nr. 74, November 10, 2010.

The second case involved the JV established by ARM Limited (ARM), Giesecke & Devrient GmbH (G&D), and Gemalto NV (Gemalto).[29] The purpose of the JV was to develop and market a TEE (Trusted Execution Environment) solution. The key participant in this JV was ARM whose main business was to license semiconductor intellectual property for application processors. Although Mofcom did not provide any data, it concluded that ARM had relatively strong market power there. The JV's activities, i.e. developing and marketing TEE, were downstream to ARM's business, and depended on ARM's TrustZone technology. Mofcom was concerned that ARM might discriminate against other developers of TEE when licensing its technologies, or might downgrade the performance of other developers of TEE by manipulating its own IPRs in certain ways. These two concerns were exactly what the commitments targeted. Within the commitments ARM promised to publish the necessary information to develop TEE based on its TrustZone in a non-discriminatory and timely manner, and not to design its IPRs in such a way as to downgrade the performance of third-party TEEs.

The last case was the acquisition by Merck KGaA (Merck) of AZ Electronics.[30] The two relevant markets defined included liquid display crystals and photoresists, which are widely used in the manufacture of flat panel displays. Merck had more than 70 percent market share in China in the market for liquid display crystals, and AZ Electronics had more than 50 percent market share in China in the market for photoresists. Due to the strong market power of the two entities, Mofcom became concerned with the strengthened market position after the merger based on the possibility of cross-subsidy between the two markets, thereby limiting the even limited competition. Previous experience suggests a structural remedy under similar situations because of the tremendous market shares of the participants in the merger. Possibly in order to trade for a non-divesture, Merck mainly agreed (i) not to bundle the two products, (ii) to license its patents for liquid display crystals based on nonexclusive and no-sublicensing terms, which should be reasonable and non-discriminatory.

It has been widely acknowledged that SEPs are in most cases substitutable with market power.[31] There is no doubt that Mofcom maintains a rather strict approach to mergers involving SEPs. In relation to non-SEPs, it is observed that Mofcom is more tolerant. The three cases discussed in this section exhibited a common feature of de facto standards. Since there were no SSOs involved in the cases the IPRs concerned there are not standards in accordance with Mofcom's definition of SEPs. However, the merged entities nevertheless held a very strong market position, though Mofcom did not make it clear in the decisions. Due to their strong market presence, de facto standards can make those merged entities control the markets in a

---

[29] Case *ARM/G&D/Gemalt*, Mofcom [2012] Nr. 87, December 6, 2012.
[30] Case *Merck/AZ Electronics*, Mofcom [2014] Nr. 30, April 30, 2014.
[31] Cotter (2014).

similar way as SEP owners. In all the cases, Mofcom expressed its concern that the merged entities may artificially increase the costs of other operators switching to a substitutable technology. As a result, the commitments were imposed in the same way as those in SEPs-related mergers, i.e. the FRAND obligations. Special attention might be paid to the *Merck/AZ Electronics* case where Mofcom imposed FRAND remedies even without a proper market definition of IPR licensing. However, the commitments in this case might have been proposed by Merck in order not to be divested by Mofcom.

23.4.2. *Other Patents*

Mofcom is more tolerant when a merger does not concern SEPs or de facto standards. In addition to the six cases discussed above, Mofcom touched upon IPRs in a group of nine cases. Due to the difference in remedies, the nine cases can be divided into two categories, cases with structural divesture and ones with functional divesture. It should be noted that in all these cases IPRs only played a minor role.

The first group is comprised of six cases, i.e. *Pfizer/Wyeth*,[32] *Panasonic/Sanyo*,[33] *United Technologies/Goodrich*,[34] *Baxter International/Gambro*,[35] *Thermo Fisher Scientific/Life Technologies*,[36] and *NXP/Freescale*.[37] In all six cases Mofcom imposed remedies of structural divesture. It is worth exploring the common features of these cases, and seeing why Mofcom required structural divesture. There were some common features shared by all six cases. First, they were all horizontal mergers. Second, the mergers took place in high tech industries. Third, entry barriers for these markets were all prohibitively high for potential competitors. Fourth and most importantly, post merger the market shares of the merged entities were all much higher than those of their competitors.

Table 23.1 reveals that only in Case *Pfizer/Wyeth* did Mofcom specify the market share of the second largest competitor, i.e. about 2.7 times smaller than the merged entity. Although Mofcom did not reveal the market shares of competitors, it can be inferred that the fact that the merged entities' market shares were all above 60 percent suggests that the merged entities must be at least twice as big as the others. Due to their relatively huge size, plus high entry barriers, the horizontal mergers would create a dominant operator that would be impossibly restrained by competitors. It thus makes sense that Mofcom finally imposed structural divesture on those merged entities. It is important to note that in order to maintain competitiveness of the

---

[32] Case *Pfizer/Wyeth*, Mofcom [2009] Nr. 77, September 29, 2009.
[33] Case *Panasonic/Sanyo*, Mofcom [2009] Nr. 82, October 30, 2009.
[34] Case *United Technologies/Goodrich*, Mofcom [2012] Nr. 35, June 15, 2012.
[35] Case *Baxter International/Gambro*, Mofcom [2013] Nr. 58, August 13, 2013.
[36] Case *Thermo Fisher Scientific/Life Technologies*, Mofcom [2014] Nr. 3, January 15, 2014.
[37] Case *NXP/Freescale*, Mofcom [2015] Nr. 64, November 25, 2015.

TABLE 23.1: *Market structure in cases of structural divesture*

Cases	Combined market share of merged entities	Market share of competitors
Pfizer/Wyeth	49.4%	18.35% (second largest)
Panasonic/Sanyo	> 60%	Small
United Technologies/Goodrich	84%	Much smaller
Baxter International/Gambro	> 60%	Small
Thermo Fisher Scientific/Life Technologies	> 60%	Small
NXP/Freescale	84%	Small

divested part Mofcom in all the cases stressed that the divesture should include not only physical assets, but also IPRs.

Functional divesture is considered by Mofcom as an innovation in merger review remedies. Unlike structural divesture, where the merged entities lose a part of their business permanently, functional divesture requires the merged entities to remain separate only within a transitional period when the merged entities still compete against each other. However, once the transitional period ends and no anticompetitive effects are observed, they can continue with the merger. Functional divesture was concerned in the remaining three cases.

The first two cases took place in the industry of hard disk drives. There were only five producers (i.e. Seagate, Western Digital, Hitachi, Toshiba, and Samsung) in the relevant market. The two mergers, i.e. *Seagate/Samsung*[38] and *Western Digital/ Hitachi*,[39] significantly reduced the number of competitors from five to three. Similar to the first group of six cases, this group of cases was also characterized by horizontal mergers, and the hard disk drives industry also exhibited high entry barriers for potential competition. The cases satisfied requirements of structural divesture observed in the previous subsection. However, they had a major difference from other cases. Here the two mergers would result in two big competitors of relatively similar size that might be able to restrain each other. Consequently, Mofcom chose functional divesture. These commitments in the two cases were reviewed in 2015 and Mofcom removed some commitments.[40]

The third case was the absorption of MStar by MediaTek. The affected relevant products were digital television chips. The merged entity would control a 61 percent market share worldwide and 80 percent market share in Mainland China. Mofcom expected that no existing competitors could counter such substantial market power.

---

[38] Case *Seagate/Samsung*, Mofcom [2011] Nr. 90, December 12, 2011.
[39] Case *Western Digital/Hitachi*, Mofcom [2012] Nr. 9, March 2, 2012.
[40] Case *Western Digital/Hitachi*, Mofcom [2015] Nr. 41, October 19, 2015; and Case *Seagate/Samsung*, Mofcom [2015] Nr. 43, October 19, 2015.

It seems that a structural divesture should have been possible. However, Mofcom also observed that the relevant market was under fierce pressure of innovation. Potential competitors might be able to challenge the merged entity's market power. Moreover, television producers in Mainland China preferred diversified supply of digital television chips. After the merger they might procure those chips from other producers. This could create an opportunity for other competitors. Consequently, Mofcom imposed functional divesture as a balanced solution. Most importantly, MediaTek and MStar were prohibited from cooperation in any manner, such as technological exchange.

23.5. IPRS-RELATED ABUSES

As far as IPRs are concerned, Mofcom only received opportunities to express its opinions with regard to SEPs and de facto standards. With regard to possible anti-competitive conduct in relation to other IPRs, there is only a requirement for the transfer of IPRs in structural divesture. Then questions would arise as to how Mofcom might react if other types of anticompetitive concerns were raised. Given the close link between merger review and abusing dominance, this section examines the practice of courts, the NDRC, and the SAIC in the related domains. This may serve as a guideline for the practice of Mofcom in the future.

23.5.1. *Case* Huawei v. InterDigital *and Case* Qualcomm

Antitrust enforcement in China has developed very quickly since the AML was adopted. Enforcement of the AML outside merger review nonetheless came much later only after 2013. This included a case of private enforcement and one of public enforcement, the former being the *Huawei v. InterDigital* cases[41] adjudicated by the Guangdong courts in 2013, and the latter the *Qualcomm* case decided by the NDRC in 2015. Both cases explore in detail the correct usage of SEPs from many perspectives.

23.5.1.1. Case *Huawei v. InterDigital*

This case was only one of a number of lawsuits filed by Huawei against Inter-Digital around the world.[42] InterDigital's main business was to license patents in relation to mobile communications standards, including both SEPs and non-SEPs. Huawei, while also owning patents of mobile communications standards,

---

[41] Shenzhen Intermediate People's Court, *Huawei v. InterDigital*, Feb. 4, 2013, [2011] Shen Zhong Fa Zhi Min Chu Zi Nr. 857 and Nr. 858; and Guangdong High People's Court, *Huawei v. InterDigital*, Oct. 16, 2013, [2013] Yue Gao Fa Min San Zhong Zi Nr. 305 and Guangdong High People's Court, *Huawei v. InterDigital*, Oct. 21, 2013, [2013] Yue Gao Fa Min San Zhong Zi Nr. 306.
[42] Jones (2014).

was mainly a communications equipment producer. Within its claims Huawei submitted that InterDigital (i) set unreasonably high royalty rates for its patents, (ii) applied unreasonably discriminatory royalty rates, (iii) required free cross-license of Huawei's patents, (iv) tied SEPs with non-SEPs, and (v) refused to license its patents by filing injunction applications with the relevant US authorities. All Huawei's claims other than the fifth one were supported both by the court of first instance and by the court of appeal in the Guangdong Province of China.

The courts first defined the relevant markets based on demand-side and supply-side substitution. The relevant markets were defined as all SEPs owned by InterDigital in relation to mobile communications standards. The courts held that there were no actual or potential substitutes for SEPs. Without those patents it would be impossible for Huawei to produce the related products. Consequently, every SEP was unique, and could not be substituted. Furthermore, from the perspective of supply-side substitution InterDigital was the only supplier in the relevant market. There was no possibility for Huawei to switch to other competitors. In addition, due to the geographic limitation of IPRs in accordance with national laws, the courts defined the geographic markets as China and the United States in accordance with Huawei's claimed behavior.

Given the definition of the relevant markets as every SEP, the assessment of InterDigital's dominant market position was no surprise. As the only supplier in every relevant market, there were no actual or potential competitors alongside InterDigital. The Court quickly came to the conclusion that InterDigital was dominant in all the relevant markets.

The last step was to evaluate whether InterDigital had abused its dominant position. The courts first examined the issue of unreasonably high royalty rates. Their analysis was mainly based on a comparison with the royalties rates given to other firms, such as Apple, Samsung, RIM, and HTC. The courts found that the rates proposed to Huawei were at least seven times higher than those for other firms. Afterwards, InterDigital did not offer any evidence to justify the higher rates proposed to Huawei with cost reasons. Consequently, the courts ruled that the proposed high royalty rates constituted unreasonably high prices under the AML.

With regard to tying, the courts first looked at the tying of SEPs with non-SEPs. The courts maintained that non-SEPs were possibly substitutable. Its tying with SEPs might allow InterDigital to leverage its market power in the market for SEPs to the market for non-SEPs. This fell within the scope of tying under Article 17 of the AML. Subsequently, the courts examined Huawei's claim that it was also tying under the AML to license the SEPs on a global basis, as Huawei argued that it was licensed only for China and the United States. The courts nonetheless affirmed that such a global license could decrease transaction costs for international firms, and remain a common practice in the industry. Thus it was not an antitrust violation to license SEPs on a global basis.

In the end, the courts required InterDigital to cease its conduct regarding unreasonably high rates and tying. It is interesting to note that the courts set a specific FRAND rate. The applicable royalty rate for the licensing of the Chinese SEPs was determined at 0.019 percent, significantly lower than the royalty rate initially required by InterDigital.[43]

23.5.1.2. Case *Qualcomm*

The *Qualcomm* case was handled by the NDRC, and also related to SEPs of mobile communications standards.[44] Qualcomm was a firm similar to InterDigital, both of which had strong market power regarding mobile communications standards. Unlike InterDigital, Qualcomm was also an active player in the market for baseband chips. In the decision the NDRC held Qualcomm liable for its abusive behavior, and imposed a fine of RMB6.088 billion yuan, a record for antitrust fines in China to date.

Without having reference to the *Huawei v. InterDigital* case, the NDRC defined the relevant market for SEPs of mobile communications standards in the same manner and based on the same logic. Given there were no substitutes for the SEPs owned by Qualcomm, the NDRC first defined a relevant market for every SEP owned by Qualcomm. The geographic market was defined as nationwide because of national IPR regulations. Such a market definition resulted in the inevitable conclusion that Qualcomm was dominant in the relevant market by reason of its 100 percent market share, its ability to control the market, no countervailing buying power, and high entry barriers.

Due to Qualcomm's prominent presence in the baseband chips market, the NDRC also defined a relevant market for baseband chips, and designated Qualcomm as the dominant firm there. The major reasons were the following. First, Qualcomm's market share was well above 50 percent, and was consistently much higher than, for example, that of the second biggest player, i.e. MTK, who had only 15.5 percent market share. Second, Qualcomm enjoyed competitive advantages over technologies, product performance, and strong brand name. All of these made it an unavoidable firm for mobile equipment producers on the downstream market. Last but not least, entry barriers to this market proved extremely high. No potential competitors could enter the market in the short term. The geographic market was defined as worldwide.

Subsequently, the NDRC investigated three types of abusive behavior, namely unreasonably high royalty rates, tying SEPs with non-SEPs, and attaching unreasonable trading conditions.

---

[43] The NDRC also conducted an investigation into InterDigital, which, however, ended in a settlement as InterDigital put forward commitments (NDRC 2014).

[44] Case *Qualcomm*, NDRC [2015] Nr. 1, February 9, 2015.

With regard to unreasonably high royalty rates, the NDRC did not examine whether the royalty rates imposed by the NDRC were excessively high per se or discriminatory as the courts did in the *Huawei v. InterDigital* case. The NDRC specifically reviewed two issues. First, Qualcomm, when licensing its patents, did not offer a list of patents to its trading partners. The NDRC observed that Qualcomm also forced its partners to pay royalty fees for invalidated patents. Although Qualcomm claimed that it also frequently added new patents to the unspecified list, the NDRC did not believe that those new patents were of similar value to invalidated patents. The NDRC ruled that it was illegal to allow trading partners to pay the royalty fees for invalidated patents. Second, when licensing its patents Qualcomm constantly required cross-license free of charge from its trading partners. Since these cross-licenses were not free but paid for in a competitive market, Qualcomm's ability to do so was an abuse based on its dominance. The NDRC finally concluded that Qualcomm's pricing strategy fell within the scope of unreasonably high prices under the AML.

The second type of conduct concerned tying SEPs with non-SEPs. Similar to the conduct analyzed in the previous paragraph, this was also one of Qualcomm's longtime business strategies. The NDRC identified the anticompetitive effects of such a compulsory tying from two perspectives. First, it limited licensees' freedom of choice, and forced them to pay more than they might actually need. Second, it could foreclose the market for non-SEPs against actual and potential competitors. Consequently, this conduct must be considered as tying within the context of the AML.

The third type of conduct was raised in the market for baseband chips. Qualcomm obliged its trading partners to accept non-challenge clauses regarding the validity of its patents when purchasing its chips. This could lead to an effect of excluding market players who would like to accept the non-challenge clause on the downstream market, and thus violated Article 17.5 of the AML.

23.5.1.3. Summary

Neither the Chinese courts nor the NDRC have published any guideline on antitrust enforcement of IPRs. The aforementioned two cases only suggest the most significant issues confronted by the two authorities in relation to abuse of IPRs in China. It seems that what concerns them most is SEPs-related issues, similar to Mofcom with regard to merger review. It should be noted that neither the courts nor the NDRC impugned the actual possession of SEPs, which is not an antitrust violation per se. The focus is always on the manner of exercising SEPs.

The *Huawei v. InterDigital* case and the *Qualcomm* case share several features. Since the two cases may have precedential value for future enforcement, they are worthy of further exploration. First and foremost, both the courts and the NDRC maintained that every single SEP should constitute an individual relevant market.

Such a market definition leaves no possibility for substitutes, and hence produces a profound effect on the assessment of dominance. It becomes inevitable that SEP holders have a monopoly on the relevant market, regardless of their market size. Secondly, both the courts and the NDRC contended that it was abusive to tie SEPs with non-SEPs.

There are also some divergences between the two authorities. First, both of them condemned excessively high royalty rates, albeit from different perspectives. The courts affirmed the anticompetitive nature of InterDigital's price strategy based on a comparison with the fees paid by other competitors, and held that the fact that the price proposed to Huawei was several times higher than that actually paid by other competitors was evidence of abusing dominance. It is more striking that the courts later even set a specific level of royalty rate at 0.019 percent. By comparison, the NDRC more or less took another approach to compare Qualcomm's price with the costs. The NDRC found that the current price included both invalidated patents and a cross-license that was requested by Qualcomm free of charge. Subsequently, the NDRC maintained that licensees should not pay for invalidated patents. Neither should they cross-license their patents for free. After deducting invalidated patents and cross-licenses from the costs, it is obvious that the current royalty rates applied by Qualcomm were unreasonably high. However, unlike the courts, the NDRC did not specify the royalty rates for Qualcomm. Second, within the *Qualcomm* case the NDRC also reviewed the illegality of non-challenge clauses, which was not disputed before the courts.

Furthermore, there is also a hidden similarity between the two authorities. Both of them seemed to place a high burden of proof on the dominant firm. When evaluating the anticompetitive harm of the conduct concerned, both authorities in effect only indicated a possibility. For example, when assessing tying of SEPs with non-SEPs, the two authorities simply mentioned that it would affect competition on the market for non-SEPs. However, they did not go further in substantiating this possibility with evidence. Instead, the authorities seemed to switch the burden of proof to the dominant firms concerned. Once the dominant firms concerned could not offer evidence to prove otherwise, a conclusion of infringements was then quickly reached. It is well known that the conduct of dominant firms has the possibility to go two ways, anticompetitive and procompetitive. There is no behavior that does not limit competition at all. Therefore, an appropriate assessment of abusive conduct should always compare the procompetitive effects with the anticompetitive ones. If a mere suggestion of anticompetitive possibilities can lead to the burden of proof being switched to the other party, it would in effect lead to a situation where the defendants or the investigated firms always had to justify themselves. Given the difficulty of proving abusive conduct or in this scenario non-abusive conduct, it would in the end make all the abusive conduct listed in the AML almost per se illegal, at least from the perspective of the authorities.

Last but not least, we might interpret the practice within the two cases as a way to protect domestic producers. There are many emerging Chinese smartphone makers, such as Huawei, ZTE, Xiaomi etc. Their further growth depends to a great extent on reduced royalty fees for mobile communications standards. The two cases can exactly satisfy the needs of those firms. However, it should be pointed out that SEPs are not only owned by foreign firms, e.g. Qualcomm and InterDigital, but also by domestic companies. Huawei, for example, has already begun to be a major player in the market for SEPs. It will be interesting to see whether the authorities maintain their practice within the two cases in the future if the defendants or the investigated firms are Chinese firms.

23.5.2. Regulation of Abusing IPRs

The SAIC is in charge of non-price-related antitrust cases. As a matter of fact, the SAIC has not investigated a case related to IPRs so far. However, it publishes guidelines to systemically explain its views on antitrust enforcement of IPRs, so that the Regulation of Abusing Intellectual Property Rights to Eliminate or Restrict Competition was promulgated.[45] It is a pity that the SAIC failed to reach a common position with the NDRC and Mofcom. Therefore, the scope of the Regulation does not explicitly include price-related anticompetitive conduct and IPR issues in merger control. There follows a brief introduction to the key provisions of the Regulation in relation to abusing dominance.

23.5.2.1. Goals and Principles

Building upon previous developments, the Regulation starts by pointing out that the AML and IPR laws have shared goals to promote innovation and competition, improve efficiency, and safeguard the interest of consumers and the public interest.[46] Moreover, the Regulation is to be enacted with the purpose of clarifying the uncertainties left by Article 55 of the AML; that provision contains the very high-level principle that "abuses of IPRs" to eliminate or restrict competition are prohibited, while the use of IPRs in line with IPR laws and regulations is exempted from the AML. The Regulation makes it clear that "abuses of IPRs" refer to anti-competitive agreements, abuse of dominance, or other types of conduct that are not compatible with the AML and IPR laws.[47]

23.5.2.2. Analytical Framework

The Regulation puts forward a five-step analysis for assessing IPR-related antitrust infringements. In an investigation, the SAIC needs to identify, first, the characteristics and form of the conduct and, second, the type of relationship (i.e., horizontal or

---

[45] The Regulation, *supra* note 24.
[46] *Id.*, Article 1.
[47] *Id.*, Article 3.

vertical) between the firms involved. The Regulation specifies that an agreement is not considered horizontal where the parties are not competitors at the time the agreement is concluded, even if they become competitors afterwards. Third, the relevant market needs to be defined and, fourth, the market positions of the parties concerned are to be examined. As the fifth and last step, the SAIC needs to appraise the impact of the conduct on competition in the relevant market.[48]

Factors to assess the impact on competition include (i) the companies' market position vis-à-vis trading partners, (ii) the degree of market concentration, (iii) entry barriers, (iv) industry practices and the level of development, (v) the duration and scope of the restraints to competition, (vi) the impact on encouraging innovation and disseminating technology, and (vii) the innovation capacity and dynamics of technological change.[49]

23.5.2.3. Prohibitions

The Regulation contains ten articles that touch upon conduct of abusing dominance, covering refusal to license, exclusive dealing, tying, unreasonable licensing conditions, discrimination, patent pools, and SEPs.

The refusal to license clause has probably been the most debated provision during the consultation process.[50] This clause extends the "essential facilities" doctrine established by the SAIC Regulation on the Prohibition of Conduct Abusing Dominant Market Position to the IPR field.[51] Under this earlier regulation, a request for access to an essential facility may not be rejected by a dominant firm provided that the following factors have been taken into account: (i) feasibility of investing in, and building, other facilities by firms other than the dominant company, (ii) dependence of other trading partners on the facility for effective production and operation, (iii) the ability of the dominant firm to provide access to the facility; and (iv) the impact of such access on the operations of the dominant firm itself.[52]

The Regulation appears to go into more detail in relation to some of the factors laid out in the earlier SAIC regulation. First, the concept of "dependence of other trading partners" is further clarified as that the requested IPR should have no reasonable substitutes and must be "necessary" for the economic activities of those trading partners. In conjunction with the earlier regulation, this may mean that it is not economically feasible to self-craft or obtain from third parties an alternative IPR that functions equivalently to the IPR owned by the dominant company. Second, the assessment of the impact on the relevant market should include the effect not only on competition but also on innovation, which echoes the EU practice in the

---

[48] Id., Article 15.
[49] Id., Article 16.
[50] E.g. International Bar Association (2014).
[51] Regulation on the Prohibition of Conduct Abusing Dominant Market Position, SAIC Order [2010] Nr. 54, December 31, 2010, Article 4.
[52] Id., Article 4.

Microsoft case.[53] This approach appears to be similar to the principles employed by the EU Courts in cases such as *Magill*, *IMS*, and *Microsoft*.[54]

The Regulation further prohibits the imposition of unreasonable conditions by dominant companies. A list of examples of unreasonable conditions is provided, including (i) exclusive grantbacks for derived technology, (ii) prohibition on challenging the IPR's validity, (iii) prohibition on using the IPR after expiry, or developing or using competing technologies, (iv) obligation to pay royalties after expiry or finding of invalidity; and (v) exclusive dealing.[55]

Then, the Regulation embarks on regulating specific figures of IPR cooperation and unilateral action. Unless justified, patent pools are not allowed to (i) prevent patent pool members from licensing individually outside the pool, (ii) prevent members or licensees from developing competing technologies, (iii) impose exclusive grantbacks, (iv) prohibit licensees from challenging the validity of the pooled patents, and (v) apply different conditions to equivalent transactions.[56]

Abusing IPRs in relation to SEPs represents the second most hotly debated provision in the Regulation. The Regulation appears to set out the general principle that patent holders are not allowed to abuse their SEPs, irrespective of whether these patents are essential, because they are compulsorily set by the government, voluntarily organized by the industry, or have evolved as a de facto industry standard. The Regulation then provides specific prohibitions of SEP-related conduct, unless justifications can be provided. The first specific prohibition concerns patent assertion after the deliberate failure on the part of the SEP holder to disclose information during the standard setting process and/or after "giving up" the right to assert the SEP to be included in the standard. The second prohibition refers to violations of the principle of licensing SEPs on FRAND terms, refusing to license under reasonable conditions, licensing under unfair conditions, or tying in licensing.[57]

Finally, exclusive dealing, tying, discrimination, and investigation procedures are also touched upon by the Regulation. That said, the provisions related to those types of conduct closely reflect the wording of the AML provisions. Few details are provided that would indicate any differences as to their application to IPR issues.

23.5.2.4. Assessment

It is the first time in history that China has attempted to adopt a guideline to regulate IPRs through antitrust enforcement. Despite waves of doubts from stakeholders,[58] the Regulation is itself as cautious as it is conservative. After close to six years of researching, drafting, and consultations, the Regulation reflects a compromise, i.e.

---

[53] The Regulation, *supra* note 24, Article 7.
[54] Hou (2012).
[55] The Regulation, *supra* note 24, Article 10.
[56] *Id.*, Article 12.
[57] *Id.*, Article 13.
[58] American Bar Association (2014); US-China Business Council (2014).

mixing both foreign and domestic elements and reconciling various interests. The Regulation is a relatively cautious attempt to bridge the gap with related pre-existing laws in China, while at the same time attempting to follow international practices in the antitrust/IPR space – namely, EU and US rules, to the extent possible. While some controversial articles have been removed during the drafting process,[59] the Regulation still manages to deliver some fresh thinking, and offer a relatively comprehensive antitrust regulation of IPRs, within the SAIC's parameter of jurisdiction. The Regulation has the potential to provide welcome guidance to market players and government bodies alike on what constitutes anticompetitive and procompetitive conduct in the IPR field, and what the analytical framework behind any enforcement would be. As mentioned earlier, the SAIC failed to reach a common position with the NDRC and Mofcom. Consequently, the Regulation covers only non-price-related issues, and generated binding effect on the SAIC itself. This drawback has recently been noticed by the Anti-Monopoly Committee of the State Council, which is a coordinating authority between competition agencies and sector-specific regulators.[60] The Anti-Monopoly Committee has begun the process of drafting a guideline that demonstrates the application of the AML to IPRs in all antitrust domains. Although the drafting process is led by the NDRC,[61] it goes without saying that the SAIC Regulation will play a major role in the forthcoming guideline.

23.6. CONCLUSIONS AND FORECAST

Access to technology and development of domestic and "indigenous" technology are key factors in China's development strategy. Often, technology is protected by IPRs, and hence policies relating to IPRs can play an important role in implementing the strategy. One such policy is IPR protection, but, increasingly in Chinese government circles, antitrust law is viewed as an instrument to tackle perceived "monopolistic" strangleholds in technology development and transfer.

Based on the discussion above, we have observed that Chinese competition agencies have been taking an active role in such a process. This can particularly be seen in Mofcom's practice in the domain of merger control. Mofcom makes a distinction between standardized IPRs (including de facto standards) and non-standardized IPRs. The former, due to their strong market presence, are subject to strict review. FRAND obligations are often imposed upon owners of standardized IPRs. With regard to SEPs for mobile communications, Mofcom even sets specific limits on the usage of injunctions; it appears to be the first in the world to do so. But this does not mean that Mofcom is hostile to IPRs. In other situations, IPRs are mostly not the main concern for merger cases. While so far standards remain the only focus of Mofcom, this chapter also examines the practice of other competition

---

[59] Emch and Hou (2014).
[60] Hao (2013).
[61] USIT (2015).

authorities regarding abusing dominance in relation to IPRs. Given the fact that the Anti-Monopoly Committee is drafting a general guideline for all three competition agencies, it may reveal Mofcom's activities in the future.

REFERENCES

American Bar Association. 2014. Joint Comments of the American Bar Association Section of Antitrust Law, Section of Intellectual Property Law, and Section of International Law on the SAIC Draft Rules on the Prohibition of Abuses of Intellectual Property Rights for the Purposes of Eliminating or Restricting Competition. July 9, 2014. Available at: http://ec.europa.eu/competition/consultations/2013_technology_transfer/aba_en.pdf.

Coco, Rita. 2008. Antitrust Liability for Refusal to License Intellectual Property: A Comparative Analysis and the International Setting. *Marquette Intellectual Property Law Review*, 12, 3–47.

Cotter, Thomas F. 2014. Comparative Law and Economics of Standard-Essential Patents and FRAND Royalties. *Texas Intellectual Property Law Journal*, 22, 311–61.

Emch, Adrian and Liyang Hou. 2014. Antitrust Regulation of IPRs: China's First Proposal. *Competition Policy International Asian Column*, August 14, 2014.

European Commission. 2014. Commission accepts legally binding commitments by Samsung Electronics on standard essential patent injunctions. IP-14-490. April 29, 2014.

Hao, Qian. 2013. The Multiple Hands: Institutional Dynamics of China's Competition Regime. In Adrian Emch and, David Stallibrass (eds.), *China's Anti-monopoly Law: The First Five Years*, Alphen aan den Rijn: Wolters Kluwer, 15–34.

Hou, Liyang. 2012. The Essential Facilities Doctrine – What Was Wrong in Microsoft?. *IIC-International Review of Intellectual Property and Competition Law*, 43, 251–71.

International Bar Association. 2014. Submission to the Administration for Industry and Commerce on the Proposed Rules on the Prohibition of Abuses of Intellectual Property Rights for the Purposes of Eliminating or Restricting Competition. July 10, 2014. Available at www.ibanet.org/Document/Default.aspx?DocumentUid=088F2ACA-A837-47A2-A8AB-E4492EC96400.

Jones, Alison. 2014. Standard-Essential Patents: FRAND Commitments, Injunctions and the Smartphone Wars. *European Competition Journal*, 10, 1–36.

NDRC. 2014. NDRC Suspends the Investigation into IDC for Price-Monopoly Conduct, May 22, 2014. Available at www.ndrc.gov.cn/gzdt/201405/t20140522_612466.html.

US-China Business Council. 2014. US-China Business Council Comments on Draft Regulations on the Prohibition of Conduct that Eliminates or Restricts Competition through Abuse of Intellectual Property Rights (IPR), July 10. Available at www.uschina.org/advocacy/testimony-speeches/us-china-business-council-comments-draft-regulations-prohibition-conduct.

USIT. 2015. NDRC Drafts Anti-Monopoly Guidelines against IPR Abuse. Available at http://usito.org/news/ndrc-drafts-anti-monopoly-guidelines-against-ipr-abuse.

Wang, Xianlin. 2000. Discussion on the Establishment and Responsibilities of China's Antimonopoly Agencies. *Zhongguo Xingzheng Guanli (China Public Administration)*, 8, 60–64.

Wang, Xiaoye. 2007. Several Questions about Our Anti-Monopoly Agencies. *Dong Luncong Dongyue Tribune* 21, 30–41.

Zhang, Bingsheng. 2005. The Establishment of Our Anti-Monopoly Authorities – Questioning the Current Institutional Design. *Fa Xue (Science of Law)*, 2, 113–21.

Index

Abbott Laboratories case, drug patent settlement in, 320–1
Abbreviated New Drug Application (ANDA), 22–3, 25–6
 product hopping cases and, 27, 38–9
 reverse payment cases and, 26–7, 32–8
AB rating
 brand/generic drug regulations and, 25–6
 monopoly power screen and, 28–32
 product hopping cases and, 38–9
AB Volvo and Erik Veng, 97–100, 108–11
Acacia Technologies, 257–8
acceleration clauses, in patent drug settlements, 322–4
Acme Precision Products, Inc. v. Am. Alloys Corp., 275–7
adaptation, copyright protection and, 210
Addanki, Sumanth, 21–41
additional abusive conduct principle, intellectual property and, 108–9
AdMob, Google acquisition of, 461
advertising
 Big Data consumer targeting and, 296
 of branded drugs, 29–30
 comparative advertising defense, 199–200
 horizontal mergers in, 461
 impact on search results of, 301–3
 market definition and, 189–91, 310
 by pharmaceutical industry, 30–1
 trademark protection and, 199–200, 210–11
 Walker Process standards and, 274–5
Airtours case, 109–10
AKZO case, 109–10
Allied Orthopedic Appliances Inc. v. Tyco Health Care Group L.P., 87–9, 398–401
Almunia, J., 351–2

Alstyne, M.W., 44–5
America Invents Act, 262–3
American Infra-Red Radiant Co. v. Lambert Industries, Inc., 288–92
American Intellectual Property Law Association (AIPLA), 326–7
American Needle case, 193
Anderman, S.D., 97–106
anticompetitive practices
 cable and satellite copyright and, 345–6
 complementarity in, 142–3
 disclosure requirements and deception as, 236–8
 in drug patent cases, 321–2
 in Japanese antimonopoly regulations, 139–43, 152–3
 in Korean IFP law, 158–63, 169–70, 174–5
 Microsoft lawsuit and, 10–12
 non-patent litigation and, 288–92
 patent assertion entities and, 263–5
 patent pooling and collusion as, 367–72
 post-expiration royalties and, 238–41
 product hopping and, 87–9
 resale price maintenance as, 425–7, 433–4
 reverse payment cases, 32–8
 tied products/tying arrangements as, 404–7, 409–18
 US government investigation of, 225–9
Anti-Monopoly Bureau (Korea), 159–60
Anti-Monopoly Commission (AMC) (China), 122
 FRAND obligations and, 127–9
 future challenges for, 135–6
 Market Definition, 122–3
Anti-Monopoly Law (AML) (China)
 establishment of, 120–1, 468–70
 holdup of patents and, 256–7
 IP–antitrust provisions, 121–9, 229–31

489

Anti-Monopoly Law (AML) (China) (cont.)
  patent law and, 129–30
Antimonopoly Law (AML) (Japan)
  Article 2 provisions, 140
  Article 3 of, 139–40
  Article 19 provisions, 141–2, 150–2
  Article 21 provisions, 143–4, 147–51
  complementarity with intellectual property of, 142–3
  expired or exhausted patent restriction, 148–9
  FRAND licensing obligations and, 233–4
  future challenges for, 155
  JFTC guidelines and, 138
  patent and trademark applications and, 145
  patent infringement and, 145–6
  unilateral refusal to license in, 146–7
Antitrust Modernization Commission (AMC), 396–7
antitrust policy
  Anti-Monopoly Law in China and, 121–9
  applications of, 85–91
  Big Data regulation and, 293–313
  in China, 120–36, 467–8
  complementarity in EU concerning, 106–8
  complementarity in Japanese law on, 142–3
  cooperative commercialization and, 15–16
  coordinated vs. single firm conduct, pharmaceuticals industry, 26–39
  data-driven mergers and defenses and, 304–5
  drug patent settlements, 319–36
  economics and, 82–3
  entrant innovation and, 9–10
  in European Union law, 92–117
  exceptionalism and, 98–106
  high technology and limits of, 384–7
  holdup of patents and, 248–9
  implied immunity in, 56
  injunctions for patent infringement and, 72
  innovation economics and, 3–6, 47–9
  intellectual property intersection with, 18–19, 81–93
  in Japan, 138–55
  limits of market definition for, 194–6
  litigation costs as antitrust injury, 272 n.4
  market definition and prices in, 188–9
  network effects and, 45–6
  non-discrimination in FRAND licensing and, 69–70
  patent assertion entities and, 263–5
  in pharmaceuticals, 21–41
  profits of innovation and, 7–9
  rule-based approach to EU intellectual property involving, 116–17
  Schumpeterian competition and, 49–51

  sham litigation doctrine and, 281–8
  support for exceptionalism in EU for, 101–2
  tied products/tying arrangements and, 11–12, 404–7
  Walker Process doctrine and, 272–81
  welfare consequences of monopolization and, 82–3
Apple, Inc. v. Motorola, 251–5
Apple, Inc. v. Motorola Mobility, Inc., 290–1
Apple, Inc. v. Samsung Electronics Co., 290–1
Apple Computer
  European litigation involving, 231–3
  patent holdup violations, 251–5
  in *Samsung Electronics* case, 176
Applied Materials/Tokyo Electron merger, 163
arbitration, FRAND licensing as, 68–9
Areeda, Phillip, 284, 417–18, 438–9
ARM Limited–Giesecke & Devrient GmbH–Gemalto N.V. merger, 474–6
Armstrong, M., 44–5
asset specificity, holdup of patents and, 61
assign back. *See* grantbacks in patent law
AstraZeneca v. Commission, 101–2
Atkinson, R.D., 47–9
AT&T-DirecTV merger, 455–8, 462–4
AT&T-T-Mobile merger, 458–62
audiovisual works, EU copyright rules on, 348–9
Audretsch, D.B., 47–9
"authorized generics," drug patent settlements and, 322–4

Babies "R" Us, coercion of suppliers by, 427–34
"balancing" approach, complementarity and, 106–7
bargaining power, of IT mergers, 462–4
"baseball style" arbitration, FRAME licensing and, 68–9
Baxter, William F., 84–5
Bayer AG v. Süllhöfer, 115
below cost pricing, innovation and antitrust policies and, 4
Bement v. National Harrow, 89, 368–9
benchmarking framework, in FRAND licensing, 69–70
Big Data
  accessibility, collectivity and cheapness of, 298–9
  antitrust regulation and, 293–313
  applicability of antitrust law to, 308–12
  appropriateness of antitrust remedies in, 311–12
  case-law on, 308–10
  consumer protection issues and, 294–6, 310–11
  data-driven mergers and defenses, 304–5
  economies of scale and, 305–8

entry barriers and, 297–8, 305–8
harm to competition from, 294–5, 300–8
innovation impact of, 296–7, 301–3
limitations of, 299–300
loss of quality impact of, 301–3
monetization of, 296
network effects of, 305–8
non-durability of, 299
nonexclusive/nonrivalrous content of, 299
overview of research on antitrust and, 294–5
platform differentiation and, 300
privacy, harm from, 303–4
procompetitive benefits of, 296–300
product market for, 310
quality improvement and innovation enhancement with, 296–7
"Big Six" Hollywood producers, EC Statement of Objections to, 341, 351–2
bioequivalence principle
"AB rating" and, 25–6
FDA brand and generic drug approval process and, 22–3
Blair, Roger D., 204–20
blanket licensing, 358–9
block-booking practices. *See also* geo-blocking
copyright protection and, 411–12
sufficient economic power principle and, 415–17
Block Exemption Regulation (EU), complementarity and, 107–8
blocking of patents, patent pooling and, 361–2
Boliek, Babette E., 43–56
Bork, Robert, 406–7
"bottom-feeder" patent assertion entities, 260–1
boundary delineation, patent pooling and, 364–6
Bowman, Ward, 84–5
branded drug products
development of, 21–2
European patent settlements and, 332–5
FDA approval of, 22–3
litigation costs for challenges to, 326–7
market definition and, 189–91
monopoly power screen and, 28–32
payment for unrelated services in settlements involving, 327–8
payment issues in drug patent settlements and, 322–4
product hopping cases and, 27, 38–9
regulations governing, 25–6
reverse payment cases and, 32–8, 319–20
branding
antitrust and IP and, 183–4
interbrand/intrabrand competition, 184
market definition and, 189–91
Brennan, T.J., 19

brick-and-mortar retailers
resale price maintenance in online markets and, 427–34
scanner data from, 434 n.35
broadband markets
innovation in, 51–2
net neutrality and, 53–4
Broadcom v. Qualcomm, 225–9, 249–50
Bronner v. Mediaprint, 102
Brulotte v. Thys Co., 87, 127, 238–41
bundled products
antitrust law and, 11–12, 379–401
discounts and rebates on, 394–8
essential facilities doctrine concerning, 391
in FRAND licensing, 255–8
growth in information technology and, 384–7
innovation and, 392–4
Internet economics and, 43
loyalty discounts for, 398–401
in NDRC (China) Qualcomm case, 133–5, 480–1
patent pooling and, 362–3
portfolio effects of mergers and, 388–90
pricing in patent pools vs. conventional bundling, 366–7
burdens of proof, in resale price maintenance litigation, 435
business practices of SEPs, FRAND licensing and, 73

Cable and Satellite Directive (EU), 344–8. *See also* Green Paper of the European Commission
cross-border transmission and, 344–5
digital single market strategy and review of, 350–1
export prohibitions in, 345–6
Murphy case and interpretation of, 347–8
national law of Member States and, 353 n.64
cable modem providers, net neutrality and, 46–7
cable transmission copyright
EU Cable and Satellite Directive, 344–5
Murphy case and, 345–8
cable transmissions, European Cable and Satellite directive and, 344–8
California Air Resources Board (CARB), 287–8
California Computer Products Inc. v. International Bus. Mach. Corp, 380–4
California Motor Transport Co. v. Trucking Unlimited, 282
Canada, drug patent settlements in, 335–6
Cannon, Joseph P., 430–1
Capital One, patent challenge by, 262–5
Carlton, Dennis W., 69–70, 386–7
Carrier, Michael A., 84–5, 319–36

cartels
  Japanese antimonopoly restrictions on, 140
  monopolization and, 82–3
  patent pooling by, 367–72
  patent pooling exclusivity and, 369–72
Cascade Health Solutions v. Peacehealth case, 396–7
cash payments, in patent drug settlements, 322–4
categorization
  applications of antitrust and, 87, 89
  in Chinese IP–antitrust regulatory practices, 121–9
causal connection test, standard-setting approach to patent protection and, 288–92
causation, in drug patent settlements, 331
chain of production
  damages for SEP infringement and, 73–5
  net neutrality and, 46–7
Charter of Fundamental Rights of the European Union, 107–8
Charter-TWC merger, 455–8, 462–4
Cheng, Thomas K., 120–36
Chicago School of Law and Economics, tied products/tying arrangements and, 406–7
China. See People's Republic of China
Chinese Patent Law, 129–30
chip technology
  Broadcom v. Qualcomm case and, 226
  Korean Qualcomm case involving, 173–4, 234–5
  Qualcomm case in China and, 480–1
claim construction doctrine, market definition and, 200–1
Clayton Antitrust Act (U.S.), 248–9, 389–90, 392–4, 407–9, 418–21, 446–53
"click-and-query data," 296–7
  impact on quality and innovation of, 301–3
  network effects and, 305–8
Coase, Ronald, 364–5
code division multiple access standards (CDMA, WCDMA)
  Chinese IP regulations and, 131
  European litigation concerning, 231–3
  FRAND licensing and, 226
  Japanese litigation concerning, 233–4
  Korean investigation of Qualcomm and, 234–5
  Samsung Electric case (Korea), 176
Coditel I and II cases, 97–8, 341–6, 354
coercive business tactics
  disclosure and deception violations and, 236–8
  resale price maintenance in online markets and, 428–9
  tying arrangements and, 413–15
collusion. See also coordinated conduct; joint conduct
  applications of antitrust and, 85–91
  express collusion, 368–9
  Japanese antimonopoly restrictions on, 140
  patent pooling and threat of, 367–72
  resale price maintenance facilitation of, 433–4
Colomo, Pablo Ibáñez, 339–55
Comanor, William, 428–9
Comcast/NBU merger, 455–8, 462–4
Comcast–TWC–Charter merger, 455–8, 462–4
Comerford, Roisin, 293–313
commerce, tied products/tying arrangements effect on, 417
Communications Act (U.S.), 53
  merger review process and, 452–3
comparative advertising defense, trademark protection and, 199–200
competition
  Big Data impact on, 293–4, 300–8
  Chinese IP–antitrust regulations classification of, 124, 129–30
  Chinese merger controls and, 467–87
  cooperative commercialization and, 12–16
  copyright and EU law on, 342–4, 386–7
  digital single market strategy and EU law on, 351–5
  EU laws concerning, 342–4, 352–3, 386–7
  ex post regulation and, 54–5
  future drug product development and, 39–41
  horizontal merger effects on, 458–62
  information technology foreclosure of, 455–8
  inherency in EU case-law concerning, 95–109
  innovation and, 3–6
  Internet economics and, 43
  Japanese antimonopoly protection of, 139–40
  Korean laws concerning, 158–78
  merger analysis and, 16–18, 447–50
  monopoly power and, 215–20
  net neutrality and, 46–7
  pharmaceutical mergers and, 39–41
  in pharmaceuticals industry, 26–39
  post-expiration royalties and, 238–41
  prevention of exploitation and, 222–4
  pro- and anti- aspects of exceptionalism and, 97–106
  product hopping cases and effects of, 38–9
  reverse payment cases and effects of, 32–8
  rules-based approach in EU to, 109–10
  Schumpeterian competition, 49–51
  static analysis of, 10–12
  tied products/tying arrangements and EU law on, 404–7
  in two-sided markets, 44–5
Competition Commission of India (CCI), 335–6

competitive harm theory
  Big Data and, 294–5, 300–10
  disclosure and deception violations and, 236–8
  drug patent cases and, 321–2, 331–2
  market definition and, 188–9
  network effects on Internet and, 45–6
  patent assertion entities and, 263–5
  patent pooling and, 362–3, 369–72
  payments in drug patent disputes and, 322–4
  product hopping cases and, 27, 38–9
  resale price maintenance and, 427–35
  in Sherman Act, 222–4
  state law on drug patents and, 331–2
  tied products and, 362–3, 409–18
  Walker Process doctrine and, 277–81
complementarity
  bundling and innovation and, 392–4
  copyright protection and, 198–9
  defined, 93
  EU antitrust and intellectual property law and, 106–8
  exposition and, 106–7
  in information technology, 388–90
  in Japanese IP–antitrust intersection, 142–3
  limits in patent pooling of, 363
  patent pooling and, 359–63
compulsory licensing
  in Chinese IP–antitrust law, 125–6
  in Chinese patent law, 129–30
  EU law and, 108–11
  Japanese IP–antitrust law, 142–3
  monopoly power and, 216–18
  rule-based approach to, 110–11
Concrete Pile case, 152–3
conditional rebates, loyalty discounts as, 398–401
conduct remedies, in merger jurisprudence, 451
Consten and Grundig v. Commission, 95–109, 112–15, 342–4
consumer protection
  applications of antitrust and, 85–91
  bargaining power of IT mergers and, 462–4
  Big Data issues and, 294–6, 310–11
  Chinese IP–antitrust regulations and, 122, 126–7
  essential facilities doctrine in Chinese IP–antitrust regulations and, 125–6
  holdups of patents and, 247–8
  Internet economics and, 43
  pharmaceuticals industry and role of, 21–2
  procompetitive effects of Big Data and, 296–300
  resale price maintenance and, 429–31
  reverse payment cases and, 32–8
  tied products/tying arrangements as threat to, 404–7
  Walker Process doctrine and standing of, 277–9

  welfare effects of intellectual property and, 83–5
consumption habit defense, Chinese IP–antitrust regulations and, 126–7
contractual principles
  country of origin principle for online transmission and, 349–50
  EU copyright rules and, 68–9
  FRAND licensing as, 68–9
  patent infringement and, 225–9
  tied products/tying arrangements and, 413–15
  trade secrets law and, 207–8
Cooper, James, 294
cooperative commercialization
  antitrust policy and, 15–16
  determinants of innovation prize and, 14–15
  in Korean IP law, 160–1
  non-challenge clause in Japanese patent law and, 150–1
  outcomes of, 13–14
  rationale for, 12–15
cooperative licensing agreement, mergers and, 17–18
coordinated conduct
  applications of antitrust and, 85–91
  mergers and, 447–50
  in Microsoft/Nokia merger case (Korea), 177–8
  in pharmaceuticals industry, 26–39
  rule-based approach to cases involving, 112–15
copyright protection
  antitrust impact on information technology and, 384–7
  blanket licensing, 358–9
  duration of, 211
  EU bundling and innovation policies and, 386
  EU digital single market strategy, 339–55
  European Cable and Satellite Directive and, 344–8
  EU rules concerning, 341–4, 348–9
  exhaustion doctrine in EU copyright and, 341–2
  inherency in EU case-law concerning, 95–109
  Japanese IP–antitrust policy and, 151–2
  limits on innovation from, 48–9
  market definition and, 186–7, 189–91, 198–9
  monopoly power and, 210
  phase-out of inherency and, 97–8
  satellite transmission, 344–5
  sham litigation doctrine and, 272–3 n.5
  sufficient economic power principle and, 415–17
  two-products element in, 411–12
country of destination principle
  European competition law and, 352–3
  Murphy case on cable and satellite copyright and, 347–8

country of origin principle
  difficulties in identification of, 353 n.64
  in EC Statement of Objections, 351–2
  EU copyright rules and, 348–9
  in *Murphy* case, 345–6
  for online transmissions, 349–50
  satellite transmission copyright, 344–5
country of uplink principle, satellite transmission copyright, 344–5
Court of Justice of the European Union (CJEU), 92–3
  coordinated conduct cases and, 112–15
  digital single market strategy for copyright and, 339–41
  exceptionalism in EU law and, 98–106
  operational irrelevance of inherency and, 96–7
  particularism vs. exceptionalism in cases before, 103–4
  patent holdup injunctions and, 251–5
  rule-based approach of, 109–17
Crane, D.A., 109–10, 397
cross-border transmission
  copyright protection and, 344–5
  country of origin principle and, 349–50
  digital single market strategy (EU) and, 350–1
  EU competition law and, 351–5
cross-elasticity of supply and demand, market definition and, 188–9
cross-group externalities, in two-sided markets, 44–5
cross-licensing
  Japanese IP–antitrust rules and, 153–4
  Korean IPR guidelines and, 167–8
  monopoly power and, 219
  patent pooling and, 358–9
  rate disclosure and, 61–2
cross-price elasticities, in two-sided markets, 44–5
CSIRO v. Cisco, 74–5
customer demands, loyalty discounts in Europe and, 400–1
Czapracka, K., 97–8

Daiichi-Kosho (DK) case, 146–7
damages
  lost profit damages for patents, 185–6
  in patent holdups, 255–8
  for patent infringement, 212
  for SEP infringement, FRAND provisions for, 73–5
data-driven defenses, antitrust policy and, 304–5
Declaratory Judgment Act (DJA) jurisdiction, *Walker* Doctrine enforcement and, 279–81
decoding devices, satellite transmission copyright and, 345–6

de facto standards, Chinese control of non-SEP related mergers and, 474–6
de facto standards, Korean IPR guidelines and, 170–1
de jure standards, Korean IPR guidelines and, 170–1
delayed entry
  drug patent settlements and, 319–20, 327–8
  European investigations of, 332–5
  horizontal mergers and, 449
  risk aversion and, 33–4
Delcamp, Henry, 218–19
Dell Computers
  disclosure and deception violations by, 236–8
  patent ambush by, 249–50
demand
  benefit function of innovation and, 3–6
  for innovation, 205–6
  Lerner index of monopoly power and elasticity of, 213–14
  loyalty discounts in Europe and, 400–1
  merger horizontal transactions and, 447–50
  monopoly markups and elasticity of, 215–16
  monopoly power screen and building of, 30
  pharmaceutical reimbursement policies and, 23–5
  tied products/tying arrangements and, 410–13
  welfare consequences of monopolization and, 82–3
demand-side effects, information technology competitive dynamics, 453–5
Department of Justice (US)
  bundling and product integration cases and, 386–7
  data-driven mergers and defenses and, 304–5
  horizontal merger jurisprudence and, 458–62
  Merger Guidelines of, 17–18, 447–50
  merger jurisprudence and, 446–53
  merger review process, 452–3
  patent holdup injunctions and, 251–5
derivative work right, copyright protection and, 198–9
design protection
  disclosure and deception violations and, 236–8
  in high technology industry, 380–4
  inherency in EU case-law concerning, 95–109
  tied products/tying arrangements and, 413–15
DG Markt, 349–50
differentiated data, online markets and, 300
digital lock-in, holdup of patents and, 245–9
digital single market strategy (DSMS)
  competition law intervention and, 351–5
  EU copyright rules and, 339–41
  proposals for, 350–1

Dippin' Dots, Inc. v. Frosty Bites Distribution, Inc., 196
direct purchasers, *Walker Process* doctrine and standing of, 277–9
direct test of monopoly power, monopoly power screen and, 28–32
disclosure requirements
  exploitation of licensees and, 222–4, 236–8
  patent ambush as violation of, 249–50
discount attribution rule
  bundled products discounts and rebates and, 396–7
  loyalty discounts and, 399–400
discounting
  of bundled products, 394–8
  loyalty discounts, 398–401
  online resale price maintenance and, 431–3
  in two-sided markets, 44–5
"dispense as written" principle, 25 n.8
dispute resolution, FRAND licensing and, 68–9
distribution
  Big Data and, 298–9
  copyright protection and, 210
  European copyright protection and, 342–4
  monopoly markups and demand elasticity, 215–16
  resale price maintenance in online markets and, 428–9
  trademark protection and, 412–13
Dolby case, 162–3, 177
dollar volume effect on commerce, tying arrangements and, 417
dominance, abuse of
  Big Data impact on, 301–3
  in Chinese IP–antitrust regulations, 124–5, 222–4, 229–31
  EU cases involving, 109–11, 231–3
  exceptionalism in cases of, 102–3
  holdup of patents and, 251–5
  Huawei v. InterDigital case in China and, 131–3, 478–80
  in Korean IP law, 160–3, 222–4
  post-expiration royalties and, 240–1
  resale price maintenance in online markets and, 429–31
  SAIC Regulation (China) and, 484–5
  tied products/tying arrangements as, 404–7
Dong-A Pharmaceutical Co., 174–5
double marginalization theory
  in information technology mergers, 388–90
  royalty stacking and, 70–1
  US private litigation and government enforcement and, 225–9

downstream markets
  exploitative abuse of IP and, 222–4
  Japanese IP–antitrust regulation and, 233–4
  non-discrimination in FRAND licensing and, 69–70
  patent assertion entities and, 258–61
Drake, Keith, 36
Drexl, J., 92, 106–7
Dreyfus, C., 97–8
drug delivery mechanisms, monopoly power screen and, 29–30
drug marketing and promotion
  product hopping cases and, 38–9
  reverse payment cases and, 32–8
drug patents, settlement of disputes involving, 319–36
  Actavis ruling and, 321–2
  in Canada, 335–6
  causation in, 331
  European settlements, 332–5
  large and unjustified payment threshold in, 329
  litigation costs and, 326–7
  merits of patent principle and, 324–5
  payment for unrelated services in, 327–8
  payment issues in, 322–4
  pleading requirements in, 329–30
  pre-*Actavis* case-law concerning, 320–1
  rule of reason in, 326–9
  state law and, 331–2
  US patent settlements, 319–32
DSD case, 102–3, 112
DSL providers, net neutrality and, 46–7
duplicative assets, cooperative commercialization and avoidance of, 13–14
duration of IP protection, 211
  non-durability of Big Data and, 299
dynamic markets
  antitrust policies and, 3–6
  established manufacturers safe harbor, RPM litigation, 436–7
  incumbency advantage and, 3–6
  in information technology, 445–6
  information technology competitive dynamics, 453–5
  innovation and, 3–6, 18–19, 47–9
  modeling innovation and, 6–10
  Schumpeterian competition and, 49–51
  static analysis and, 10–12

Easterbrook, 49
Eastman Kodak Co. v. Image Technical Services, Inc., 410–13
eBay court decision, 262
  FRAND licensing and, 73

e-commerce
    digital single market strategy (EU) and, 350–1
    EU digital single market strategy, 339–41
econometric analysis
    FRAND licensing balance and, 62–4
    pharmaceuticals industry and limits of, 25 n.7, 30 n.18
economic theory
    antitrust and intellectual property and, 82–5
    Big Data and, 297–300
    innovation and, 3–19, 47–52
    intellectual property and, 83–5, 205–6
    Internet economics, 43–56
    Korean intellectual property and competition policies and, 158–9
    monopoly power and, 212–15
    resale price maintenance justification based on, 438–9
    rule-based approach in EU antitrust and avoidance of, 116–17
    tied products/tying arrangements and, 404–7
economies of scale
    Big Data and, 305–8
    information technology competitive dynamics and, 453–5
    tied products/tying arrangements and, 404–7
Edlin, Aaron, 326–7
effective price concept, European bundled products discounts and rebates guidelines and, 397–8
effects doctrine
    Korean Free Trade Commission IPR applications and, 162
    loyalty discount guidelines in Europe and, 400–1
efficiency defense
    mergers and, 450–1
    resale price maintenance and, 433–4
    tied products/tying arrangements and, 420–1
efficient competitor test, European bundled products discounts and rebates guidelines and, 397–8
Eisenach, Jeffrey A., 51–2, 445–65
Elhauge, Einer, 70–1, 397
EMI v. CBS, 96–7
Enchelmaier, S., 95
encryption technology, satellite transmission copyright, 344–5
enforcement of antitrust and IP laws
    exceptional circumstances principle and, 98–100
    in Japan, 140
    by Korean Free Trade Commission, 160–1
    mergers and, 16–18

in United States, 225–9
    Walker Process doctrine and issue of, 279–81
"Enforcement Perspectives on the *Noerr-Pennington* doctrine" (FTC), 285–7
Enhanced Patent Quality Initiative, 261–3
entrant innovation
    antitrust policy and, 9–10, 15–16
    cooperative commercialization and, 12–15, 18–19
    exclusive customer contracts and rate of, 15–16
    mergers and, 17–18
    static analysis of, 10–12
entry barriers
    Big Data lowering of, 297–8, 305–10
    Chinese SEPs mergers and role of, 470–4
    innovation and antitrust policies and, 3–6
    market definition in antitrust and, 188–9
    tied products/tying arrangements as tool for, 404–7
entry dates
    holdup of patents and, 59–62
    instantaneous vs. delayed entry, 33–4
    in reverse payment cases, 27 n.12
    reverse payment cases and, 32–8
equilibrium rate of innovation, 8–9
equivalence principle, *Walker* Doctrine enforcement and, 279–81
Erauw-Jacquery v. La Hesbignonne, 115
Ericsson v. D-Link Systems, 251–5
error-cost framework
    limits to innovation and, 49
    Schumpeterian competition and, 50–1
E.R.R. Presidents Conference v. Noerr Motor Freight, Inc., 281–8
essential facilities doctrine
    Chinese IP–antitrust regulations and, 125–6
    information technology and role of, 390–1
essential function principle
    EU antitrust and intellectual property policy and, 94–8
    inherency and, 97–106
    in *Samsung Electronics* case, 176
    tied products/tying arrangements and, 410–13
Essential Patent Claims, 74–5
European Commission (European Union)
    antitrust and intellectual property in law of, 92–117
    Big Data cases and, 308–10
    bundled products discounts and rebates and, 397–8
    bundling and information technology and antitrust policies in, 386
    Cable and Satellite Directive, 344–8
    competition law and copyright in, 342–4

complementarity antitrust and intellectual property law in, 106–8
coordinated conduct cases in, 112–15
copyright rules in, 341–4, 348–9
data-driven mergers in, 304–5
digital single market strategy, copyright licensing, 339–41, 350–1
drug patent settlements in, 332–5
emergence of inherency in case-law of, 95–109
essential facilities doctrine in, 390–1
exceptionalism in antitrust and intellectual property policy and, 98–106
GE/Honeywell merger opposed by, 388–90
high tech design protection cases in, 380–4
inherency in antitrust and intellectual property policy of, 94–8
license pricing and terms in IPR policy, 112
loyalty discounts in bundled products guidelines, 398–401
patent holdup injunctions and, 251–5
product design protection in, 380–4
regulatory context of competition law in, 352–3
remedy for IPR, cases involving, 111–12
resale price maintenance rules in, 425 n.1
rule-based approach to antitrust-IP intersection and, 109–17
European Court of Justice (ECJ), 341–2
loyalty discounts and rulings by, 400–1
Murphy case and, 345–8
European Telecommunications Standards Institute (ETSI)
FRAND licensing and, 62–4
Huawei v. InterDigital case in China and, 131–3
UMTS standard, 60, 225 n.10
Evans, D.S., 3–6, 10–12, 16–18, 50–1, 384–7
event study research, reverse payments cases and, 36–8
ex ante rules
complementarity theory and, 93
Internet regulation and, 54–5
patent disclosure and deceptive practices and, 64–7
rule-based approach to intellectual property and, 109–10
exceptionalism, theory of
antitrust law and, 102–3
complementarity as rejection of, 106–7
defined, 92–3
EU antitrust and intellectual property cases and, 98–106
lack of specificity concerning intellectual property rights, 102
limits of, 105
open-ended content of, 104–5

opposition in EU courts to, 102–5
particularism and, 103–4
pro- and anti- aspects of, 97–106
refusal to license cases and, 98–100
support in EU for, 101–2
excessive pricing
Chinese litigation involving, 229–31
disclosure and deception violations and, 236–8
European litigation concerning, 231–3
FRAND obligations and, 224–35
Japanese litigation concerning, 233–4
Korea Free Trade Commission investigation of, 234–5
US litigation and enforcement involving, 225–9
exclusionary practices
antitrust and, 18–19
in Chinese IP–antitrust regulations, 126–7
disclosure requirements and deception and, 236–8
exclusion payment, in reverse payment cases, 32–8
information technology foreclosure of competition and, 455–8
network effects on Internet and, 45–6
patent pooling and, 369–72
SAIC Regulation (China) concerning, 484–5
in US antitrust law, 222–4
exclusive customer contracts
antitrust policy and, 15–16
loyalty discounts and, 399–400
exclusive rights
EU exceptionalism and exercise of, 101–2
Murphy case on cable and satellite copyright and, 347–8
exclusive-supply contracts, loyalty discounts and, 399–400
exercise of right, in Japanese IP–antitrust law
establishment of, 143–4
output and price restrictions and, 148
restrictive licenses and, 147–51
tying and package licensing and, 149–50
unilateral refusal to license and, 146–7
exhaustion doctrine
cable and satellite copyright and, 347–8
EU copyright licensing and, 341–2
Japanese IP–antitrust law, 142–3
market definition and, 201
existence v. exercise dichotomy
EU antitrust and intellectual property policy and, 94–8
inherency in EU case-law and, 95–109
phase-out in EU case-law of, 97–8

exoneration, rule, EU IPR remedy cases and, 111–12
expert analysis, FRAND licensing and, 58–9
expired or exhausted patents. *See also* post-expiration royalties
  exploitation of post-expiration royalties and, 238–41
  Japanese IP–antitrust law concerning, 148–9
  Korean *Qualcomm* case involving, 173–4
  post-expiration royalties, 222–4
exploitative abuse
  applications of antitrust and, 85–91
  Chinese litigation involving, 229–31
  disclosure violations and deception as, 236–8
  excessive pricing as, 224–35
  future challenges in antitrust law concerning, 241–2
  of intellectual property, 222–42
  in US antitrust law, 222–4
ex post regulations
  complementarity theory and, 93, 106–8
  FRAND licensing and, 66–7
  Internet economics and, 54–5
  rule-based approach to EU antitrust and, 116–17
express collusion, pooling of patents and, 368–9
extra-territoriality
  country of origin principle for online transmission and, 349–50
  EU copyright law and, 342–4, 348–9
  in *Murphy* case, 345–6
  pay TV operators in EU and, 351–2
Ezrachi, Ariel, 302

Facebook, WhatsApp merger with, 308–10
Fair, Reasonable and Non-Discriminatory (FRAND) licensing
  balance in policies of, 62–4
  business model of SEPs and, 73
  Chinese IP–antitrust regulations and, 127–9, 229–31
  Chinese merger remedies involving, 473–4
  damages for SEP infringement, 73–5
  disclosure and deceptive violations and, 238
  dispute resolution and, 68–9
  emphasis on fair and reasonable in, 64–7
  EU litigation involving, 111–12, 231–3
  evolution of interpretations of, 64–71
  exceptionalism and particularism concerning, 103–4
  excessive pricing and obligations of, 224–35
  exploitative abuse issues and, 222–4, 241–2
  holdup of patents and, 247, 250–8
  Huawei v. InterDigital case in China and, 131–3, 478–80
  impetus for, 59–62
  injunctions as tool for, 71–2, 251–5
  Japanese litigation concerning, 233–4
  Korean IP and competition policies and, 158–9, 168–9, 173–4, 234–5
  litigation of SEPs and, 71–5
  in Microsoft/Nokia merger case (Korea), 177–8
  non-discrimination prong of, 69–70
  patent ambush cases and, 249–50
  patent assertion entities and, 263–5
  patent pooling and, 372
  post-expiration royalties and, 238–41
  royalty stacking and, 70–1
  in *Samsung Electronics* case, 176
  sham litigation doctrine and, 288–92
  standard setting organizations and, 58–9, 168–9, 219–20, 224–5
  US private litigation and government enforcement and, 225–9
fair use. *See also* usage
  copyright protection and, 210
  market definition and, 187
Farrell, Joseph, 65–6
Federal Communications Commission (FCC)
  AT&T-T-Mobile merger and, 458–62
  on bargaining power of mergers, 462–4
  Internet regulation and, 53
  merger remedies and, 451
  merger review process, 452–3
  net neutrality and, 53–4
  regulatory authority of, 56
Federal Trade Commission (US) (FTC)
  Big Data cases and, 308–12
  Dell computers case and, 249–50
  disclosure and deception violations and, 236–8
  drug patent settlements and, 320–1
  established manufacturers safe harbor, RPM litigation, 436–7
  FRAND licensing and, 58–9, 227
  horizontal mergers and, 460–1
  Japanese IP–antitrust policies compared with, 233–4
  merger jurisprudence and, 446–53
  merger review process, 452–3
  monopoly power screen and direct test of, 28–32
  patent ambush cases and, 64–7
  patent assertion entities and, 260–2
  reverse-payment drug patent settlements and, 319–20
  sham litigation doctrine and, 285–7, 290–1
feedback loops, Big Data management and, 294, 305–8
field of use restriction, Japanese law on, 147–51
First, Harry, 222–42

First Amendment rights, sham litigation doctrine and, 287–8
first consent principle, in Microsoft/Nokia merger case (Korea), 177–8
first sale doctrine. *See* exhaustion doctrine
Fisher, Franklin, 10–11, 47–9
fixation condition, copyright protection and, 210
fixed fees, in two-sided markets, 44–5
Food and Drug Administration (FDA)
  "authorized generics" and, 322–4
  brand and generic drug approval, 22–3, 25–6
  labeling of drugs by, 30–1
  patent litigation and, 286
  pharmaceuticals industry and, 21–2
Football Association Premier League Ltd. (FAPL) case, 112–15
foreign enterprises, Korean Free Trade Commission IPR applications for, 162
formularies
  monopoly power screen and role of, 31
  in pharmaceuticals industry, 23–5
Framework Act on Intellectual Property (Korea), 160
franchises, trademark protection and, 412–13
FRAND licensing. *See* Fair, Reasonable and Non-Discriminatory (FRAND) licensing
fraudulent intent, *Walker Process* doctrine and claims of, 274–7
free competition, reducing, in Japanese antimonopoly laws, 141–2
free riding behavior
  new entrants safe harbor, RPM litigation, 437–8
  resale price maintenance as, 427–31
FTC v. Actavis, 261–2, 321–2. *See also New York v. Actavis* case
  litigation cost issues in, 326–7
  merits of patent issue in, 324–5
  patent pooling and collusion in, 368–9
  payment for unrelated services in, 327–8
  payment issues in, 322–4
FTC v. Cephalon, Inc., 276–7, 324–5, 327–8
FTC v. Rambus, 61, 64–7. *See also Hynix Semiconductor Inc. v. Rambus Inc.*; *Rambus, Inc. v. Infineon Techs, AG*
FTC v. Watson Pharmaceuticals, 320–1
functional remedies, in Chinese merger litigation, 476–8
future competition, horizontal mergers and, 461
future profits of innovation
  determinants of, 7–9
  net present value, 205–6
future rate of innovation, incumbency advantage and, 3–6

Gans, Joshua S., 3–19
gatekeeping practices, in pharmaceuticals industry, 21–2
GE/Honeywell merger, 388–90
Genentech Inc. v. Hoechst GmbH, 115
General Electric, Shenhua Group joint venture with, 474–6
generic drugs
  "authorized generics," 322–4
  European patent settlements and, 332–5
  FDA approval of, 22–3
  Korean GSK anticompetitive patent decision and, 174–5
  litigation costs in defense of, 326–7
  market definition and, 189–91
  monopoly power screen and, 28–32
  payment for unrelated services in settlements involving, 327–8
  payment issues in drug patent settlements and, 322–4
  product hopping cases and, 38–9, 87–9
  regulations governing, 25–6
  reverse payment cases and, 26–7, 38–9, 89–90, 319–20
geo-blocking. *See also* block-booking practices
  copyright protection and, 339–41
  digital single market strategy (EU) and, 350–1
  European competition law and, 352–3
Georgia Pacific Corp. v. U.S. Plywood Corp., 67
Gifford, David, 379–401
Gilbert, Richard, 69–70, 365
Gilead Sciences, 215–16
Gilliam, Terry, 209–10
Ginsburg, Douglas H., 326–7
GlaxoSpain case, 342–4
good will defense, tied products/tying arrangements and principle of, 418–20
Google
  AdMob acquired by, 461
  DoubleClick acquisition by, 304–5, 308–10
  European litigation involving, 231–3
  FTC investigation of, 227, 233–4
  ITA Software acquisition by, 458
Google/Motorola merger (China), 470–4
grantbacks in patent law
  China IP–antitrust case-law concerning, 127–9, 133–5
  Japanese IP–antitrust law and, 150
Greenfield, Leon, 271–92
Greenlee, Patrick, 397
Green Paper of the European Commission. *See also* Cable and Satellite Directive (EU)
  on EU copyright rules, 348–9

GSK anticompetitive patent decision, Korean ISP law and, 174–5
GSK v. Commission case, 112–15
Guidelines for Patent and Knowhow Licensing Agreements (Japan), 142–3
Guidelines for the Use of Intellectual Property under the Antimonopoly Act (Japan), establishment of, 138
Gundlach, Gregory, 430–1

Handgards, Inc. v. Ethicon, Inc., 282–3
Handgards doctrine
  basic principles of, 271–2
  improper patent assertion and, 288–92
  litigation costs as antitrust injury in, 272
  sham petitioning activity and, 282–3
Hartford-Empire case, 371–2
Hart-Scott-Rodino Antitrust Improvements Act, 452–3
Hatch-Waxman litigation, 326–7
Hazlett, T.W., 49
Hemphill, Scott, 326–7
Henry v. A.B. Dick Co., 407–13
Herfindahl-Hirschman Index (HHI), 447–50
Heyer, Ken, 386–7
high technology. See information technology
Hoechst Marion Roussel case, drug patent settlement in, 320–1
Hoffmann-LaRoche v. Centrafarm, 95–109
Hokkaido Press case, 145
holdup of patents. See also licensee holdout
  damages as factor in, 255–8
  in FRAND disputes, 250–8
  future challenges concerning, 265–6
  injunctions for, 71–2, 251–5
  leveraging of royalty payments with, 59–62
  in Microsoft/Nokia merger case (Korea), 177–8
  overview of litigation concerning, 245–66
  patent ambush and, 64–7
  by patent assertion entities, 247, 258–65
  patent pooling and avoidance of, 360
  poor patent quality and, 261–3
  reverse holdup/holdout theories and, 62–4
  royalty stacking and, 70–1
  in standard-setting context, 249–58
  US litigation and enforcement involving, 227–9
Horizontal Cooperation Guidelines (European Commission), 231–3
horizontal mergers
  bargaining power of, 462–4
  Chinese SEPs mergers as, 470–4
  competitive effects of, 458–62
  US jurisprudence concerning, 447–50
Hovenkamp, E., 358–73, 393–4, 417–18, 438–9

Hovenkamp, H., 48–9, 52, 293–313, 326–7, 358–73, 393–4
Huawei v. InterDigital case, 131–3, 229–31, 256–7, 478–80
Huawei v. ZTE case, 103–4, 107–8, 111–12, 251–5
Hylton, Keith N., 81–91
Hynix Semiconductor Inc. v. Rambus Inc., 288–92. See also FTC v. Rambus; Rambus, Inc. v. Infineon Techs, AG
hypothetical monopolist test, merger horizontal transactions and, 447–50

IBM
  bundling and innovation at, 393
  product design protection and, 380–4, 386–7
IBM v. United States, 418–20
Illinois Tool Works Inc. v. Independent Ink, Inc., 87, 124–5, 416–17
Image Technical Services v. Eastman Kodak, 125–6
immunity from antitrust
  Internet providers, implied immunity of, 56
  Noerr-Pennington doctrine, 273
  sham litigation doctrine and, 281–8
"improvement" patents, 361–2
IMS Health case, 98–105, 110–11, 386, 391
incentive theory of copyright, 210
incremental value principle, FRAND licensing and, 65–6
incumbency
  accelerated product development and, 16
  antitrust policy impact on, 9–10
  cooperative commercialization and, 12–15
  exclusive customer contracts and, 15–16
  innovation and antitrust policies and, 3–6
  innovation prize and, 7–9
  mergers and, 17–18
  net neutrality and, 46–7
  network effects on Internet and, 45–6
  Schumpeterian competition and, 49–51
  static analysis of, 10–12
India
  drug patent settlements in, 335–6
  exploitative abuse of IP in laws of, 222–4
indirect network effects, in two-sided markets, 44–5
infant industries defense, tied products/tying arrangements and, 420
information technology
  bargaining power of mergers in, 462–4
  Big Data regulation and, 293–313
  blocking of patents in, 361–2
  bundled products and antitrust in, 379–401
  Chinese merger litigation concerning, 476–8
  competition foreclosure, incentive and ability for, 455–8

competitive market dynamics of, 453–5
conduct remedies concerning antitrust in, 451
discounts and rebates on bundled products and, 394–8
double marginalization, efficiencies, and portfolio effects in, 388–90
FRAND licensing and, 224–5
holdups of patents and, 245–9
Huawei v. InterDigital case in China and, 478–80
limits of antitrust involving, 384–7
merger review process for, 452–3
mergers in, 445–65
net neutrality and, 53–4
product design and, 380–4
Schumpeterian competition and, 49–51
tied products/tying arrangements and, 404–7
InfoSoc Directive, 341–2
infringement litigation
European "infringement by object" principle, 332–5
market definition and, 200–1
misuse doctrine and, 409
patent assertion entities use of, 258–61
patent protection and, 206–7
remedies and investment incentives and, 212
Walker Process doctrine and, 273–4
inherency, theory of
defined, 92
in EU antitrust and intellectual property, 94–8
exceptionalism compared with, 97–106
formative EU case-law on, 95–109
operational irrelevance, 96–7
phase-out in EU case-law of, 97–8
injunctions for patent infringement
copyright protection and, 210
FRAND licensing and, 71–2
holdups of patents and, 247–8, 251–5
in Korean IPR guidelines, 169
patent assertion entities and, 262
trade secrets law and, 207–8
innovation
antitrust law as limit on, 47–9, 384–7
Big Data impact on, 296–7, 301–3
in broadband markets, 51–2
bundled products and, 392–4
Chinese IP–antitrust regulations and, 122
costs of incumbency and, 8–9
dynamics modeling, 6–10
essential facilities doctrine in Chinese IP-antitrust regulations and, 125–6
horizontal merger impact on, 458–62
incentivization of, 83–5
intellectual property law as limit on, 47–9

in Japanese antimonopoly laws, 139–40
Korean IP law and protection of, 163
mergers' impact on, 16–18, 450
patent assertion entities and, 258–61
patent pooling and rates of, 372–3
in pharmaceuticals industry, 21–2
post-expiration royalties and, 238–41
Schumpeterian competition and, 49–51
size of prize of, 7–9
tied products/tying arrangements impact on, 404–7
innovation economics
antitrust/intellectual property limits on innovation and, 47–9
Internet and, 47–52
net present value and, 205–6
theory of, 3–19
In re Aggrenox Antitrust Litigation, 329–30
In re Buspirone Patent Litig./In re Buspirone Antitrust Litig., 286
In re Cardizem CD Antitrust Litigation, 320–1
In re Cipro Cases I & II, 331–2
In re Ciprofloxacin Hydrochloride Antitrust Litigation, 320–1
In re DDAVP Direct Purchase Antitrust Litig., 277–9
In re Effexor XR Antitrust Litigation, 329
In re K-Dur Antitrust Litigation, 320–1
In re Lamictal Direct Purchaser Antitrust Litigation, 329
In re Lipitor Antitrust Litigation, 329–30
In re Loestrin 24 FE Antitrust Litigation, 329
In re Niaspan Antitrust Litigation, 331
In re Robert Bosch case, 251–5, 291 n.110
In re Tamoxifen Citrate Antitrust Litigation, 320–1
Institute of Electrical and Electronics Engineers (IEEE)
FRAND licensing and, 61–2
IPR policy of, 71–5
patent holdup injunctions and policy update of, 251–5
institutional culture, in pharmaceuticals industry, 21–2
Insung Industry case, 172–3
integration of products. *See also* interoperability standards; technology transfer agreements
design protection and, 380–4
limits of antitrust principles concerning, 384–7
Microsoft lawsuits and level of, 10–12, 380–4
Intel Corporation, disclosure and deception violations by, 236–8
Intellectual Property Basic Act (Japan), 142–4

intellectual property law/intellectual property
  rights (IPL/IPR)
  abuse of dominance cases, 109–11
  additional abusive conduct principle and, 108–9
  Anti-Monopoly Law in China and, 121–9
  antitrust and, 18–19, 81–91
  in China, 120–36, 467–8, 478–86
  complementarity in EU concerning, 106–8
  complementarity in Japanese law on, 142–3
  coordinated conduct cases in EU and, 112–15
  copyright protection in Europe and, 342–4
  duration of protection, 211
  economics of, 83–5, 205–6
  essential facilities doctrine in EU law and, 390–1
  European law and, 92–117, 342–4, 352–3
  European Telecommunications Standards
    Institute policies and, 62–4
  exceptionalism in EU concerning, 98–106
  exemptions in Japanese law concerning, 143–4
  exploitative abuses of, 222–42
  FRAND licensing commitment to, 59–62
  history of tying arrangements and, 407–9
  information technology and limits of, 384–7
  innovation collaboration and, 52
  in Japan, 138–55
  Japan Fair Trade Commission "General
    Designation" and, 141–2
  Korean laws concerning, 158–78
  legitimate business defenses in violation of,
    418–21
  license pricing and terms and, 112
  limits to innovation and, 47–9
  litigation costs and, 326–7
  market definition and, 185–91, 196–201
  market miscalculation in, 194
  monopoly power and, 204–20
  in pharmaceuticals, 21–41
  pro- and anti- aspects of exceptionalism and,
    97–106
  refusal to license cases, 110–11
  remedies and investment incentives, 111–12, 212
  rule-based approach to, 109–17
  SAIC Regulation (China) and, 483–6
  sufficient economic power principle and, 415–17
  tied products and, 404–22
*Intellectual Ventures I LLC v. Capital One Fin.
  Corp.*, 262–5
intent, in *Walker Process* doctrine, 275–7
interbrand/intrabrand competition, market
  definition and, 184, 192–4
InterDigital corporation, 131–3
Interim Provisions on the Administration of
  National Standards Involving Patents
  (China), 128

intermediary platform, Internet two-sided markets
  and, 43–7
International Salt case, 409, 415–21
International Technology Transfer Guidelines,
  142–3
Internet. See also online markets
  broadband innovation and, 51–2
  Chinese laws concerning, 129–31
  data differentiation on, 300
  design protection and development of, 380–4
  economics of, 43–56
  ex post regulation of, 54–5
  holdups of patents and innovation on, 245–9
  implied immunity from antitrust and, 56
  innovation economics on, 47–52
  net neutrality and, 46–7
  nonexclusive/nonrivalrous content on, 299
  regulation and economics of, 52–6
  Schumpeterian competition and, 49–51
  streaming, EU transmission copyright and, 353
    n.64
  tied products/tying arrangements and, 404–7
  two-sided markets and network effects of, 43–7
"Internet of Things," holdups of patents and, 245–9
Internet Service Providers (ISPs)
  Chinese regulation of, 130–1
  as multi-homing agent, 44–5
  net neutrality and, 53–4
interoperability standards
  design protection and, 380–4
  FRAND licensing and, 59–62, 224–5
  holdup of patents and, 245–9
  monopoly power and, 219–20
inter partes review, patent assertion entities and,
  262–3
Interpretation of the Internet Rules (MIIT
  Interpretation) (China), 129–31
Investigative New Drug (IND) applications, 22–3
investment in innovation
  antitrust impact on information technology and,
    384–7
  holdup of patents and, 245–9
  incentives in IP for, 212
  net present value and, 205–6
  patent assertion entities and, 258–61
investors in pharmaceuticals, reverse payment
  settlements as reward for, 36–8
ITA Software, Google acquisition of, 458
ITT Promedia case, 102

Japan
  complementarity in IP–antitrust law of, 142–3
  excessive pricing litigation in, 233–4
  future IP–antitrust issues in, 155

Index

intellectual property and antitrust in, 138–55
non-challenge clause in patent law of, 150–1
patent and trademark application in, 145
patent infringement lawsuits in, 145–6
research and development restrictions in, 151
restrictive licenses regulations in, 147–51
royalty payment and calculation policies in, 151–2
tying and package licensing in, 149–50
unilateral refusal to license in IP–antitrust law of, 146–7
Japanese Society for Rights of Authors, Composers and Publishers (JASRAC), 151–2
Japan Fair Trade Commission (JFTC)
Antimonopoly Law (AML) Article 3 and, 139–40
Antimonopoly Law (AML) Article 19 and, 141–2
complementarity in rulings by, 142–3
FRAND licensing obligations and, 233–4
guidelines of, 138
Microsoft (NAP) case and, 153–4
Microsoft (Tying) case and, 149–50
patent and trademark application, 145
resale price maintenance rulings and, 148–9
Twentieth Century Fox Japan (TCFJ) case and, 148
Japan Patent Office (JPO), 145
Javico case, 342–4
JEDEC standards body, 64–7
Jefferson Parish Hospital No. 2 v. Hyde, 126–7, 384, 389–90, 410–13, 418–20
joint conduct
applications of antitrust and, 85–91
FRAND licensing negotiations and, 67
patent pooling and, 359–63, 371
in pay-for-delay and reverse payment cases, 26–7
royalty stacking and, 70–1
judicial economy principle, rule-based approach in EU antitrust and, 116–17

Kahn, Alfred, 54–5
Kaplow, Louis, 84–5, 195
Katz, M.L., 49–51
Kimble v. Marvel Entertainment, 87, 127, 238–41
King Drug Company of Florence v. Smithkline Beecham Corporation (Lamictal), 322–4
Kirkwood, John, 425–40
Klein, Benjamin, 429–33
knowledge, imputation of, in *Walker Process* doctrine, 275–7
Kobayashi, Bruce H., 326–7
Kobe, Inc. v. Dempsey Pump Co., 290
Kobec case, 172
Korah, V., 116–17

Korea
cases of IPR abuse in, 171–8, 222–4
competition law and intellectual property in, 158–78
drug patent settlements in, 335–6
future of IPA abuse regulation in, 178
KFTC IPR guidelines and, 161–71
post-expiration royalties cases in, 240–1
Korea Intellectual Property Office (KIPO), 160
Korean Fair Trade Commission (KFTC)
anticompetitive patent dispute agreements and, 169–70
cases of IPR abuse and, 171–8
cooperation with government IP agencies, 160
determination of illegality by, 163
Dolby unfair licensing practices case and, 177
drug patent settlements, 335–6
enforcement activities of, 158–9
GSK anticompetitive patent decision, 174–5
IPR abuse guidelines, 161–71
licensing terms in IPR guidelines, 164–7
Microsoft/Nokia merger and, 177–8
Monopoly Regulation and Fair Trade Act regulations and, 162–3
nonpracticing entities patent rights, 170
patent pooling and cross licensing and, 167–8
patent rights guidelines, 163–4
post-expiration royalties cases and, 240–1
proposed amendments to IPR guidelines, 170–1
Qualcomm case and, 173–4
Samsung Electronics SEP abuse case, 176
scope of application, 162
SK Telecom unfair licensing practices and, 175
standards-based approach to intellectual property and, 168–9
structure and activities of, 159–60
Krehl v. Baskin-Robbins Ice Cream Co., 412–13
Kudrle, Robert T., 379–401

Lambrecht, Anja, 297–300
Lampe, Ryan, 218–19
Lang, Temple, 108–9
language, in tied products/tying arrangements litigation, 410–13
Lao, Marina, 427–8, 431–3
large and unjustified payment threshold, in drug patent settlements, 329
Larouche, Pierre, 68–9
Layne-Farrar, Anne, 58–75
Lear v. Adkins, Inc., 127
Lee, Hwang, 158–78
Leegin case, 425–7

legality principle
  determination of illegality in Korean IP law and, 163
  discount attribution rule and, 396–7
  EU coordinated conduct cases and, 115
legal uncertainty, rule-based approach to intellectual property and, 116–17
legitimate expectations principle, patent holdup injunctions and, 251–5
Lemley, Mark A., 68–9, 183–202
LePage's Inc., 394–8
Lerner, Abba, 213–14
Lerner, Andres, 294, 305–8
Lerner, Josh, 62–4, 218–19
Lerner index of monopoly power, 213–14
  information technology competitive dynamics and, 453–5
Les Laboratoires Servier, 332–5
Leslie, Christopher R., 279–81, 404–22
leverage theory
  applications of antitrust and, 87
  holdup of patents and, 245–9
  product hopping cases and, 87–9
  royalty holdups and, 59–62
  tied products/tying arrangements and, 404–7
liability
  production integration cases and, 383
  rule-based approach to intellectual property and, 109–12, 116–17
licensee holdout, SEP holders and, 62–4
licensees, in patent pooling, 363
licensing. *See also* cross-licensing; grantbacks in patent law; package licensing; tied products/tying arrangements; unfair trade practices
  Chinese IP–antitrust regulations and, 125–7, 229–31
  in Chinese Patent Law, 129–30
  compulsory licensing, 108–9
  copyright, EU competition law and, 341–4
  disclosure requirements and deception concerning, 236–8
  EU digital single market strategy and, 339–55
  exceptionalism in refusal to license cases, 98–100
  FRAND licensing, 58–75
  "holdups" to leverage royalty payments, 59–62
  innovation and, 14–15
  IPR license pricing and terms, 112
  Japanese IP agreements concerning, 152–3
  Korean IPR abuse case-law and, 171–8
  Korean IPR guidelines and, 164–7
  Korean limitations on scope of, 165–6
  Murphy case on cable and satellite copyright and, 347–8
  patent assertion entities and, 258–61
  patent pooling and, 359–63
  post-expiration royalties and economics of, 238–41
  restrictive licenses regulations in Japan, 147–51
  territorial licensing, cable and satellite transmission copyright, 345
  unilateral refusal to license, Japanese IP–antitrust laws and, 146–7
likelihood of confusion test, trademark registration and, 199–200
Lim, Daryl, 245–66
limited-license theory, in EU antitrust and intellectual property, 92 n.1
limited period of protection, Japanese IP–antitrust law, 142–3
Line Materials case, 361–2, 368–9
litigation costs
  as antitrust injury, 272 n.4
  rule of reason in drug patent settlements and, 326–9
Litton Systems v. Am. Tel. & Tel. Co, 285–7
Liyang Hou, 467–87
Llobet, Gerard, 65–6, 73–5
Loewe's case, 415–17
long-term therapies, monopoly power screen and, 31–2
loss of quality theory, Big Data and, 301–3
lost profit damages, market definition and, 185–6
"lottery-ticket" patent assertion entities, 260–1
loyalty discounts, for bundled products, 398–401
Lundbeck case, 102–3
Lundbeck Company, 332–5

Magill case, 97–106, 110–11, 116–17, 390–1
Manne, G.A., 49
Manning, Kenneth C., 430–1
manufacturers
  online resale price maintenance impact on, 427–34
  safe harbor for, in resale price maintenance, 436–7
marginal cost of production
  antitrust impact on information technology and, 384–7
  Big Data creation and, 298–9
  Lerner index of monopoly power and, 213–14
  market definition and, 194
  monopoly markups and demand elasticity, 215–16
  patent pooling exclusivity and, 369–72
  welfare consequences of monopolization and, 82–3
Mariniello, Mario, 66–7
market-by-market assessment, pharmaceutical mergers and, 39–41

Index

market definition/market power. *See also* two-sided markets
  antitrust and, 85–91, 188–9
  Big Data and, 310
  boundary delineation in patent pooling and, 365–6
  broadband innovation and, 51–2
  bundling and innovation and, 392–4
  case-law on IP and, 192–4
  Chinese control of non-SEP related mergers and, 474–6
  Chinese IP–antitrust regulations and, 122–3
  Chinese merger control and assessment of, 472–3
  copyright protection and, 186–7, 189–91, 198–9
  design protection cases and, 384
  in drug patent cases, 321–2
  established manufacturers safe harbor, RPM litigation, 436–7
  exhaustion and first sale doctrines, 201
  FRAND licensing and, 61
  future challenges for, 201–2
  holdup of patents and, 248–9
  horizontal transactions in mergers and, 447–50
  implicit definition and, 187–8
  information technology and antitrust policy and, 384–7
  information technology competitive dynamics, 453–5
  injunctions on patent infringement and, 71–2
  innovation and, 3–6
  intellectual property doctrine and, 185–91, 196–201
  Japanese antimonopoly protection of, 139–40
  in Korean IFP law, 158–9, 162–3
  limitations of, 194–6
  merger horizontal transactions and, 447–50
  Microsoft lawsuit and, 10–12
  miscalculation in IP of, 194
  monopoly power and, 215
  net neutrality and, 53–4
  overview of, 183–202
  patent ambush and, 64–7
  patent assertion entities and, 263–5
  patent pooling and, 367–72
  in pharmaceuticals industry, 21–2
  product differentiation and, 188–94
  Qualcomm case in China and, 480–1
  rate of innovation and, 3–6
  regulatory limits to innovation and, 47–9
  resale price maintenance in online markets and, 427–8, 435
  scope of patent rights and, 200–1
  standard essential patents and, 471–2
  sufficient economic power principle and, 415–17
  trademark registration and, 185, 199–200
market share
  information technology foreclosure of competition and, 455–8
  innovation and antitrust policies and role of, 4–5
  Japanese antimonopoly protection of, 139–40
  loyalty discounts for bundled products and, 398–401
  merger horizontal transactions and, 447–50
  Schumpeterian competition and, 49–51
Masimo Corp. v. Tyco Health Care Group, 398–401
Mattel suit, 186–7
McDonough case, 429–31
MCI v. AT&T, 125–6
McKenna, Mark P., 183–202
Measures for Compulsory Licensing of Patent Implementation (China), 129–30
media rating services, horizontal mergers of, 460–1
medical best practices, off-label drug use and, 30–1 n.17
MedImmune test, *Walker* Doctrine enforcement and, 279–81
Mediplorer case, 148–9
Merck KGaA–AZ Electronics merger, 474–6
Mercoid Corp. case, 409
mergers
  antitrust impact on, 13, 16–18
  bargaining power of IT mergers, 462–4
  Chinese IPR and control of, 467–87
  data-driven defenses of, 304–5
  efficiencies in, 450–1
  Facebook/WhatsApp merger, 308–10
  holdup of patents and, 256–7
  horizontal mergers, 447–50
  horizontal transactions and, 447–50
  as industry maneuvers, 18–19
  information technology competitive dynamics and, 453–5
  in information technology sector, 445–65
  IPR-related abuses in China and, 478–86
  market definition and, 195
  Microsoft/Nokia merger case (Korea), 177–8
  non-SEP related mergers, Chinese control of, 474–8
  in pharmaceuticals industry, 39–41
  portfolio effects doctrine and, 388–90
  remedies in jurisprudence on, 451
  review process for, 452–3
  SEPs-related mergers in China and, 470–4
  US jurisprudence concerning, 446–53
  vertical transactions in, 450
Micro Leader Business v. Commission, 98–100

Microsoft (NAP) case, 153–4
Microsoft (Tying) case, 149–50
Microsoft Corp v. Motorola, 61, 70–2, 226–7
Microsoft lawsuits, 10–12
  damage to company from, 49
  design protection in technology and, 380–4
  essential facilities doctrine in, 390–1
  EU litigation of, 386, 404–7
  exceptionalism in, 98–100
  limits of antitrust principles in, 384–7
  SAIC Regulation (China) concerning, 484–5
  standards-based approach in, 110–11
  tied products/tying arrangements in, 404–7
Microsoft/Nokia merger, 170–1, 177–8, 470–4
Miller International, 342–4
Ministry of Commerce (China) (Mofcom)
  cases reviewed by, 469–70
  establishment of, 468–70
  General Electric-Shenhua Group joint venture and, 474–6
  market power assessment in merger control by, 472–3
  remedies for mergers and, 473–4
  SEPs-related mergers and, 470–4
misleading promotions, online resale price maintenance facilitation of, 433–4
misuse doctrine
  bundling and innovation and, 392–4
  tying arrangements and, 408
mixed bundling, discounts and rebates and, 394–8
modularity, information technology competitive dynamics, 453–5
monitoring costs, online resale price maintenance and lowering of, 431–3
monopoly power
  Chinese Patent Law and, 129–30
  competition and, 215–20
  compulsory licensing and, 216–18
  cooperative commercialization and, 13–14
  copyright protection and, 210
  cross-licensing agreements and, 219
  design protection and, 380–4
  disclosure and deception violations and, 236–8
  duration of IP protection and, 211
  economics of antitrust and, 82–3
  exploitative abuses of IP and, 222–4
  intellectual property and, 204–20
  Japanese restrictions on, 139–40
  Lerner index of, 213–14
  market power and, 215
  markups and demand elasticity, 215–16
  Microsoft lawsuits and, 10–12, 380–4
  network effects and, 45–6
  patent pooling and, 218–19
  patent protection and, 206–7
  in pharmaceuticals industry, 26–39
  properties of, 215
  rule of reason and, 28
  Schumpeterian competition and, 49–51
  screen, application of, 28–32, 38–9
  standard setting organizations and, 219–20
  trademark registration and, 210–11
  trade secrets law and, 207–8
  tying arrangements and, 406–7
  welfare consequences of, 82–3
Monopoly Regulation and Fair Trade Act (MRFTA) (Korea)
  determination of illegality in, 163
  excessive pricing investigations and, 234–5
  interagency cooperation and, 159–60
  IPR abuse cases and, 173
  IP regulation under, 160–1
  Korean Free Trade Commission IPR guidelines and, 162–3
  post-expiration royalties cases and, 240–1
Morton Salt Co. v. G.S. Suppiger Co., 409
Moser, Petra, 218–19
Motion Picture Patents v. Universal Film Mfg. Co., 407–8
Motorola
  European litigation involving, 231–3
  FTC investigation of, 227, 233–4
  patent holdup disputes and, 251–5
MPEG-LA pool licenses, 363
MPHJ Technology Investments, 260
multi-homing agents
  Big Data and, 299
  two-sided markets and, 44–5
Murphy case, 339–44
  EU competition law and, 354
  exhaustion doctrine in interpretation of, 347–8
  facts and ruling in, 345–6
  scope of, 351–2

Nalebuff, Barry, 397
Nash bargaining
  joint negotiations with standard setting organizations, 67
  license fees and costs of innovation, 14–15
National Development and Reform Commission (NCRC) (China)
  establishment of, 120–1, 468–70
  holdup of patent violations and, 256–7
  investigation of Qualcomm by, 131, 133–5, 480–1
  IP–antitrust regulatory practices of, 121–9, 229–31
  IPR-related abuses in mergers and, 478–86
  post-expiration royalties cases and, 240

Index

National Disease and Therapies Index (NDTI) dataset, 31
Nature of the Firm (Coase), 364–5
Negotiated Data Solutions (N-Data)
  FTC investigation of, 227–9, 233–4
  patent ambush in, 249–50
net neutrality
  FCC and, 53–4
  Internet economics and, 46–7
net present value, innovation economics and, 205–6
Netscape, Microsoft lawsuit and, 10–12, 380–4, 386–7
network effects
  antitrust and role of, 45–6
  antitrust impact on information technology and, 384–7
  of Big Data, 305–8
  information technology competitive dynamics and, 453–5
  Internet two-sided markets and, 43–7
  mergers and, 389–90
  net neutrality and, 46–7
  Schumpeterian competition and, 50–1
network goods, innovation and antitrust policies and, 4–5
New Chemical Entity (NCE), FDA process for, 22–3 n.4
New Drug Application (NDA), 22–3
  reverse payment cases and, 26–7
new entrants, resale price management and safe harbor for, 437–8
new products defense, tied products/tying arrangements and, 420
New York v. Actavis case. *See also FTC v. Actavis*
  anticompetitive conduct in, 26–7
  product hopping in, 87–9
  reverse payment principle in, 89–90
Nicsand, Inc. v. 3M Co., 399–400
Nielson/Arbitron case
  antitrust policy and, 308–10
  horizontal merger effects in, 460–1
Nine West case, 436–7
no-authorized-generic (no-AG) promises, drug patent settlements and, 322–4
Nobelpharma case, 275–7
no-contest/no-challenge clauses, NDRC Qualcomm case in China and, 135, 480–1
Noerr-Pennington doctrine, immunity from antitrust and, 271–92
Nokia/Alcatel merger, 470–4
  Chinese review of, 256–7
non-assertion of patent (NAP) obligations,
  Japanese IP–antitrust rules and, 153–4, 233–4

non-challenge clause, in Japanese patent law, 150–1
non-discrimination in FRAND licensing, 69–70
  damages in patent holdups and, 255–8
  Korean focus on, 234–5
nonexclusionary content
  of Big Data, 299
  grant-back obligations and, 150
  patent pooling and, 369–72
  standard setting organizations and, 219–20
nonobviousness requirements
  in patent protection, 206–7
  product hopping and, 88–9
nonpracticing entities (NPEs)
  FRAND licensing and, 73
  Korean IPA regulations and, 158–9, 178
  Korean patent protections for, 170
nonprice competition, privacy as, 303–4
non-replicable assets, information technology competitive dynamics and, 453–5
nonrivalrous content and products
  Big Data and, 299
  patent pooling and, 368–9
Notice on Institutional Establishment (China), 468–70
Novel Corporation, complaints against Microsoft by, 380–4, 386–7
novelty requirement in patent protection, 206–7
Nungesser and Eisele case, 112–15, 342–4

Obama, Barack, 257–8
OEMs
  design production in computers and, 382–3
  disclosure and deception violations by, 236–8
off-label drug use
  manufacturers' research sponsorship for exploring, 32 n.19
  monopoly power screen and, 30–1
Ohlhausen, Maureen, 294–5
Oihoo v. Tencent, 131
Okuliar, Alexander P., 294–5
online markets. *See also* Internet
  Big Data role in, 294–7
  country of origin principle for transmissions, 349–50
  EU copyright rules and, 348–9
  EU digital single market strategy and, 339–41
  European competition law and transmission of content in, 352–3
  harm to competition from Big Data and, 300–8
  horizontal mergers and, 458–62
  limits of Big Data on, 299–300
  mobile online advertising and, 461

online markets. (cont.)
  Murphy case on cable and satellite copyright and, 344–8
  platform differentiation in, 300
  resale price maintenance on, 425–40
open access remedies, information technology and, 451
open-ended content of exceptionalism, 104–5
open source licensing arrangements, 358–9
Ordover, J., 10
originality condition, copyright protection and, 210
Ortho Diagnostic Systems, Inc. v. Abbott Laboratories, Inc., 394–8
Ostrom, Elinor, 364–5
Ottung case, 115
output
  complementarity of products and, 388–90
  express collusion and limitations on, 368–9
  Japanese patent protections on, 148
  patent pooling exclusivity and, 369–72
  resale price maintenance impact on, 434 n.35
  Schumpeterian competition and, 49–51
overuse of resources, health care financing and, 24 n.5
Owen, Brude, 46–7
ownership rights
  copyright protection and, 210
  limits on innovation and, 48–9

Pachinko Patent Pool case, 154–5
package licensing
  Huawei v. InterDigital case in China and, 131–3
  Japanese IP–antitrust policy and, 149–52
Padilla, Jorge, 67, 73–5
pan-EU licensing, country of origin principle for online transmission and, 349–50
parallel trading, European competition law and, 342–4
Paramount Bed case, 145–6
Parke David case, 95–109
Parker, G.G., 44–5
particularism, exceptionalism and, 103–4
Patent Act (Japan), 143–4
Patent Act (Korea), Monopoly Regulation and Fair Trade Act regulations and, 162–3
Patent Act (US), patent pooling and, 367–72
patent aggregators, 260–1
patent ambush
  disclosure violations and, 222–4, 236–8, 245–9
  FRAND licensing and issues of, 64–7
  future challenges concerning, 265–6
  in standard setting context, 249–50
patent-as-property theory, 89
patent assertion entities (PAEs), 73

antitrust responses to, 263–5
holdups by, 247, 258–65
judicial and legislative responses to, 261–3
market function of, 257–8
standard essential patents switch to, 257–8
Patent Litigation Toolkit, 261–3
patent protection. *See also* holdup of patents; pooling of patents; standard essential patents (SEPs)
  abuse of patent actions, Korean IP law on, 164
  anticompetitive dispute settlement in Korea and, 169–70, 174–5
  applications of, 85–91
  boundary delineation and, 365–6
  in Chinese IP–antitrust regulations, 121–30
  disclosure and deception violations and, 236–8
  drug patent settlements, 319–36
  duration of, 211
  exclusion payment, in reverse payment cases, 33 n.22
  expired or exhausted patents, Japanese restrictions on, 148–9
  FDA brand and generic drug approval and, 22–3
  four-part test for infringement of, 73
  grantback obligations, 127–9, 133–5, 150
  inclusion in standards and, 59–62
  inherency in EU case-law concerning, 95–109
  injunctions for infringement of, 71–2, 145–6
  in Japanese IP–antitrust policies, 145–6, 150–1
  Korean IPR guidelines concerning, 163–4, 168–70, 174–5
  limits on innovation from, 48–9
  litigation costs of, 326–7
  market definition and, 36, 185–6, 200–1
  monopoly power and, 206–7
  non-assertion of patent obligation and, 153–4
  non-challenge clause in Japanese law and, 150–1
  nonpracticing entities in Korea and, 170
  patent pools, Chinese IP–antitrust regulations and, 127–9
  in pharmaceuticals industry, 21–2
  post-expiration royalties and, 238–41
  restrictive licensing in Japan and, 147–51
  reverse payment cases and, 26–7, 32–8
  royalty stacking and, 70–1
  scope of patent rights, 200–1
  sham litigation doctrine and, 271–2, 281–8
  SK Telecom unfair licensing practices case (Korea), 175
  standards-based approach to, 168–9
  sufficient economic power principle and, 415–17
  tying arrangements and, 407–13
  Walker Process doctrine and, 272–81
  welfare effects of, 83–5

patent terminators, *inter partes* review and emergence of, 262–3
patent-term split agreements, reverse-payment drug patent settlements and, 319–20
Patented Medicine Notice of Compliance Regulations (PM(NOC)) (Canada), 335–6
pay-for-delay cases, anticompetitive joint conduct and, 26–7
payments in drug patent settlements
  cash payments, 322–4
  causation and, 331
  European settlements, 332–5
  large and unjustified payment threshold and, 329
  merits of patent and, 324–5
  pleading requirements, 329–30
  for unrelated services, 327–8
payment systems
  digital single market strategy (EU) and, 350–1
  payment card market, two-sided markets and, 43–7
  pharmaceuticals industry and role of, 21–2
pay-per-click business model
  "click-and-query data," 296–7
  impact on quality and innovation of, 301–3
  network effects and, 305–8
pay TV services, EU investigation of, 351–2
People's Republic of China
  case-law involving IP–antitrust interface in, 131–5
  de facto standards for non-SEP related mergers in, 474–6
  future of IP–antitrust interface in, 135–6, 486–7
  holdup of patent violations in, 256–7
  Internet regulation in, 129–31
  IP–antitrust interface in, 120–36
  IPR-related abuses and merger control in, 478–86
  market power assessment in merger control by, 472–3
  merger control and IPR in, 467–87
  non-SEP-related mergers in, 474–8
  patent-involved mergers in, 476–8
  patent law in, 129–30
  post-expiration royalties cases in, 240
  remedies for merger violations in, 473–4
  SEPs-related mergers in, 470–4
perfectly competitive market, applications of antitrust and, 85–91
peripheral devices, design protection and product integration and, 380–4
per-period profits, antitrust policies and innovation and, 12, 15–16
per se rule
  bundling and innovation and, 392–4

coercion in tying arrangements and, 413–15
design protection in high tech products and, 384
EU coordinated conduct cases and, 112–15
patent holdup injunctions and, 251–5
portfolio effects of mergers and, 389–90
resale price maintenance and, 425–7, 434–9
tying arrangements and tied products and, 409–18
Petit, Nicolas, 92–117
Pfizer/Wyeth merger, 40 n.36
pharmaceuticals industry
  antitrust and intellectual property issues in, 21–41
  brand/generic regulations, 25–6
  competitive effects and monopolization in, 26–39
  coordinated vs. single firm conduct, analysis of, 26–39
  drug patent settlements in, 319–36
  econometric analysis of, 25 n.7, 30 n.17
  exceptionalism in EU case-law concerning, 101–2
  FDA brand and generic drug approval and, 22–3
  Korean ISP law and, 174–5
  market definition in, 189–91
  mergers in, 39–41
  monopoly markups and demand elasticity, 215–16
  monopoly power screen in, 28–32
  product development and improvement in, 21–6
  product hopping cases in, 38–9, 87–9
  reverse payment cases in, 32–8, 89–90, 319–20
  sham litigation doctrine and, 286
  third-party payors and principal/agent problem for, 23–5
Philips case, blocking of patents in, 361–2
physician as decision-maker, pharmaceuticals industry and role of, 21–2, 31
pipeline products, pharmaceutical mergers and development of, 39–41
platform properties
  bargaining power of IT mergers and, 462–4
  Big Data and differentiation in, 300
  EU bundling and innovation policies and, 386
  intermediary platforms, 43–7
  third-party platforms, 298–9
  two-sided platforms, 301–6
pleading requirements, in drug patent settlements, 329–30
pooling of patents
  blocking of patents and, 361–2
  boundary delineation and defense, cost reduction relating to, 364–6
  Chinese regulations concerning, 127–8

pooling of patents (cont.)
  collusion threats and, 367–72
  complements vs. substitutes, 359–63
  exclusivity and, 369–72
  express collusion and, 368–9
  FRAND agreements and, 372
  innovation rates and, 372–3
  Japanese licensing agreements and, 152–5
  Korean IPR guidelines and, 167–8
  limitations on complement/substitute distinction, 363
  market divisions and, 371–2
  monopoly power and, 218–19
  pricing vs. conventional bundling, 366–7
  royalty stacking and holdup, avoidance of, 360
  rule of reason and, 358–9, 371
  SAIC Regulation (China) concerning, 484–5
  technology sharing and, 358–73
  tied products and, 362–3
portfolio effects doctrine
  mergers and, 388–90
  pharmaceuticals industry, 39–41
Posner, R., 109–10, 322–4
post-entry behavior
  innovation and antitrust policies and, 4
  profits of innovation, 7–9
post-expiration royalties
  Chinese licensing restrictions and, 127
  prevention of exploitation and, 222–4, 238–41
post hoc standards, rule-based approach to EU antitrust–IP intersection and, 109–17
PowerReviews, 304–5
predatory behavior, entrant innovation and, 10
pre-installed software, Microsoft lawsuit and presence of, 10–12
Presidential Council on Intellectual Property (PCIP), 160
presumption of illegality
  merger policy and, 445–6
  for online resale price maintenance, 427–34, 438–9
  in reverse payment cases, 37 n.35
price/cost test, loyalty discounts and, 399–400
price protection clause, in tied products/tying arrangements, 421
pricing. *See also* resale price maintenance (RPM)
  bargaining power of IT mergers concerning, 462–4
  bundling and innovation and discrimination in, 393–4
  coercion in tying arrangements through, 413–15
  complementarity of products and, 388–90
  discounts and rebates on bundled products, 394–8

excessive pricing, FRAND obligations and, 224–35
Japanese patent protections on, 148
loyalty discounts and, 398–401
market power and, 85–91
net neutrality and discrimination in, 53–4
non-discrimination in FRAND licensing and, 69–70
in patent pools, 366–72
patent protection and, 83–5
in pharmaceuticals industry, 23–5
resale price maintenance impact on, 434 n.35
Schumpeterian competition and, 49–51
two-sided markets and, 43–7
tying arrangements and discrimination in, 406–7
welfare consequences of monopolization and, 82–3
Prieger, James, 55
principal/agent problem, in pharmaceuticals industry, 23–5
privacy, Big Data threats to, 303–4, 310–12
private litigation
  excessive pricing abuses and, 225–9
  patent assertion entities and, 258–61
  post-expiration royalties and, 238–41
private monopolization, Japanese prohibition of, 139–40, 222–4
private property rights, resource pooling and, 364–6
prize of innovation
  antitrust policy impact on, 9–10
  determinants of, 14–15
  size of, 7–9
procompetitive effects
  of Big Data, 296–300
  data monetization and targeted advertising, 296
  in Japanese antimonopoly laws, 139–42
  Korean intellectual property and competition policies and, 158–9
  Korean limitations on scope of licensing and, 165–6
  of mergers, 388–90
  network effects and, 45–6
  patent pooling and, 369–72
  pharmaceutical antitrust cases, 28
  of resale price maintenance, 425–7, 431–3
  reverse payments and, 32–8
  of tied products/tying arrangements, 409–18
product development and improvement
  Big Data and, 296–7
  FDA brand and generic drug approval process and, 22–3
  in high technology industry, 380–4

incumbent accelerated product development, 16
innovation and, 6–10
net present value of innovation and, 205–6
pharmaceutical mergers and, 39–41
in pharmaceuticals industry, 21–2
product differentiation
  antitrust and IP and, 183–4
  antitrust impact on information technology and, 384–7
  bundled products discounts and rebates and, 394–8
  market definition and, 188–94
product hopping cases
  anticompetitive unilateral conduct and, 26–7
  competitive effects of, 38–9
  leveraging theory and, 87–9
  monopoly power screen in, 28–32, 38–9
  patent-as-property theory and, 89
  rule of reason in, 28
product incompatibility, Chinese regulation of Internet services and, 130–1
product integration
  design protection and, 380–4
  Microsoft lawsuits and level of, 10–12, 380–4
product subsidization, Big Data monetization and, 296
Professional Real Estate Investors (PRE) case, 271–2
  Handgards doctrine and, 282–3
  non-patent anticompetitive litigation and, 288–92
profits
  antitrust policy impact on, 15–16
  duration of IP, 211
  Lerner index of monopoly power and, 213–14
  net present value of innovation and expectation of, 205–6
  patent pooling and boundary delineation and, 364–5
  period-by-period profits, antitrust policies and innovation and, 12
promotional allowances, resale price maintenance in online markets and, 427–8, 431–3
Protégé International v. Commission, 98–100, 102
providers of Internet information services (PIIS), Chinese regulation of, 130–1
public display, copyright protection and, 210
public interest
  essential facilities doctrine in Chinese IP-antitrust regulations and, 125–6
  Japanese antimonopoly protections for, 139–40
  patent protection and, 271–2
  production integration and design protection cases and, 383–4

Publicis/Omnicom merger, 304–5
public performance
  copyright protection and, 210
  exhaustion doctrine and, 341–2
public tendering process
  Japanese patent infringement cases and, 145–6
  in Korean cases of IPR abuse, 171–8

QPX airfare pricing and shopping system, 458
Qualcomm
  Chinese investigation of, 131, 133–5, 229–31, 256–7, 480–1
  European investigation of, 231–3
  Japanese litigation involving, 233–4
  Korean ruling on SEP abuse by, 173–4, 234–5, 240–1
  patent ambush violations by, 249–50
  patent holdup injunctions, 251–5
Qualcomm Inc. v. Broadcom Corp., 225–9, 249–50
quality control in innovation
  Big Data impact on, 296–7, 301–3
  patent holdups and, 261–3
  tied products/tying arrangements justification using, 418–20

Rainbow Scape case, 172
Rambus, Inc. v. Infineon Techs, AG, 222–4, 236–8, 249–50. *See also FTC v. Rambus*; *Hynix Semiconductor Inc. v. Rambus Inc.*
Ramirez, Edith, 460–1
rate determination and disclosure, FRAND licensing and, 61–2, 68–9
reasonable apprehension test, *Walker* Doctrine enforcement and, 279–81
reasonable royalty framework
  damages in patent holdups, 255–8
  excessive pricing and, 224–5
  FRAND licensing and, 67
  Korean IPR licensing conditions and, 166–7
  market definition and, 186
  patent disclosure and deceptive practices and, 65–6
  patent holdup injunctions and, 251–5
  patent infringement and, 206–7
rebates
  of bundled products, 394–8
  conditional rebates, 398–401
  to third-party payors, pharmaceuticals practice of, 23–5
refusal to license cases
  exceptionalism in, 98–100
  Japanese IP licensing agreements and, 152–3
  Korean IPR licensing guidelines and, 165
  rules-based approach to, 110–11

refusal to license cases (cont.)
    SAIC Regulation (China) and, 484–5
Regibeau, P., 116–17
Regulation of Abusing Intellectual Property Rights to Eliminate or Restrict Competition (China), 483–6
Regulation on the Prohibition of Conduct Eliminating or Restricting Competition by Abusing Intellectual Property Rights (China), 471–2
regulatory policies
    broadband innovation and, 51–2
    EU competition law and, 352–3
    ex post regulations, Internet economics and, 54–5
    innovation and antitrust policies and, 3–6
    Internet economics and, 52–6
    merger jurisprudence in US and, 446–53
    net neutrality and, 53–4
    in pharmaceuticals industry, 21–2
Reitman, David, 397
remedies in antitrust
    Big Data and appropriateness of, 311–12
    Chinese remedies in merger control, 473–4
    in EU IPR case-law, 111–12
    in IP protection, 212
    merger remedies, 451
rent extraction
    antitrust policy and, 15–16
    cooperative commercialization and, 12–15
    EU cases involving, 233–4
    exploitative abuse of IP and, 222–4
    in information technology, 386–7
    innovation and antitrust policies and, 3–6
    merger analysis of, 16–18
    patent infringement and, 225–9
    simpliciter protections and, 238
reproduction rights, copyright protection and, 210
resale price maintenance (RPM)
    burdens of proof and production in litigation concerning, 435
    in Chinese IP–antitrust regulations, 122–4
    future legal issues concerning, 440
    Japanese expired or exhausted patent restriction, 148–9
    in Japanese IP–antitrust regulations, 141–2
    justifications for, 438–9
    legal precedent concerning, 425–7
    new entrants safe harbor and, 437–8
    online markets and, 425–40
    presumption of illegality concerning, 427–34
    rule of reason and, 425–7, 434–9
    safe harbor provisions concerning, 435–8

resale services, online resale price maintenance facilitation of, 433–4
research and development (R&D)
    entrant investment in, 6–10
    in high tech industries, 386 n.23
    Japanese IP–antitrust policies and, 151
    merger impact on, 17–18
    pharmaceutical mergers and, 39–41
    in pharmaceuticals industry, 21–2
    Schumpeterian competition and, 49–51
resource allocation, product development and improvement and, 6–10
resource pooling, boundary delineation in, 364–6
restrictive licenses, Japanese regulation of, 147–51
reverse free riding, resale price maintenance and, 430–1
reverse holdup/holdout theories, FRAND licensing and, 62–4
reverse payment cases
    anticompetitive joint conduct and, 26–7
    antitrust law and, 89–90
    in Canada, 335–6
    cash payments, 322–4
    causation and, 331
    competitive effects in, 32–8
    drug patent settlements, 320–1
    European settlements, 332–5
    FTC v. Actavis ruling and, 321–2
    illegality presumption in, 37 n.35
    in India, 335–6
    in Korea, 335–6
    large and unjustified payment threshold in drug patent settlements, 329
    merits of patent and, 324–5
    monopoly power screen in, 28–32
    pleading requirements in, 329–30
    rule of reason analysis, 28
    state law and, 331–2
    stock market rewards for settlement of, 36–8
    US drug patent settlements, 319–20
Review Guidelines for Unfair Trade Practices (Korea), 162–3
Rhone-Poulenc case, 284
right of communication to the public
    EU copyright licensing and, 341–2
    EU transmission copyright and, 353 n.64
Ringtone case, 154–5
risk aversion and discounting
    antitrust impact on information technology and, 384–7
    drug entry dates and, 32 n.21
    instantaneous vs. delayed entry and, 33–4
Ritz Camera & Image, LLC v. SanDisk Corp., 277–9

Robinson-Patman Act, 386 n.23
robust functionality doctrine, trademark registration and, 199–200
Rockett, K., 116–17
Royal Information Technology Corporation (RITCO) case, 173
royalties. *See also* post-expiration royalties
  Brulotte rule and, 87
  Chinese IP–antitrust regulations and, 121–9, 133–5, 229–31
  damages for SEP infringement and, 73–5
  disclosure and deception violations and, 236–8
  European litigation concerning, 231–3
  excessive pricing abuses and, 225–9
  "holdups" of licensing leverage payments, 59–62
  in *Huawei v. InterDigital* case, 478–80
  Japanese IP protections and, 151–3, 233–4
  Korean IPR licensing guidelines and, 164–5
  patent pooling, avoidance of stacking and, 360
  Qualcomm case in China and, 480–1
  royalty-free licensing, 58 n.1
  stacking of, 70–1, 224–5, 245–9, 251–5, 360
rule-based approach
  abuse of dominance cases in EU and, 109–11
  in antitrust cases involving IPRs, 116–17
  coordinated conduct cases, 112–15
  to EU antitrust–IP intersection, 109–17
  in EU IPR remedy cases, 111–12
  IPR license pricing and terms, 112
rule of reason
  burdens of proof and production in RPM litigation and, 435
  in drug patent settlements, 321–2, 326–9
  patent pooling challenges and, 358–9, 371
  pharmaceutical monopoly behavior and, 28
  resale price maintenance and, 425–7, 434–9
  tied products/tying arrangements and, 421–2
Rules on Regulating the Market Order of Internet Information Services (MIIT Internet Rules) (China), 129–31
Rysman, M., 43–7, 60

safe harbor principle
  bundled products discounts and rebates and, 396–7
  in Chinese IP–antitrust regulations, 123–4
  established manufacturers safe harbor, RPM litigation, 436–7
  FRAND licensing and, 66–7
  in Japanese antimonopoly laws, 139–40
  Korean intellectual property and competition policies and, 158–9
  new entrants safe harbor, RPM litigation, 437–8

  resale price maintenance litigation and, 435–8
  Walker Process doctrine and, 279–81
sales contracts, tied products/tying arrangements and, 413–15
Samsung Electronics case, standard essential patents abuse and, 176
satellite transmission
  copyright aspects of, 344–5
  EU Cable and Satellite Directive and, 344–8
  Murphy case facts and ruling and, 345–6
Scarlett Extended case, 107–8
Schering-Plough Corp. v. FTC, 320–1
Schmalensee, R., 3–6, 10–12, 16–18, 67, 384–7
Schmidt, H., 97–106
Schumpeter, Joseph, 3, 49–51
scope of patent principle
  drug patent settlements and, 320–1
  Korean licensing regulations and, 160, 165–6
  market power and, 200–1
  rejection in drug patent cases of, 321–2
Scott Morton, F. and Shapiro, C., 73
Segal, I. *See* Segal and Whinston innovation model
Segal and Whinston innovation model, 5–6
  antitrust evaluation procedures and, 18–19
  dynamics modeling and, 6–10
  static analysis of, 10–12
self-interest seeking, holdup of patents and, 61
SEPs. *See* standard essential patents
sham litigation doctrine
  basic principles of, 271–2
  drug patent settlements and, 320–1
  non-patent anticompetitive litigation and, 288–92
Shampine, Allan L., 69–70
Shapiro, C., 44–6, 68–9, 219, 326–7
Shapley Value method, FRAND licensing, 67 n.39
Shelanski, H.A., 49–51, 54–5
Shenhua Group, General Electric joint venture with, 474–6
Sherman Act
  applications of, 85–91
  bundled pricing and, 394–8
  disclosure and deception violations and, 236–8
  essential facilities doctrine and, 390–1
  holdup of patents and, 248–9
  market definition and, 193
  Microsoft lawsuits and, 380–4
  patent pooling exclusivity and, 369–72
  sham litigation doctrine and, 281–8
  smartphone wars and, 225–9
  tying arrangements and, 408
  Walker Process doctrine and, 271–2
Sibley, David, 397

Siegel v. Chicken Delight, 412–13, 415–17
Simcoe, Tim, 60
single firm conduct
 applications of antitrust and, 85–91
 monopolization and, 82–3
 in pharmaceuticals industry, 26–39
single-homing agents, two-sided markets and, 44–5
Sirena v. EDA, 96–7
SK Telecom case, 175
Sky UK, EC Statement of Objections to, 341, 351–3
slotting fees, online resale price maintenance and, 431–3
small but significant nontransitory increase in price (SSNIP)
 IP–antitrust regulations and, 183
 market definition and, 188–9
 merger horizontal transactions and, 447–50
smallest salable patent practicing unit (SSPPU)
 damages in patent holdups and, 255–8
 damages for SEP infringement and, 73–5
 FRAND licensing and, 61–2
smartphone technology
 holdups of patents and, 245–9
 horizontal mergers and, 461
 litigation involving, 225–9
 patent pooling and, 364–6
SmithKline Corp v. Eli Lilly & Co., 394–8
social loss
 welfare consequences of monopolization and, 82–3
 welfare effects of intellectual property and, 83–5
Société Technique Minière, 354
soft law instruments, complementarity and, 106–8
Sokol, D. Daniel, 293–313
Sot Lelos kai Sia EE and Others v. GlaxoSmith Kline, 116–17
South Africa, exploitative abuse of IP in laws of, 222–4
specific intent, in *Walker Process* doctrine, 275–7
specific subject matter
 EU antitrust and intellectual property policy and, 94–8
 inherency and, 96–106
Spulber, Daniel F., 56
SSOs. *See* standard setting organizations (SSOs)
standard essential patents (SEPs)
 business models for, 73
 China IP–antitrust regulation, 131–5, 229–31
 Chinese mergers related to, 470–4
 damages for infringement of, 73–5
 disclosure and deceptive practices and, 64–7
 dispute resolution SSOs and, 68–9
 enforcement of antitrust laws concerning, 241–2

 European litigation on excessive royalties for, 231–3
 exceptionalism and particularism concerning, 103–4
 FRAND licensing and, 60, 62–4, 173–4, 224–5, 250–8
 holdup disputes and, 250–8
 Huawei v. InterDigital case and, 478–80
 injunctions for protection of, 71–2, 251–5
 Japanese IP–antitrust policies and, 145–6, 233–4
 Korean IPR guidelines and, 158–9, 168–71, 173–4, 234–5, 240–1
 litigation issues for, 71–5
 non-discrimination in FRAND licensing and, 69–70
 patent pooling and, 358–9, 372
 post-expiration royalties cases and, 240–1
 relevant markets for, 471–2
 royalty stacking and, 70–1
 SAIC Regulation (China) concerning, 484–5
 sham litigation doctrine and, 288–92
 switch to patent assertion entities by, 257–8
 unilateral refusal to license violations, Japanese law on, 146–7
Standardization Administration of China, 128
standards-based approach to intellectual property
 causal connection test and, 288–92
 coordinated conduct cases and, 115
 exceptionalism and, 103–4
 FRAND disputes and, 250–8
 holdup of patents and, 245–58
 Korean IPR guidelines and, 168–9
 patent ambush and, 249–50
 rule-based approach vs., 109–10
 US litigation and enforcement involving, 227–9
standard setting organizations (SSOs)
 Chinese IP–antitrust regulations and, 127–9
 commitments in FRAND licensing to, 59–62
 disclosure and deception violations and, 236–8
 dispute resolution with SEPs and, 68–9
 emphasis on fair and reasonable in, 64–7
 FRAND licensing and, 58–9, 224–5
 holdup of patents and, 62–4, 245–9
 injunctions for patent infringement and, 71–2
 joint negotiations and, 67
 mergers of standard essential patents and, 471–2
 monopoly power and, 219–20
 patent ambush and, 249–50
 patent disclosure and deceptive practices and, 64–7
 patent pooling and, 358–9
 sham litigation doctrine and, 288–92
State Administration of Industry and Commerce (SAIC) (China), 120–1

essential facilities doctrine and, 125–6
establishment of, 468–70
FRAND obligations and, 127–9
future challenges for, 135–6
IP-antitrust regulatory practices of, 121–9, 483–6
IPR-related abuses and, 478–86
standard essential patent mergers and, 471–2
State Intellectual Property Office (SIPO) (China), 128
Chinese Patent Law and, 129–30
state law, US drug patent settlements and role of, 331–2
Statement of Objections (European Commission), 351–2
state of mind requirement, *Walker Process* doctrine and, 275–7
static analysis of innovation, 10–12, 18–19
Steiner, Robert L., 437–8
stock market, reverse payments cases and impact of, 36–8
Straus v. Victor Talking Machine Co., 290 n.105
structural remedies
in Chinese merger litigation, 476–8
in merger jurisprudence, 451
Stucke, Maurice E., 301–3
substantial similarity principle, copyright protection and, 198–9
substitute products
exclusivity in patent pooling and, 369–72
limits in patent pooling of, 363
market definition in antitrust and, 188–9
monopoly power constraints from, 85–91
patent pooling and, 359–63
suction effect of loyalty discounts, 398–401
sufficient economic power principle, tied products/tying arrangements and, 415–17
sunk costs, in high tech industries, 386 n.23
supply
benefit function of innovation and, 3–6
essential facilities doctrine and feasibility of, 125–6
market definition and elasticity of, 188–9
resale price maintenance in online markets and coercion of, 428–9
supracompetition
market definition and, 188–9
patent holdup injunctions and, 251–5
Supreme Court (US)
eBay court decision, 73
exploitative abuse of IP in rulings of, 222–4
Handgards doctrine and, 282–3
loyalty discounts on bundled products and, 398–401

merits of patent in drug patent settlements and, 324–5
patent assertion entities and, 262
patent pooling cases and, 361–2
pay-for-delay cases before, 26–7
per se rule and, 393–4
post-expiration royalties cases and, 238–41
quality control and good will in tying arrangements rulings of, 418–20
resale price maintenance rulings by, 425–7
rule of reason in drug patent settlements and, 326–9
sham litigation doctrine and, 281–8
tied products/tying arrangements in rulings by, 404–9
Walker Process claims before, 273–4
surcharge payments, in Japanese antimonopoly laws, 141–2
Swanson, 65–7, 69–70

Taft, William Howard, 222–4
technological tie-ins, 413–15
sufficiency defense for, 420–1
technology transfer agreements. *See also* information technology; telecommunications industry
blocking of patents and, 361–2
collusion and patent pooling and, 367–72
complementarity and, 106–8
exclusivity in patent pooling and, 369–72
holdups of patents and, 245–9
in Japan, 142–3
patent assertion entities and, 258–61
patent pooling and, 358–73
in *SK Telecomm* case, 175
Teece, D.J., 12–13
telecommunications industry
bargaining power of mergers in, 462–4
Chinese regulation of, 131–5, 229–31
Chinese SEPs mergers and, 470–4
ex post regulation and, 54–5
foreclosure of competition in, 455–8
FRAND licensing and, 224–5
holdups of patents and, 245–9
merger review process for, 452–3
net neutrality and, 53–4
Telefonica/Vodafone/EverythingEverywhere joint venture, 304–5
television copyright
block-booking and, 411–12
EU laws concerning, 97–8
territorial protection
country of origin principle for online transmission and, 349–50

territorial protection (cont.)
  EU copyright law and, 342–4, 348–9
  in *Murphy* case, 345–6
  pay TV operators in EU and, 351–2
third-party developers, exclusivity in patent pooling and, 369–72
third-party payors
  monopoly power screen and, 31
  in pharmaceuticals industry, 23–5
third-party platforms
  Big Data collection by, 298–9
  patent pooling and, 359–63
3G telecommunications standard
  Chinese IP regulations and, 131
  FRAND licensing and, 226
  Samsung Electric case (Korea), 176
3M Company, 399–400
  bundled discounts from, 394–8
tied products/tying arrangements
  anticompetitiveness of, 392–4
  antitrust law and, 11–12, 404–7
  case-law applications of, 409–18
  Chinese IP–antitrust regulations restrictions on, 126–7
  coercion or conditions relating to, 413–15
  commerce effects of, 417
  copyright protection and, 411–12
  defenses to antitrust violations from, 418–21
  EU bundling and innovation policies and, 386
  in high tech industry, design protection and, 380–4
  history of intellectual property and, 407–9
  Huawei v. InterDigital case in China and, 131–3, 478–80
  infant industries/new products defenses for, 420
  innovation and, 392 n.39
  intellectual property rights and, 404–22
  Japanese IP–antitrust policy and, 149–50
  limits of antitrust in high technology concerning, 384–7
  misuse doctrine and, 409
  NDRC Qualcomm case in China and, 133–5, 480–1
  patent pooling and, 362–3
  portfolio effects of mergers and, 388–90
  price protection clause and, 421
  Qualcomm case in China and, 480–1
  quality control and good will as defenses for, 418–20
  rule of reason in, 421–2
  SAIC Regulation (China) concerning, 484–5
  sufficiency defense and, 420–1
  sufficient economic power principle and, 415–17
  trademark protection and, 412–13
  two separate products principle and, 410–13
  unwanted products, 362–3
Tiercé Ladbroke SA v. Commission case, 98–100
tiered reimbursement
  brand/generic drug preferences and, 25–6
  in pharmaceuticals industry, 23–5
Time Insurance v. AstraZeneca, 324–5, 331
Tinder online dating site, 299–300
Tirole, Jean, 62–4, 218–19
TomTom/TeleAtlas merger, 304–5
top-down negotiation, damages in patent holdups and, 255–8
trademark protection
  European competition law and, 342–4
  exceptionalism in EU cases involving, 98–100
  inherency in EU case-law concerning, 95–109
  in Japanese IP–antitrust policies, 145
  market definition and, 185, 199–200
  monopoly power and, 210–11
  sham litigation doctrine and, 272–3 n.5
  sufficient economic power principle and, 415–17
  two separate products principle and, 412–13
Trade Related Aspects of Intellectual Property Rights (TRIPS), 216–18
trade secrets law, 207–8
TrafFix Devices, Inc. v. Marketing Displays, Inc., 185
transaction costs
  merger horizontal transactions and, 447–50
  merger policy and, 445–6
  patent pooling and reduction of, 364–6
  in two-sided markets, 44–5
Treaty on the Functioning of the European Union (TFEU), 95–109
  complementarity and, 106–7
  coordinated conduct cases and, 112–15
  copyright protection and, 342–4
  drug patent settlements and, 332–5
  exceptionalism in cases involving, 102–3, 105
  exploitative abuse in IP law and, 222–4
  loyalty discounts and, 400–1
  Murphy case and, 345–6
  phase-out of inherency and, 97–8
  rule-based approach to antitrust–IP intersection in, 109–17
true network effects, in two-sided markets, 44–5
Trusted Execution Environment (TEE), in Chinese merger policies, 474–6
Tsai, Joanna, 326–7
Tucker, Catherine, 297–300
12 Monkeys (film), 209–10
Twentieth Century Fox Japan (TCFJ) case, 148

two separate products principle
  tied products/tying arrangements and, 410–13
  trademark protection and, 412–13
two-sided markets
  Internet network effects and, 43–7
  net neutrality and, 46–7
two-sided platforms
  Big Data impact on, 301–6
  network effects and, 305–8

unfair trade practices
  in Japanese IP law, 222–4, 233–4
  in Korean IP law, 160–3, 165–6, 175, 177
unilateral conduct
  applications of antitrust and, 85–91
  in Korean IP law, 162–3
  monopolization and, 82–3
  patent pooling and collusion and, 367–72
  in pharmaceuticals industry, 26–39
  refusal to license, Japanese IP–antitrust laws and, 146–7
  resale price maintenance in online markets and, 428–9
unilateral effects theory
  horizontal mergers and, 458–62
  mergers and, 447–50
United Food & Commercial Workers Local 1776 v. Teikoku Pharma USA (Lidoderm), 329–30
United Shoe Machinery case, 392–4
United States
  drug patent settlements in, 319–32
  exploitation in antitrust law of, 222–4
  FRAND excessive pricing investigations in, 225–9
  merger jurisprudence in, 446–53
  post-expiration royalties cases in, 238–41
United States v. Bazaarvoice, 304–5, 458–62
United States v. General Electric, 222–4, 368–9
United States v. Jerrold Electronics Corp., 420
United States v. Terminal R.R. Ass'n, 390–1
Unitherm case, 276–7, 279–81
Universal Mobile Telecommunications Standard (UMTS) (3G), development of, 60, 225 n.10
Unocal case, 236–8, 287–8
unreasonable damage principle, essential facilities doctrine in Chinese IP–antitrust regulations and, 125–6
unreasonable licensing, SAIC Regulation (China) concerning, 484–5
unwanted tied product principle, patent pooling and, 362–3
upward pricing pressure (UPP)
  information technology competitive dynamics and, 453–5
  unilateral effects theory and, 449

usage. *See also* fair use
  Big Data creation from, 298–9
  Japanese IP agreements concerning, 152
  US Antitrust Guidelines for the Licensing of Intellectual Property, Korean IP law and, 163
usefulness requirement in patent protection, 206–7
user consent, Big Data and role of, 311–12
U.S. v. Microsoft (Microsoft III), 87–9
  design protection in, 380–4
  limits of antitrust in, 384–7
utility, welfare consequences of monopolization and, 82–3

"value-added services," Big Data enhancement of, 296–7
vertical integration
  market definition in IP and, 192–4
  net neutrality and, 46–7
vertical leveraging
  Chinese merger control and, 473–4
  in General Electric-Shenhua Group joint venture, 474–6
  in mergers, 450
  regulatory policy and, 56
  resale price maintenance in online markets and, 428–9, 435
Vestager, Margrethe, 352–3
VITA standards body, 61
VME bus technology, standards for, 61

Wakui, Masako, 138–55
Walker Process doctrine
  basic principles of, 271–2
  current standards, 274–5
  direct purchasers' standing under, 277–9
  enforcement issues with, 279–81
  history of claims involving, 277
  improper patent assertion and, 288–92
  infringement litigation and, 273–4
  knowledge and intent requirement in, 275–7
Walker Process Equipment, Inc. v. Food Machinery & Chemical Corp., 273
Wang, Wenche, 204–20
Wathelet (AG), 115
wealth transfer
  patent assertion entities and, 258–61
  patent holdups and, 247–8
  welfare consequences of monopolization and, 82–3
welfare consequences of monopolization, 82–3
WhatsApp
  Big Data applications on, 299–300
  FTC investigation of, 308–10

Whinston, M. *See* Segal and Whinston innovation model
"wholly exceptional circumstances," exceptionalism in IPR and, 102
Willing, R., 10
willing licensee standard, in *Samsung Electronics* case, 176
Windsurfing International case, 95, 97
"winner takes all" markets
   merger analysis and, 16–18
   Microsoft lawsuit and, 10–12

Wong-Ervin, Koren, 70–1
Woods, Lebbeus, 209–10
Wright, J.D., 49, 326–7

Yoo, Christopher S., 46–7, 53–4, 56

zero-pricing, in two-sided markets, 44–5
ZF Meritor, LLC v. Eaton Corp., 399–400